Developments in Petroleum Science, 45

thermal modeling of petroleum generation: theory and applications

to Yvonne, Conan, and
the memory of Ashley

Developments in Petroleum Science, 45

thermal modeling of petroleum generation: theory and applications

COLIN BARKER
Geosciences Department, University of Tulsa,
600 S. College Avenue, Tulsa Oklahoma 74104, U.S.A.

ELSEVIER
Amsterdam – Lausanne – New York – Oxford – Shannon – Tokyo 1996

ELSEVIER SCIENCE B.V.
Sara Burgerhartstraat 25
P.O. Box 211, 1000 AE Amsterdam, The Netherlands

ISBN: 0-444-82030-2

This book is printed on acid-free paper.

Printed in The Netherlands

PREFACE

Petroleum exploration is always limited by the lack of adequate subsurface control, and exploration problems are usually problems of extrapolation – to greater depth, to laterally equivalent rocks, or back through time. Models are widely used as a way of describing complex geological systems so that they can be treated quantitatively and used as the basis for extrapolations and predictions. Models typically consider a simplified geological system that can be described mathematically, and it is very important to know what simplifying assumptions have been made, when these assumptions are valid, and under what conditions their use may not be appropriate. This requires an understanding of the concepts involved in building the model and how the model operates.

The controlling factors in petroleum generation appear to be time and temperature, but what is usually available is age and depth. Clearly age provides time. Temperature is much more difficult to constrain, but values are commonly obtained from burial histories that summarize depth of burial through time for all the rock units at the modeled site. Temperatures are then obtained either directly, using geothermal gradients, or by combining heat-flow values with the conductivities of individual rock units. In the last few years several commercial computer programs have become available for modeling petroleum generation. These models assume first-order chemical kinetics and combine time and temperature throughout the history of the rock. They are widely used to establish the time and depth when source rocks became thermally mature and generated oil and gas. For any particular exploration site, using such a modeling program requires the geologist to provide a variety of input parameters. Usually the first step in thermal modeling is to assemble sufficient data for the program to build a realistic burial history. Subsequent calculations require information on the type of organic matter (or activation energies for its maturation reactions), as well as rock properties, details of unconformities, heat flow data, etc. The exploration geologist needs to know how these data can be obtained, and if data are not available how reliable are the default values that most programs offer.

There is a tendency to believe that thermal modeling provides "The Answer", but models are best used as a tool for probing the system and evaluating the sensitivity of the conclusions to possible uncertainties in the values of the input parameters. In a sense, models permit experimental petroleum geochemistry and let the user answer the "what if?" questions. What if the geothermal gradient had been higher in the past? What if less rock had been eroded at an unconformity? What if the organic matter type had been different?

This book starts with an overview of petroleum geochemistry. Chapter 1 describes the main factors involved in the generation of oil and gas, and then reviews the techniques available for evaluating the current status of source rocks. This first chapter also introduces some of the "tools" needed later, such as vitrinite reflectance and the way correlation relates an oil to a specific source rock. Basic physical chemistry of kinetics is covered in Chapter 2. The third chapter deals with depth of burial through time and how this is represented on burial history diagrams. It describes construction of the diagrams, including complications due to compaction, unconformities, faulting, and folding. A discussion of thermal regimes and the present-day temperatures they produce occupies Chapter 4, while Chapter 5 reviews the methods that are available for obtaining paleotemperatures. The thermal effects of time-temperature combinations can be calibrated by reference to well-documented areas, as is done in the so-called Lopatin method (Chapter 6), or can be treated more rigorously using first order-kinetics in the Tissot-Espitalie method (Chapter 7). The various uses of the thermal models for defining organic matter maturity through time and space form the contents of Chapter 8. Extended models that use 2-D and 3-D information to handle abnormal pressures, migration, and quantities generated are explained next (Chapter 9), and the book concludes with examples where various models have been applied in a wide range of different geological environments (Chapter 10).

This is not a book for experts in developing mathematical models, and is intended for geologists without strong chemical, geochemical, or mathematical backgrounds. It is worth stressing what this book does not do. It does not provide a manual that will explain how to use specific commercial programs. It does not teach you how to write a program to model the petroleum generative system. What it does do is provide students, exploration geologists, and others who would like to use the available models, with a general idea of how the models work, what they can do, and what their limitations are. It also provides the information necessary to obtain the input data required by the commercial models.

University of Tulsa
Tulsa, October 15th, 1996

COLIN BARKER

ACKNOWLEDGEMENTS

This book, like so many others, started out as a set of lecture notes and has benefited from the input of all the students who had early versions of the course inflicted on them. The course was taught to university graduate students, and also internationally to a wide variety of petroleum industry geologists. Their questions, comments, and insights have helped mould this text. I especially appreciate the time spent by Ted Beaumont, Tom Ho, Mike Horn, Indu Meshri, and Radomir Petrovich in reviewing early drafts of the manuscript. The final version is much improved by their contributions – but I take responsibility for any remaining errors!

I would like to express my appreciation to those organizations who gave their permission to use the figures and tables for which they hold the copyright.

CONTENTS

CHAPTER 1

TIME AND TEMPERATURE IN PETROLEUM GENERATION

1.1. INTRODUCTION

Organic matter is a common constituent of sedimentary rocks, ranging in amount from less than 0.01% in some red beds, clean sands and carbonates, to more than 90% in coals. It is usually extremely complex (which from an optimistic point of view means that it has high information content). Studies of organic matter are complicated because the material is thermally unstable, and its composition changes in response to increasing depth of subsurface burial. These changes lead to the important processes of coalification and the generation of petroleum. In this book the term petroleum is used to include both crude oil and natural gas. The qualitative aspects of petroleum generation are now well understood, and have been described by Barker (1979), Tissot and Welte (1978, 1984), Hunt (1995), Waples (1985), Killops and Killops (1993), Bordenave (1993), and Engel and Macko (1993).

Effective quantitative modeling of the petroleum generation process requires a knowledge of the factors controlling the conversion of organic matter into oil and gas, and the roles of various organic matter types. It is now widely accepted that time and temperature have the dominant roles in controlling generation, but the specific details will depend on the nature of the organic material itself. Some types of organic matter are more thermally resistant and require greater thermal exposure to generate hydrocarbons, while other types may generate only gas and no oil. Also, for oil-prone organic materials the chemical composition of the oil will depend on the organics in the source rock. These basic concepts are reviewed in this chapter.

Organic materials show irreversible changes in composition in response to passing time and rising temperature, and in a sense behave like maximum reading thermometers. This behavior gives them several important roles in the modeling process. First, they can be used as indicators of past thermal history. Vitrinite reflectance, biomarker transformations, and coal rank have all been used in this way. In addition, breaks in the expected depth profiles of thermal indicators may be used to infer the amount of rock section removed by erosion. These losses at unconformities are often important in reconstructing burial histories.

Present day organic maturities are calculated by the thermal models, and so measured and calculated values can be compared to provide a check on the validity of the thermal and burial reconstructions used in the model. This first chapter provides a general overview of the basic concepts involved in understanding organic matter behavior.

1.2. CHEMICAL TERMINOLOGY

The simplest organic compound is methane in which one carbon atom is linked tetrahedrally to four hydrogen atoms. Even in more complex organic compounds carbon always forms four bonds, and these commonly link the carbon to hydrogen, oxygen, sulfur, or to other carbon atoms. This ability to form carbon-carbon bonds results in the formation of large groups of carbon atoms that can be linked into chains and rings. An enormous number and variety of organic compound structures is possible.

1.2.1. Hydrocarbons

Hydrocarbon molecules contain only carbon and hydrogen atoms. The carbon atoms may be in arranged in straight chains, branched chains, rings, or in combinations of these structures.

Straight Chains: Hydrocarbons with chain structures are members of a group known as "*alkanes*", or "*paraffins*". If the chain is unbranched they are called "*normal*", and written "*n-alkane*" or "*n-paraffin*". Chains with 1 to 80 or more carbon atoms occur in oils and rock extracts:

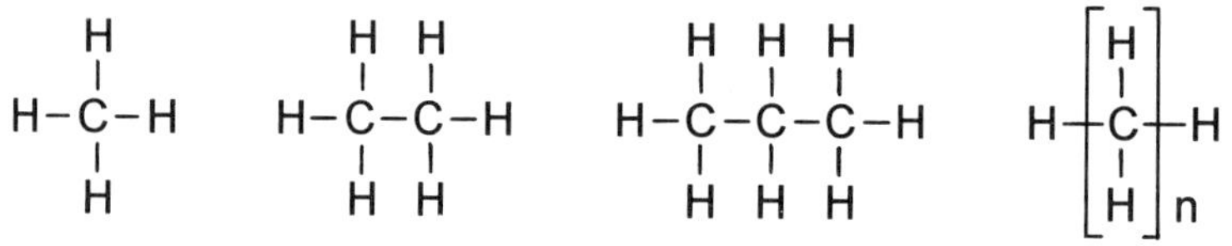

methane ethane propane C_nH_{2n+2}

Branched Chains: Branching can occur at one or more of the carbon atoms in a chain, and the branches may be made up of one or more carbon atoms. If a branch with a single carbon atom is located on the second carbon in a chain, it is called an "*isoalkane*" or "*isoparaffin*". It is often convenient to number the carbon atoms in a molecule so that the position of the branch(es) can be specified uniquely. In this system the carbon with the

branch is identified by assigning it a number starting from 1 at one end of the chain, so an "isoalkane" would be a "2-methyl alkane" (Barker, 1996).
Isomers: It is usually possible to arrange a given number of carbon and hydrogen atoms in a variety of different ways. These different structures are known as "*structural isomers*". C_5H_{12}, for example, could be a normal alkane, or a 4-carbon chain with a branch (2-methylbutane), or a 3-carbon chain with 2 branches (2,2-dimethylpropane):

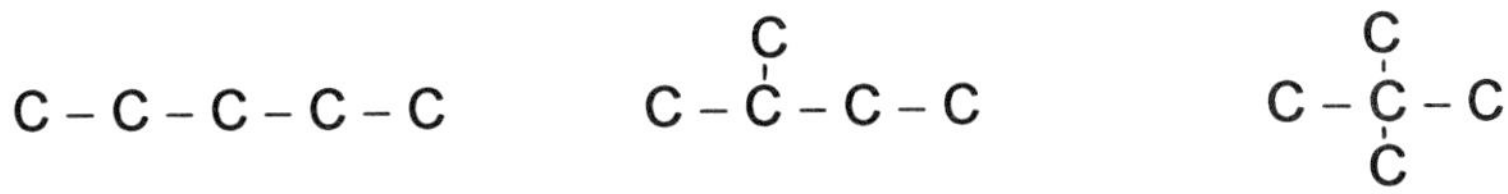

normal pentane (n-pentane) | isopentane (2-methylbutane) | neopentane (2,2-dimethylpropane)

When a molecule contains a carbon atom that is bonded to 4 different groups, the carbon atom is said to be *"asymmetric"* and gives a "*chiral center*" where 2 different isomeric structures can form. These isomeric structures are related to each other in the same way that an asymmetrical object is related to its mirror image (Figure 1.1), or the way a right hand is related to a left hand. Such compounds are said to be "*optically active*" because they rotate the plane of polarized light.
Naphthenes: The most common rings are those containing five or six carbon atoms. These rings can be fused into groups of 2, 3, or more, and there may be one or more side chains (Figure 1.2). The compounds in this very large group are usually called "*naphthenes*", and they account for the major fraction of medium gravity crude oils. Alternative names are "*cycloalkanes*" and "*cycloparaffins*". A wide range of multi-ringed naphthenes is found in source rock extracts and oils and many of the

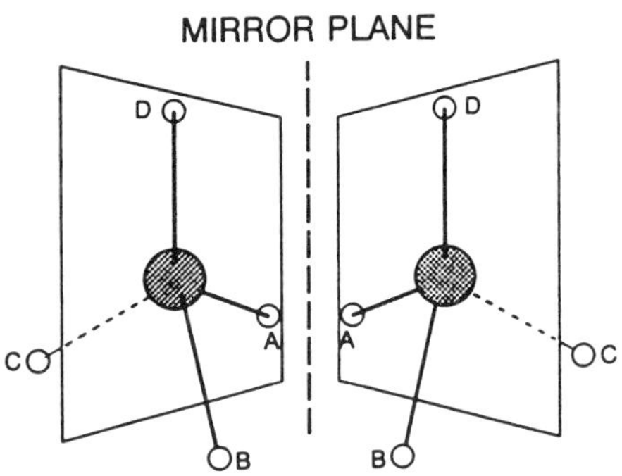

Figure 1.1. Optical isomers. The spatial configurations of the two forms are related as though the dotted line represented a mirror.

compounds can be related to biochemical precursors synthesized by living systems. Among these chemical fossils, now more commonly called "*biomarkers*", the steranes (which are four-ring naphthenes) and the hopanes (which are five-ring naphthenes) are particularly important geochemically (Figure 1.3).

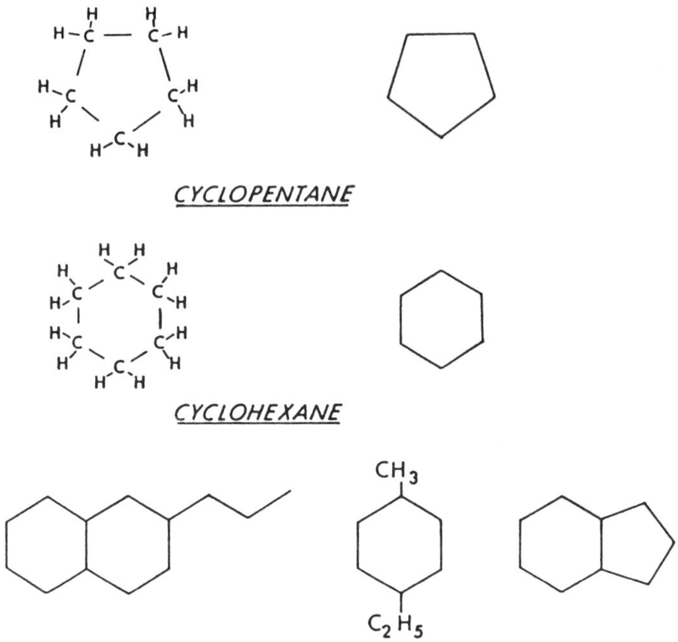

Figure 1.2. Saturated ring compounds showing the arrangement of carbon atoms and conventional representations. Selected examples show how rings may be combined with other rings and with side chains.

Figure 1.3. Representative examples of a sterane (ergostane) and a hopane (moretane).

Aromatics: Aromatic hydrocarbons are molecules that contain one or more benzene rings. The six-membered benzene ring can be written with 3 double bonds and 3 single bonds but actually behaves as though the extra electron density of the double bonds is spread uniformly over the ring. These rings are very stable structures. Aromatic rings can be fused either to other aromatic rings (to give "*polynuclear aromatics*") or to saturated rings, and any of the rings may have side chains (Figure 1.4) Again, there is an enormous number of possible structures. Aromatic compounds have relatively low hydrogen contents and this is well illustrated by comparing benzene (C_6H_6) with the C_6 normal alkane and the C_6 naphthene, which have 14 and 12 hydrogens, respectively (Table 1.1).

Table 1.1
Relative hydrogen contents of aromatic and saturated hydrocarbons with six carbon atoms.

COMPOUND TYPE	STRUCTURE	CONVENTIONAL REPRESENTATION	FORMULA	H/C
PARAFFIN	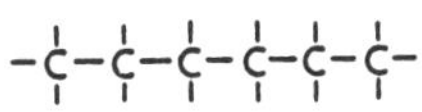		C_6H_{14}	2.33
NAPHTHENE	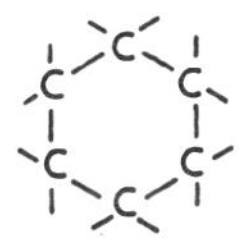	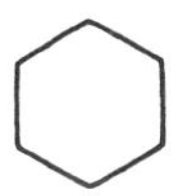	C_6H_{12}	2.00
AROMATIC	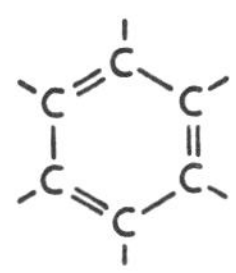		C_6H_6	1.00

1.2.2. Compounds with Heteroatoms

In addition to carbon and hydrogen many molecules contain other elements and these are called "*heteroatoms*". The most important heteroatoms in petroleum geochemistry are nitrogen, sulfur and oxygen, which has led to the designation of this group of compounds as "*NSO's*". Significant amounts of nickel and vanadium may be present in some oils.

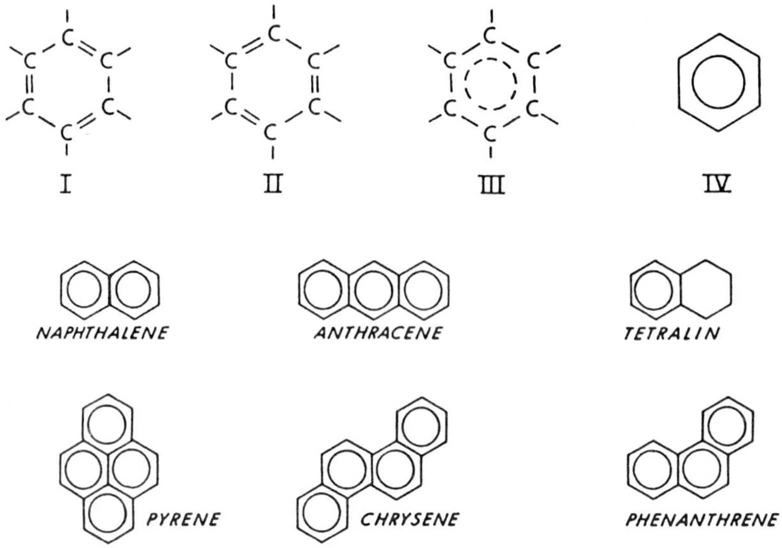

Figure 1.4. Some conventional ways of representing benzene (I to IV) and examples of fused ring (*"polycyclic"*) compounds. Note that in the tetralin structure the hexagon without a circle represents a saturated, non-aromatic ring.

1.2.3. Isotopes

Carbon has two stable isotopes, carbon-12 and carbon-13, which differ slightly in mass but have essentially the same chemical properties. The relative amounts of the stable isotopes vary somewhat in carbon-bearing materials, and these variations give useful geochemical information. Usually the relative abundances of the isotopes are given in terms of a $\delta^{13}C$ value. This has units of parts per thousand, or *"per mil"*, and is defined by the expression:

$$\delta\,^{13}C\text{‰} = \left[\frac{(^{13}C/^{12}C)_{\text{sample}} - (^{13}C/^{12}C)_{\text{standard}}}{(^{13}C/^{12}C)_{\text{standard}}}\right] \times 1000.$$

The most widely used standard is a belemnite carbonate from the Pee Dee Formation, South Carolina, and it is referred to as the *"PDB standard"*. Other isotope ratios are defined in an analogous way ($\delta^{18}O$, δD, $\delta^{34}S$).

1.3. ANALYTICAL METHODS

Most naturally-occurring carbonaceous materials are either very complex mixtures containing a large number of closely related compounds (such as crude oil), or are high molecular weight, polymeric materials which are not soluble in normal organic solvents. In most cases, the organic matter in rocks is present in low concentrations. It is only relatively recently that techniques have been developed for separating and analyzing such organic materials, and it is this more than any other single factor that has contributed to the rapid development of organic geochemistry. With this has come an understanding of the generation, migration, and accumulation of oil and gas and opened up the possibility of modeling these processes.

1.3.1. Total Organic Carbon and Total Organic Matter

The total organic carbon (*"TOC"*) in a sample is found by burning a weighed amount of the acid-treated rock to convert the organics to carbon dioxide, which can then be measured. Results are expressed as weight percent. The quantity of carbon dioxide produced from the organic matter gives a measure of the organic carbon content of the rock, not the organic matter content (*"TOM"*). Since organic matter includes hydrogen and NSO's, it will always be a higher percentage of the rock than the organic carbon content, and for ancient rocks a conversion factor of 1.22 is almost universally used.

1.3.2. Kerogen Concentrates

The complex, high molecular weight fraction of the organic matter in rocks is insoluble in all organic solvents, and is by definition "*kerogen*". It can be separated from rocks by dissolving away the mineral matter (Durand and Nicaise, 1980). This is easy for carbonates, which readily dissolve in HCl, but for silicates it is necessary to use hydrofluoric acid (HF).

1.3.3. Solvent Extraction of Bitumen

Soluble organic materials are known as "*bitumen*", and can be removed from finely-ground rocks by solvent extraction in a system called a Soxhlet extractor. This is designed so that 10 - 100 g of finely ground rock is repeatedly extracted with fresh solvent. Extraction typically takes 8 - 24 hours. Removal of the solvent by evaporation then leaves a *"heavy hydrocarbon extract"* containing compounds with 15 or more carbon atoms and identified as the *"C_{15+} extract"*. Other extraction schemes are possible.

1.3.4. Liquid Chromatography

Chromatography was first used to separate plant pigments (hence the "chroma") and relies on the fact that the different components in a mixture will migrate over the surface of a suitable substrate at different rates. If, for example, a hydrocarbon mixture is poured onto a vertical column packed with a suitable powdered solid (as shown schematically in Figure 1.5), and then pure solvent is continuously added, the components in the mixture will be washed down the column. Separation occurs because the various components move at different rates depending on their interaction with the packing – the greater the interaction, the more time they spend adsorbed, and the slower they move. Liquid chromatography of this type is commonly used as the first step in analysis, and separates crude oils and Soxhlet extracts into saturate, aromatic, and NSO fractions.

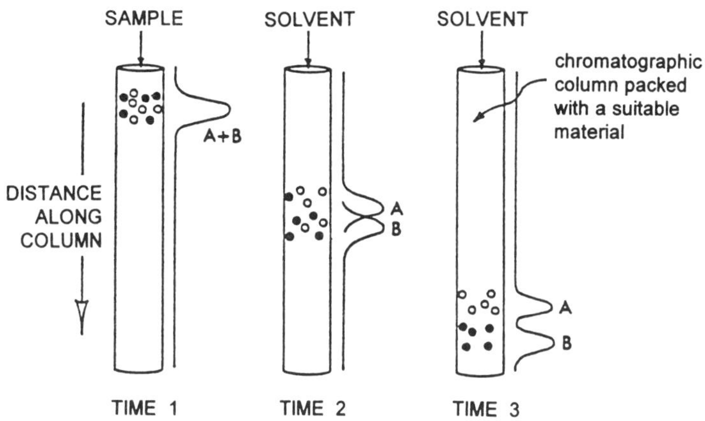

Figure 1.5. Schematic illustration of the principle of chromatographic separation. Compound A interacts more strongly with the column packing than compound B, so that A's rate of movement is impeded and B moves through the column faster. This leads to separation of A and B.

1.3.5. Gas Chromatography (GC)

When a gas stream is used to carry the components over the column packing (or, more often, a coating), the technique is called *"gas chromatography"*. This is a sensitive and widely used procedure for separating complex volatile mixtures, such as the saturate or aromatic fractions obtained by liquid chromatography. A typical gas chromatogram is shown in Figure 1.6. In whole oil chromatograms, or chromatograms of total rock extracts, there is often a prominent "hump" that is due to the thousands (millions?) of closely-related branched, cyclic, aromatic, and NSO compounds that are not completely resolved by the column. Resolution can be improved by steadily increasing the column temperature during the analysis in a procedure called "*temperature programming*". This also permits the analysis of compounds with a very wide range of boiling points, such as those typically found in crude oils. In general, the lighter components come off the column first and many gas chromatographic separations approximate a boiling point separation. Gas chromatograms are commonly reproduced in the literature and provide both qualitative information (from the horizontal scale, which is the "*retention time*"), and quantitative information (from the vertical scale, or "*peak height*").

1.3.6. Mass Spectrometry (MS)

Mass spectrometry is an older technique than gas chromatography and much of the early development was related to the analysis of refinery streams. Analysis by mass spectrometry depends on creating and separating ions with different mass-to-charge ratios (m/z), although the charge is usually 1. In a typical analytical instrument, there is a sample inlet; a method for generating the ions and then separating them; an ion collector connected to an amplifier; and a recording or data handling device. The ion source, separator and collector are mounted in a vacuum (Figure 1.7). Ions are commonly formed by electron bombardment (or other methods, such as chemical ionization) and ions with different masses are subsequently separated, either by deflection in a magnetic field or by the paths they follow through a quadrupole system. By sweeping across the range of masses, a plot of relative intensity against mass number can be obtained and is called a "*mass spectrum*" (Figure 1.7). In a gas chromatogram, a single compound gives a single peak, whereas in a mass spectrometer one compound generally gives a large number of peaks. This leads to both the power and the limitation of mass spectrometry. Mass spectrometry is an extremely powerful technique because for a single compound, the different mass fragments and their relative abundances can be interpreted to identify the compound uniquely, including details of its stereochemistry. The limitations come because mass spectra are linearly additive and in a complex mixture the peaks overlap and cannot be interpreted to give concentrations of individual components.

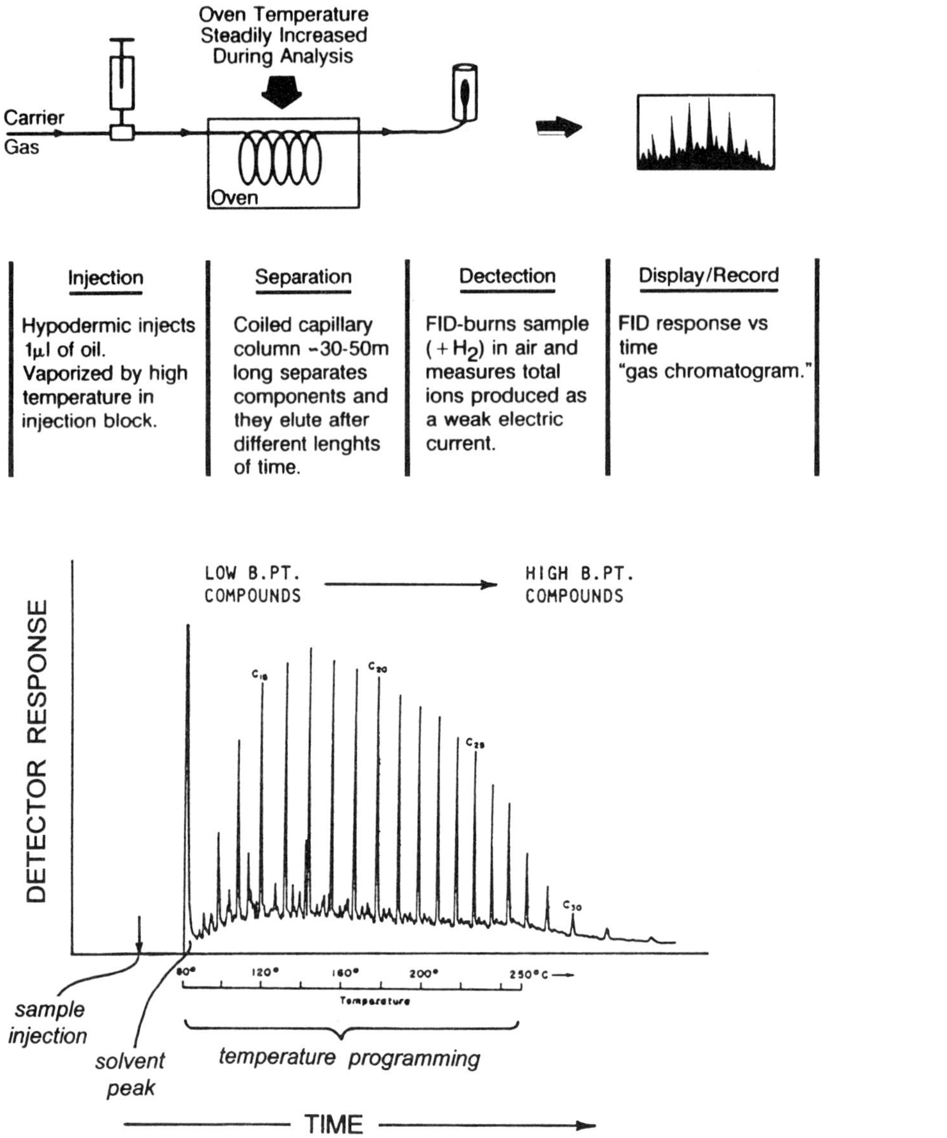

Figure 1.6. Schematic layout of a gas chromatograph (top) showing sample injection; separation through the column; detection of eluting components; and display. An example of a gas chromatogram (bottom) shows prominent peaks that are normal alkanes (labeled C_{15}, etc.), and a hump of unresolved compounds. Note that the time scale may be reported increasing from left to right, or from right to left.

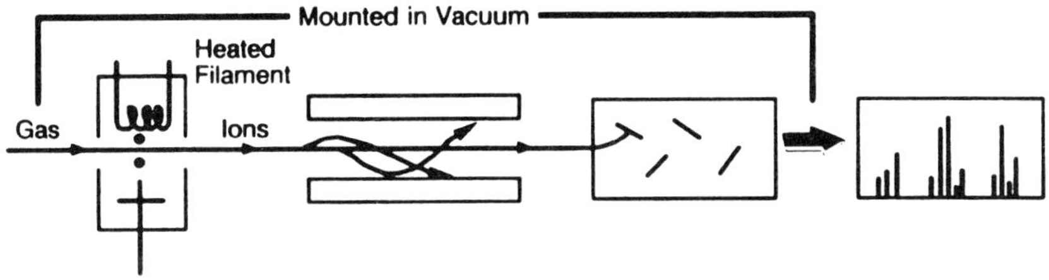

Ionization	Separation	Detection	Display/Record
Ions formed by electron impact (or in other ways such as chemical ionization). Ions accelerated into separator.	In quadrupole MS ions oscillate through four parallel rods. Only one mass gets through. Separation may be done using magnetic deflection.	Ions collected commonly amplified by an electron multiplier.	Detector response vs mass "mass spectrum."

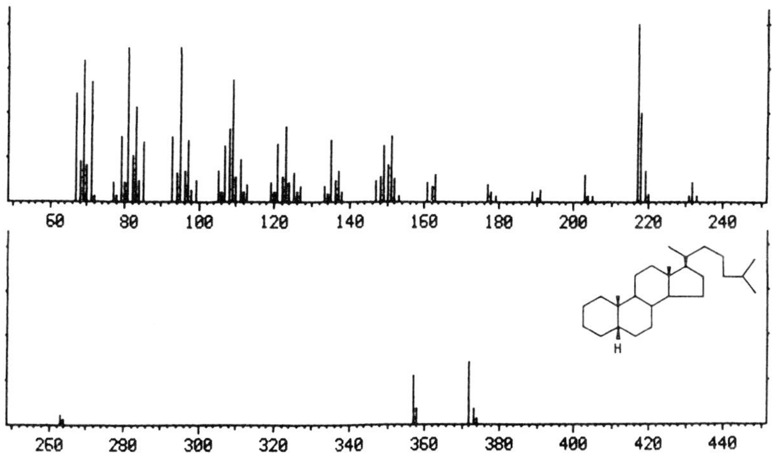

327. 5β(H)-Cholestane — Rhead et al., 1971

M.Wt.372 $C_{27}H_{48}$ Base peak m/z 217 — 70 eV - GC-MS

Shales, oils, source rocks, marine input — I and C

Figure 1.7. Schematic layout of a mass spectrometer (top) showing the formation of ions; separation of ions; and detection and display as a mass spectrum. Mass spectrum produced by 5β(H)-cholestane (bottom). Note the prominent peak at m/z 217 (from Philp, 1985).

1.3.7. GCMS – The Combination of Mass Spectrometry with Gas Chromatography

Gas chromatography is an excellent tool for separating complex mixtures, but it is often difficult to identify the compound responsible for a given peak. In contrast, mass spectrometry is a very powerful technique for identifying individual compounds but is less useful with mixtures. Clearly, a combination of the two analytical techniques could use the best aspects of both, and the development of GCMS systems has provided a powerful tool for the detailed characterization of organic materials and it is being used widely by petroleum geochemists.

1.3.8. Overall Analytical Scheme

Figure 1.8 shows a generalized overall analytical scheme for the detailed chemical characterization of an oil or source rock extract. In practice, not all the techniques would be used, partly because of cost and partly because the various techniques give overlapping information. Some redundancy, of course, provides a useful check.

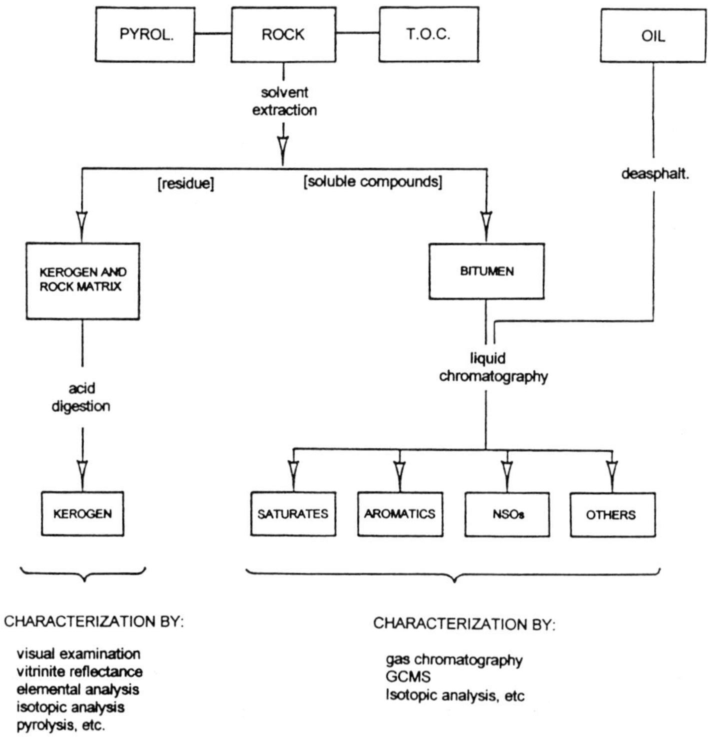

Figure 1.8. Generalized analytical scheme for rocks and oils.

1.4. RECENT SEDIMENTS

Petroleum is generated from the organic matter in rocks. This organic matter was incorporated at the time the rocks were being laid down as recent sediments, so that conditions in the depositional environment control how much organic matter and what type of organic matter is incorporated.

1.4.1. Total Organic Content

A wide variety of factors combine to control the total amount of organic matter that ends up in sediments:

Productivity: High organic productivity creates large quantities of organic matter and this is a requirement for high organic contents in sediments. However, high biological productivity alone does not guarantee high organic content for the sediments because in addition to generation there must be preservation.

Grain size: In clastic sediments, fine-grained muds have higher organic contents than the coarser-grained sands. This is a consequence of the winnowing process that removes fine-grained clays and organics from the sands. The clays and organic materials later settle out together in quieter environments that become reducing.

Sedimentation rate: At high sedimentation rates there is heavy clastic dilution of the organic materials and values of total organic content are low. Values are also low for very slow sedimentation rates because the organic materials remain close to the sediment surface for a long time and are degraded by bacteria. In general the middle range of sedimentation rate leads to higher organic contents.

Oxidation/Reduction: Organic matter is not stable under oxidizing conditions but can be preserved when the environment is reducing (Demaison and Moore, 1980) (Figure 1.9). Organic matter preserved under reducing conditions is generally of higher quality in the sense that it is a better petroleum generator. Oxidizing conditions are often indicated by the nature of the inorganic constituents, for example, by the presence of iron oxides. Since these give the red color to red beds, rocks of this type generally have low organic matter contents and are usually excluded from consideration as source rocks.

1.4.2. Organic Compounds

Biopolymers: The total amount of organic material incorporated into

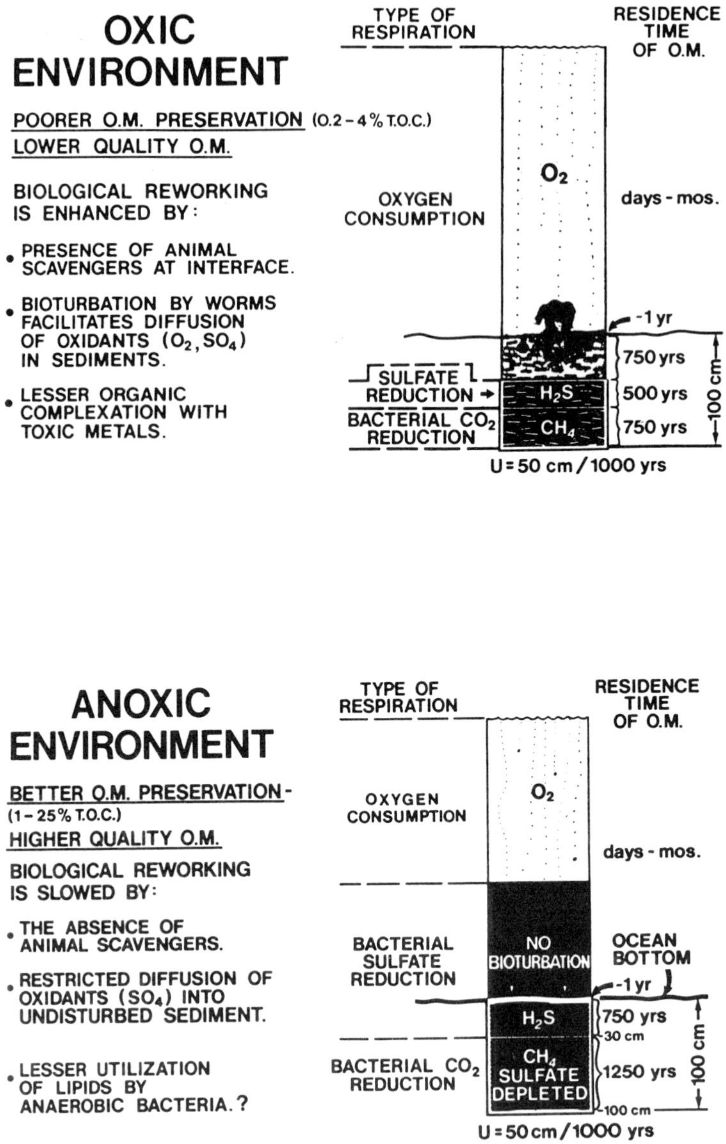

Figure 1.9. Preservation of organic materials in oxic and anoxic environments (Demaison and Moore, 1980).

sediments is important, but so is its chemical composition. The nature of the organic compounds which become part of the sediment depends on the type of organisms living in the overlying water column and on the relative contribution of organic debris transported from terrestrial environments (Figure 1.10). Much of the organic matter is made up of high molecular weight biopolymers, including carbohydrates, proteins, and lignin. As these are broken down, the monomers can be reassembled to make the complex, dark-colored *"humic material"*. With increasing size and complexity this becomes insoluble and is then, by definition, kerogen. In addition, organic debris in the form of wood, leaf coatings, spores, and pollen grains can contribute to the kerogen.

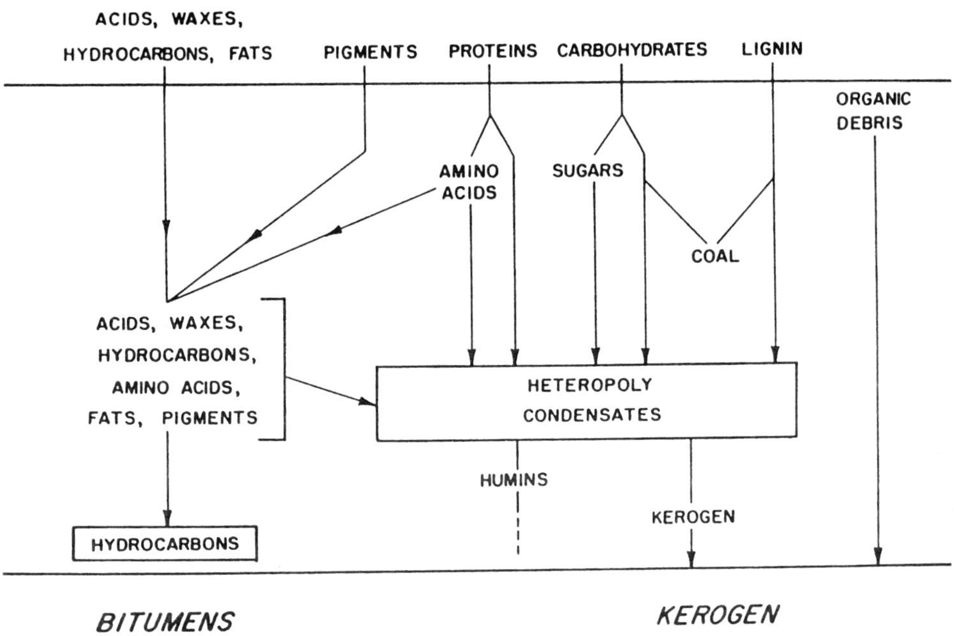

Figure 1.10. Summary of the fate of organic matter in recent sediments. The major types of organic materials made by living systems are given above the top line, and the materials found in ancient sediments (rocks) summarized below the bottom line. In between the major changes produced in the water column and in recent sediments are summarized.

Bitumen: Along with the biopolymers, organisms synthesize an enormous number of different compounds and these are transferred to the water/sediment system when the organisms dies. Although there are almost no compounds between C_2 and C_{10}, many of the larger molecules are important because they occur (often in modified form) in crude oils and source rock extracts. The long chain n-alkanes with more than 27 carbon atoms are especially prominent in terrestrial organic matter and these chains commonly show an odd carbon predominance, the most abundant chains being those with 27, 29, or 31 carbon atoms. Long chain waxes are important in the cuticle of higher land plants and are an important source of the long chain alkanes. Compounds containing naphthene rings, and nitrogen-, sulfur-, and oxygen-bearing functional groups, are also quantitatively important in organisms and in recent sediments.

Biomarkers: When an organism dies it may leave a record of its hard parts as a classical fossil, but it will also leave in the sediment all of the biochemically-synthesized molecules as "*chemical fossils*". These are now more commonly called "*biological markers*", and this term is often contracted to "*biomarkers*". A biomarker is defined as "any organic compound detected in the geosphere whose basic skeleton suggests an unambiguous link to a known contemporary natural product" (Mackenzie, 1984).

Acyclic isoprenoids: As part of their metabolic processes organisms build complex molecules that are formally related to linked isoprene units. These molecules can be joined together to form extended chains ("*acyclic isoprenoids*") or rings. The most common chain compounds are pristane (C_{19}) and phytane (C_{20}) which can be considered as derived from 4 isoprene units linked "*head-to-tail*" (Figure 1.11). This leads to chains with a branch on every fifth carbon atom. Pristane and phytane are present (and relatively abundant) in living organisms, recent sediments, ancient sediments, and crude oils. Other acyclic isoprenoids are known but are generally less abundant.

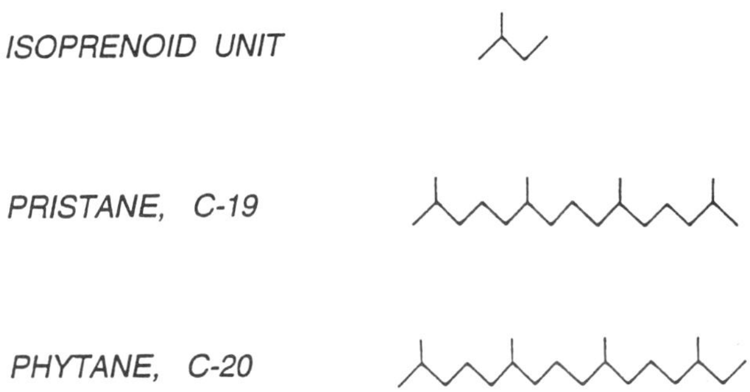

Figure 1.11. The pristane (C_{19}) and phytane (C_{20}) alicyclic isoprenoid hydrocarbons showing the characteristic branching on every fifth carbon atom in the chain produced by linking isoprenoid units.

Cyclic isoprenoids: Two linked isoprene units are called a monoterpane, four units give diterpanes, and six units give triterpanes. The terpanes can be closed into ring structures in a variety of different ways, although individual rings almost always have five or six carbon atoms. The configurations most frequently used in petroleum geochemistry involve the steranes with four rings (3 six-membered, 1 five-membered), and the hopanes with five rings (4 six-membered, 1 five-membered). In sediments and organisms these compounds commonly contain oxygen-bearing groups and double bonds. Cholesterol provides an example of a sterol that may subsequently be modified to give a sterane (in this case cholestane). Note that the carbon skeleton remains unchanged (Figure 1.12). Triterpanes are believed to be derived from bacterial triterpenoids (Ourisson et al, 1984). Steranes are derived from the sterols which occur in algae and most higher plants but appear to be rare in procaryotic organisms (Volkman, 1986, 1988).

Cholesterol

Cholestane

Figure 1.12. Comparison of the cholesterol molecule found in living systems and the related cholestane molecule (*"biomarker"*) found in rock extracts and oils.

1.4.3. Isotopic Composition

The isotopic composition of plant material is strongly influenced by the isotopic composition of the carbon source used in biosynthesis. Carbon dioxide in the atmosphere is richer in ^{12}C (i.e., "lighter") than the bicarbonate in sea water, and as a consequence terrestrial plants tend to be richer in ^{12}C than marine plants (such as algae). This results in more negative $\delta^{13}C$ values for terrestrial organic materials, including crude oils derived from them (Sofer, 1984).

1.5. PETROLEUM GENERATION

1.5.1. Introduction

The organic matter incorporated into the muds of recent sediments is buried with them as they convert to shales or micritic carbonates. Average shales contain approximately 1 wt% organic matter. About 90% of this is the high molecular weight, insoluble, polymeric material "*kerogen*", and the rest is solvent soluble and is the "*bitumen*" fraction. Both the compositions and relative amounts of kerogen and bitumen change with burial, and thermodynamic calculations show that in a reducing environment, methane and carbon are the stable end products produced by the influence of increased temperature on all types of organic matter. This implies a redistribution of hydrogen to give one product rich in hydrogen (CH_4) and one with no hydrogen (C). For example, the nC_{21} alkane would ultimately redistribute its hydrogen:

$$C_{21}H_{44} \implies 11CH_4 + 10C.$$

With the increasing temperature that accompanies increasing depth of burial, the kerogen gets progressively richer in carbon by becoming more condensed and developing more aromatic structures, while the extractable fraction gets richer in hydrogen as the amount of alkanes increases and average chain length decreases. This redistribution of hydrogen can be summarized by the scheme shown in Figure 1.13.

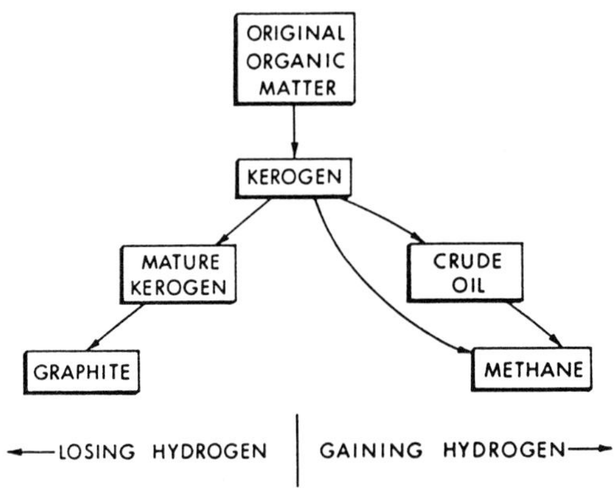

Figure 1.13. Schematic illustration of the redistribution of carbon and hydrogen that occurs when organic matter is buried.

The quantity of bitumen increases because it is formed from the kerogen. We will see later that part of the bitumen fraction migrates and accumulates as crude oil, so that the formation of bitumen from kerogen is the process of petroleum generation. Since the kerogen accounts for the bulk of the organic matter in rocks and has a major role in forming bitumen, changes in kerogen composition with depth provide an important indication of the progress of petroleum generation.

1.5.2. Changes in Kerogen Composition

Various graphical representations have been developed for displaying changes in the chemical composition of kerogen as a function of depth. The simplest one is a plot of %C, %H, etc., against depth. LaPlante (1974) used this for kerogens separated from shales in several Gulf Coast wells, including the one from South Pecan Lake, Louisiana, shown in Figure 1.14. In this case there is a systematic increase in carbon content from about 68% at 4,000 ft (1220 m) to 82% at 15,000 ft (4575 m). The carbon increase is balanced by a decrease in oxygen content. Hydrogen content increases down to about 10,000 ft (3050 m) but then decreases. Similar hydrogen trends were found in other wells studied, and LaPlante suggested that this decrease in the hydrogen content occurs because it is being used to make the hydrogen-rich hydrocarbons of the bitumen fraction. Oxygen is lost mainly as carbon dioxide.

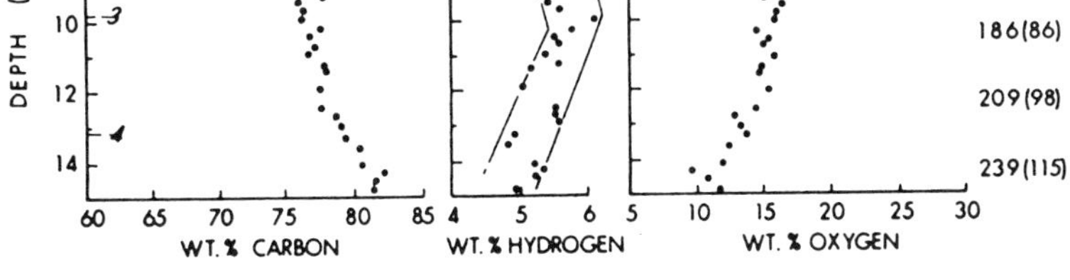

Figure 1.14. Changing composition of kerogen with depth in the South Pecan Lake field, Cameron Parish, Louisiana (LaPlante, 1974) showing the variation in weight percentages of carbon, hydrogen, and oxygen.

Percentages of carbon, hydrogen and oxygen can be plotted on C-H-O ternary diagrams, but a more convenient alternative method has been developed for coal research. This uses a plot of atomic H/C ratios versus atomic O/C ratios. Such diagrams are called "*van Krevelen diagrams*", and Tissot et al (1974) were the first to show that they are suitable for use with kerogen data. Different types of organic matter follow different tracks across the diagram depending on their original elemental composition (Figure 1.15), and the three major tracks are those of kerogen Types I, II, and III. Type III is mainly wood-derived, terrestrial organic matter, Type I is common in lacustrine algae, while Type II corresponds to marine algae and is the main oil-generating track. Low hydrogen, degraded organic matter is sometimes referred to as Type IV kerogen (or "*inertinite*").

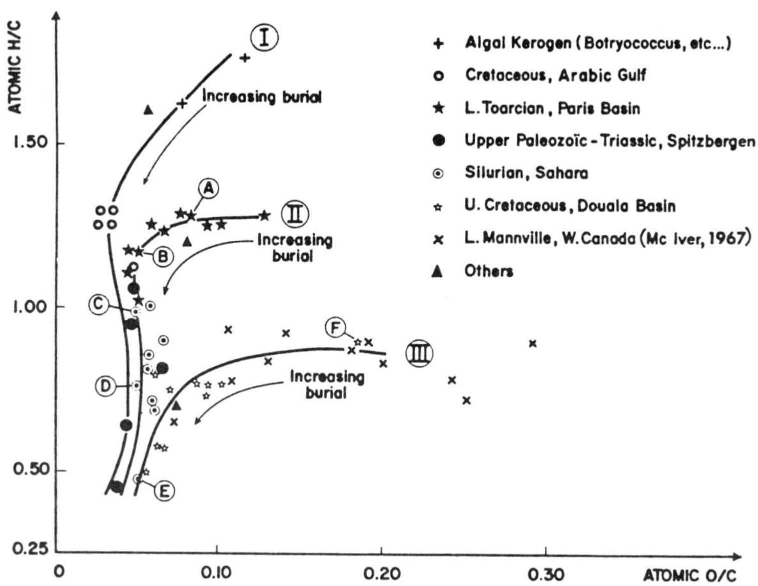

Figure 1.15. Examples of kerogen evolution paths from Tissot et al (1974). Path I includes algal kerogen and excellent source rocks from the Middle East; Path II includes good source rocks from North Africa and other basins; Path III corresponds to less oil-productive organic matter, but may include gas source rocks. Evolution of kerogen composition with depth is marked by arrows along each path.

1.5.3. Changes in Physical Characteristics of Kerogen

Kerogen color: With increasing subsurface temperatures kerogen begins to generate petroleum compounds, and its chemical composition changes in a process called "*maturation.*" Changes in the chemical composition of the kerogen during maturation lead to changes in its physical appearance. Shallow (immature) samples that have not been deeply buried generally contain kerogen that is light yellow in color, but with increasing depth there is a progressive darkening through golden yellow to brown, and finally to black. Several schemes have been introduced for rating the color intensities in terms of a "*Thermal Alteration Index*", or "*TAI*" (Figure 1.16).

Vitrinite reflectance: As kerogen matures it becomes "shinier", and this increase in reflectivity can be a useful measure of organic matter maturity. For comparative purposes it is necessary to select only one type of organic matter and vitrinite (a woody, Type III kerogen) is universally used because of its well documented composition and predictable behavior (Dow, 1977) (Figure 1.17). Measurements of "*vitrinite reflectance*" (R_o) are made in oil – hence the "o" subscript.

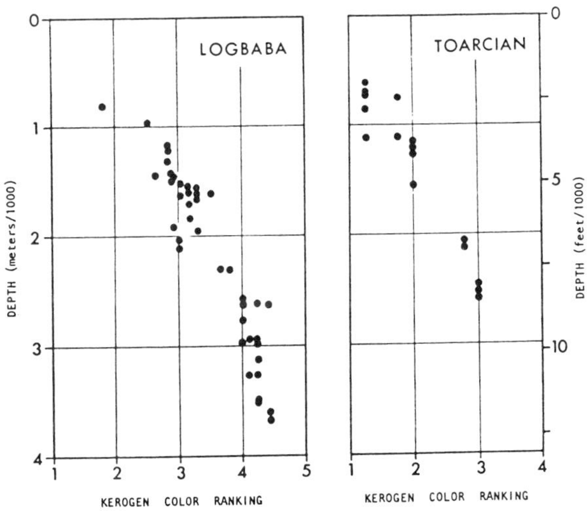

Figure 1.16. Darkening of kerogen color with depth in the Toarcian shales of the Paris basin and the Logbaba series of the Douala basin (data from Correia, 1971).

1.5.4. Generation of Hydrocarbons and the Factors Involved

Introduction: Since kerogen accounts for over ninety percent of the organic matter in shales, small changes in its composition can produce large changes in the nature of the lower molecular weight, solvent extractable bitumen that is generated from it. The trend of increasing carbon content in the kerogen with progressive burial is balanced by an increasing hydrogen content in the bitumen, which becomes lower in molecular weight and richer in alkanes as thermal maturation proceeds. The net effect is the preferential transfer of hydrogen from the kerogen to the bitumen (Figure 1.13).

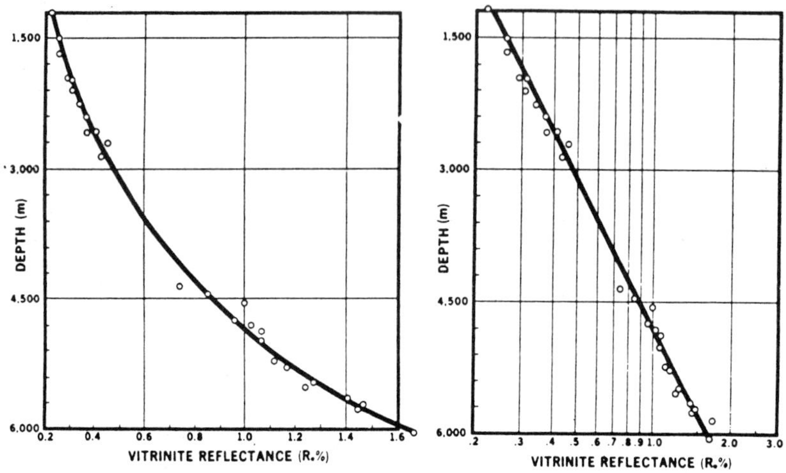

Figure 1.17. **A**: Kerogen maturation profile of a typical Louisiana Gulf Coast well showing vitrinite reflectance increase with depth. Both depth and average R_o values are plotted on linear scales. **B**: The same maturation profile as shown in A, but with R_o values plotted on a logarithmic scale. This is the simplest case of a continuously subsiding basin. (Data from Dow).

The formation of bitumen is particularly important because this is the process of petroleum generation. The C_2-C_{10} hydrocarbons, which are absent in living organisms and recent sediments, but can be such an important part of many crude oils, must be made from some other carbonaceous material. Kerogen is the most likely source because of its availability and its response to changing temperatures.

The chemical compositions of kerogen and bitumen change systematically as depth increases. This is also the direction of increasing time, temperature, and pressure, and these factors have to be evaluated for their influence on the petroleum generation process. It has also been suggested that radioactivity, bacteria and catalysts have a role in the maturation of organic matter but, except for the bacterial generation of shallow dry gas, their influence seems to be minor. In most cases pressure also seems to be a minor factor, and in the following sections emphasis is placed on the effects of temperature and time.

Temperature: Temperature is the most important parameter controlling the generation of petroleum. This was demonstrated very clearly by Philippi (1965) in a classic study of Neogene sediments from the Los Angeles and Ventura basins of California. For rocks of different depths but similar ages he determined the amount of organic matter and the amount and composition of the bitumen in the C_{15+} range. His curves showing the ratio (extractable C_{15+} hydrocarbons)/(total organic carbon) as a function of sample depth are reproduced here as Figure 1.18. Both show a rather low initial ratio which increases rapidly at 10,000 ft (3050 m) in the Los Angeles basin and at 16,000 ft (4880 m) in the Ventura basin. What makes this study so remarkable is that the increase in the ratio occurs at the same temperature – 120 °C – in both basins. This is compelling evidence that temperature has a critically important role in the generation of the extractable hydrocarbons. Other factors, such as pressure, have only a minor role, if indeed they have any role at all. For example, in the Los Angeles basin the pressure is 5000 psi at the onset of generation but it is 60 % higher, at 8000 psi, when generation starts in the Ventura basin.

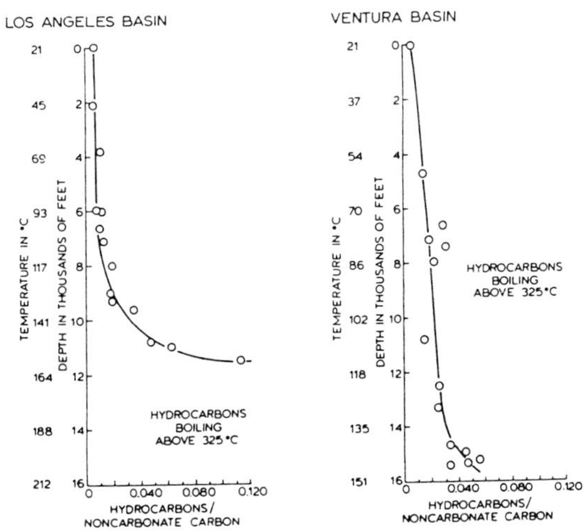

Figure 1.18. Variation of the (hydrocarbon/non-carbonate carbon) ratio with depth and age in shales from the Los Angeles and Ventura basins, California (Philippi, 1965).

Supporting evidence for the role of temperature in petroleum generation comes from the study of sedimentary rocks intruded by igneous dikes. Because of the steep thermal gradient away from the dike when it is intruded, the organic matter in the rocks close to the intrusion has been exposed to high temperatures, while the organic matter farther away has been heated only a little above the normal temperature of the country rock. Immediately adjacent to the intrusion the organic matter is destroyed and only graphitic material remains. Further out the extractable material is present in the greatest quantities but then decreases smoothly with distance from the contact, eventually reaching values characteristic of the host shale. In this situation the role of temperature is unambiguous.

Further information on the effect of temperature on organic matter is provided by laboratory heating experiments ("*pyrolysis*"), which can duplicate many of the features observed for subsurface samples. In order to produce appreciable reaction in a reasonable length of time, the temperatures used must be higher than those encountered in the subsurface.

Time: When bitumen/kerogen ratios are plotted against depth for different basins, they show that the onset of generation occurs at quite different temperatures (Figure 1.19). Clearly temperature alone is not the only factor influencing the onset of generation. In general generation starts at greater depths in younger basins and shallower depths in older basins, suggesting that the effect of time (age) modifies the effect of temperature. As a first approximation the rate of reaction roughly doubles for every 10 °C (18 °F) rise in temperature. However, the quantitative relationship between time and temperature is at the core of thermal modeling and will be addressed in more detail in Chapter 2.

Summary: Petroleum generation is temperature-dependent and time-dependent, but varies exponentially with temperature, and linearly with time. These thermal processes are non-biogenic, and operate to remove the characteristics of biogenically-derived molecules. Petroleum generation is unidirectional and irreversible, and is suspended by an appreciable temperature decrease (such as occurs on uplift). Catalysis, radioactivity, and pore fluid pressure do not appear to have significant roles. The effects of the generation process are summarized in Figure 1.20 which shows that biologically derived molecules form an important part of the bitumen in shallow samples, but become quantitatively less important in deeper samples.

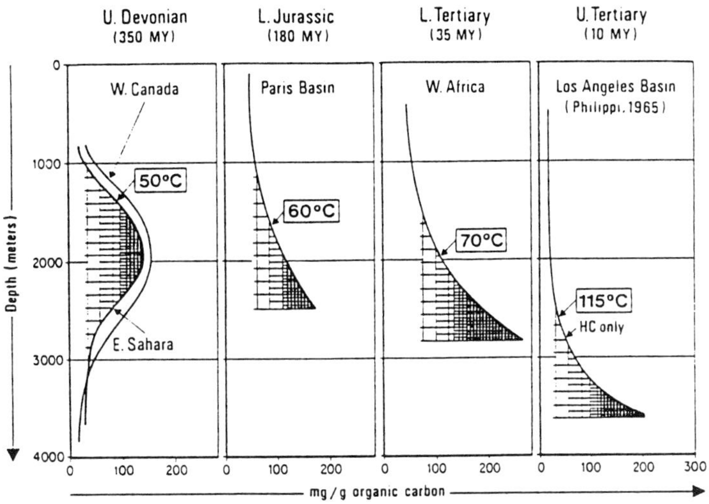

Figure 1.19. Comparison of the depths and temperatures for the onset of oil generation in several source rocks of different ages (Tissot et al, 1975).

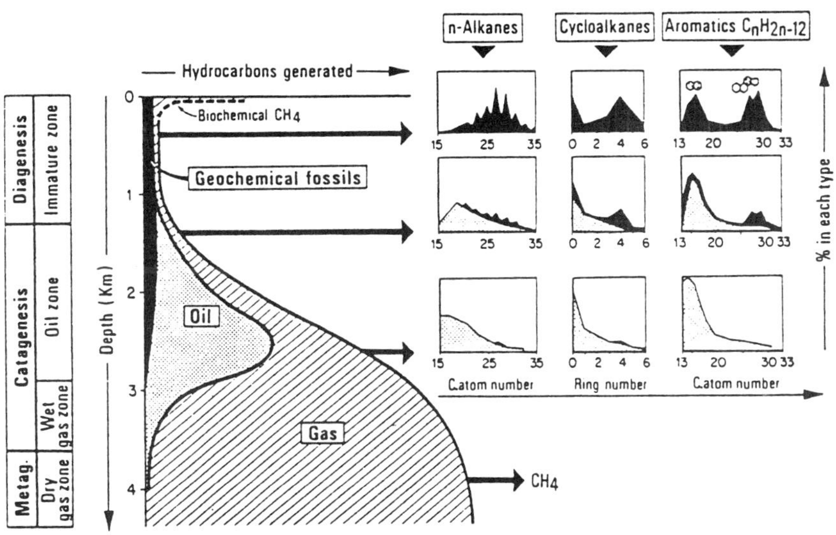

Figure 1.20. General scheme for hydrocarbon formation as a function of burial of the source rock. The evolution of hydrocarbon composition is shown for three different depths. The depth scale given is only approximate and will depend on the geothermal gradient. The values given correspond to an average for Mesozoic and Paleozoic source rocks. Actual depths will also depend on organic matter type and burial history. (From Tissot et al, 1974).

1.6. BIOGENIC GAS

There is an important exception to the thermal generation of petroleum, and this is the bacterial formation of methane (Rice, 1993). Bacterial, or "*biogenic*", methane is generated under anoxic, sulfate-free conditions by anaerobic bacteria. The bacteria cannot tolerate subsurface temperatures above approximately 75 °C, but in practice they run out of living space due to compaction, and also consume the available food, so that generation of biogenic gas is usually complete by about 350 m (Rice and Claypool, 1981).

Biogenic gas is essentially pure methane with no higher hydrocarbons ("*dry gas*"), and it is isotopically very light, with $\delta^{13}C$ values more negative than -55°/₀₀. The characteristics of dry, shallow, isotopically light gas that are used to recognize a bacterial origin have been reported for the huge gas fields of W. Siberia, and it appears that Urengoi, one of the largest gas fields in the world with 286 Tcf, contains gas that was bacterially generated (Grace and Hart, 1986). Many other smaller, but commercial, biogenic gas accumulations are known (Table 1.2; Rice and Claypool, 1981). However, since biogenic gas is not generated by thermal processes, its formation cannot be modeled by the techniques discussed in this book.

Table 1.2
Biogenic gas fields of Pleistocene age in offshore Gulf of Mexico

FIELD (AND DEPTH)	$\delta^{13}C_1$ (°/₀₀)	C_1/C_{1-5}
Ship Shoal 271 (1818 - 1876 m)	-59.9 to -56.9	0.989 to 0.987
East Cameron 245 (458 - 83 m)	-69.4 to -65.9	0.999
West Cameron 513 (2224 - 2312 m)	-57.2 to -56.9	0.997 to 0.972
West Cameron 533 (917 m)	-64.3	0.999
West Cameron 543 (2498 - 2573 m)	-58 to 56.7	0.980 to 0.978
West Cameron 587 (1355 - 1562 m)	-58.7 to -58.1	0.981
West Cameron 639 (1414 -1512 m)	-63.2 to -57.9	0.996 to 0.985
West Cameron 643 (1391 - 1399 m)	-55.4 to -55.3	0.974
High Island A302 (2721 - 2786 m)	-65.1 to -61.7	0.999 to 0.997
High Island A309 (1390 - 2208 m)	-61.5 to -57.4	0.986 to 0.982
High Island A330 (1456 - 2430 m)	-66.5 to -64.5	0.997
High Island A343 (1829 - 1916 m)	-67.0 to -66.8	0.999 to 0.998
High Island A370 (1270 - 2393 m)	-65.3 to -60.6	0.998 to 0.960

1.7. PETROLEUM MIGRATION

1.7.1. Introduction

Petroleum can accumulate in any rock with adequate permeability and porosity, but in most cases the reservoirs are either sandstones or carbonates. In general, the processes that produce sediments which later become rocks with high permeability and porosity act to remove the organic matter by winnowing and oxidation. Reservoirs do not retain sufficient organic matter to produce the volumes of oil and gas that accumulate in them. The idea that petroleum is generated in organic-rich, generally fine-grained source rocks, and that part of it subsequently migrates and accumulates in a reservoir, is the so-called *"Source Rock Concept"* (Figure 1.21). Clearly, migration processes have a key role in linking the source rock to the reservoir, and as a consequence, an understanding of the migration mechanisms is important for many aspects of modeling and exploration (Figure 1.22).

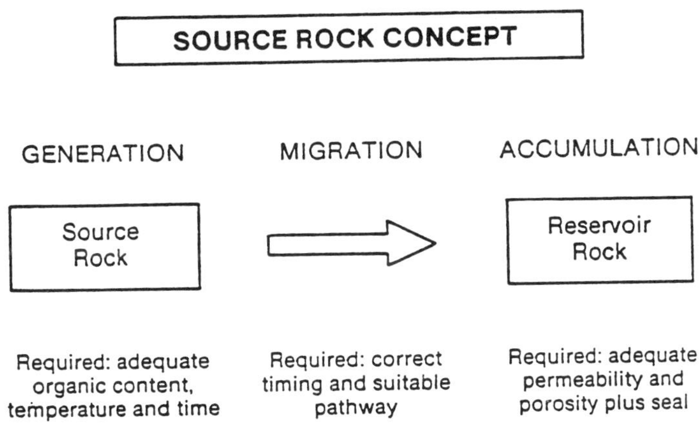

Figure 1.21. The "Source Rock Concept".

1.7.2. Migration in Solution

The data for natural systems do not seem to be compatible with any of the solubility mechanisms that have been suggested for oil migration (Vandenbroucke, 1993). It is difficult to understand how a complex mixture of hydrocarbons can dissolve in water, be transported, and then come back out of solution without being fractionated in terms of solubility –

and there is no evidence that this happens. In addition, there is the problem of finding adequate amounts of water and moving it at the right time. Also, the process and timing of exsolution are poorly documented.

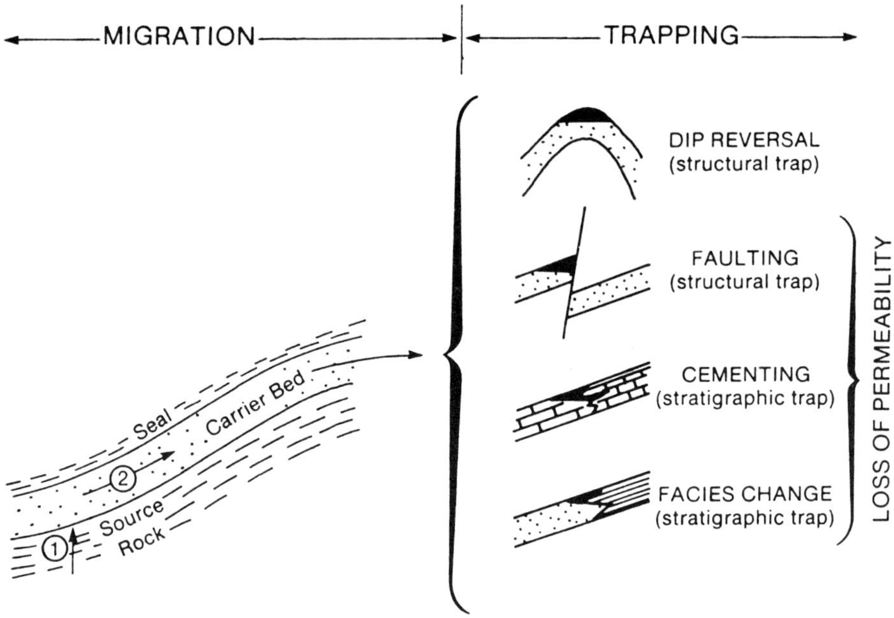

Figure 1.22. Schematic illustration of expulsion–(1); lateral migration–(2); and trapping. Vertical migration routes through faults can be important in some geological settings.

1.7.3. Migration of a Petroleum Phase

When hydrocarbons are generated in source rocks in quantities sufficient to saturate the pore water and the adsorptive capacities of the pore surfaces, then a separate petroleum phase forms as a discontinuous network (Figure 1.23). Some directed driving force is needed to cause movement of the hydrocarbons out of the rock. Here it is important to recognize that the system is not isothermal, and that the petroleum is there only because the rising temperature that accompanies increasing depth of burial drives the generation process. Deeper burial increases pressure and drives compaction, which tends to squeeze oil out of the rock. In addition, the rising temperature causes water and other fluids in the pores to expand and create internal high pressures within the source rocks, and the generation process itself also tends to overpressure the system.

Overpressuring plays an important role because it produces a pressure gradient from inside the source rock to its effective boundary, probably a near-vertical fracture. The pressure gradient that is generated is greater than the hydrostatic gradient, and oil can be expelled downwards and out of the bottom of the source rock just as well as upwards and out of the top. Thus it is quite possible for a reservoir to lie immediately under a source rock.

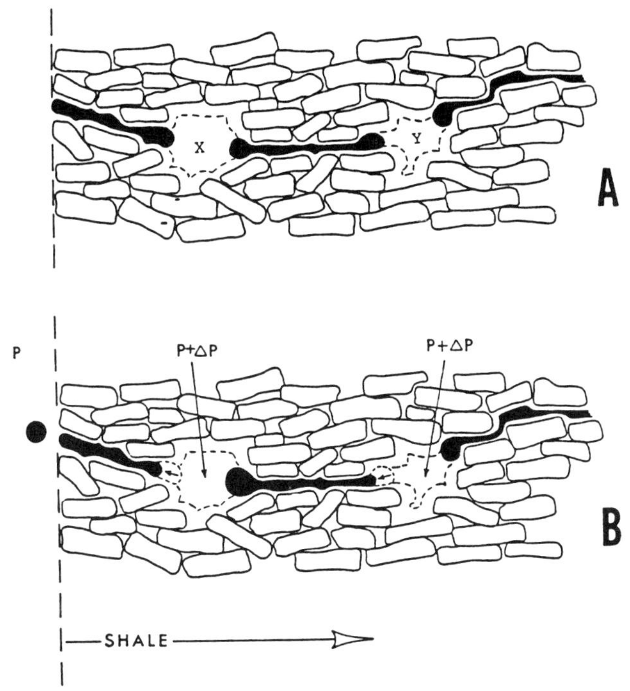

Figure 1.23. Possible distribution of bitumen (oil) as a pore center network in a shale adjacent to a sand. **A**: distribution of oil and water at approximately 3000 m and 85 °C. **B**: distribution after burial to approximately 450 m and 115 °C that has caused continued generation and the thermal expansion of water which generates overpressures and led to the expulsion of part of the hydrocarbons. (From Barker, 1980).

Not all types of compounds migrate out of source rocks with equal ease. Correlation of oil in the reservoir with rock extracts (i.e., the material left behind in the source rock) shows marked differences in the relative amounts of the various types of compounds. Although there is generally little fractionation within a compound type, such as the alkanes, when alkanes are compared with NSOs there are marked differences. The functional groups in the NSO compounds favor adsorption by mineral surfaces and kerogen (see Figure 1.24), and so these compounds are preferentially retained as the less strongly adsorbed alkanes are being expelled (Tissot and Pelet, 1971).

Migration of gas differs in some important aspects from oil migration because a free gas phase is compressible, and pressure relationships are different. In addition gas molecules (C_1 through C_4) are smaller, which means that they have higher solubilities in water. They also migrate more rapidly, and diffusion can be a significant transporting mechanism (Leythaeuser et al, 1980, 1982).

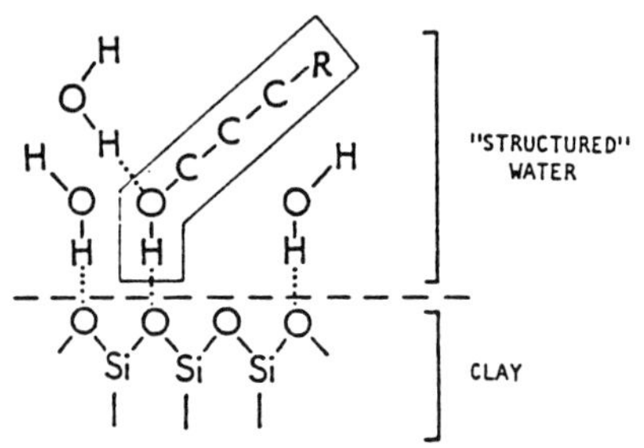

Figure 1.24. Interaction of the –OH group of the R–C–C–C–OH alcohol with a clay surface leading to adsorption. Hydrogen bonding is involved in the same way as it is for the –OH groups in water molecules. NSO molecules are adsorbed in this way and are retained in the source rocks while hydrocarbons are preferentially expelled.

Efficiency and Timing of Expulsion: The fact that extractable bitumens are present in rocks which have been effective sources shows that the migration process is not 100% efficient, and that much of what was generated remains in the source rock and never accumulates in a reservoir. Expulsion efficiency can be quite low, but appears to be a function of organic matter content of the source rock, with very organic-rich rocks expelling more than 80% of the generated bitumen (Mackenzie and Price et al, 1987; Espitalie et al, 1988). This follows from a consideration of the migration mechanism proposed above and laboratory simulations (Lafargue et al, 1994). The quantity of bitumen needed to fill the available pore space depends on the nature of the rock, while the availability of bitumen depends on the organic matter amount and type. Momper (1978) has claimed that it takes an average of 850 ppm of bitumen to fill up the pore system before anything can migrate out (analogous to filling a pipeline before products appear at the far end). In a very organic rich rock, 850 ppm will be generated earlier than in a lean one (Figure 1.25). Not only do rich rocks start to expel earlier, but they also expel a greater percentage of what is generated. The time delay between onset of generation and onset of migration will depend on the organic content of the source rock.

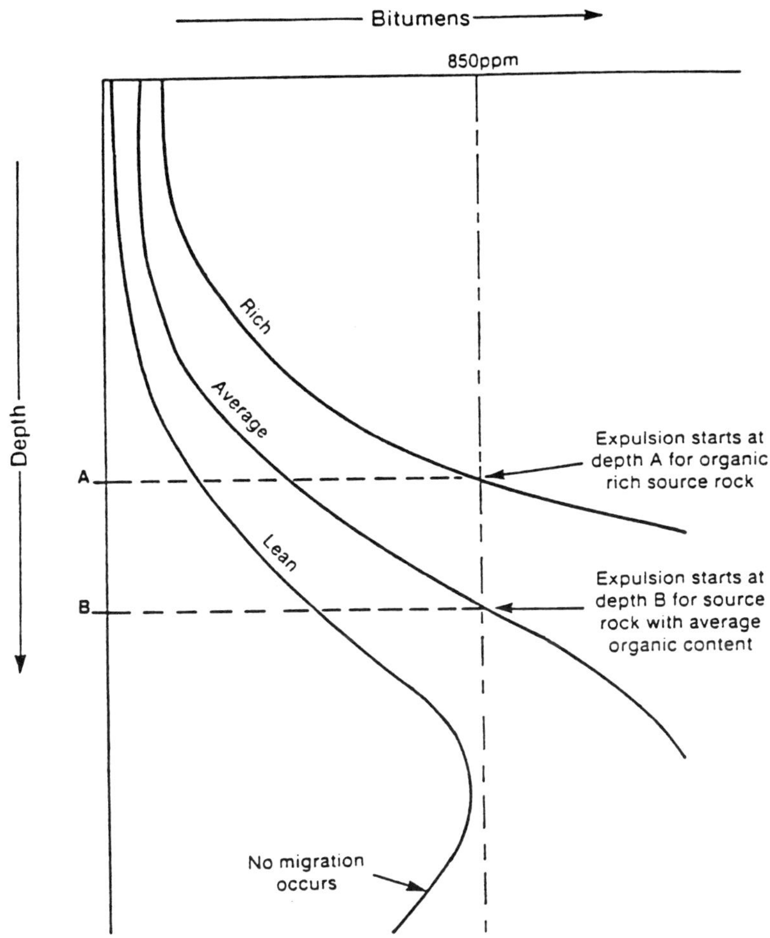

Figure 1.25. Bitumen content of source rocks as a function of depth. If some threshold value for the amount of bitumen (850 ppm in this example) is needed to initiate expulsion, then rich source rocks will start expelling before average ones. Rocks with low organic contents may never generate enough bitumen to drive migration.

If the rock does not generate enough bitumen to induce migration nothing will be expelled, and with rising temperature, the organics will be cracked to gas. This may be expelled later (due to the volume increase during cracking), so that a lean rock will be a gas source regardless of the type of organic matter it contains.

1.7.4. Transport through Carrier Beds

The first step of migration is called "*expulsion*" (Figure 1.22, (1)) and ends when the generated petroleum leaves the source rock. The next step involves transporting the oil droplets to a reservoir over distances that may be more than 100 km horizontally (Figure 1.22, (2)), and thousands of meters vertically (Demaison and Huizinga, 1991). This requires an adequate conduit through the rock system that may be provided by laterally continuous permeable rock strata, by faults, or in some cases by unconformity surfaces where permeability has developed. The behavior of faults as avenues for migration is complex and often difficult to predict since they can range from conduits, through partial conduits, to very effective seals. In discussing migration routes, it is important to stress the poor control that is available, even in very well drilled basins (see discussion in Section 2.1), and to recognize that it is the migration path available at the time the source rock is generating (not the present one) that is important.

Buoyancy is the main force acting on oil droplets in carrier beds and operates vertically upwards, so oil droplets will tend to move up. A drop may be stopped initially at a narrow constriction, but as more oil accumulates below it the pressure differential rises until finally the oil moves through. If the vertically rising oil meets a rock which has very small pores (like a shale) or is effectively impermeable (such as an evaporite, or cemented carbonate), then the pressure differential generated from buoyancy may never become high enough to continue moving the oil upwards. The overlying rock is then acting as a seal. Notice that the same shale can act as both a source rock and a seal. This happens because oil is forced out of the shale down a pressure gradient generated by internal overpressuring where pressure differentials are high enough to cope with small pores, but the oil cannot reenter the shale because the forces generated by buoyancy are inadequate for the drop to negotiate these small pores. If the base of the seal is sloping, then there is a component of buoyancy updip and petroleum will tend to move in that direction, assuming, of course, that there is a permeable pathway.

Direction of Migration: The overall route followed by the migrating petroleum is upward, but the actual route will depend on the distribution of adequate permeability. Permeable rocks are probably the most common route in many structurally simple settings and tend to introduce a significant lateral component to oil migration. Some examples in continental interior basins are discussed later (Section 10.2).

The upward movement of petroleum under the influence of buoyancy does not eliminate the possibility of moving from a younger source rock into an older reservoir. The giant Hassi Massaoud field in Algeria, for example, is a Cambrian sandstone reservoir but contains oil from a Silurian source rock. In this case, the Silurian shales lap up onto a Cambrian high and the oil has migrated upwards to accumulate in the reservoir. A similar situation occurs in rifted areas, like the North Sea, where petroleum from source rocks in the grabens migrates up faults to accumulate at the tops of the rotated fault blocks. "Buried hills" that represent considerable relief on an old exposed surface can accumulate oil from younger source rocks deposited between the highs. Also, younger rocks lapping onto an unconformity can sometimes lead to oil accumulations in older reservoirs.

Migration Distance: Oil will continue to move under the influence of buoyancy until it encounters a permeability barrier (or is lost to the surface). If there is continuity of migration pathways, then petroleum can move very long distances, ultimately moving the dimensions of the basin. Geochemical correlation between oils and source rocks is providing an increasing number of well-documented cases where migration has moved oil into a reservoir more than 100 km from the source rock. Examples of long-distance migration in the United States have been given by Momper (1978), and include data from Williams (1974) for the Williston basin, where mature source rocks are located in the deeper parts of the basin and are nearly 150 km from some reservoirs. In the Denver basin, oils have migrated as much as 125 km from the area of mature source rocks in the western part of the asymmetric basin up onto the shallow eastern flank Momper and Williams (1985) have documented very long distance migration of Phosphoria-sourced oils now reservoired in the Powder River Basin. This is a particularly interesting study because the postulated migration pathway does not exist today due to the presence of the Casper Arch and the Big Horn Mountains. However, the uplift developed relatively late (and defined the basin), but at the time the source rocks started generating, there was a simple updip migration pathway into what is now the Powder River Basin.

Examples of very long distance migration (>100 km) are most common in intracratonic basins because these are structurally simple with only minor, basement-controlled tectonics, so they preserve long-distance continuity of carrier beds (e.g., Clayton and Swetland, 1980 for the Denver basin, and Burrus et al, 1996 for the Williston basin). In contrast, the big Tertiary deltas are broken up by growth faults, and rifted margins are broken up by horst and graben development. In both these settings, migration has a dominant vertical component and lateral movement is less significant. Thus migration distance is a strong function of geological setting. This is a point that we will return to later because many of the widely used thermal models only calculate source rock maturity, and in some cases the reservoirs (which are, of course, the exploration targets) can be more than 100 km from mature source rocks.

Traps--Termination of Migration: Migration ends (at least temporarily) when petroleum reaches a trap where the buoyancy forces are no longer capable of moving it. The detailed geologic configurations of traps need not concern us (see Levorsen (1967) and North (1985)), but information about the time of trap formation and the efficiency of trapping is important.

Clearly, traps must develop before migrating petroleum arrives. This requires that the trapping configuration forms before the time of petroleum generation (but not necessarily at a time greater than the age of the source rock). (This is shown schematically in Figure 8.6). In practice, the thermal models provide a way of establishing the time of generation, so that it is often possible to predict whether the reservoirs were available to trap migrating oil and gas.

1.7.5. Remobilization

When a sequence of structural traps occurs along a carrier system with oil moving through it, the migrating crude starts to accumulate in the first reservoir that has closure, and continues to accumulate there until the trap is full. As more hydrocarbons flow through the carrier system towards this trap, it must spill because it cannot hold any more. Thus the next updip trap starts to fill. This is an important sequence of events with exploration significance, because it implies that if a trap in a thermally immature area is drilled and is not full to the spill point, then the downdip trap must be full while all the updip traps will have no production. The Media and Eagle Mesa fields in the Entrada sandstone of the San Juan basin provide a good example, and are shown in Figure 1.26 (Vincelette and Chittum, 1981). This situation often develops on the flanks of intracratonic basins where long-distance continuity of permeable carrier beds is coupled with downdip petroleum generation in the deeper, axial parts of the basin. The oil will flow updip toward the basin margins where rocks are still geochemically immature. Traps spill off the bottom at the spill point, but if there is any water washing or biodegradation, it will affect the oil most at the oil-water contact. Thus the most degraded oil will move preferentially updip, leading to a sequence of API gravities that decrease towards the basin margin.

When remobilization involves reservoirs containing both oil and gas, some important separations can occur. Initially, a full reservoir will

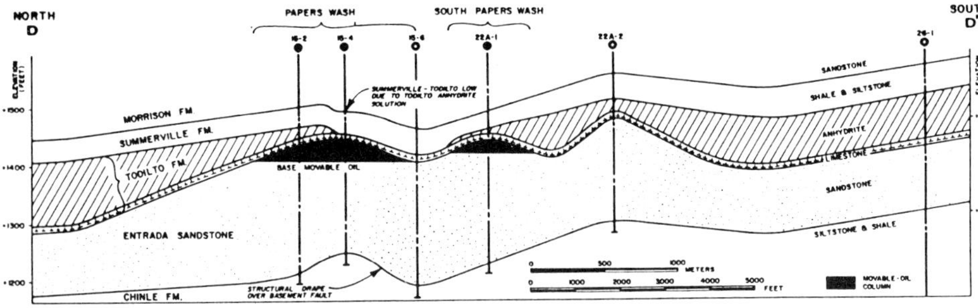

Figure 1.26. Structural cross section of the Papers Wash field in the San Juan basin (from Vincelette and Chittum, 1981).

contain a gas cap at the highest part of the structure and an oil leg which extends down to the level of the spill point. If more petroleum--either oil or gas--is introduced into this reservoir, the capacity of the trap will be exceeded and spilling will result. But, it will be the oil from the bottom which spills regardless of the size of the gas cap (Figure 1.27). In a geologic situation where a sequence of structural traps are related to a single carrier, the end product of this "fill'n spill" process will be downdip traps that are full of gas, then one with a gas cap over the oil, then one or more traps full to the spill point with oil, then minor oil, and from there updip, no production. This sequence is illustrated in Figure 1.27 and was called "*differential entrapment*" by Gussow (1954).

Many examples of differential entrapment have been described in the literature in addition to those originally given by Gussow (1954, 1968) (e.g., Gill, 1979). Most of these examples are for fields located on the flanks of continental interior basins, but differential entrapment can occur in other settings. Barker (1979), for example, has suggested that it can occur in deltas due to spilling from rollover anticlines up growth faults, and suggested that the Niger delta might provide examples. Differential entrapment in a rift setting has been described by Thomas et al (1985). Their interpretations of oil and gas distributions in the central graben of the North Sea involves generation in the Oseberg "*kitchen*" (see Figure 1.28) with migration westward through the sequence of traps at North Alwyn (oil and gas), Brent (oil and gas), Statfjord (oil), 34/7 structure (oil), and Gullfaks (oil). Eastward migration has led to oil and gas in the Oseberg field, oil in Brage, and finally spilling of oil into Troll. The gas in Troll was generated in the more northerly Troll kitchen and arrived through a different migration route.

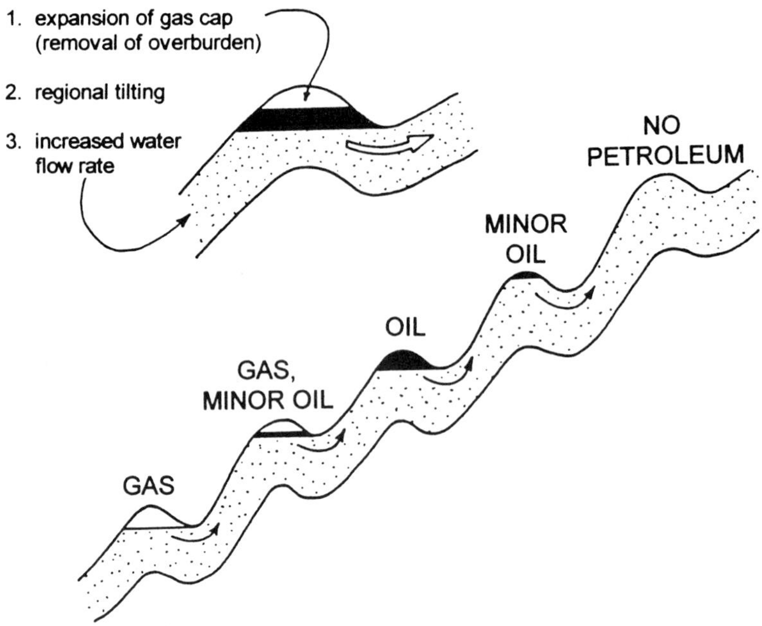

Figure 1.27. Schematic representation of the possible results of differential entrapment.

1.8. MATURATION AND ALTERATION OF PETROLEUM IN RESERVOIRS

1.8.1. Introduction

The chemical composition of a crude oil is not fixed, but changes in response to changing conditions in much the same way that kerogen composition changes. A considerable body of information is now available documenting the evolution of crude oil composition, and these changes are best considered under two headings. First, "*maturation*": this term is used for all processes which occur as an inevitable consequence of increasing age and depth of burial. Second, "*alteration*": this is caused by any external influence which acts to change the composition of the crude oil, and usually involves flowing water. All crude oils are subject to maturation, but the effects of alteration may vary widely from one crude oil to another, or from one part of a reservoir to another. In many cases the effects of alteration are absent.

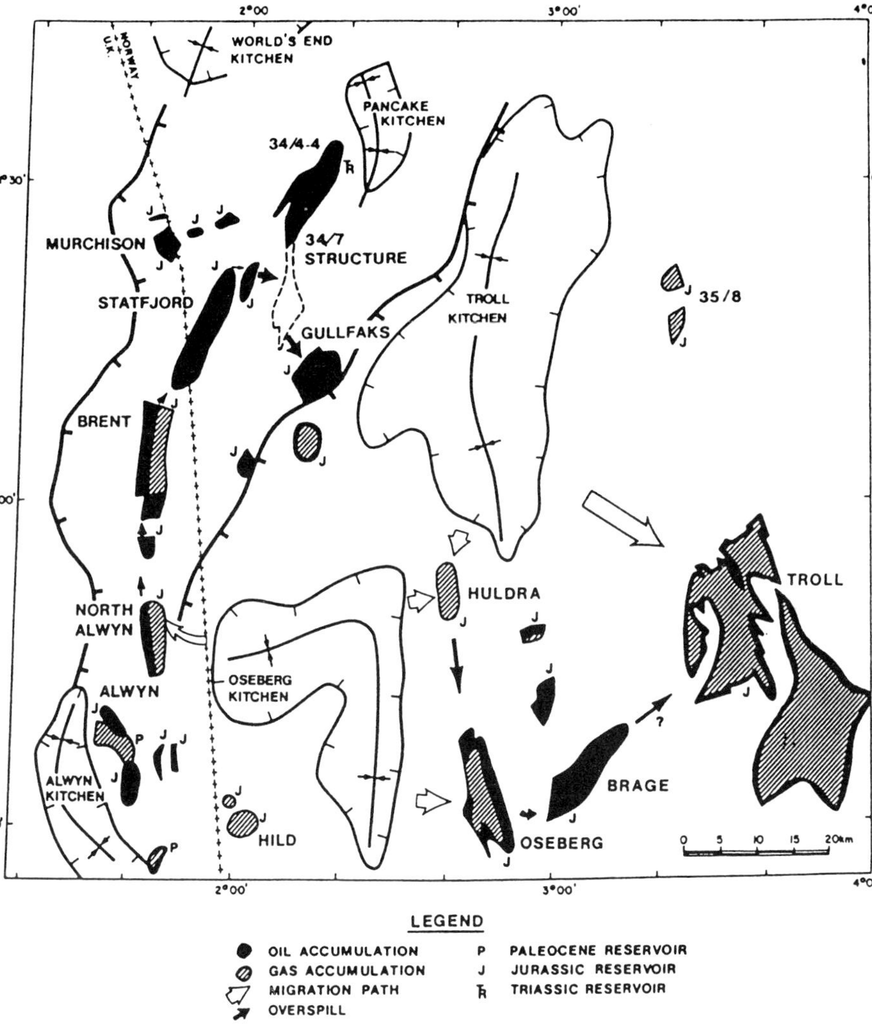

Figure 1.28. Inferred migration pathways in part of the North Sea showing the effects of differential entrapment (Thomas et al, 1985).

1.8.2. Maturation

Temperature: Temperature has a major role in maturation. Crude oils are not equilibrium mixtures and as temperature increases they readjust towards thermodynamic equilibrium at an increasing rate in a process of continuous irreversible change, usually called "*thermal maturation*". This readjustment takes the form of a redistribution of hydrogen (Figure 1.29)

ultimately giving methane and a solid carbon-rich residue in a process similar to that discussed for the thermal maturation of kerogen. As oils get more thermally mature, they become lighter (API gravity rises), have reduced amounts of NSO compounds, and become isotopically heavier.
Natural Deasphaltening: The steady increase in the concentration of light ends which accompanies thermal maturation may lead to precipitation of asphaltenes in a process known as "*natural deasphaltening*". This produces marked changes in the chemical and physical properties of the remaining crude oil because the solid precipitate takes with it sulfur, nitrogen, and metallic compounds, leaving the producible oil lighter and of higher quality. However, the solid precipitate can degrade reservoir quality by reducing both permeability and porosity (Lomando, 1992). Although it is often disseminated throughout the reservoir, in some cases the precipitated material accumulates close to the oil-water contact as a tar layer.
Floor for oil: The conversion of oil to gas with increasing temperature implies that there is some definite temperature above which oil will not be found. On the other hand, it takes a certain amount of thermal exposure for oil generation to begin, so oil is only available over a limited temperature range. This is called the "*oil window*", or the "*liquid window*", and has as its lower boundary the "*floor for oil*". Gas formed by cracking oil is called "*thermal gas*". This dry gas is generally associated with a graphitic residue, has high contents of carbon dioxide and hydrogen sulfide, and is frequently overpressured. Thermal gas is isotopically heavy, which serves to distinguish it from the isotopically light dry gas made by bacteria.

1.8.3. Alteration

Water washing: Several external factors operate to produce changes in the composition of crude oils. Most of these are associated with water because this acts as the transporting medium for materials being removed from the crude oil or being brought into contact with it. Water flowing past crude oil preferentially removes the most soluble compounds, and these include the lower molecular weight hydrocarbons, particularly the aromatics. The water-washed oil is depleted in gasoline range hydrocarbons and becomes much heavier (Lafargue and Barker, 1988; Palmer, 1993).
Bacterial degradation: Under some subsurface conditions crude oils can be degraded by bacteria (Connan, 1984; Palmer, 1993). The most severe bacterial degradation, or "*biodegradation*", is caused by aerobic bacteria that require free oxygen. They also need nutrients and cannot tolerate temperatures above about 75 ^{o}C. These conditions are often satisfied in

situations where penetrating surface waters bring oxygen and nutrients into the subsurface and remove toxic byproducts.

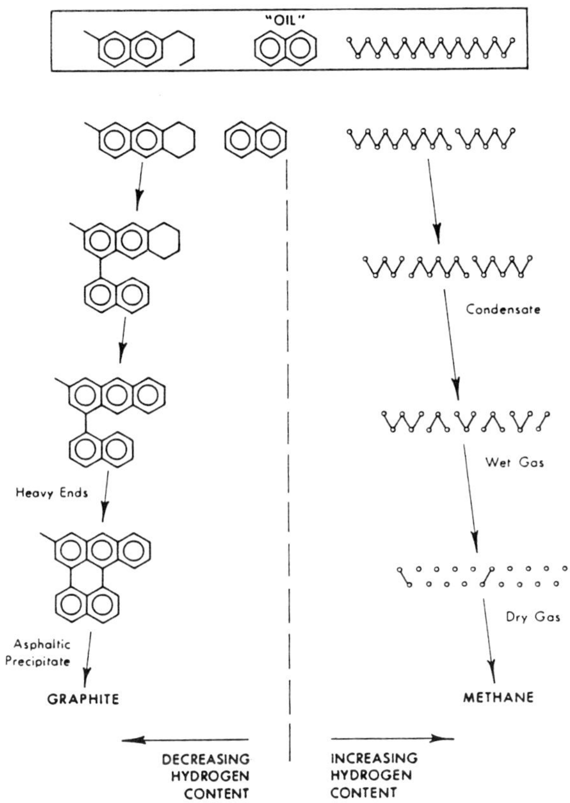

Figure 1.29. Schematic representation of the thermal maturation of crude oil showing the parallel development of increasingly aromatic heavy molecules and progressively lighter alkanes (Barker, 1985).

The bacteria eat n-alkanes preferentially, and will remove them completely before consuming branched chains (including pristane and phytane) and side chains from naphthene rings. This leads to minor changes in the biomarkers, but subsequent degradation causes more drastic modifications of first the steranes and later the hopanes (Table 1.3). Bacteria do not appear to consume most large aromatic molecules or sulfur compounds. The overall effect of biodegradation is to produce a very heavy, viscous, high sulfur crude (Figure 1.30). Biodegradation is very common and is the major process controlling crude oil composition.

Table 3
Scale of biodegradation proposed by Volkman et al (1983)

LEVEL	CHEMICAL COMPOSITION/ALTERATION	BIODEGRADATION
1	Typical oil; abundant n-alkanes	Not degraded
2	Light end n-alkanes removed	Minor
3	>90% n-alkanes removed	Moderate
4	Alkylcyclohexanes removed isoprenoids reduced	Moderate
5	Isoprenoids removed	Moderate
6	C_{14} - C_{16} bicyclic alkanes removed	Extensive
7	50% biological configuration steranes removed	Very extensive
8	Steranes altered; demethylated hopanes abundant	Severe
9	Demethylated hopanes predominate no steranes	Extreme

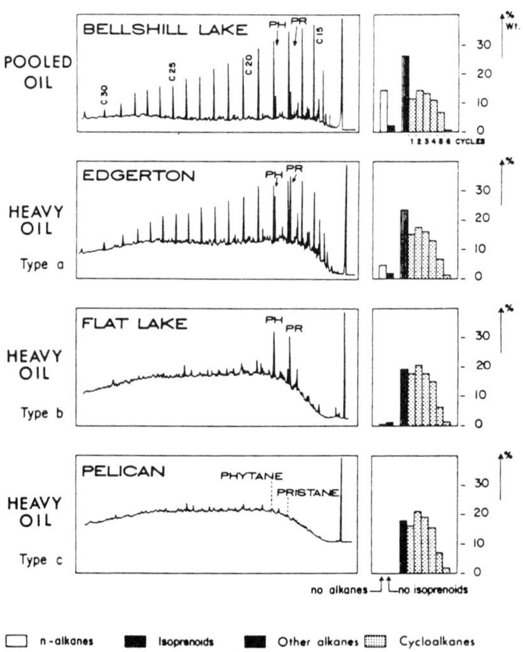

Figure 1.30. Gas chromatograms and compositional data for Canadian oils showing the progressive changes produced by bacterial degradation. Unaltered oil is at the top, and the most severely degraded oils are at the bottom (Deroo et al, 1977).

1.9. CLASSICAL SOURCE ROCK EVALUATION

1.9.1. Introduction

One of the most important uses of geochemistry in petroleum exploration is to identify the source rocks that generated petroleum. Qualitative techniques have been widely used for many years, and these form the basis for the development of the quantitative methods that are the main focus for most of this book.

For many years explorationists have assumed that black shales would be source rocks. There is at least some truth in this because red beds, for example, generally indicate oxidizing conditions and imply low organic matter contents. However, some source rocks are not black and some black shales contain little organic matter – the color being due to pyrolusite (MnO_2) or other inorganic material. It now seems that an evaluation of a shale, or other rock type, for source rock character must provide answers to the following questions (Barker, 1979):

(1) Does the rock have sufficient organic matter?
(2) Is the organic matter capable of generating?
(3) Has this organic matter generated petroleum?
(4) Has the generated petroleum migrated out?
(5) Is the rock oil-prone or gas-prone?

These questions are discussed individually below.

Correlation of source rock extract to a reservoired crude oil provides the only direct indication that generation and migration have occurred. It is assumed that the generation of petroleum drives the migration process (see discussion on migration), and that a certain minimum quantity of bitumen is needed before expulsion can occur. This threshold value is set by the need to saturate adsorbers in the rock and to fill (at least partially) the pore system. A rock will not be an effective petroleum source rock until the threshold has been exceeded.

1.9.2. Quantity of Organic Matter

The requirement that source rocks generate commercially significant quantities of oil and gas implies that there is probably some lower limit for the organic carbon content in effective source rocks. A relationship between high organic matter content in rocks and the occurrence of petroleum is well documented, both globally and regionally. In an extensive Russian study, Ronov (1958) analyzed nearly 26,000 samples from both petroliferous and nonpetroliferous areas. The average organic content in the latter was below 0.5%, whereas in petroliferous areas it was nearly three times as high. These studies (and others) led to the widely accepted use of 0.5% TOC as a lower cut-off for effective source character. However, the lower limit for an effective source rock will depend on the nature of the rock framework, and many authors now use values of 1.0% (or even higher) as their lower limits, but the values may be as high as 2% TOC (Lewan, 1988). A lower limit for organic carbon also appears to be consistent with the idea that it is necessary to generate a certain amount of bitumen before migration can start. This is an important concept because it means that a lean rock with less than 1.0% TOC will never generate enough oil to initiate migration, and therefore it is not possible to offset low organic contents with large volumes of rock.

1.9.3. Type of Organic Matter

The organic matter in potential source rocks has to be capable of generating petroleum. Organic matter type has an important influence on the nature of the petroleum products (gas versus oil), and specific details are discussed below. Recycled material, which has already been buried once and is now no longer capable of generating, should be excluded from the total when establishing the percentage of organic matter in the rock. Most clastic rocks contain at least a few particles of recycled material. Degradation during transport generally produces a low-hydrogen, high-oxygen material that plots below the Type III track on a van Krevelen diagram, and this non-generating organic matter is sometimes called Type IV kerogen or "*inertinite*".

1.9.4. Generation Status ("*Maturity*")

Introduction: Both physical and chemical characteristics of kerogen show the combined effects of time and temperature and indicate present thermal maturity. With increasing time-temperature exposure kerogen color intensifies and darkens; vitrinite reflectance increases; palynomorph translucency decreases; thermal stability (as indicated by pyrolysis) increases; and the elemental composition of kerogen shows a progressive increase in percent carbon and decrease in H/C ratio. These changes in the kerogen lead to changes in both the amount and chemical composition of the bitumen. A wide range of techniques has been developed for evaluating the maturity of source rocks using these changes in kerogen and bitumen. The more important ones are discussed below.

Kerogen: Physical characteristics: Kerogen color darkens and intensifies as increasing maturity drives the kerogen towards high carbon contents, and ultimately to black graphite. This sequence can be assigned a Thermal Alteration Index (TAI) value. Unfortunately, several schemes are being used, and while all of them start with 1.0 for the palest yellow, immature kerogen, the black end member is assigned a value of 6, 7 or even 10. Since color ranking is subjective, values are typically given as 2+, 3-, $1^1/_2$, etc., and values reported with more than one decimal place are not justified. Apart from inevitable operator bias in assigning a numerical ranking to a color, there are additional problems because color intensity varies with the thickness of the organic particles. This problem can be reduced somewhat by recording color for a single type of kerogen particle. Spores are often used and lead to a *Spore Coloration Index* (*SCI*) scale. There is also a *Conodont Alteration Index* (*CAI*) with a scale of 1 - 5 (Epstein et al, 1977).

The single most important maturity indicator is the reflectivity of vitrinite (R_o), because this increases steadily and predictably as maturity increases. Most other maturity indicators have been calibrated to give corresponding R_o values. In spite of its importance, the interpretation of vitrinite reflectance data is often not unambiguous, and reported values can be significantly affected by both experimental procedures and geologic processes acting in the subsurface.

Clearly for R_o to be used as a maturity indicator, the sample must contain vitrinite. Because vitrinite is a wood-derived kerogen that originates on the land surface, it is transported mainly by rivers and accumulates with clastic rocks. Carbonates, which commonly develop in clear waters indicating lack of clastic input, frequently do not have vitrinite. Cambro-Ordovician and older rocks will not have vitrinite because higher land plants had not evolved at that time.

Reflectivity is usually determined for at least 30 pieces of vitrinite in a sample. A large number of measurements is needed because there are natural variations in vitrinite reflectance due to minor variations in chemical composition, and because particles are cut randomly during polishing and reflectivity varies somewhat with orientation. An ideal, single vitrinite population in a low maturity sample will produce a histogram with a single, narrow bell-shaped curve, but at higher maturities (1.5% R_o or more) the peak broadens considerably because vitrinite becomes increasingly anisotropic. Real samples commonly show multiple populations in the histogram, such as those given in Figure 1.31. Peak A may be due to primary vitrinite with B and C being oxidized and/or recycled material. It is also possible that B is the primary population, C is oxidized and A is due to caved material from shallower in the well. As a final possibility, populations A and B could be cavings with C, the primary population. In the absence of any other information it may be difficult or impossible to decide among these alternative interpretations. For this reason, it can be very dangerous to depend on vitrinite reflectance data obtained from a single sample of cuttings. Core samples are better, but the most reliable estimates of maturity come from vitrinite reflectance profiles down a well.

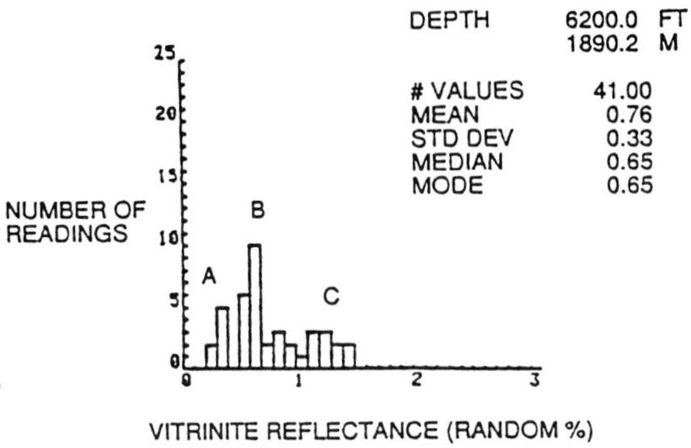

Figure 1.31. Vitrinite reflectance data for 41 particles showing several populations (from Dow and O'Connor, 1982).

In geologically simple settings, vitrinite reflectance starts at about 0.2% near surface and increases steadily with depth (Figure 1.32), but vitrinite reflectance is a poor measure of thermal maturity below about 0.4% reflectivity (Robert, 1988). As vitrinite becomes increasingly anisotropic at high maturities, histograms spread and recognition and resolution of multiple populations is more difficult. The rate of increase in reflectivity increases with depth, and plots of log R_o against depth are close to straight lines (Dow, 1977).

Simple R_o profiles can be modified by several geologic processes:

(1) *Thrust faulting* can move deeper, more mature samples with higher R_o values to shallower depths. In a well that crosses a thrust fault, the vitrinite reflectance profile will show an abrupt break from high values to lower ones with increasing depth (Figure 1.33). Normal faulting produces the opposite trend.

(2) *Erosional unconformities* show that part of the section has been removed and imply that material with the lowest range of R_o values has been lost. Subsequent sedimentation will deposit vitrinite with an R_o value of about 0.2% onto the erosional surface, and the profile will show an abrupt increase at the unconformity (Figure 1.33).

(3) *Igneous intrusions* cause locally high temperatures (though generally of short duration). Vitrinite reflectance is largely an indication of past thermal history, so that close to dikes and sills the values will be increased due to the elevated thermal regime. Intrusions, however, have

only a rather localized effect and if a dike or sill is D units thick, then thermal effects can be expected for about 2D units on either side (Dow, 1977; Wang et al, 1989) (Figure 1.33).

(4) *"Bitumen soaking"* occurs in rocks that have high organic contents and hydrogen-rich organic matter (as indicated by high pyrolysis HI values). This suppresses R_o values (Jones and Edison, 1978; Nuccio and Johnson, 1984a; Price and Barker, 1985).

(5) Caving of materials from shallower in the section may introduce contaminating vitrinite particles with lower reflectivities.

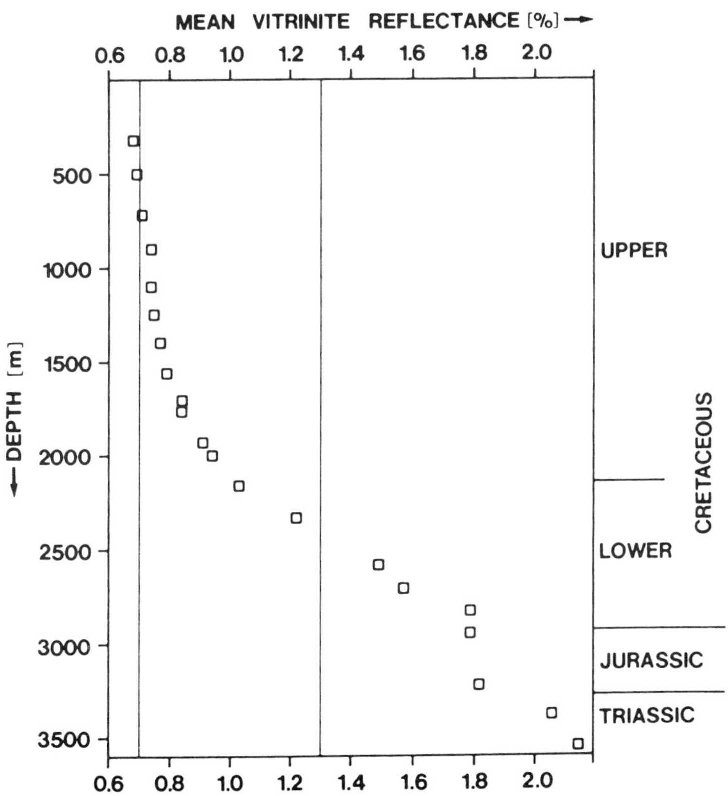

Figure 1.32. Profile of vitrinite reflectance with depth for a well in the Elmsworth field , Deep Basin of western Canada (Welte et al, 1984). Erosion of younger rocks has removed the part of the profile with vitrinite reflectance values below 0.65%.

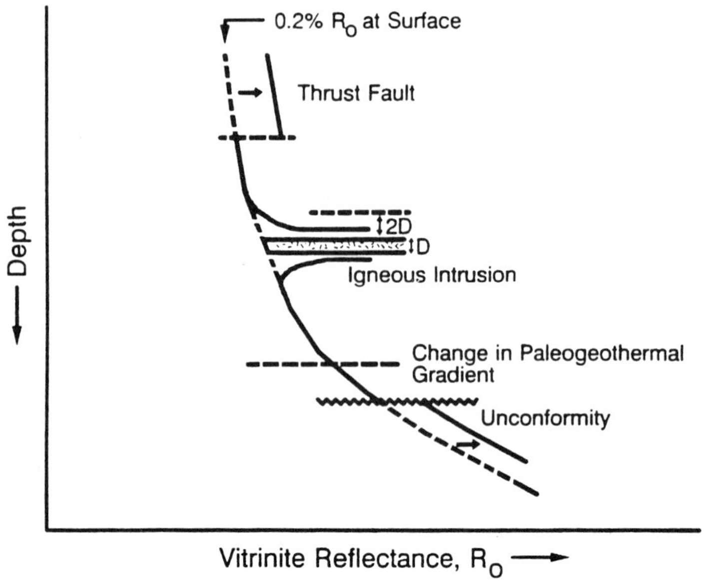

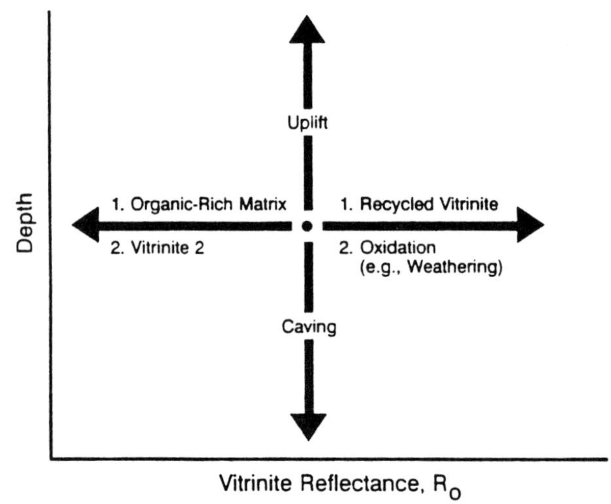

Figure 1.33. Schematic diagram showing some of the geologic processes that can disturb vitrinite reflectance profiles (top), and factors that may change the measured reflectance for a single vitrinite particle (bottom).

Vitrinite is just one of the many types of organic matter that may be present in a rock. It has its own characteristic kinetic parameters (E, A) and these may differ from those for other organic materials, so that on burial, rates of reaction including petroleum generation can be different. While R_o values have been calibrated to give a measure of thermal stress (essentially time-temperature influence), they may not always give a precise indication of maturation status for other types of organic matter.

Kerogen: Chemical Composition: Changes in the physical characteristics of kerogen are caused by changing chemical composition. The simplest indication of these chemical changes is the variation of elemental composition with maturity. Carbon, hydrogen and oxygen weight percentages (LaPlante, 1974; Durand and Monin, 1980); weight percent carbon; or atomic hydrogen/carbon ratios can be used. However, van Krevelen diagrams, where hydrogen/carbon ratios are plotted against oxygen/carbon ratios, are most widely used, and they have the added advantage of presenting kerogen composition in a way that not only shows maturity but may also indicate organic matter type (Tissot et al, 1974) (Figure 1.34).

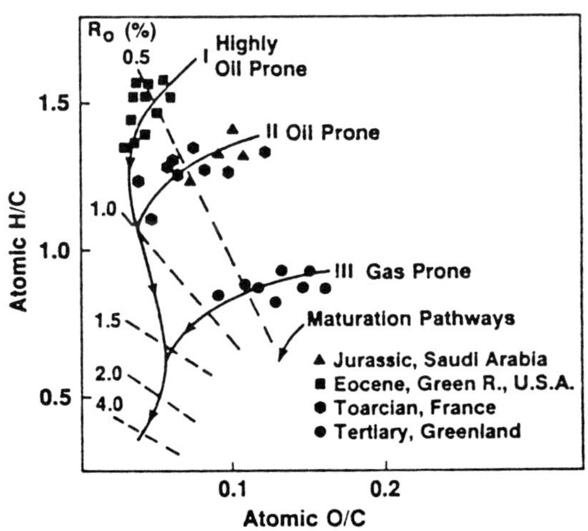

Figure 1.34. An example of elemental data for kerogen plotted on a van Krevelen diagram showing various hydrocarbon-generative types. Equivalent vitrinite reflectance values are superimposed. Mixed organic matter types that plot between the main tracks are common. (From Peters, 1986).

Bitumen: Amount: Although kerogen accounts for the major part of organic matter in rocks, it is the bitumen which eventually migrates and accumulates in the reservoir as petroleum. The most direct way of establishing the amount and composition of generated material is to analyze this solvent-extractable bitumen fraction. With increasing maturity, both the amount of bitumen and its chemical composition change. Either (or both) of these parameters can be used to help assess source rock maturity. Since bitumen is generated mainly from kerogen, maturity is usually established from bitumen/kerogen ratios, but it is worth remembering that the absolute concentration of bitumen is what initially controls the onset of migration. It is possible to have a mature rock (as indicated by R_o, percent carbon, etc.) that has generated insufficient bitumen to initiate migration. Strictly, the term *"bitumen "* refers to all extractable organic material, but it is often restricted to just C_{15+} when being used with source rock data.

Bitumen: Composition: The overall changes in bitumen composition with increasing maturity are shown by the gas chromatograms for the total solvent extract (Figure 1.35). The trend is for normal alkanes to increase relative to other components. Initially, these other materials are simply diluted, but with rising temperature, many biologically derived compounds break up, either to smaller fragments or to more complex residues that are not part of the soluble bitumen fraction. The small hump in the gas chromatogram that is due to multiring biomarkers steadily diminishes with maturity, and the main hump in the overall chromatogram also becomes less prominent. Alkane/isoprenoid ratios, such as n-C_{17}/pristane, are initially less than 1.0 but continues to rise with maturity, until at very high maturity levels there are essentially no isoprenoid chains left (Figure 1.36). As depth and temperature increase, the C_{15+}-bitumen/kerogen ratio decreases, reflecting the trend from oil-like materials towards gas.

Bitumen: Biomarkers: Rock extracts contain an enormous variety of compounds, including the complex but characteristic molecules derived directly from organisms that are generally called biomarkers (Mackenzie, 1984a; Philp, 1985). These molecules have variable, but limited, thermal stabilities and change in structure and composition as the host sediment is buried. These changes can be used to indicate maturity level.

Organisms synthesize compounds with very specific structures because these molecules must fit like a key in a lock if they are to work with the biochemical processes. In studying these molecules it is important to report spatial relationships in detail and the nomenclature can become complex. First, it is necessary to specify the carbon where branches are located using the standard numbering system that is available for all multi-ringed compounds (Figure 1.37). Then the orientation of the bonds

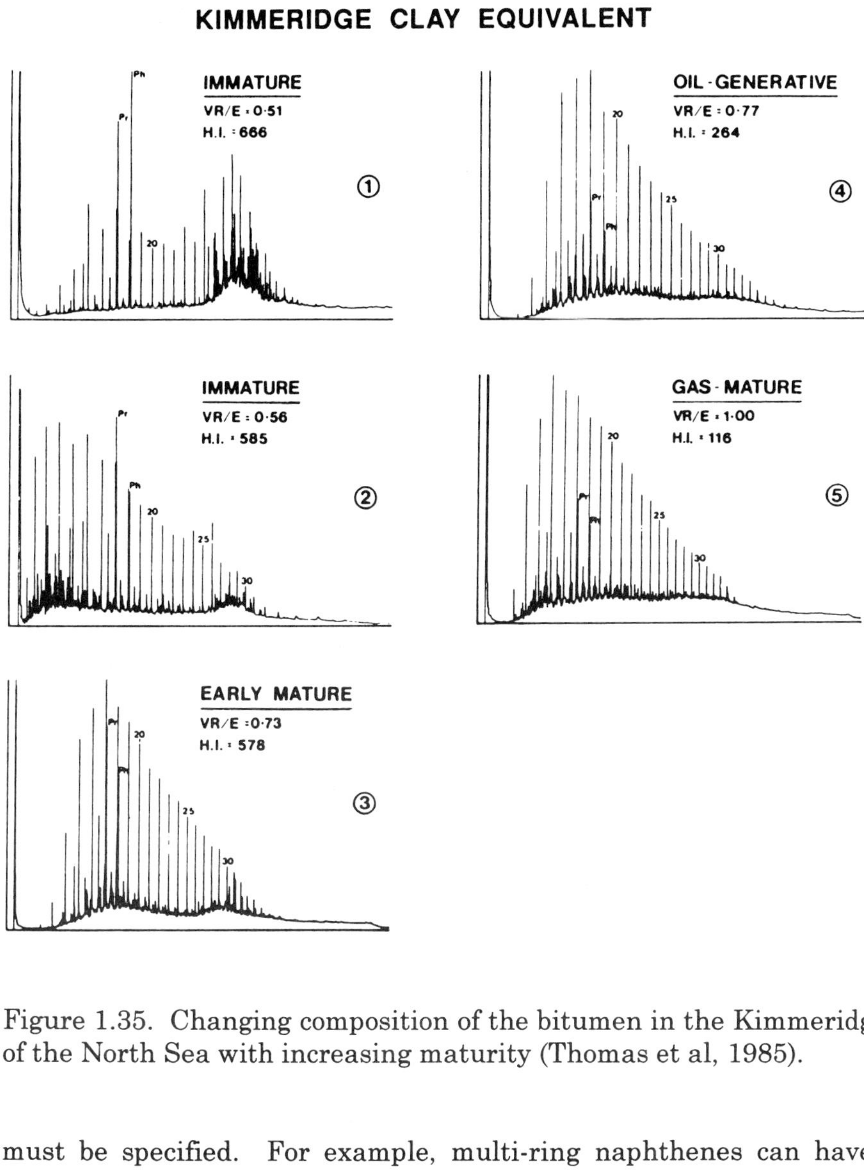

Figure 1.35. Changing composition of the bitumen in the Kimmeridge Clay of the North Sea with increasing maturity (Thomas et al, 1985).

must be specified. For example, multi-ring naphthenes can have short chains (or hydrogens) that are located either above or below the plane of the main part of the molecule. In fully describing such a molecule it is necessary to have some convention for indicating this stereochemistry. Commonly used conventions are shown in Figure 1.38. In addition, at each asymmetric carbon atom, or chiral center, it will be possible to have a right-handed or a left-handed form. Asymmetry occurs when four different groups are bonded to the carbon (see Figure 1.1), so that carbon No. 20 in the sterane structure of Figure 1.39, which has a hydrogen, a six-carbon chain, a two-carbon chain, and a four-ring group around it, is asymmetric. Both left-handed and right-handed forms can exist and they are designated *"R"* and *"S"* (from the Latin *"rectus"* for right, and *"sinister"* for left) (Mackenzie, 1984; Waples and Machihara, 1991; Peters and Moldowan, 1993).

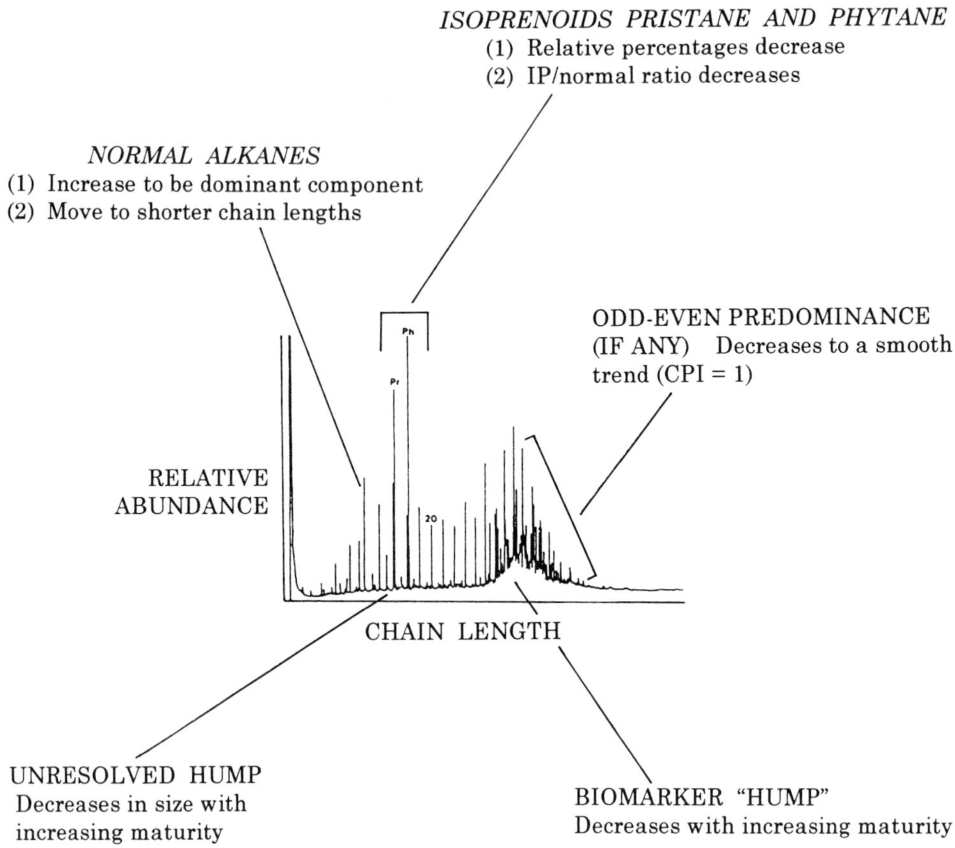

Figure 1.36. Changing character of bitumen gas chromatogram with increasing maturity.

Although the C-20 carbon atom in the C_{29}-sterane is a chiral center (Figure 1.39), and can assume either an S or an R form, organisms only make the 20R form, and initially the concentration of 20S is zero. With increasing maturity the ratio (20S)/(20R+20S) steranes rises from 0 to 0.50-

0.55 (Table 1.4 and Figure 1.40). An exactly analogous isomerization occurs at the C-22 carbon in the homohopanes. Similar modifications occur in the stereochemistry of other types of molecules, and this "*configurational isomerization*" is very commonly used to indicate maturity.

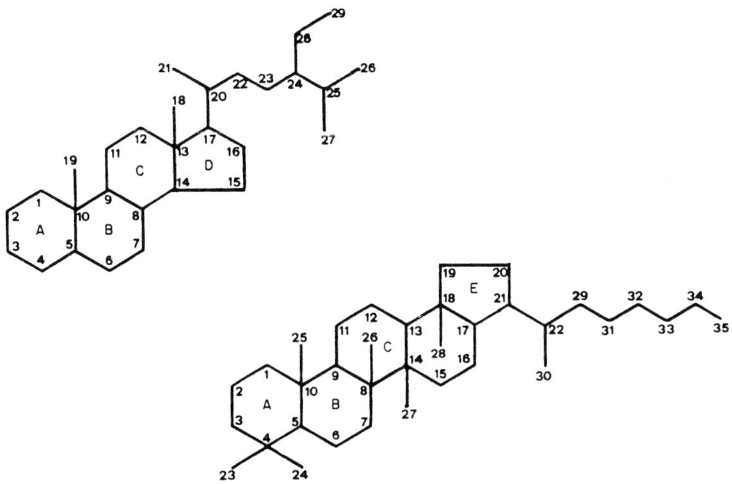

Figure 1.37. The standard numbering scheme for carbon atoms in steranes and hopanes.

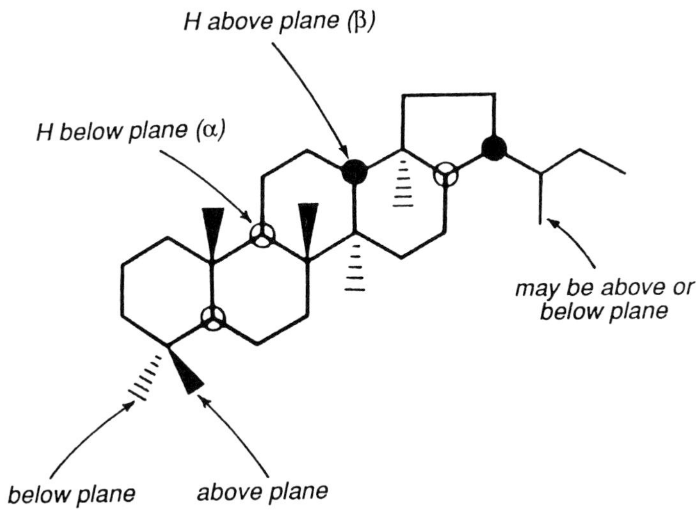

Figure 1.38. Some widely used conventions for representing the spatial relationships of bonds.

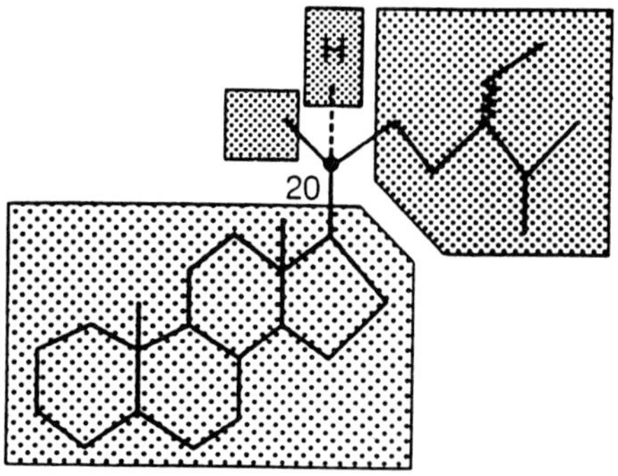

Figure 1.39. Sterane molecule showing the chiral center at C-20 and the four different groups bonded to it.

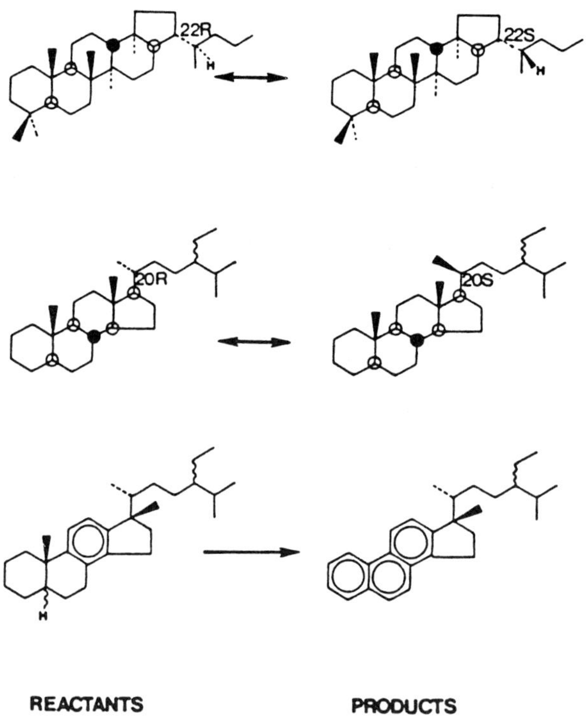

Figure 1.40. The three most commonly used, temperature-dependent geochemical biomarker reactions used for assessing the extent of thermal maturation in sediments and crude oils (from Mackenzie, 1984).

Table 1.4
Reactions of biomarkers used for determining thermal maturity (Mackenzie, 1984).

Reaction	Ratio (2)	Starting Value of (2)	End Value of (2)
Configurational isomerization	$\frac{\text{RR + SS}}{\text{total pristane}}$	0	0.5
	$\frac{\text{22R}}{\text{(22R + 22S)} - 17\alpha\text{(H)-hopanes}}$	0	0.6
	$\frac{\text{24R}}{\text{(24R + 24S)-steranes}}$	variable	0.5
	$\frac{\text{20S}}{\text{(20R + 20S)-steranes}}$	0	0.5–0.55
	$\frac{17\alpha\text{(H)},21\beta\text{(H)-hopanes}}{\text{total hopanes}}$	variable	0.9
	$\frac{5\alpha\text{(H)},14\beta\text{(H)},17\beta\text{(H)-steranes}}{\text{total steranes}}$	variable	0.8
Aromatization	$\frac{\text{triaromatic steroid hc.}}{\text{mono + triaromatic steroid hc.}}$	0	1.0
Apparent C–C bond cleavage	(e.g.) $\frac{C_{20}\text{ triaromatic steroid hc.}}{C_{20} + C_{28}\text{ triaromatic steroid hc.}}$	variable	1.0
	$\frac{\text{Aetio}}{\text{DPEP + Aetioporphyrins}}$	variable	1.0

Some molecules show systematic changes in chemical composition in response to increasing thermal exposure, and several of these changes are widely used to establish source rock maturity. Steranes are four-ring naphthenes, but compounds in which one of the rings is aromatic occur, and with increasing maturation there is a trend toward triaromatic steranes in which all three six-member rings become aromatic (Figure 1.40). Since there are no triaromatic steranes in living systems or recent sediments, this provides a convenient maturity indicator (Table 1.4). Aromatic molecules themselves show compositional trends with maturity and a pair of ratios involving relative concentrations of methylphenanthrenes have been proposed as maturity indicators by Radke et al (1982).

When a reaction involving the biomarkers has gone to completion (i.e., reached equilibrium), it is no longer controlled by time and temperature and cannot be used as an indicator of maturity. For some molecules, such as the unstable triterpanes called the moretanes, or the isomerization of pristane, this happens early. For others, it happens rather later, but in general, biomarkers are not useful for establishing maturation much above an equivalent vitrinite reflectance of about 1.0% (Figure 1.41). They are very useful, however, because they cover the important phases of oil generation and expulsion.

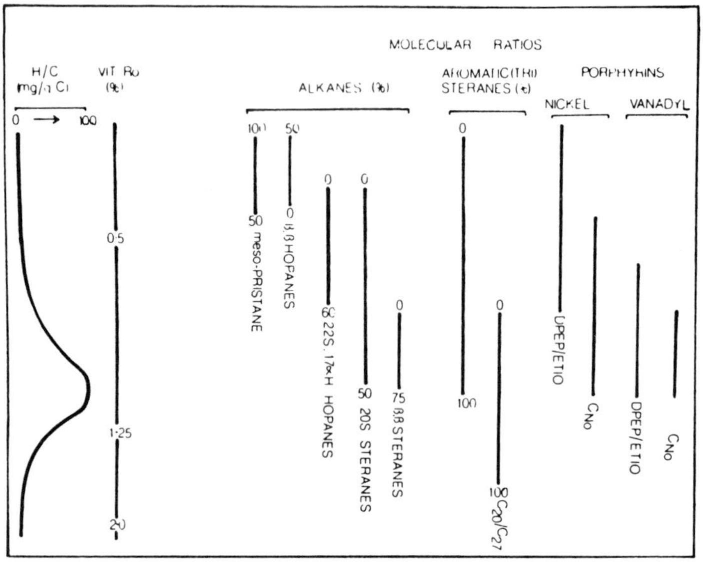

Figure 1.41. Ranges of individual molecular measurements of thermal maturation plotted against the downhole hydrocarbon generation curve and equivalent vitrinite reflectance values (Mackenzie, 1984).

Pyrolysis: A variety of pyrolysis techniques is being used for evaluating source rock generation status, as well as providing information about organic matter type (Horsfield, 1984; Larter, 1984; Barker and Wang, 1988). These methods offer some important advantages over the more conventional kerogen and bitumen evaluation techniques because they use very small samples, typically 20-100 mg, and only minor sample preparation is needed. Results are obtained in less than 1 hour and require little interpretation time. In one widely used pyrolysis procedure, small rock samples are heated at a steadily increasing temperature in an inert gas (usually helium) while the total evolved hydrocarbons are monitored as a function of temperature. Hydrocarbon release curves are called "*pyrograms*" and generally show a first peak around 200 °C which is produced by hydrocarbons already present in the rock due to generation in the subsurface (Peak 1), and a second peak at higher temperature which can be assigned to the thermal breakdown products of the kerogen (Peak 2) (Figure 1.42). Peak 1 is a bitumen peak but differs from the commonly used Soxhlet extract bitumen because it also includes the C_{15-} fraction. Peak 2 provides an indication of remaining generation potential (but does

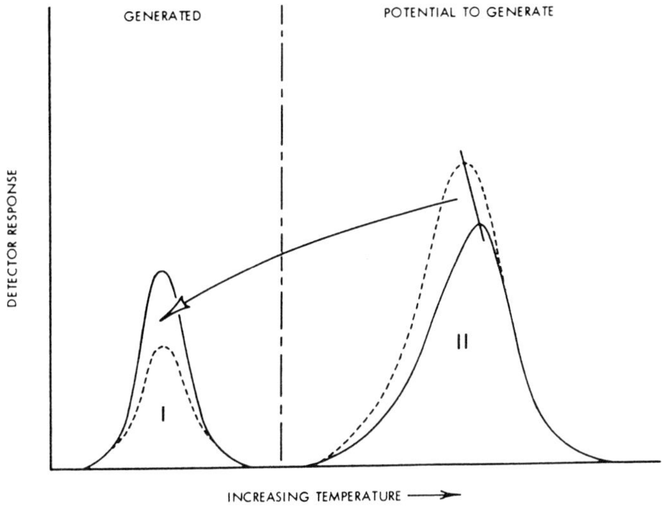

Figure 1.42. Schematic representation of the way in which the low-temperature pyrolysis peak (Peak 1, S1) increases with depth at the expense of the high-temperature peak (Peak 2, S2). Dashed line: response for shallow sample. Solid line: response for deeper sample.

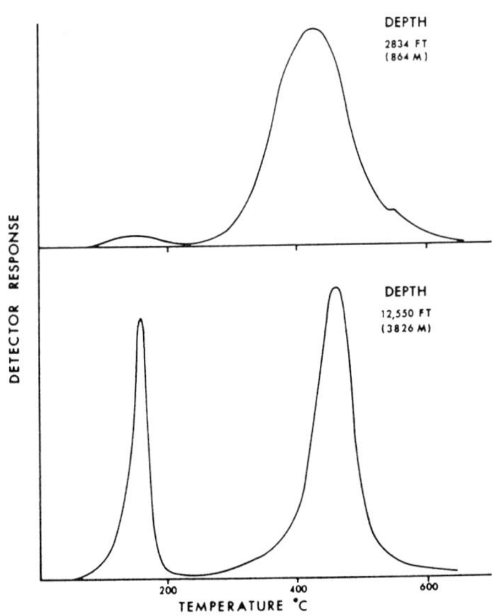

Figure 1.43. Yield of hydrocarbons as indicated by detector response with increasing heating of Miocene shale samples from a well in the Gulf of Mexico (Barker, 1974).

depend on organic matter type). As rocks are buried more deeply in the subsurface and subjected to higher temperatures, increasing amounts of hydrocarbons are generated and Peak 1 gets larger. This increase occurs at the expense of Peak 2, which not only decreases in size, but also moves to higher temperature because the less thermally stable material is broken down first during natural maturation, leaving a thermally more stable kerogen residue in the rock (Figure 1.43) (Barker, 1974). Both the relative sizes of the peaks and the temperature of the maximum in the hydrocarbon release curve for Peak 2 can be used to indicate maturity (Figure 1.44).

Relative peak sizes are given by the *"Generation Index"* (GI) defined by Barker (1974), although it is now more often called a *"Transformation Ratio"* (TR) (Espitalie, et al, 1977).

$$\text{GI or TR} = \frac{\text{amount generated}}{\text{total potential for generation}}$$

$$= \frac{\text{Peak 1}}{\text{Peak 1} + \text{Peak 2}}$$

This ratio steadily increases with maturity (Figure 1.44) and a value of 0.1 corresponds to the onset of generation and 0.4 marks the end of the oil window.

The temperature corresponding to the maximum value for Peak 2 is called "T_{max}". Values increase with maturity (Figure 1.44). While transformation ratios are relatively "transportable," in the sense that different instruments give comparable values, values for T_{max} vary with heating rate (increasing with increasing rate) and may differ from one instrument to another. It is important to recognize this when using published data or comparing data from different sources.

RockEval Pyrolysis: A commercial pyrolysis instrument called *"RockEval"* is being widely used for routine analysis of large numbers of rock samples. The system was developed initially by IFP-Fina and is now sold under license. Like many other pyrolysis systems, the hydrocarbons released from heated samples are monitored as a function of increasing temperature, but some models include another type of detector (called a thermal conductivity detector) that can detect the carbon dioxide collected over a preselected temperature range. The two hydrocarbon peaks (Peak 1, Peak 2) and the carbon dioxide peak (Peak 3) are usually called S1, S2, and S3, respectively (Table 1.5).

The temperature range chosen for carbon dioxide collection was selected to include the main stage of carbon dioxide generation from kerogen, but to avoid thermally decomposing inorganic carbonates. The quantity of carbon dioxide then becomes a measure of the oxygen content of the kerogen and S3/TOC has been called the "*Oxygen Index*." Similarly, Peak 2 is a measure of the hydrogen content of the kerogen so that S2/TOC is a "*Hydrogen Index*." A plot of Oxygen Index against Hydrogen Index produces a pseudo-van Krevelen diagram (Espitalie et al, 1977) (Figure 1.45). The position of a point on this diagram shows not only the maturity of the organic matter, but also organic matter type.

Table 1.5
Geochemical parameters and pyrolysis information used to establish source rock maturity, generating potential, and type of hydrocarbon product. Values are from Peters (1986) but there are variations among values quoted by other authors.

GENERATING POTENTIAL	TOC (wt%)	S1 mg HC/g rock	S2 mg HC/g rock
Poor	0.0 - 0.5	0.0 - 0.5	0.0 - 2.5
Fair	0.5 - 1.0	0.5 - 1.0	2.5 - 5.0
Good	1.0 - 2.0	1.0 - 2.0	5.0 - 10
Very good	>2.0	>2.0	>10

TYPE	HI	S2 / S3
Gas	0 - 150	0 - 3
Gas and oil	150 - 300	3 - 5
Oil	>300	>5

MATURATION	TR	Tmax	Ro
Top of oil window	~0.1	~435 - 445	~0.6
Bottom of oil window	~0.4	~470	~1.4

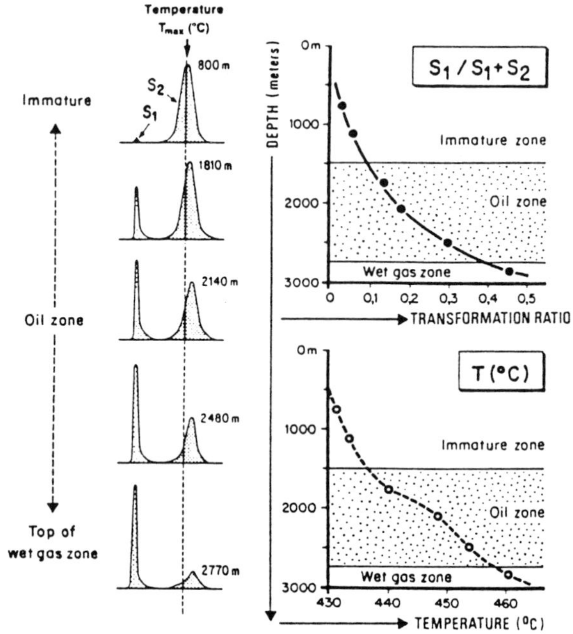

Figure 1.44. Characterization of source rock maturity by pyrolysis showing changes in Transformation Ratio and Tmax over a depth range of 1900 m (Espitalie et al, 1977).

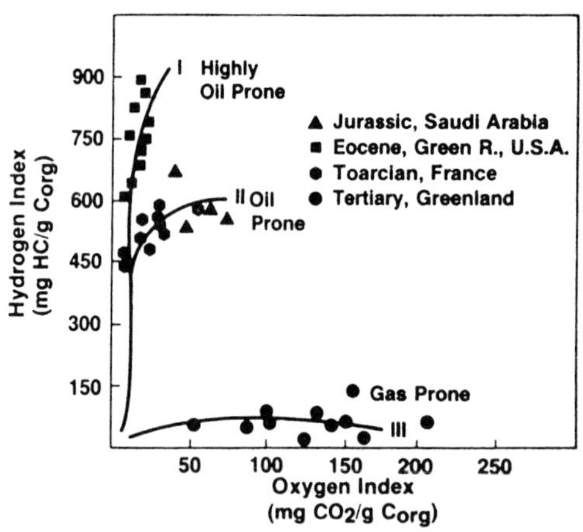

Figure 1.45. Plot of pyrolysis data as Hydrogen Index (HI) versus Oxygen Index (OI). This gives a pseudo van Krevelen diagram and can be used to define the three main kerogen types (Peters, 1986).

There has been some criticism of the validity of the assumption that all the carbon dioxide measured in S3 is in fact coming from decomposing organic matter (Peters, 1986). While hydrogen indices are useful, reliable, and widely used, oxygen indices should be viewed with caution.

Summary: Although many of the maturity indicators have been related to R_o values, there have been other attempts to relate all the commonly used parameters (Heroux et al, 1979). Not all of the techniques discussed for evaluating maturity are equally effective or applicable to all samples. In a comprehensive program it is advisable to have some preliminary analysis to screen samples and eliminate those with very low organic contents, which have no potential as source rocks. Pyrolysis is often used in this way because it only needs small samples and is quick and cheap. Total organic carbon contents can also be used to eliminate lean samples.

It is good practice to combine at least one kerogen-based method with at least one bitumen-based method (Figure 1.46) in evaluating maturity. Bitumen is mobile and a rock may be partially depleted due to loss by migration out, or conversely, may be contaminated by migration in. In the latter case it is commonly a migrating crude oil (or drilling fluid) that contaminates the rock, and this may introduce an apparent mature character into the bitumen of an immature rock, and also raise bitumen/kerogen ratios.

SOURCE ROCK EVALUATION

KEROGEN	Chemical	%C, O/C vs H/C, IR
	Physical	TAI, R_o
BITUMEN	Amount	ppm
	Composition	CPI, % normals, biomarkers
PYROLYSIS	O.M. type	response/TOM, GC
	O.M. maturity	$S_1/(S_1+S_2)$, Tmax

Figure 1.46. Summary of the techniques available for source rock evaluation.

1.10. NATURE AND AMOUNT OF PETROLEUM PRODUCTS

1.10.1. Introduction

In many areas, particularly those that are hostile or remote from consumers, the economics of transportation make oil much more desirable than gas. The ability to use source rock character to predict the nature of the hydrocarbons generated is important in exploration. At intermediate depths, the type of organic matter exercises primary control over whether oil or gas is the main petroleum product, but any type of organic matter will generate gas if buried deeply enough. It follows that being able to recognize organic matter type is a prerequisite for predicting the nature of the petroleum products. A range of techniques is available for characterizing organic matter type and clear distinctions can usually be made at intermediate maturity levels. Unfortunately, at high maturity all types of organic materials begin to appear similar -- they all become black, tracks on a van Krevelen diagram converge, etc. While this may show that the rock is currently gas generating or overmature, it does not help in deciding whether it followed a gas-prone or oil-prone track as it was buried. There is no easy solution to this dilemma.

1.10.2. Microscopic Characterization

Microscopic examination of kerogen concentrates often provides considerable information about the types of organic materials present. However, it can be difficult to get estimates of weight percentages of the various components. Several classification schemes have been developed and a variety of terms introduced, some with overlapping meanings (Figure 1.47).

Amorphous: Frequently, much of the organic material seen under the microscope lacks any recognizable structure even at high magnification. This "*amorphous kerogen*" is often algal-derived, though it is very unwise to assume that this is the case in the absence of supporting data. All types of organic matter can be physically degraded to the extent that they no longer retain any identifiable structure.

Algal: Recognizable remnants of algae are commonly observed in separated organic matter, especially in lacustrine sediments.

Herbaceous: This term covers a broad range of organic materials, including spores, pollen grains, cuticle, leaf epidermis, cellular structures of plants, and other recognizable discrete cell material.

Woody: Woody kerogen is derived from the structural support of higher land plants and often has a compact, angular morphology.

Coaly: Included in this category are angular to subangular fragments of naturally charred plant material (fossil charcoal, or "*fusain*") and recycled organic material.

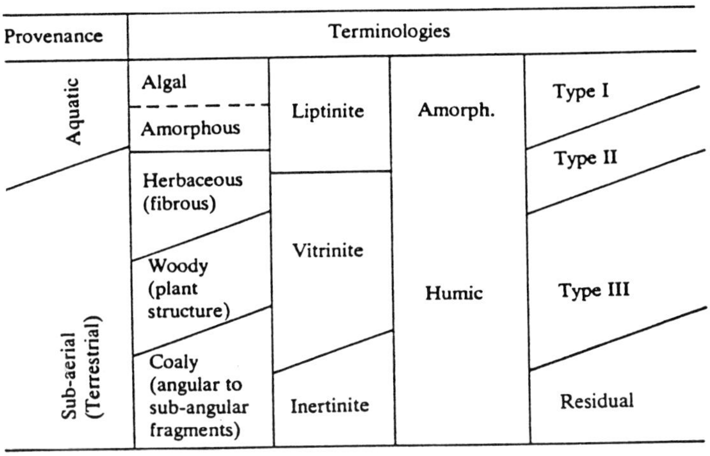

Figure 1.47. Approximate equivalents of various terms used in kerogen descriptions (after Cornford).

Often, a large percentage of the organic material in a rock will have a recognizable structure, and its source can be inferred with confidence. Spores and pollen grains provide obvious examples. In addition, since woody and herbaceous materials are terrestrially derived, they are transported into depositional environments where they accumulate with clastic rocks. Carbonates develop in clear waters, indicating a lack of clastic input, and are usually dominated by amorphous and/or algal kerogens.

Organic matter type becomes increasingly difficult to characterize as the size of the particles diminishes (especially below 1 μm), and as the organic material becomes increasingly mature and darker.

1.10.3. Elemental Analysis

Van Krevelen diagrams show both the type of organic matter and its stage of petroleum generation (Figure 1.48). As organic matter moves down the diagenetic tracks defined by Type I or Type II kerogens, it generates first oil, then gas. The gas is formed at decreasing rates until eventually the material is metamorphosed. Type III kerogen, in contrast, generates little, if any, oil before it forms gas. While the microscopic examination of kerogen provides information on single particles, bulk chemical methods (such as elemental analysis displayed on a van Krevelen diagram) present weighted averages of all the types of organic materials in the sample.

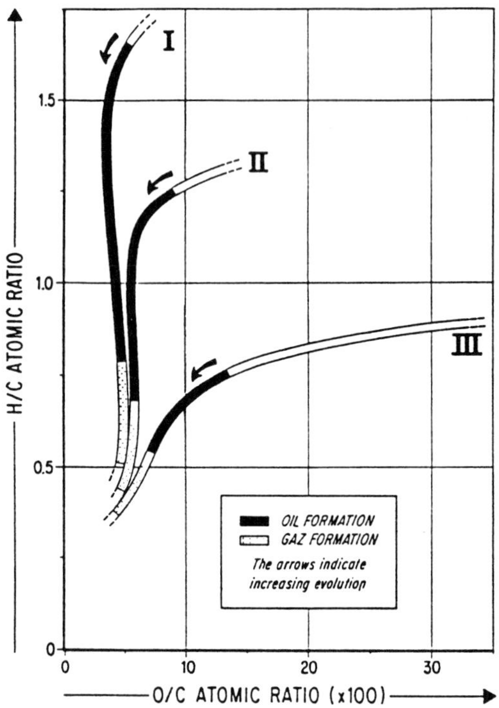

Figure 1.48. Changes in the elemental compositions of different kerogen types with increasing maturity (Tissot and Welte, 1984).

1.10.4. Pyrolysis Techniques

The information provided by pyrolysis can be interpreted in terms of organic matter type. Different types of organic matter have different thermal convertibilities (see below) and since pyrolysis is causing the thermal breakdown of the kerogen, the quantity of hydrocarbon products will depend on organic matter type. The ratio of (pyrolitic hydrocarbons)/(total organic carbon) is a measure of organic matter type at a given maturity level. In addition, pyrolysis products can be trapped and analyzed gas chromatographically to provide information about organic matter type.

1.10.5. Convertibilities

The quantity of petroleum generated from a given weight of organic matter varies with organic matter type. The hydrogen-rich, oil-prone kerogens are able to convert more of their carbon to hydrocarbon products and generally produce $2^1/_2$ times as much gas (if deeply buried) as the hydrogen-poor, gas-prone (Type III) kerogen (Figure 1.49). The Type I oil-prone kerogen is somewhat more prolific than the Type II kerogen, as would be expected from its slightly higher hydrogen content.

Lean rocks (<0.5 wt% organic carbon) that contain oil-prone kerogen may never generate enough bitumen to migrate as oil. With increasing burial, this bitumen may crack to gas and this gas may subsequently migrate. In this case, the organic matter type does not have the major role in controlling the nature of the hydrocarbon products.

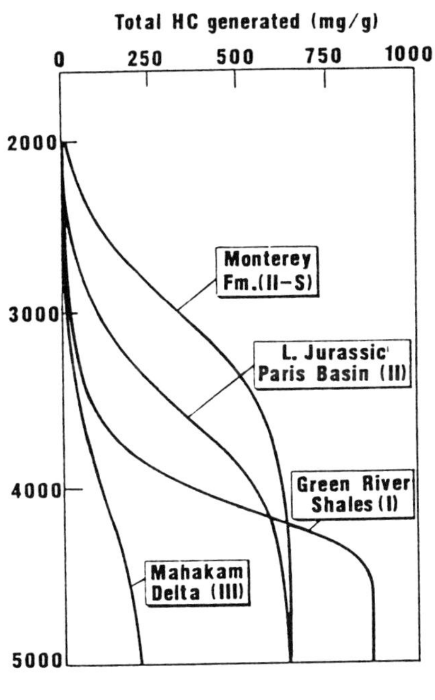

Figure 1.49. Calculated values of the quantities of petroleum products formed by the thermal evolution of the three main kerogen types (Tissot et al, 1987)

1.11. CORRELATION

1.11.1. Introduction

Geochemical correlation techniques provide methods for deciding whether oils are related (i.e., generated from a common source rock), and for establishing the source rock of a particular oil type. It is particularly important to know whether oils in an area belong to a single family, or are members of different families (Figure 1.50). A single oil type requires only one source rock, while the presence of several oil families shows that there must be more than one source rock in the area. Any area with multiple source rocks is more attractive for exploration, but each source rock will have to be modeled separately. Thermal maturation and biodegradation can make oils from a single family look very different, and it is essential that correlation tools are capable of recognizing similarities, even in severely altered and degraded oils.

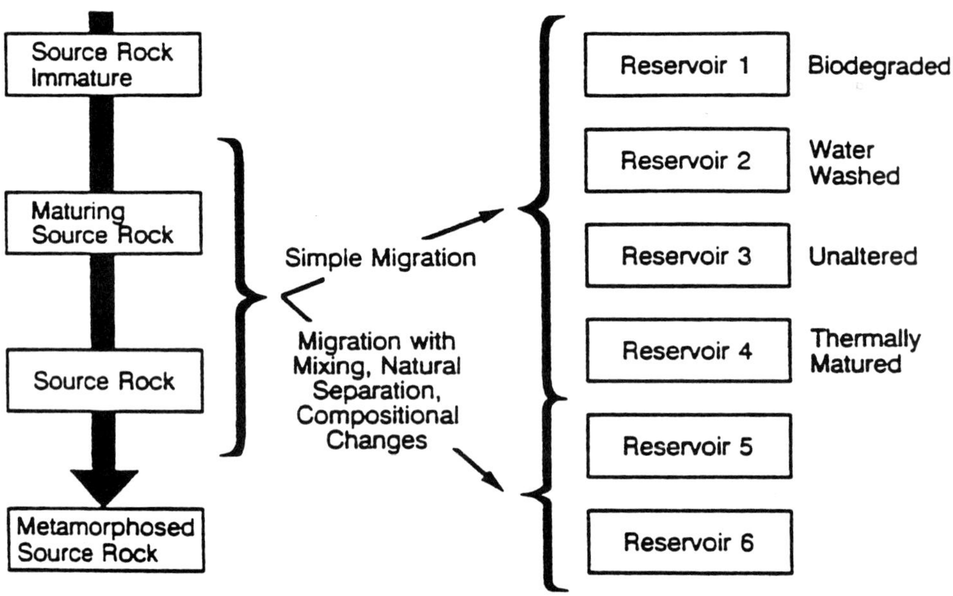

Figure 1.50. Illustration of the way in which the simple Source Rock Concept can be complicated in nature.

When an oil can be correlated to a specific source rock it often provides some constraints on possible migration pathways. Structures on pathways are prime exploration targets. In addition for three-dimensional modeling it is important to establish permeability distributions and a knowledge of likely routes and their lithologies can be helpful. Also, when the source rock is identified its organic matter type can be characterized to provide the appropriate E and A values required by the thermal models. Organic richness can also be measured directly. Finally, a sample of source rock from a well-documented location, and a knowledge of its current maturity, can provide useful checks on the validity of the models developed for the area.

1.11.2. Oil-to-Oil Correlation

A wide variety of techniques has been used in correlating crude oils, and they all try to show that the oils are sufficiently alike to be regarded as genetically related. Physical properties (API gravity, color, pour point, etc.) are not always reliable, and single parameters such as percent sulfur, carbon isotope values, etc. should be used with caution, especially if the oils are degraded or altered. It is unwise to rely on a single parameter to demonstrate a relationship between oils, and the most convincing correlation studies are those which have used many different types of chemical data.

Correlation is generally based on either using a lot of data to summarize the essential character of the whole crude, or on using ratios of specific components. A whole oil gas chromatogram is one of the simplest ways of "*fingerprinting*" an oil and comparisons of gas chromatograms is frequently a helpful first step in correlating. In addition, particular compound types can be monitored in the gas chromatographic trace using a mass spectrometer, and traces of the mass 191 fragment from hopanes or the mass 217 fragment from steranes are commonly used along with other biomarker fragments. These are called "*fragmentograms*".

Many different compounds and compound ratios have been used in correlating, but generally it is not possible to decide in advance what particular correlation parameters will work best in a specific area. In some cases carbon isotope values may be very different for the various oil families and classification is easy, while in some nearby basin all oil types may have essentially the same $\delta^{13}C$ values or overlapping ranges. Similarly for pristane/phytane ratios. In one area all oils may have ratios close to 1.0, while another area may show a range of over 10. For these reasons correlation only gives the minimum number of oil types in an area. It is always conceivable that analysis of some additional parameters will show subgroups within the groups previously defined. For example, in the Williston basin carbon isotope values for two of the three major oil types overlap, but n-alkane data clearly show differences (Williams, 1974).

When one or more of the crudes is altered or degraded, correlation becomes more difficult. It is necessary to use compounds, or compound types, that are resistant to the degradation process. For example, n-alkanes would be useless in correlating biodegraded oils because they are the first compounds to be removed by the bacteria. In contrast, sulfur compounds, multi-ringed naphthenes (including many of the biomarkers), and large aromatic molecules are resistant to bacterial degradation and can be used to correlate related crude oils. The particular choice depends on the degree of degradation.

1.11.3. Oil-to-Source Rock Correlation

Oils are correlated to source rocks by comparing the composition of the bitumen in the rock ("*source rock extract*") with the composition of the oil (Curiale, 1993). Although there are potential problems due to non-uniform source rocks, compositional changes during migration, mixed oils, overmature source rocks, and degraded oils, in many cases oil-source rock correlation is possible and reliable.

If there has been long distance lateral or vertical migration of petroleum, a well that penetrates the oil accumulation will not provide samples of the source rock. Even in these cases it is still sometimes possible to deduce the nature of the source rock from details of the oil composition. The overall character of the oil may be helpful. Waxy crudes with high pour points and low sulfur contents are frequently derived from lacustrine source rocks, or from nearshore environments like deltas. Certain biomarkers also indicate these environments. High concentrations of perhydrobeta-carotene suggest a lacustrine source and gammacerane appears to be a high salinity indicator (lacustrine or sabkha). C_{27}-C_{28}-C_{29} sterane ternary diagrams are also helpful because specific areas on the plots correspond to different environments (e.g., open marine vs. terrestrial). High pristane/phytane ratios usually relate to a somewhat more oxidizing (i.e., nearshore) environment. Oleanane is derived from angiosperms, which did not evolve until the late Cretaceous, so its presence in an oil limits the age of the source rock to the Tertiary. Middle Ordovician oils have a very characteristic n-alkane distribution and low concentrations of acyclic isoprenoids. An increasing number of age-specific, or environment-specific biomarkers are being reported and can be useful in inferring source rock character from oil composition.

CHAPTER 2

INTRODUCTION TO THERMAL MODELS

2.1. INTRODUCTION

Traditional methods for evaluating organic matter type, organic richness, and the thermal maturity of potential source rocks, require samples for analysis. Subsurface sample coverage is always very limited, even in mature exploration areas, and the major problem in exploration involves extrapolating away from the few available control points. This may require extrapolation to greater depths or to adjacent prospects. In addition, the samples that are available represent only a single point in time – the present – and they may give little information about the timing of past events. Extrapolation back in time to reconstruct geologic and thermal events is often of major importance in exploration.

With improved understanding of the processes of petroleum generation and basin evolution has come the ability to develop theoretical models that describe these processes through time. The few available data points can then be used to constrain the models. Even in a well-explored, well-drilled basin substantial extrapolations are usually needed. For example, the Williston basin in the north-central U.S. and south-central Canada, is about 250 miles (402 km) in diameter but is only 17,000 ft (5185 m) deep (Figure 2.1). Assuming an average depth of 8,500 ft leads to a basin volume of 9.94 x 10^{15} ft^3. Roughly 8000 wells have been drilled in this basin, but most are relatively shallow and we can assume an average depth of about 5000 ft. If cuttings, cores, and electric logs provides detailed rock information for a 1 ft diameter cylinder around each well, then the volume of rock documented by well control totals 1.26 x 10^8 ft^3. This leads to a factor for extrapolation from known rocks to basin volume of 7.9 x 10^7*. The value of nearly 10^8 is for a well-explored basin, but in frontier areas, or poorly explored basins, this factor can be very much higher. A combination of well data with a detailed geologic model and high quality seismic data can give a good physical picture of an area. Unfortunately, neither organic

* Or, as Rick Bass put it in "Oil Notes" (1989), "like trying to map the State of Colorado based only on what you see for eight inches from, say, any fireplug in the state".

carbon contents nor permeabilities can be obtained from this information. This is an important point to consider, especially when using 2-D and 3-D models that require detailed knowledge of permeability distribution.

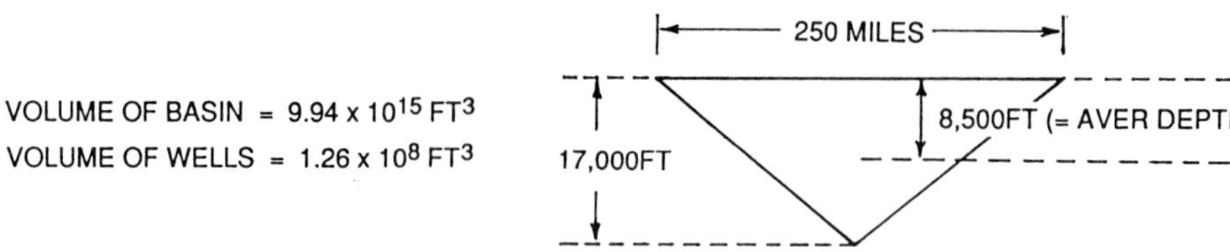

Figure 2.1. Extrapolation from known rocks in wells to total basin volume using the Williston basin, north-central U.S.A. as an example.

2.2. KINETICS

Neither crude oil nor the organic matter in rocks is at thermodynamic equilibrium under any subsurface conditions of pressure and temperature. Even though they are thermodynamically unstable, organic materials commonly occur in a wide variety of sedimentary rocks because they are converting to the stable products of graphite and methane at a rate which is slow, even on a geological time scale. The approach to equilibrium can be described by the principles and laws of chemical kinetics, and these provide a quantitative treatment of the relationship between the two factors of time and temperature that control the conversion of organic matter to crude oil and gas, i.e., the process of petroleum generation.

Petroleum generation can be represented by a simple overall reaction of the form A = B + C, or more specifically:

$$KER \implies BIT + RESIDUE \qquad (2.1)$$

(Kerogen) (Petroleum Products) ("Bitumen") (Graphitic residue)

At time $t = 0$ $\quad$ $Ker = V_o$ and $Bit = 0$

while at some later time $t = t$ $\quad$ $Ker = (V_o - V_t)$ and $Bit = V_t$

The rate at which *Ker* is converted to *Bit* at constant temperature depends on the amount of *Ker* remaining and is given by:

$$dV / dt = k (V_o - V_t)^n \qquad (2.2)$$

where (V_o-V_t) is the quantity of reactant remaining; k is a rate constant; and n is the *"order of reaction"*. When n = 1 the reaction is said to be *"first order"*, for n = 2 it is *"second order"*, and so on. Reactions involving two reactants, such as M + N = P + Q, or 2X = Y + Z, are frequently second-order reactions. Fractional orders are also possible. All the reactions of interest in petroleum generation and oil cracking are treated as first order, so that in Equation 2.2, n = 1.

The value of the rate constant k in Equation 2.2 remains unchanged only for the special case where the reaction is occurring at a fixed temperature. However, in general petroleum generation occurs under non-isothermal conditions with temperature varying through time. The temperature dependence of k is given by the so-called Arrhenius equation:

$$k = A.e^{[-E/RT]} \qquad (2.3)$$

where A is the frequency factor (also called the *"Arrhenius constant"* or *"pre-exponential factor"*); E is the activation energy; R is the gas constant (which is 0.001987 kcal/mol °K, or 0.008314 kJ/mol °K); and T is absolute temperature.

Activation energy can be considered as the height of the energy barrier that must be passed over in going from a metastable state (such as kerogen) to a more stable product (such as oil) (Figure 2.2). Here more stable simply means of lower free energy. At any given temperature, the reactant molecules will have a Gaussian distribution of thermal energies. Although most will be close to average, a few will be particularly low, while a few will be at the very high end of the range (Figure 2.3). Those with high energies will be able to pass the activation energy barrier and convert to products. If the temperature is raised, a greater percentage of the molecules will have sufficient energy to pass the barrier, and this leads to a faster reaction rate at the higher temperature. The actual increase in rate for a particular temperature increase depends on the height of the activation energy barrier. In the case of the reactions that occur during the generation of bitumen from kerogen, the height of the activation energy barrier is such that increasing the temperature by 10 °C roughly doubles the reaction rate. As a consequence, one widely used approximation is that the rate of oil generation roughly doubles for every 10 °C rise in temperature.

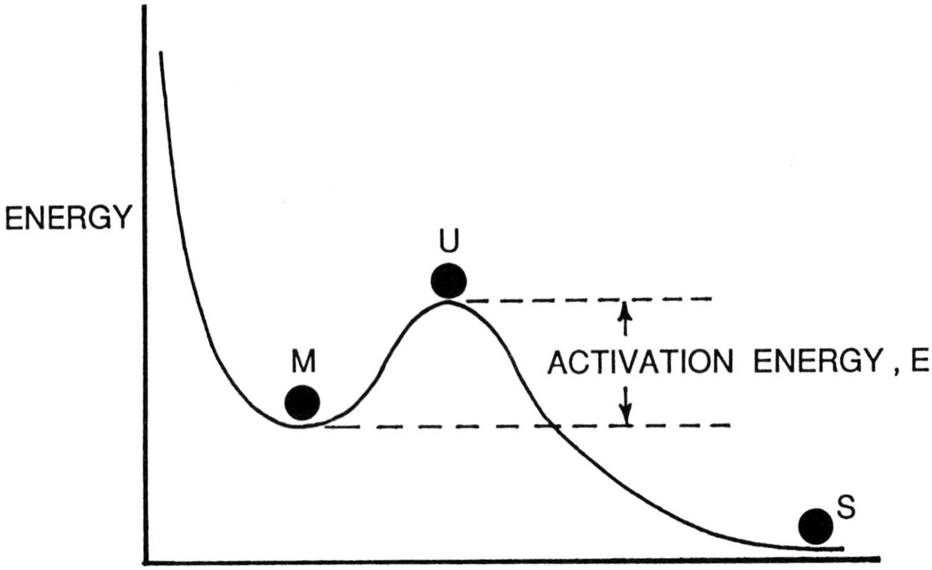

Figure 2.2. Schematic illustration of the concept of activation energy. S : stable; M : metastable; U : unstable. (The horizontal scale has no units).

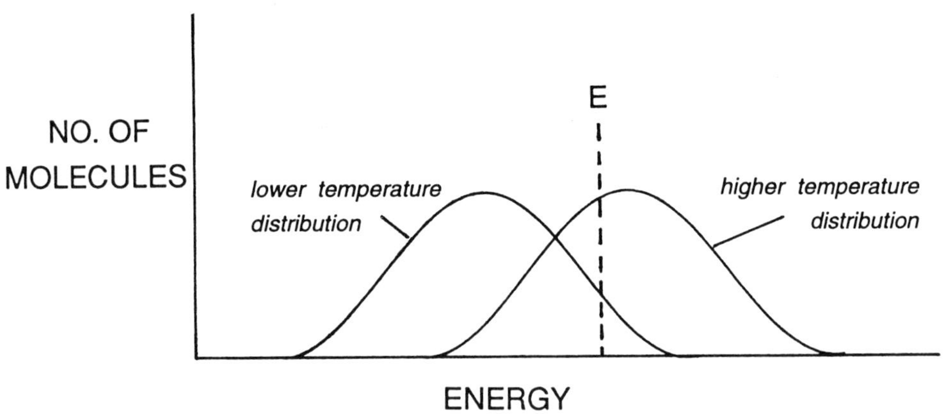

Figure 2.3. Energy distribution in molecules at higher (right), and lower (left) temperatures.

The reactions that generate bitumen by breaking smaller units off the large, complex kerogen molecules involve bond cleavage. Different bonds have different energies, and so require different energies to break them. Table 2.1 lists some generalized values for common bonds. Values are commonly given in kcal/mol, but may also be expressed in kJ/mol with a conversion factor of 4.184. It is not possible to give a unique value for each type of bond because the bond strength varies with position in the molecule. For example, it takes less energy to break a C_{20} chain from the kerogen than to break off a methyl group (C_1). Note in particular the low energy for C - S bonds, and the very high energies for aromatic C = C bond cleavage. Activation energies for oil generation are in the general range from 40 to 60 kcals/mol.

Table 2.1
Approximate bond energies. Actual values vary with position in the molecule, and may be different from one molecule to another. Data from March (1985).

BOND	kcal/mol	kJ/mol
C – H	96 - 99	401 - 414
C – C	83 - 85	347 - 356
C – N	69 - 75	289 - 314
C – O	85 - 91	356 - 381
C – S	66	276
C = C	146 - 151	460 - 464
N – H	93	389
S – H	82	343

The rate of reaction increases with temperature, so that the instantaneous rate of generation of hydrocarbon products from kerogen increases. As the reaction proceeds kerogen is used up faster and faster, and the amount remaining becomes smaller and smaller, until eventually there is none left to react. Because of this the amount of bitumen being generated eventually decreases, and falls to zero when all the kerogen is used up. The rate of consumption of the kerogen is given by

$$V = V_o . \exp(-kt) \qquad (2.4)$$

where V_o is the initial quantity of kerogen, and V is the amount remaining after time t. Kerogen is rapidly used up by reaction at high temperatures and Bredehoeft et al (1994) have calculated the fraction of kerogen remaining as a function of time for various reservoir temperatures (Table 2.2). For a 10 m.y. old reservoir at a depth of 5 km the difference between a geothermal gradient of 22 and 28 °C/km makes the difference between essentially no kerogen remaining and 87% remaining.

Table 2.2
Fraction of the kerogen remaining as a function of time at different reservoir temperatures (Bredehoeft et al, 1994). v/v_o is the volume of kerogen/initial volume of kerogen.

Geothermal gradient (°C/km)	22	25	28
Temperature, 5 km (°C)	125	140	155
Kinetic rate constant (l/sec)	4.5×10^{-16}	5.0×10^{-15}	4.7×10^{-14}
v/v_o 10,000 years	1.00	1.00	0.99
v/v_o 100,000 years	1.00	0.98	0.86
v/v_o 1 million years	0.99	0.85	0.23
v/v_o 10 million years	0.87	0.21	0.00

Combining Equation 2.4 with Equations 2.2 and 2.3 shows that for the bitumen dV/dt initially increases, but then goes through a maximum and decreases to zero (Figure 2.4). Although the time and temperature scales are very different, the same response is produced in both the natural system and in non-isothermal laboratory pyrolysis. Gretener and Curtis (1982) have pointed out that at low temperatures (less than 50 °C) time alone will not mature a source rock – even if it is Precambrian. In contrast, at temperatures exceeding 130 °C maturation is so rapid that the influence of time is minimal. It is only in the window between approximately 70 and 100 °C that time has a significant role. Barker (1989, 1991) has also argued that high temperatures lead to rapid stabilization of maturity (as indicated by R_o) after a geologically short heating time.

When activation energies are higher, the conversion of kerogen to oil occurs at higher temperatures (for a particular value of A), and this is

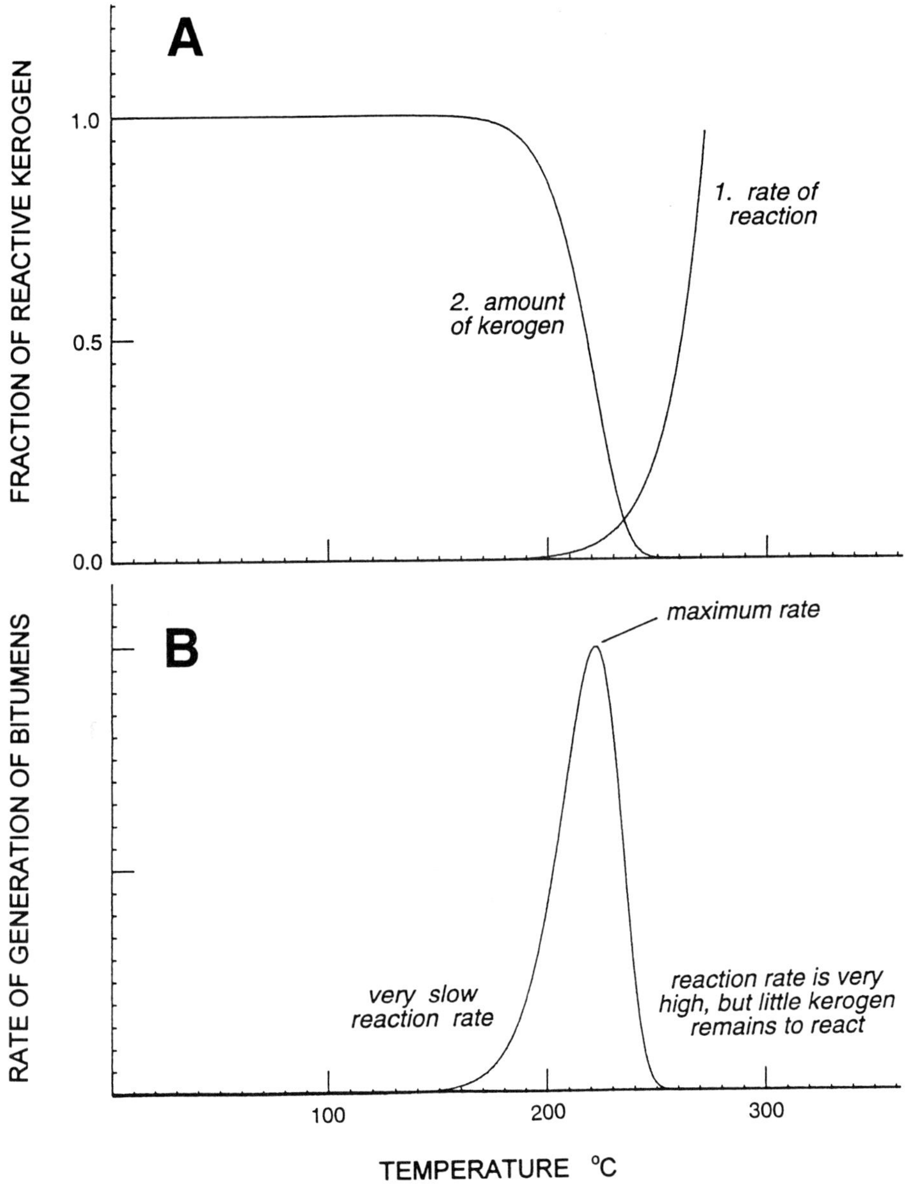

Figure 2.4. (A) Curve 1 showing increasing reaction rate with temperature, and Curve 2 showing how the amount of kerogen decreases as it is used up by reaction. (B) Rate of formation of bitumen products as a function of temperature.

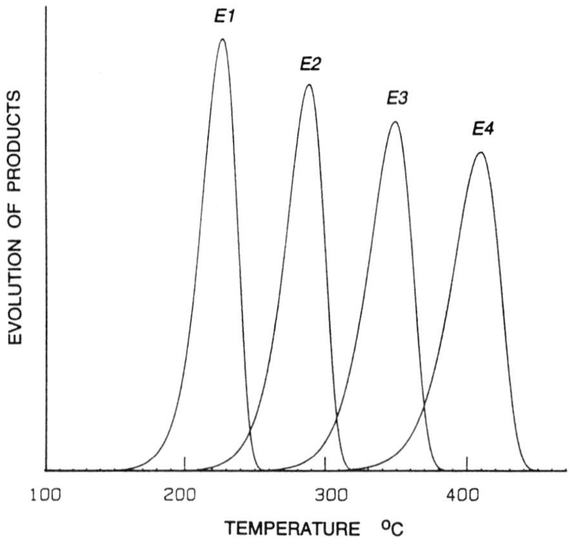

Figure 2.5. Bitumen release curves for kerogens with different values of activation energy: E1 = 40 kcals/mol; E2 = 45 kcals/mol; E3 = 50 kcals/mol; E4 = 55 kcals/mol. The value for A remains unchanged at 10^{16}, and heating rate is 0.4 °C/sec.

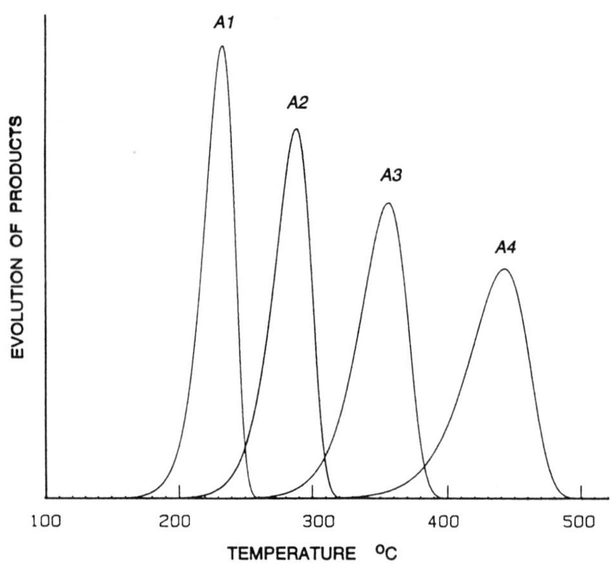

Figure 2.6. Bitumen release curves for kerogens with different values for frequency factors: A1 = 10^{12}; A2 = 10^{14}; A3 = 10^{16}; A4 = 10^{18}. The value for E remains unchanged at 45 kcal/mol, and heating rate is 0.4 °C/sec.

reflected in the maximum in dV/dt moving to higher temperatures. Figure 2.5 shows this shift in temperature and also shows that the peaks broaden. In nature rising temperature is usually associated with increasing depth of burial. A broader peak in temperature (Figure 2.5) implies that petroleum generation will be extended over a greater range in depth for cases where the kerogen is characterized by higher activation energies. When organic matter is rapidly buried in nature, heating rate is high and generation is pushed to higher temperatures because there is less heating time. Again, the peaks are broader, corresponding to a greater depth range and a wider temperature interval.

Because the Arrhenius equation (Equation 2.3) is exponential, the higher the activation energy the greater the effect of a change in temperature on the reaction rate. Table 2.3 gives increases in reaction rates for various 10 °C increments calculated using activation energies between 25 kcal/mole and 55 kcal/mol.

Table 2.3
Calculated increases in reaction rates for four different activation energies and selected 10 °C temperature increments.

Temperature interval	Activation energies (kcal/mol) 25	35	45	55
40 - 50	3.46	5.69	9.35	15.35
90 - 100	2.53	3.66	5.31	7.69
140 - 150	2.05	2.74	3.65	4.86

Frequency factors (the A term in Equation 2.3) are difficult to determine and range over roughly four orders of magnitude, from about 10^{12} to 10^{16} sec^{-1} (according to Burnham, 1994), or 10^{14} to 10^{18} sec^{-1} (according to Ungerer and Pelet, 1987). Pelet (1994) suggested that it might be appropriate to adopt a single, agreed-upon value as an interim measure. A values have the opposite effect from E on the temperature dependence of dV/dt, and as values for A increase (at constant E), the peaks move to lower temperatures (see Figure 2.6). The frequency factor is temperature-dependent and varies as T^n, but theoretical studies of simple reactions have shown that n = 1 or ½, so that for the geologically significant temperature range of 200 °K (from 350 °K to about 530 °K) the small variation of A with temperature can be ignored (Mackenzie and McKenzie, 1983).

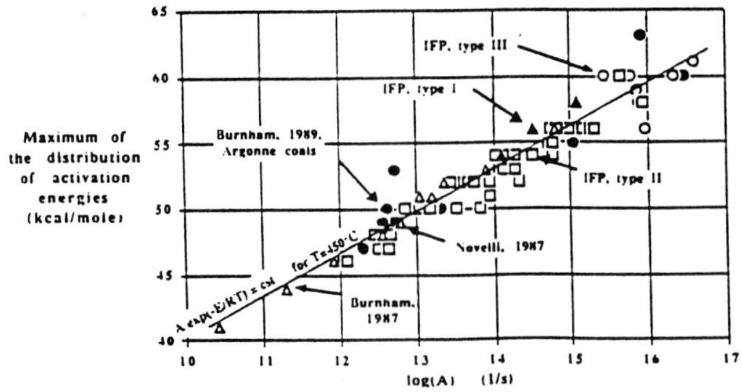

Figure 2.7. Comparison of pairs of values for activation energies and frequency factors obtained from RockEval pyrolysis (Ungerer, 1990).

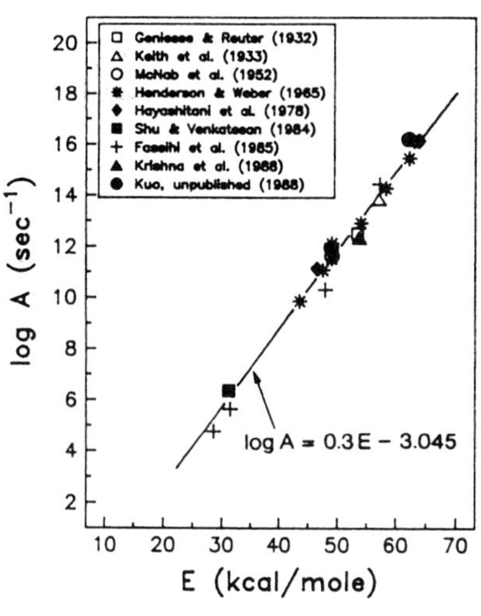

Figure 2.8. Correlation between activation energy (E) and frequency factor (log A) for cracking of crude oils with different compositions. The straight line is the statistical fit to the data ($R^2 = 0.964$). Data from Kuo and Michael (1994) (see their paper for references to specific data points).

E and log A appear to be linearly related (but see Nielsen and Dahl, 1991), and Figure 2.7 shows a plot of E and A pairs taken from a wide variety of publications that illustrates this empirical relationship. The values in Figure 2.8 were obtained for the thermal cracking of crude oils.

The relationship summarized by the Arrhenius equation (Equation 2.3) provides one method for obtaining E and A values. The log form of the equation is:

$$\ln k \quad = \quad -E/RT + \ln A \qquad (2.5)$$

and this is in the form of the equation for a straight line if 1/T is plotted against ln k. The slope of the line will be E/R and the intercept will be ln A. R is a constant, so that values for both activation energy and frequency factor can be found. When wide temperature ranges are involved, it is possible that the reaction mechanism may change with temperature. Each reaction mechanism will then have its own characteristic E and A values, and as a consequence lines on 1/T versus ln k plots will show two or more straight line segments, with each segment corresponding to a specific reaction mechanism. This is a particular concern when geological and laboratory data are combined. Laboratory experiments are run at much higher temperatures than those that normally occur in nature, in order to get sufficient product in a short time. Sajgo and Lefler (1986) obtained a good single line fit to geologic data points for the sterane C-20 isomerization in the Pannonian basin (Figure 2.9), and their extrapolated value of 90.14 kJ/mol (21.6 kcal/mol) is in excellent agreement with the value of 91 kJ/mol (21.7 kcal/mol) obtained by Mackenzie and McKenzie (1983) when North Sea data were added. However, Figure 2.10 shows laboratory hydrous pyrolysis data from Rullkotter and Marzi (1989) combined with the natural data set of Mackenzie and McKenzie. Although quite a good line can be fitted to all the data points, it is clear that the geologic data alone leads to a much lower value for E (91 kJ/mol) than the value of 169 kJ/mol obtained from the combined data sets.

Petroleum generation cannot be adequately described by a single reaction. The formation of the smaller bitumen molecules from the large, complex kerogen molecules involves breaking a wide variety of bonds. These include C–C bonds at various positions in chains (middle versus end), C=C aromatic bonds, as well as C–S, C–N and C–O bonds. In addition, there are isotopic effects because ^{13}C-^{12}C bonds are stronger than ^{12}C-^{12}C bonds. Each reaction is characterized by a unique pair of E and A values, and petroleum generation is best treated as a set of parallel, first order reactions (Figure 2.11). This becomes especially important when the cracking of oil to gas is considered.

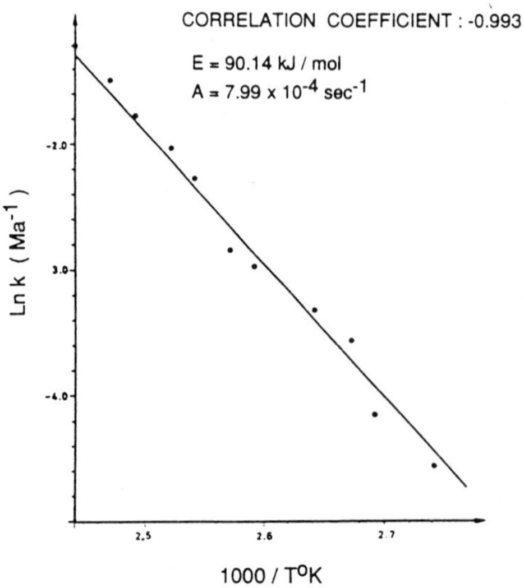

Figure 2.9. Arrhenius plot of sterane isomerization at C-20 for samples from the Hod-1 well in the Pannonian Basin. Geological ages of samples were used in the calculations (Sajgo and Lefler, 1986).

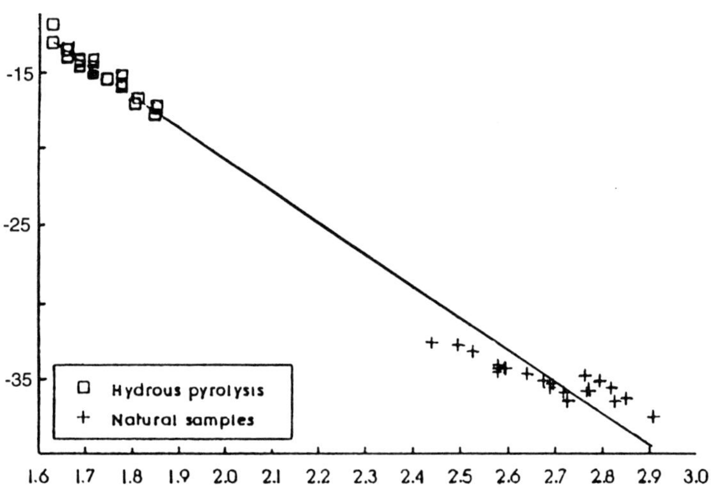

Figure 2.10. Arrhenius plot of sterane isomerization at C-20 for natural samples from the Pannonian Basin and the North Sea, together with sterane isomerization data based on hydrous pyrolysis (Rullkotter and Marzi, 1989).

Some early models attempted to simplify the kinetic treatment of petroleum generation by treating it as if it were a single reaction, and using only one pair of E and A values. Such values are called *"global"*, and do not properly describe the system. Combined values are generally too low, often leading to values of about 15 kcal/mol. Figure 2.11 shows the rate of generation of products (in this case methane) for eight parallel first order reactions together with the combined curve. The value of A remains unchanged. Although the individual reactions have activation energies that range from 48 kcal/mol up to 62 kcal/mol, the overall, pseudo-activation energy (the "global" value) is only 20 kcal/mol. Methods for obtaining suites of activation energies for describing petroleum generation and cracking are discussed later (Chapter 7). Figure 2.12 shows the suite of E values used to describe the thermal behavior of a Type II kerogen from the Toarcian shale in the Paris basin (Tissot et al, 1987). In some cases a simplified approach is taken by assuming a Gaussian distribution for the activation energies. The E values shown schematically in Figure 2.13 are derived from an immature kerogen and correspond to the pyrolysis curve given. If this kerogen had been buried and heated, the reactions with the lowest values of E would have occurred first, and in the case of the lowest value of all would have gone to completion (Figure 2.13). The reactions corresponding to the high E values would not even have started. Clearly, partial heating changes the distribution of activation energies for the remaining kerogen. The pyrolysis response obtained from the more mature sample would lack the contributions from the low temperature reactions, and the peak would be shifted to higher temperatures. This is, of course, exactly what is observed in pyrolysis of mature samples, and is why pyrolysis-derived T_{max} values can be used as an indicator of maturity.

2.3. THERMAL MODELS

Improved understanding of the kinetics of petroleum generation has stimulated the development of models that attempt to describe the generation process quantitatively. Kerogens that generate oil and gas have extremely complex structures, and the structure varies from one source rock to another but has not been characterized in detail for any sample. Even if complete structural information were available for kerogen, so many reactions would be involved in its thermal breakdown to petroleum that no computer could handle all the simultaneous reactions. The calculations would demand a knowledge of the E and A values, as well as

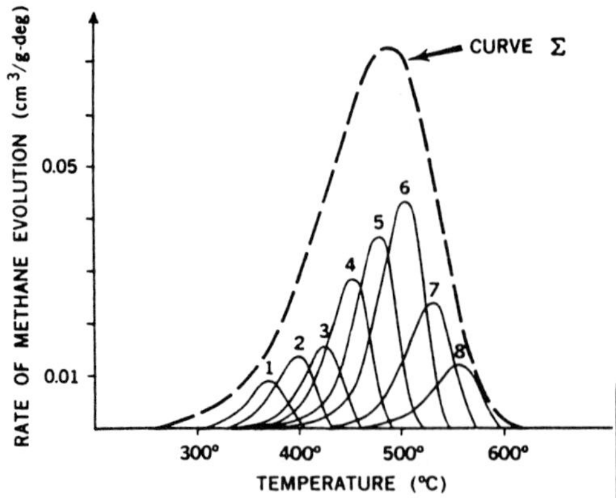

PARALLEL FIRST ORDER REACTIONS

1.	48 KCAL/MOL	10^{13} SEC^{-1}
2	50	
3	52	
4	54	
5	56	
6	58	
7	60	
8	62	
SUM	20	2×10^{2} SEC^{-1}

Figure 2.11. Bitumen release curves for 8 individual, first-order reactions with different activation energies, and the curve for total bitumen evolved. (Data from Juntgen and Klein, 1975).

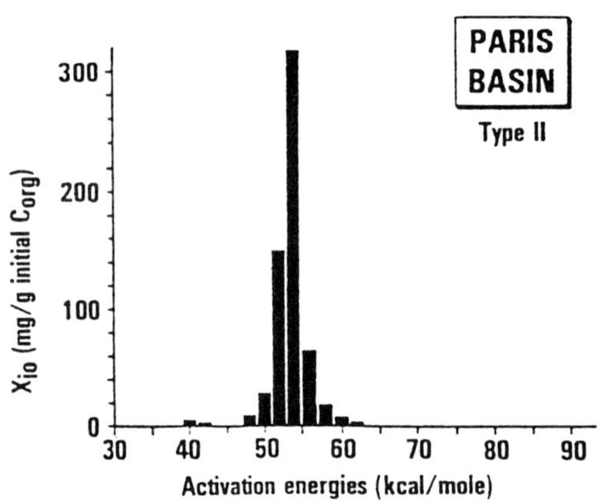

Figure 2.12. Distribution of activation energies deduced from the pyrolysis behavior of Type II kerogen form the Toarcian shale in the Paris Basin (Tissot et al, 1987).

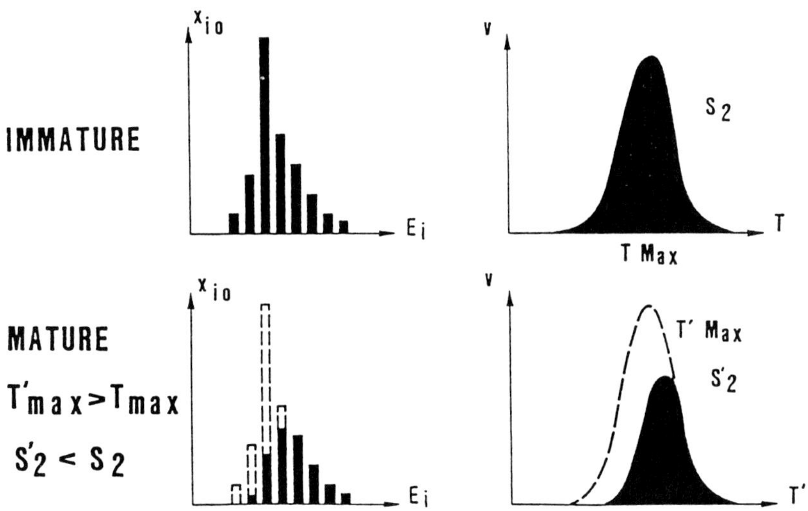

Figure 2.13: Distribution of activation energies in an immature kerogen (Top), and the distribution of activation energies in the kerogen remaining after partial maturation of the kerogen (Bottom). The pyrolysis curves generated from these two kerogens are also shown. (Tissot et al, 1987).

the relative amounts, for each of the reacting materials. In attempting to circumvent these problems two different approaches have been used, and in a sense both of them are empirical. The *"Lopatin method"* relies on geological calibration using data from well-understood basins. The other group of methods is better grounded in chemical kinetics, but is calibrated using the kinetic parameters A and E derived from laboratory pyrolysis studies.

In 1971 Lopatin introduced a method based on the assumption that the rate of petroleum generation doubled for every 10 °C increase in temperature, but that the role of time in thermal maturation was linear. He calculated a time-temperature increment (TTI) for every 10 °C temperature interval throughout the burial history. The cumulative values were then calibrated by reference to well-understood basins and provided the basis for estimating petroleum generation status in other areas. The method was clearly described and recalibrated by Waples (1980) and has been widely used. It is discussed in Chapter 6. The Lopatin method has some serious limitations and is being rapidly replaced by more appropriate models.

Models that apply the principles of chemical kinetics to petroleum generation were first introduced by Tissot and Espitalie in 1975, and are described in Chapter 7. They are more complex and require additional information in the form of E and A values for each of the relevant reactions. In practice only a very limited number of E and A pairs can be accomodated (usually less than 10), and clearly these are not capable of describing all possible reations. Even in the simpler case of thermal cracking of oil it is not possible to accumulate and handle data for every reaction that might occur. The solution has been to simulate the thermal breakdown of the geologically occurring organic materials by laboratory pyrolysis experiments. These methods provide a better numerical description of the processes occurring in the subsurface, and with the widespread availability of commerical computer models, and increasing computing power, the methods are now generally accessible. Several versions that make different assumptions are available and will be discussed in detail later.

The geological input required in using both modeling methods is the same. They need a burial history through time coupled with information for the temperature history through time. Methods for generating and displaying this information are discussed in the next three chapters.

CHAPTER 3

BURIAL HISTORY

3.1. INTRODUCTION

Previous chapters have shown that petroleum generation is a process controlled by both temperature and time, and that the quantitative relationship between these two can be described by the equations for first order reaction kinetics. To model the timing of petroleum generation it is necessary to have information about both time and temperature. In practice what is usually available is information on geologic age and depth of burial. Clearly geologic age provides time. Temperature data are much more difficult to constrain through the geologic history of the section, and two approaches are most commonly used. Either (1) geothermal gradient is combined with depth, or (2) rock thickness is combined with thermal conductivity and heat flow. In either case, a burial history is a critically important element in developing thermal models.

Burial histories are normally reconstructed at one (or more) discrete locations, and diagrams are plotted that show the depth of any significant rock unit as a function of time. Figure 3.1 shows a typical plot of depth (linear) against geologic age. (Here present day is put on the right, as is most commonly done, but present day may also occasionally be on the left.) The line shows the depth history for an element of rock that is 125 m.y. old. From its initial location at the surface 125 Ma, it was buried at a constant rate until 100 Ma, when it had reached a depth of m (segment (1)). Thus, the initial burial rate is $m/25$ m.y. (in appropriate units). Over the next 25 m.y. it was only buried an additional n (where $n<m$), so that the average rate of burial for segment (2) was less than the rate for segment (1). Segment (3) from 75 to 50 Ma was a time of uplift, and was followed by an interval of 25 m.y. when there was no change in depth (segment (4)). Finally, reburial (segment (5)) carries the rock to its present-day depth corresponding to point (6). It is worth noting that the only part of this diagram that can be established by direct measurement is the present day depth, and that all the rest of the diagram has to be deduced from presently available data.

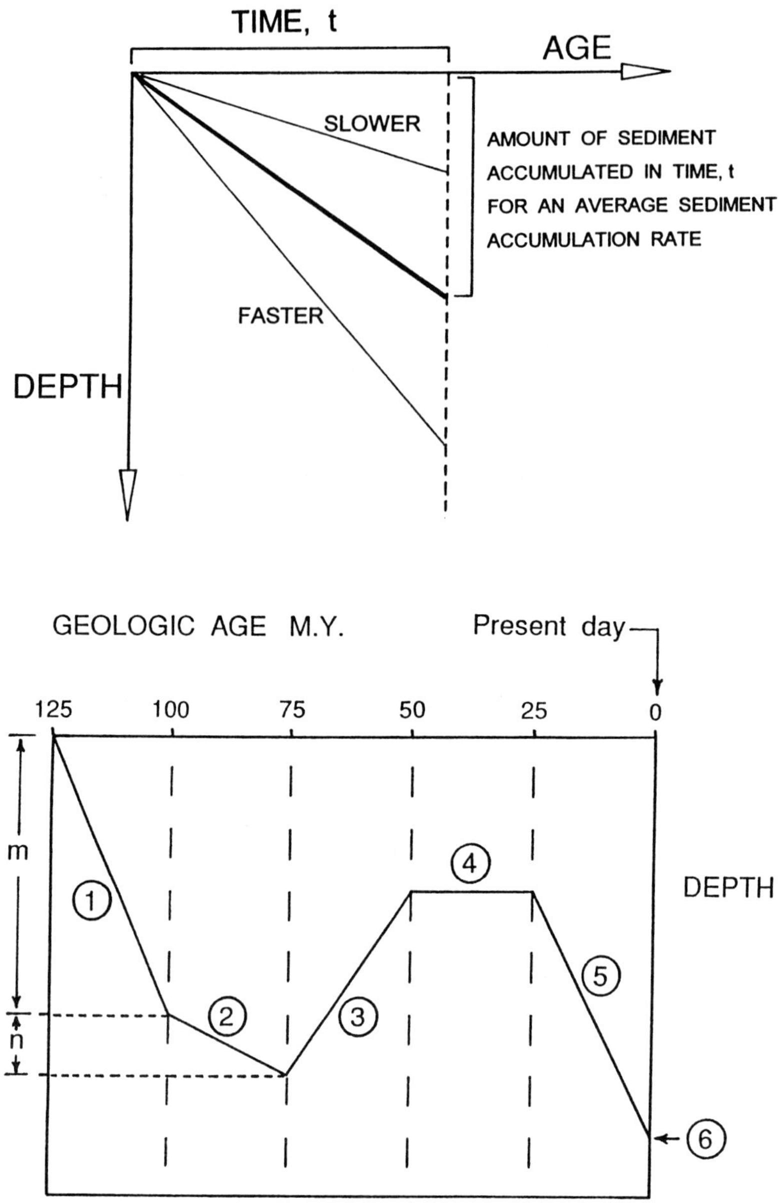

Figure 3.1. (Top) Increases in depth through time for slow, average, and fast rates of sediment accumulation. (Bottom) Rapid burial–(1): slower burial–(2); uplift–(3); burial hiatus–(4); and reburial–(5) to the present day depth–(6) for a single rock element on a burial history plot.

3.2. MULTIPLE LAYERS

Any real rock has a significant vertical thickness, unlike the idealized case discussed in Figure 3.1. Also, in most modeled areas there is interest in a package of rocks that may contain multiple potential source rocks, reservoirs, and seals. Figure 3.2 schematically illustrates this situation. It shows a rock unit A starting to accumulate at 150 Ma, with sedimentation continuing until 125 Ma. Thickness increased steadily during this time due to a uniform rate of burial. We will assume that A is a sandstone, and that as the sand supply diminished, the type of sediment accumulating at this location changed and a new lithology B developed. B steadily buried the underlying sand A, but the sand thickness remained constant (neglecting compaction for the moment), and the vertical distance from the top to the bottom of the sand does not change. As a consequence the lines on the diagram representing the top and bottom of the sand remain parallel throughout the entire burial history. Similarly, after accumulation of unit B was completed at 100 Ma, its thickness remained unchanged through its subsequent burial history. This is true for all the units in the package being considered, and for the simple case shown in Figure 3.2 all the lines will be parallel, corresponding to constant thicknesses for all the rock units. Casual examination of unit C makes it look as though it is thicker from 75 Ma to the present than it was, say, at 80 Ma. This is an optical illusion, and vertical distance between the lines actually remains the same over the complete age range. Notice that the sediment accumulation rate for Unit D was fast from 80-75 Ma, but slowed down after 75 Ma. The burial history for the Paleozoic section of the Anadarko basin shown in Figure 3.3 provides an illustration of varying rates of sediment accumulation and a final period of uplift from 60 Ma to the present.

3.3. CONSTRUCTION OF SIMPLE BURIAL HISTORY

In some simple situations the information needed for constructing a burial history plot can be obtained easily from single well data. The procedure is explained by reference to the 5000 m well shown schematically in Figure 3.4. This well penetrated four major units (I to IV), and the age at the base of each was 20, 30, 65 and 85 m.y., respectively. Thus the base of shale I at 500 m is 20 m.y. old. This means that at 20 Ma this shale was being deposited at the surface. Today it is 500 m deep, and in the absence

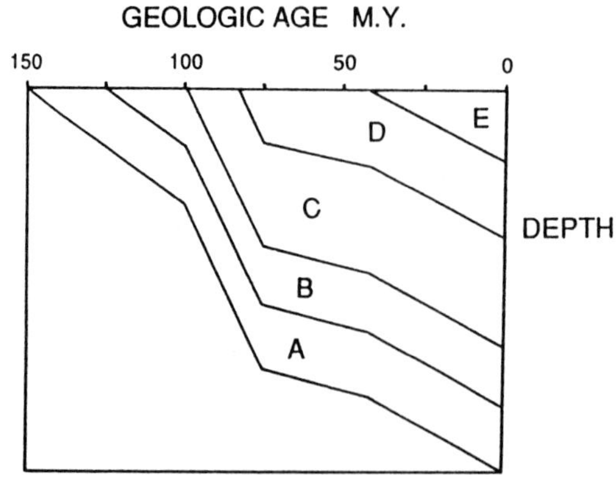

Figure 3.2. Schematic burial history plot for 5 sedimentary units with A the oldest and E the youngest. Unit D shows a decrease in sediment accumulation rate at 75 Ma.

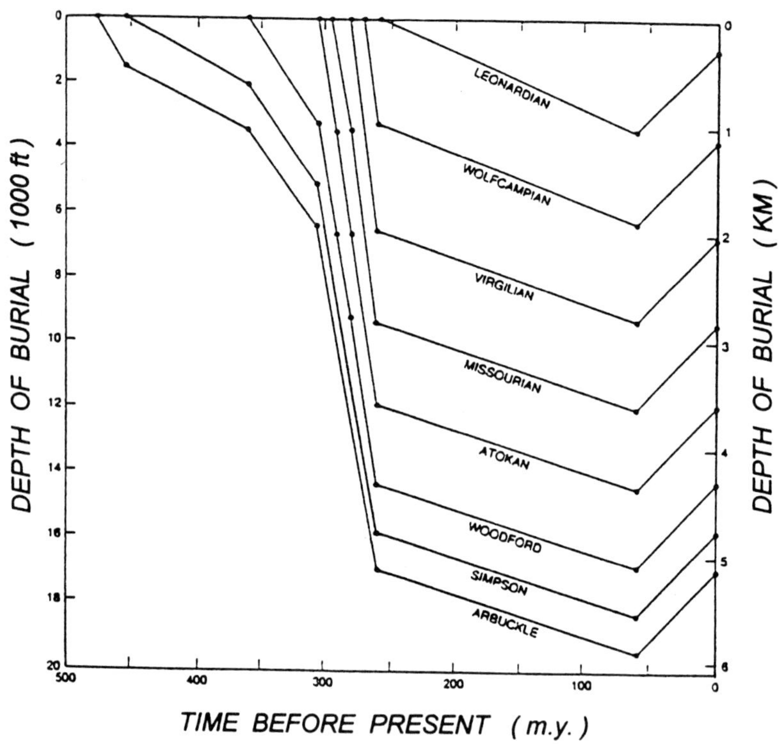

Figure 3.3. Simple burial history from the Anadarko basin, Oklahoma showing decompacted burial curves for the Paleozoic section in the area of Caddo, Canadian, and Grant Counties (Schmoker, 1986).

of more detailed information, we assume steady burial of the shale from time of deposition to the present and draw in the straight line AB. If compaction is not important and there are no unconformities in the section, all the units below shale I must have been buried at the same rate, and so the burial histories for units II, III and IV between 20 Ma and the present are represented by lines parallel to AB (i.e., CD, etc.) The base of unit II is 30 m.y. old and must have been at the surface at 30 Ma. But its position at 20 Ma has already been established, and the second step is to draw line EC. Again, all the units below must have been buried at the same rate from 30 to 20 Ma, and all the lines below ED must be parallel to it. This procedure is followed step-by-step back through time for each unit down the well until the plot is completed.

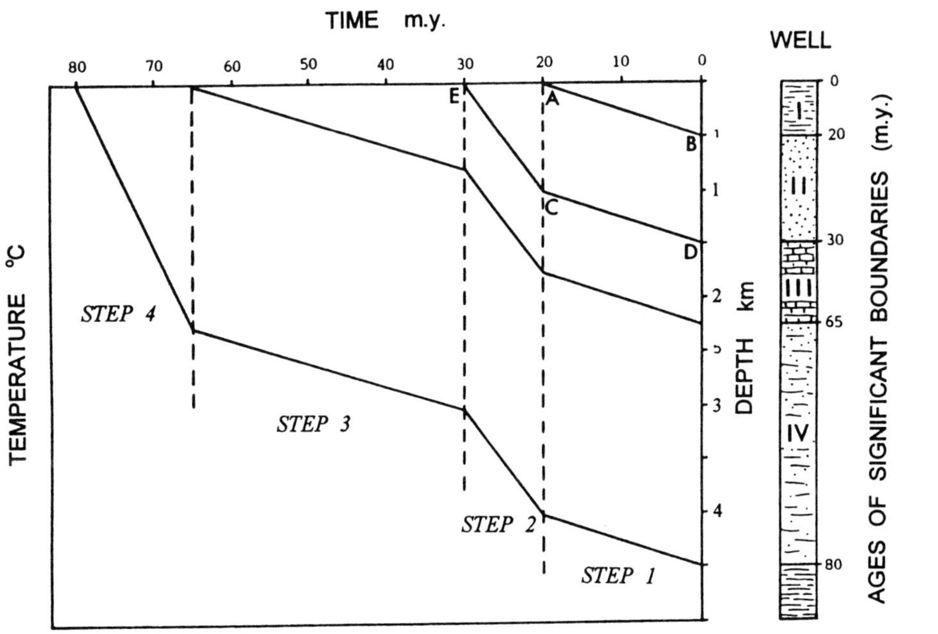

Figure 3.4. Burial history derived from the well information shown.

3.4. SEDIMENT ACCUMULATION RATES

Lines on a burial history diagram show how depth below the surface for a given unit changes through time. The rate of burial depends on the rate of accumulation of the overlying beds, but these rates are extremely variable and cover at least 11 orders of magnitude. Sadler (1981) concluded from a study of 25,000 values that accumulation rates are inversely related to the time span for which they are determined, and also that short term rates are a poor guide to long term accumulation regimes. In general, sedimentation rates are typically a few tens to a couple hundred meters per million years (Table 3.1), and the m/m.y. unit (equivalent to 1 mm/yr, or 1 mm/1000 years) is sometimes called a *"Bubnoff unit"* (Fischer, 1969). Average burial rates over some appropriate time interval are easily recovered from burial history plots. For example, from the burial history plot for part of the Oklahoma Paleozoic section shown in Figure 3.3 it appears that the Simpson accumulated at 24.4 m/m.y. since deposition started at 475 Ma and ended at 455 Ma, and in this 20 m.y. interval 1600 feet (equal to 488m) of sediment accumulated. Similarly, accumulation rates for the Woodford and Atokan can be calculated as 6.4 m/m.y. and 17.7 m/m.y., respectively. Incremental sediment accumulation rates over suitable time intervals can be plotted as a function of geologic age (e.g., Pisciotto, 1981; Mudford and Best, 1989; Mudford, 1990). Note, however, that the sedimentation rates obtained from burial history plots by some authors are derived from present-day compacted values, and these show less vertical thickness accumulated over a given time interval. Actual, near-surface sediment accumulation rates are higher, how much higher depends on the lithology and the degree of compaction.

In clastic sequences accumulation rates are usually higher for poorly sorted materials and sands than they are for clays. In particular, pelagic sediment accumulation rates are low. In the Baltimore Canyon off the East Coast of the U.S., pelagic sediments have accumulated at 25 - 45 m/m.y., compared with 75 - 320 m/m.y. for nearer shore clastic deposition (Lawrence et al, 1990). Sediment accumulation rates in carbonates can be very variable, and there may be continuing rapid growth of reef builders up towards a steadily rising sea level, and then the maximum accumulation rates equal the rate of sea level rise. However, the apparent changes in sea level are commonly caused by a subsiding basement. Although carbonate accumulation rates can be very high, they are easily inhibited by changing conditions. As a consequence, short-term accumulation rates in excess of 12,000 m/m.y. have been reported for Holocene reef growth (Macintyre et al, 1977), whereas average carbonate platform growth is closer to 200 m/m.y. (Bosscher and Schlager, 1993). Schlager (1981) suggested that 1000 m/m.y. was a conservative estimate of average growth potential for carbonates. Many of the fastest sediment accumulation rates are for evaporites, where values are often in the range from 5 - 10,000 m/m.y. (Table 3.1).

Table 3.1
Selected sediment accumulation rates for a variety of lithologies and geologic settings.

LOCATION (REF)	LITHOLOGY	TIME INTERVAL M.Y.	RATE M/M.Y.
INDIVIDUAL LITHOLOGIES			
Oxford clay (1)	Shale	15	23.6
Kimmeridge clay (1)	Shale	12	88.5
Baltimore Canyon (2)	"Clastics"	5-66	75-320
"	"Carbonates"	9-57	150-1100
"	"Pelagics"	67-127	25-45
Woodford Shale (3)	Shale	67-127	25-45
Atoka (3)		55	17.7
Simpson (3)		20	24.4
Florida (4)	Reef and debris	0.007	3000
Rodrigues bank (4)	Lagoon	0.003	1000
Andres Island (4)	Open sea bank	0.005	1000
Sabkha Faishak (4)	Tidal flats	0.002	700
Andres Island (4)	Bank sediment	120	35
Suniland Field, Florida (4)	Bank-shallow shelf	120	30
Golden Lane, Mexico (4)	Bank sediment	20	80
Persian Gulf (4)	Shallow marine and tidal flat	200	30
Oklahoma, Arbuckle Group (4)	Tidal flat - lagoonal	100	30
Songliao basin, China (9)	Lacustrine clastics		46
Gulf of Gabes, Tunisia (8)	Evaporite		6000
Plio-Pleistocene Lisan salt, Dead Sea (8)	Evaporite		4-5000
Mid-Devonian Muskeg Fm, W. Canada (8)	Evaporite		10,000
Permian Zechstein, North Sea (8)	Evaporite		10,000

(see references, next page)

Table 3.1 (cont.)
Selected sediment accumulation rates for a variety of lithologies and geologic settings.

LOCATION (REF)	LITHOLOGY	RATE, M/M.Y.
BASIN FILL		
Ellenberger, W.Texas (5)	Carbonate highstand system tract	15
Pinnacle reef, Michigan (5)	"	13
L. Keg River, Canada (5)	"	14
Swan Hills, Canada (5)	"	122
Canning Basin, Australia (5)	"	30
M. San Andres, New Mexico (5)	"	280
L. Clear Fork, Midland basin (5),	"	365
Bekes basin, Hungary (7)	Basinally restricted	79
"	Retrogradational	390
"	Early progradational	1100
"	Late progradational	590
Cratonic basins (6)		1-24
Ancient miogeosynclines (6)		12-31
Modern continental terraces (6)		24-57
Ancient eugeosynclines (6)		29-37
Modern continental rises (6)		15-44
Foreland basins (6)		43-927
Trench fill (6)		1-500
Successor basins (6)		58-200
Rift valley and aulacogens (6)		71-400

(1) Ebukanson and Kinghorn (1990); (2) Lawrence et al (1990); (3) calculated from data in Schmoker (1986); (4) Wilson (1975); (5) Sarg (1988); (6) Schwab (1976); (7) Mattick et al (1994); (8) Bordenave (1993); Xue and Galloway (1993).

The completeness of the geologic record is often a matter of considerable uncertainty (Anders et al, 1987), and the relationship of present day rock thicknesses to original rock thickness may not be clear. This is rarely taken into account in constructing burial history diagrams for thermal modeling of petroleum generation, though in some cases it may be very significant.

Basin fill is made up from a variety of sediment types, and these would have accumulated at very different rates. The single most important factor influencing the way basins fill is tectonics, because this controls not only the rate of basin subsidence but also the availability of sediment supply from adjacent uplift and erosion. Schwab (1976) tried to summarize the rate of basin fill in terms of plate tectonic setting, and has found that on average plate interior basins (intracratonic basins) accumulate sediments at 1 - 24 m/m.y., whereas rates are much higher on plate margins, where rates can exceed 70 m/m.y., and are usually several times faster. His numbers summarize the rate at which sediments accumulate in a basin, and this includes erosional loss at unconformities (negative accumulation rate). No compaction corrections were made, and the numbers are not rates of sedimentation but rather net sediment accumulation rates. His average accumulation rates are included in Table 3.1.

Several of the factors significant for thermal modeling are related to sediment accumulation rates:

(1) When sediments are rapidly loaded by a superimposed sediment pile they tend to compact (see below), but if the sediment has low permeability, which would be the case for a mud/shale, then the water cannot be expelled fast enough for the fluid pressure to remain normal, and it rises to create an abnormally-pressured zone (Mann and Mackenzie, 1990; Mello and Karner, 1996). Such an undercompacted zone with high water content has a low thermal conductivity, and this leads to high geothermal gradients (see Section 4.1.4). In addition, undercompacted sections do not lose vertical thickness in the same way as normally compacted sections. Note, however, that abnormal pressures can be generated in other ways (such as the thermal maturation of organic matter; see Section 9.4), and so abnormal pressures are not always associated with rapid sediment accumulation.

(2) If a sediment pile has accumulated very quickly, there may have been insufficient time for it to reach thermal equilibrium. In such non-equilibrium situations the geothermal gradient tends to be low, and it cannot be calculated from heat flow and thermal conductivity. For example, in the Neogene foredeep that developed adjacent to the Apennine thrust belt and filled with turbidites, Mattavelli and Novelli (1990)

reported sedimentation rates up to 1000 m/m.y. and low geothermal gradients that ranged from 19 to 25 °C/km and averaged 22 °C/km. In the case where sedimentation rates are negative because sediments are being rapidly eroded, the rocks might not be able to cool fast enough to remain in thermal equilibrium, and the temperatures would be too high for a given depth. This can be locally important in the case where ice is being removed as glaciation ends.

(3) The rapid influx and accumulation of mineral matter dilutes any organic materials that are accumulating with the sediments, and average organic matter contents are low. In deltas, for example, they are rarely above 1 wt%. Interestingly, low sedimentation rates can also lead to low organic contents because the organic materials are exposed near the surface for a relatively long time, and there is plenty of opportunity for bacteria to consume them. This trend is reinforced by the fact that deep quiet waters where pelagic sediments accumulate slowly are usually of low energy and anoxic. In contrast, near-shore environments where coarser sediments accumulate rapidly are more oxidizing. Organic matter is not stable in oxidizing environments and tends to be removed by conversion to carbon dioxide, while the organic matter of the deep pelagic muds survives (Stein, 1990). Katz (1995) showed that in rifts high subsidence rates favor the development of oil-prone source rocks, while for low subsidence rates gas-prone source rocks are favored.

(4) For a given geothermal gradient, sediment accumulation rate controls the heating rate. Table 5.1 (of Section 5.2.1) shows burial rates of 10 to 500 m/m.y. combined with geothermal gradients in the range 25 - 45 °C/km. This corresponds to heating rates of approximately 0.25 - 22.5 °C/m.y. Heating rates are typically 0.3 - 1.0 °C/m.y., but may reach 10 °C/m.y. or higher. Several simplified models have used heating rate in estimating organic maturation.

3.5. COMPACTION

3.5.1. Introduction

During burial, sediment pore water is squeezed out as the load imposed by overlying, younger sediments increases. This results in compaction, decrease in porosity, and a reduction in thickness of the compacting unit (Rieke and Chilingarian, 1974) (Figure 3.5). Sediment thickness is the key parameter used in constructing burial diagrams, and the present thickness measured in a well for deep, compacted units can be

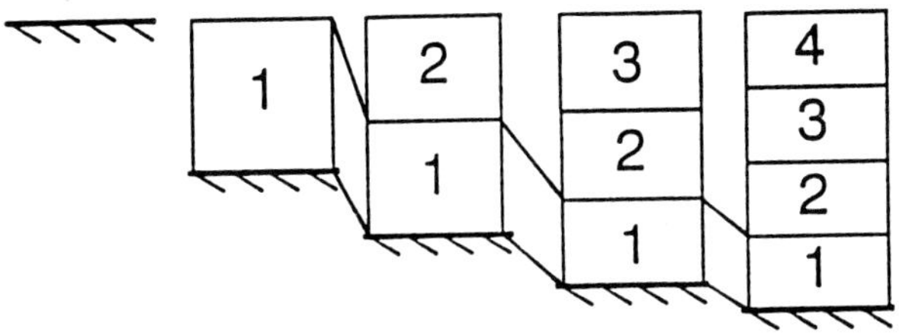

Figure 3.5. Reduction in thickness for a compacting sediment showing deeper individual units decreasing in thickness as they are loaded.

as much as 40% less than the original thickness soon after deposition. Because of this, it is necessary to correct the thickness of rocks through time to account for the changing porosity produced by compaction (i.e., to *"decompact"* them). Compaction is one of the reasons why lines on burial reconstructions might not be parallel.

3.5.2. Porosity

Depth Relationships: Most rock types (except evaporites that have no initial porosity) compact in response to the weight of superimposed material. In general, compaction occurs rapidly at first, but then more slowly as depth of burial increases. The main effect is to squeeze out water and diminish porosity, and usually the progress of compaction is established by monitoring porosity. Typical generalized porosity reduction (i.e., compaction) versus depth curves are shown in Figure 3.6, and Figure 3.7 shows some actual data points from the COST B2 well off the East coast of the U.S. Porosity (or compaction) is usually related to depth by some sort of logarithmic relationship, and one widely used expression (first suggested by Athy for shale in 1930) has the form:

$$\varphi_z = \varphi_o e^{-cz} \qquad (3.1)$$

where φ is porosity, z is depth and c is a constant. Because different rock types compact at different rates, the value of c is lithology-dependent; selected values are given in Table 3.2. Luo et al (1993) have argued that Equation 3.1 should not be used at depths less than a few tens to a few hundreds of meters. Often it appears that over the depth range of interest in modeling, porosity does not decrease to zero but to some low, but

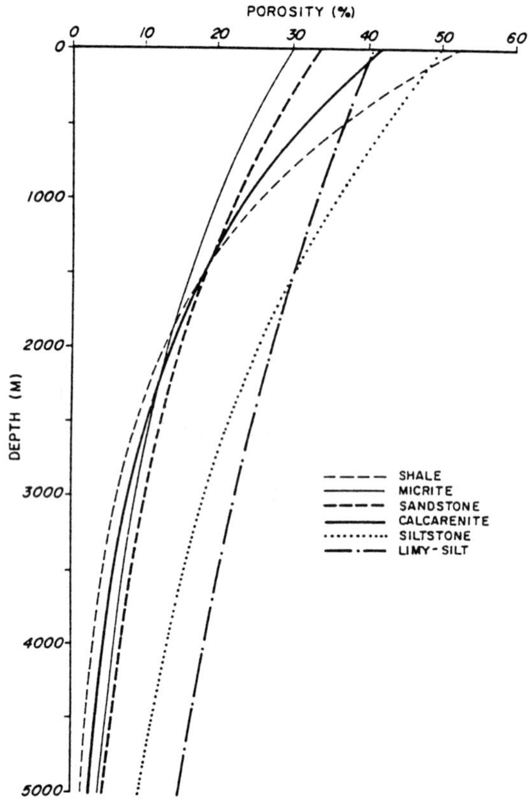

Figure 3.6. Generalized porosity reduction (i.e., compaction) versus depth curves for various lithologies on the south Australian margin (Hegarty et al, 1988).

unchanging, value. To accommodate this irreducible porosity, Bethke et al (1991) in their study of the Illinois basin used the expression:

$$\varphi = \varphi_0 e^{-cz} + \varphi_1 \tag{3.2}$$

where φ_1 is a constant with a value of 0.05, corresponding to an irreducible porosity of 5% for all lithologies. The same value was used in a study of the Denver basin (Lee and Bethke, 1994).

Other forms for the porosity-depth relationship have been suggested. Baldwin and Butler (1985), for example, used:

$$\text{BURIAL DEPTH (in km)} = 6.02 \times S^{6.35} \qquad (3.3)$$

where S is the *"solidity"*, or complement of porosity. Schmoker and Gautier (1988) have argued that the processes leading to porosity reduction are time-dependent and that porosity at any depth can be expressed by an equation of the form:

$$\varphi = A(M)^B \qquad (3.4)$$

where A is a lithology-dependent constant, and M is a measure of thermal maturity. They suggested using either TTI (see Chapter 6) or vitrinite reflectance. Other researchers have emphasized the mineralogy of the compacting sediments, especially for sandstones, where the presence of ductile grains can be an important factor (Smosna, 1989). In practice, values for porosity usually show enough scatter (e.g., Figure 3.7) that they can be fitted equally well by several different equations. Even multiple, straight line segments have been used successfully.

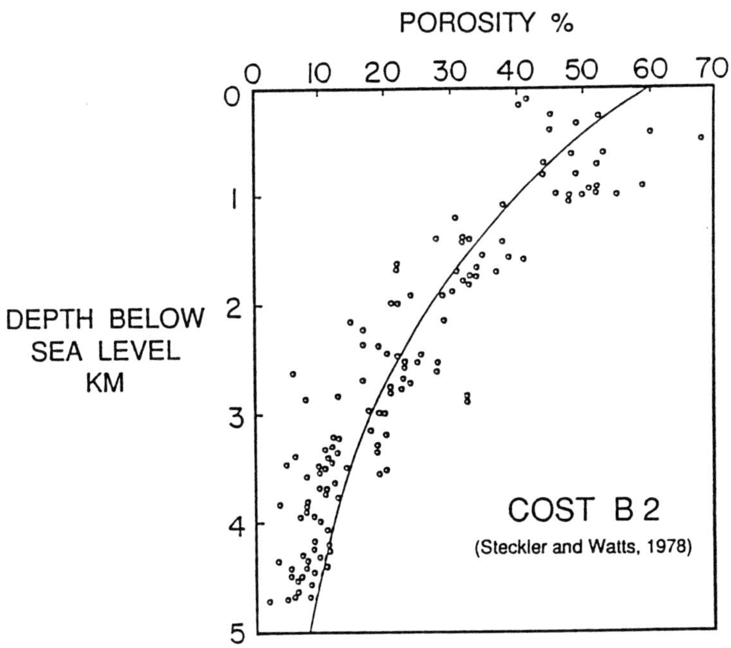

Figure 3.7. Porosity values measured for sediments in the COST B2 well, offshore Eastern U.S. (Steckler and Watts, 1978).

Table 3.2
Selected values of initial porosities and compaction constants reported for various lithologies.

LITHOLOGY*	REF	INITIAL POROSITY ϕ_o	COMPACTION CONSTANT, C 10^{-4}/m
Shale	1	0.52	1.4
	4	0.75	9.0
	5	0.55	8.5
	7	0.63	5.1
Sandstone	1	0.34	2.5
	2	0.45	3.0
	3	0.46	2.8
	4	0.40	3.0
	5	0.40	5.0
	7	0.49	2.7
Siltstone	1	0.50	3.0
	2	0.50	4.0
	4	0.50	6.0
Shaley sandstone	7	0.56	3.9
Limy silt	1	0.41	5.0
Carbonate	2	0.50	7.0
	5	0.40	5.5
Calcarenite	1	0.42	1.8
Micrite	1	0.30	2.4
	2	0.60	5.0
Chalk	7	0.70	7.1
Evaporite	4	0.04	2.0
	6	0.55	8.5

* Lithology terms are very loosely used and are not clearly defined.

(1) Hegarty et al (1988); (2) Moussavi-Harami and Brenner (1992); (3) Mudford and Best (1989); (4) Forbes et al (1992); (5) Bethke et al (1991); (6) Lee and Bethke (1994); (7) Sclater and Christie (1980).

The exponential relationships commonly used to describe compaction imply that the reduction in porosity continues smoothly with depth. Compaction is actually driven by the effective stress, and this can fall to low values if pore fluid pressure rises and produces abnormal pressures (see discussion in Section 9.4). When pore fluid pressure reaches the rock load of 1.0 psi/ft, formation fluids support the overlying rock column and the pressure driving compaction drops to zero. These sediments do not compact, resulting in undercompaction and porosity values that are anomalously high for the depth. (The consequences of this for thermal conductivity are discussed later.) Nakayama (1987) suggested adding an extra term to the exponent of Equation 3.1 to account for abnormal subsurface pressures. This term is given by:

$$\delta = (1 - P/P_{ob}) / (1 - P_e/P_{ob}) \qquad (3.5)$$

where P is the actual pore-fluid pressure, P_e is the hydrostatic pressure, and P_{ob} is the overburden pressure. When the pore-fluid pressure is equal to the hydrostatic pressure (i.e., the section is normally pressured), δ is equal to 1 and the porosity reduction equation is the same as Equation 3.1. When pore fluid pressure, P, exceeds the normal hydrostatic pressure, P_e, δ is less than 1.0 and the rate of porosity reduction with depth diminishes.

The variation of porosity with depth for carbonates is commonly more complex than in clastics. Because of the significant solubility of carbonates, cementation and dissolution often play a more important role than mechanical processes. Figure 3.8 shows inferred porosity variations for the core and flank of a bioherm in the Wegener Halvo Formation of East Greenland (Scholle et al, 1991) where there are marked differences over a relatively short distance.

Stylolites in carbonates and sandstones may indicate considerable loss of material. They develop as irregular dark bands of residual insoluble material (mainly clays and organic matter) as the host rock dissolves. Details of formation depend on the sizes and shapes of mineral grains, temperature, pore fluid compositions, and other factors. It is usually assumed that maximum amplitude can be taken to be the minimum amount of material removed. Mossop (1972) used this method to calculate loss of vertical thickness in the Upper Devonian Redwater Atoll of Alberta, Canada, and found a 13% volume reduction (174 ft, 53 m) in the rim carbonates but a 24% reduction (319 ft, 97 m) in the lagoonal carbonates (see Table 3.3). The 145 ft (44 m) of differential compaction agreed well with the 150 ft (46 m) of differential downwarping obtained by mapping.

Stylolite amplitudes in the Kekiktuk Formation sandstone of the North Slope of Alaska suggest a thickness reduction of about 13% (Bloch et

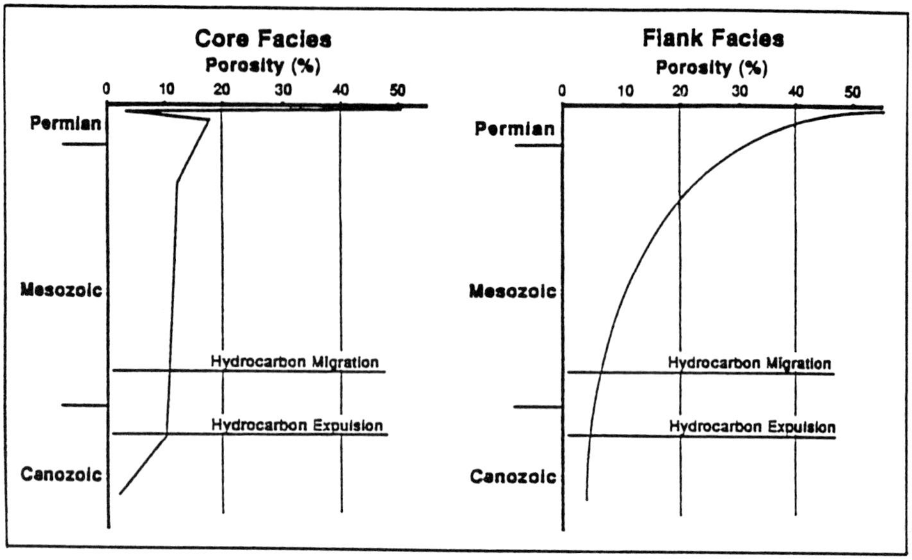

Figure 3.8. Inferred porosity variations at locations on the core and flank of a bioherm in the Wegener Halvo Formation of East Greenland (Scholle et al, 1991).

al, 1990). More accurate estimates can be obtained either from the offset of preexisting oblique markers across the stylolite surface (Pettijohn, 1957), or from estimates of the concentration of heavy mineral residues along stylolite seams (Heald, 1955). These approaches showed that the estimates obtained from Kekiktuk stylolite amplitudes were too low by a factor of between 3 and 10. In the Salem limestone of west-central Indiana, USA, stylolite amplitudes lead to a value for the shortening of section that is only 43% of what actually occurred (Finkel and Wilkinson, 1990).

The coalification process that leads from peats to high-rank coals causes a substantial reduction in thickness. Although it has been difficult to quantify this compaction, the peat appears to be reduced in thickness by a factor of approximately 10, but estimates range from 3.7 to 30 (Ryer and Langer, 1980; Elliott, 1985). Compaction is most rapid in the early stages that involve peat and lignite. Most literature descriptions of coal have been restricted to individual seams, and there is less information on coal as a constituent of the total basin fill. In general, individual coal seams are relatively thin, but in the central Powder River Basin, Wyoming, the Wyodak seam in the Fort Union Formation has a maximum thickness of 60 m, implying an initial peat thickness of about 600 m. Even though

Table 3.3
Calculated results for rim compaction in the Upper Devonian Redwater Reef, Alberta, Canada (data from Mossop, 1972).

	THICKNESS REDUCTION	
	feet	% original thickness
"Non-Duvernay" Rim	174	13.1
Duvernay Rim	321	17.4
Lagoon (Imp Eastgate 1-22)	319	23.6

individual beds are thin, the cumulative amount of coal in an area can be quite substantial.

In some situations mineral reactions can be significant in reducing the vertical thickness of sediments. This is potentially quite important in the case where gypsum ($CaSO_4.2H_2O$) dehydrates and converts to anhydrite ($CaSO_4$) because it can lead to as much as 39% decrease in volume (Jowett et al, 1993). Although the gypsum-anhydrite transition depends on temperature, it also depends on water salinity. Jowett et al (1993) showed that transition depth varied with heat flow and the nature of the overlying rocks and could vary from 400 m to as much as 4000 m.

Decompaction: Well data are obtained from rocks that have been compacted, and in using this information to reconstruct burial histories it is necessary to undo the effects of compaction and recognize that compacted rocks had greater vertical thickness when they were at shallower depths. The need for *"decompaction"* was recognized by Perrier and Quiblier (1974), and they proposed methods for making the required corrections. An alternative method was suggested by van Hinte (1978) who treated a compacting rock as being made up of mineral grains and water (which occupies the porosity) and pointed out that, as porosity decreases with depth of burial, the water content decreases but the volume of mineral grains remains unchanged. The principle of van Hinte's method is summarized in Figure 3.9. The original sediment had a thickness T_o that can be considered as being made up of porosity, $(\phi_o)T_o$, and grains, $(1 - \phi_o)T_o$, where ϕ_o is the original porosity. After compaction the thickness of the rock has been reduced, so that it now has a thickness of T_n and a porosity of ϕ_n. Because compaction only expels water, the volume of the grains remains unchanged and we can equate the terms for original and current values to give:

$$(1 - \phi_o)T_o = (1 - \phi_n)T_n \tag{3.6}$$

so that the original thickness, T_o, is given by:

$$T_o = T_n \{ (1 - \phi_n) / (1 - \phi_o) \}. \tag{3.7}$$

Current values for T_n and ϕ_n can be measured, and it is only necessary to know the original porosity, ϕ_o, to calculate the initial thickness. Values for ϕ_o for different sediments are reasonably well known and selected values are given in Table 3.2.

The example in Figure 3.9 is taken from Angevine et al (1990). Here the thickness measured today (i.e., compacted) is 100 m and the porosity is 0.20, so that the rock can be regarded as 80 m of mineral grains and 20 m of water. If initial porosity for this lithology was 50% (i.e., 0.5), then the original rock still had 80 m of mineral grains but also had 80 m of water for a total decompacted thickness of 160 m. D'Heur (1991) provides an example of using this method in the West Ekofisk area of the North Sea. Here initial porosity in the coccolith ooze was very high and was taken to be 70%. The reservoir is currently at about 11,500 ft (3507 m) with present-day porosities of 33% and 19% for compacted chalks that are 452 and 366 ft (138 and 112 m) thick, respectively. Both decompaction calculations suggest an initial, uncompacted thickness of about 1000 ft (305 m). A variety of decompaction models has been proposed (e.g., Liu and Roaldset, 1994), and computer programs for decompacting sediments are given by Gildner (1990) and Springer (1993).

Corrections for compaction are usually based on information about the relationship of porosity to depth. This assumes that porosity is reduced solely by mechanical forces that squeeze out the water and drive the grains closer together, and that loss of porosity by cementation, or development of secondary porosity, are negligible (Houseknecht, 1987). (Often petrographic examination of the rock will show whether there has been significant cementation.) The considerable scatter in porosity values that is commonly reported suggests that the basic assumptions may not always be justified. Local fluctuations due to cementation, dissolution of unstable grains, or low permeabilities that lead to overpressure, may all be reflected in anomalous porosity values. Porosity values measured by different techniques also contribute to the scatter. However, an overall trend can usually be established, and it is particularly helpful if a fairly uniform

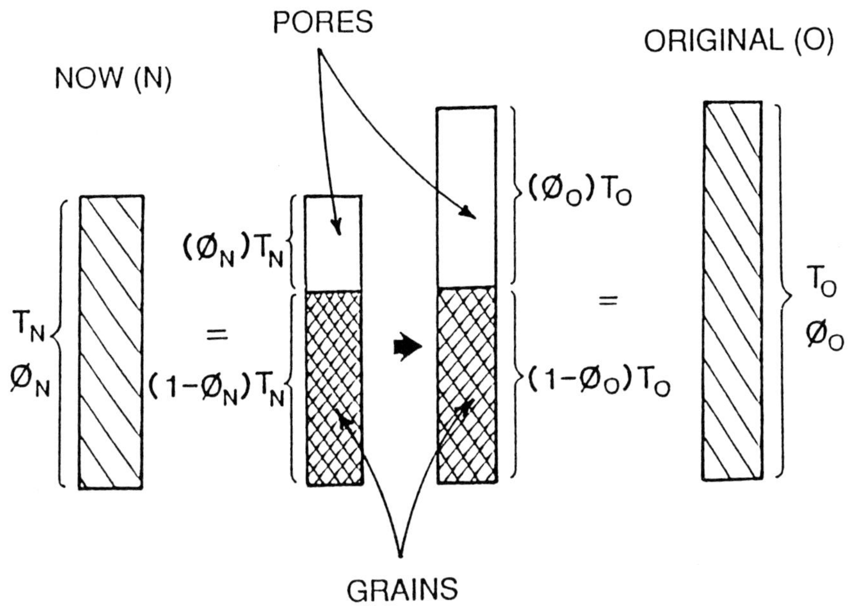

COMPACTION CORRECTION (VAN HINTE 1978)

since $(1-\emptyset_N)T_N = (1-\emptyset_O)T_O$

$$T_O = \frac{(1-\emptyset_N)T_N}{1-\emptyset_O}$$

EG. $T_N = 100$ m

$\emptyset_N = 0.20$ $\quad$ $\emptyset_O = 0.50$

$(1-\emptyset_N)T_N = 0.8(100) = 80$ m of sediment

$$T_O = \frac{(1-\emptyset_N)T_N}{1-\emptyset_O} = \frac{80}{0.5} = 160 \text{ m.}$$

Figure 3.9. Summary of the principle of the method suggested by van Hinte (1978) for decompacting sediments (from Angevine et al, 1990).

lithology (say, shale) can be sampled at many different depths in the well.

Porosity can be measured directly if rock samples are available, or deduced from electric-log response using either sonic logs or density logs. Sonic logs determine the speed at which sound is propagated through rocks. This generally increases with depth, and the transit time over some fixed depth interval decreases with depth. Values obtained from sonic logs can be related to porosity (Magara, 1978; Stam et al, 1987), but the sonic log response is sensitive to barium and should be used with caution in sections where mud weight has been increased by adding barite. As low-density water is squeezed out by compaction, bulk rock density rises and this increase can be obtained from density logs (Xiao and Suppe, 1988). Hegarty et al (1988) have shown that combining both density log and sonic log data optimizes porosity curves for the southern Australian margin (Figure 3.10).

3.5.3. Significance of Compaction

Lines representing the tops and bottoms of compacted units on burial history diagrams do not remain parallel because the thickness of the rock unit decreases with burial. The greatest decrease generally occurs in the

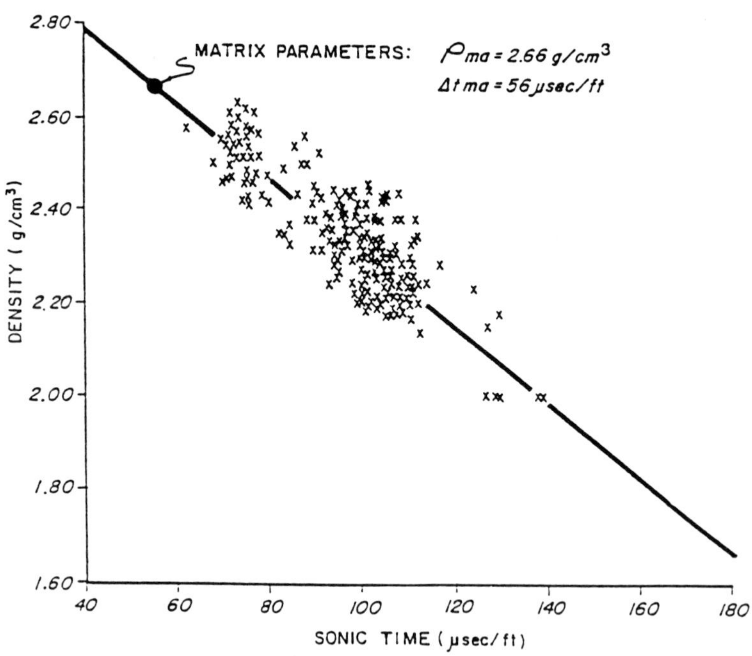

Figure 3.10. Combination of electric log-derived density and porosity values for the southern Australian margin (Hegarty et al, 1988).

first 2500 m of burial. Failure to decompact in reconstructing burial diagrams influences units ***deeper*** than the one compacting. This apparently counter-intuitive conclusion is a consequence of the way burial diagrams are constructed from present day thickness values that are compacted back through time to uncompacted thicknesses. In the example shown in Figure 3.11, depth d_a is measured in the well, and age t_a is constrained, so that burial history for the base of Unit III has to be the straight line shown. Similarly d_b is measured and t_b is constrained to give a straight line for the base of Unit II. The compacting unit has a present day measured thickness of $d_c - d_b$, but this has been greater in the past when it was less compacted than today. However, the burial history for the top of Unit I is controlled by the straight line shown for the base of Unit II, so that the only way the extra thickness can be accommodated on the burial diagram is by moving the base of Unit I to greater depth as shown. As a consequence, units deeper in the section will be moved down to greater depths of burial.

Compaction corrections can be quite significant in thermal reconstructions. For example, if a shale was initially 1000 m thick, a 40% reduction in thickness by compaction would lead to a decompacted burial profile in which underlying units were buried as much as 400 m deeper.

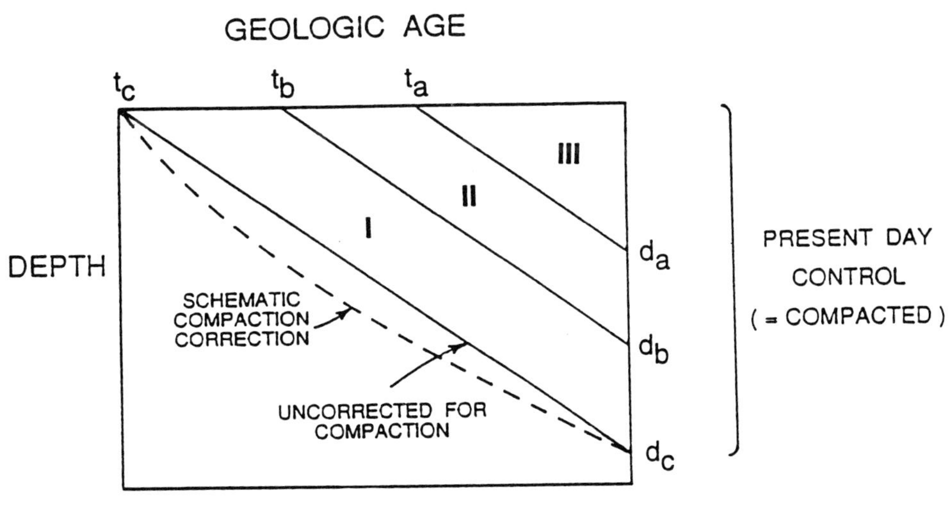

Figure 3.11. Schematic burial history showing reconstructions from present day compacted values for Unit I with (dotted line) and without (solid line) decompaction.

This corresponds to 10 °C for an average geothermal gradient of 25 °C/km, and leads roughly to a doubling of generation rate. Clearly, the effect of decompaction is to move all of the rock units in a section to greater depths and to higher temperatures, and as a consequence generation of petroleum starts earlier. Dykstra (1987) has provided examples to illustrate the differences between compacted and decompacted sections. In the Sauk A-57 well, offshore Nova Scotia, Canada, the calculated time of generation is about 8 m.y. earlier when the decompacted burial history curve is used. For the deepest rock unit penetrated by the Inigok well, Alaska (Horizon E, Figure 3.12), the corrected curves give the time of the oil window as 284 to 230 Ma, whereas for the uncorrected curves it is considerably later at 240 to 115 Ma. Notice that the effects of timing on the oil window position are greatest for the deeper units because the effect of temperature increases exponentially with depth.

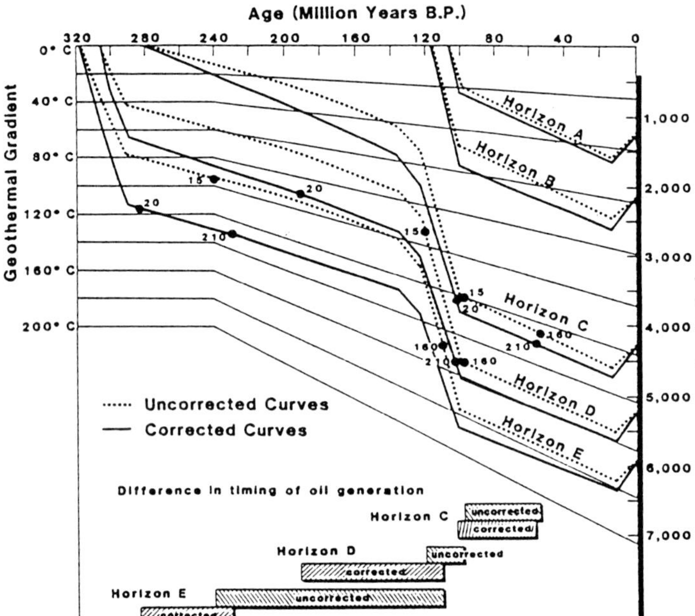

Figure 3.12. Burial history model for multiple units in the Inigok No. 1 well, Alaska showing curves corrected and uncorrected for compaction, and the effects on the timing of the oil window (Dykstra, 1987).

3.6. UNCONFORMITIES

3.6.1. Principle

Unconformities are breaks in the geologic record where material has been lost by erosion, generally due to uplift. In a few cases unconformities can be due to a time of non-deposition, and locally may be caused by scour. Most vertical sections of any appreciable depth have unconformities.

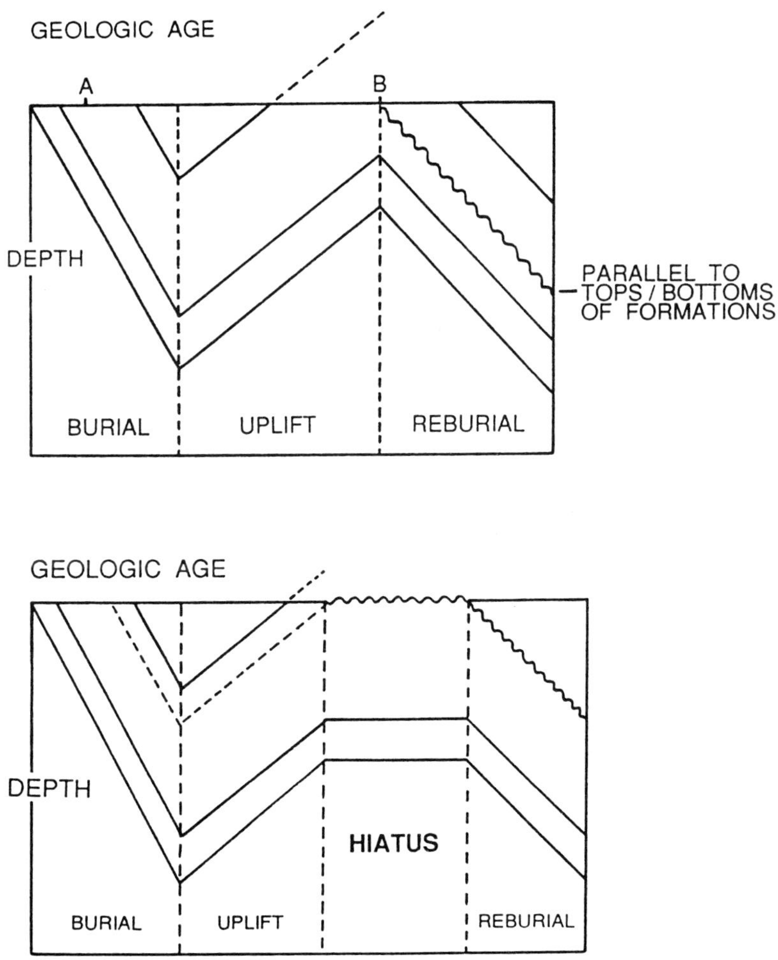

Figure 3.13. Schematic burial histories showing examples with (bottom) and without (top) a hiatus between uplift and reburial.

Burial diagrams for sections with unconformities are shown in Figure 3.13. After initial burial there is uplift that leads to thinning by erosion and loss of section. On reburial the vertical thickness of the remaining, partially eroded, rock does not change, so lines representing unconformities on burial diagrams are parallel to lines representing the tops and bottoms of lithologic units (neglecting for the moment the effects of compaction). If there is a time period when there is neither erosion nor sediment accumulation, part of the burial history diagram would be horizontal as shown in Figure 3.13 ("*hiatus*"). Complete reconstruction of a section with an unconformity requires a knowledge of ages immediately above and below the unconformity, the amount of vertical section lost by erosion, and constraints on the timing of maximum depth of burial before uplift. This is illustrated in Figure 3.14 where the thickness of rock

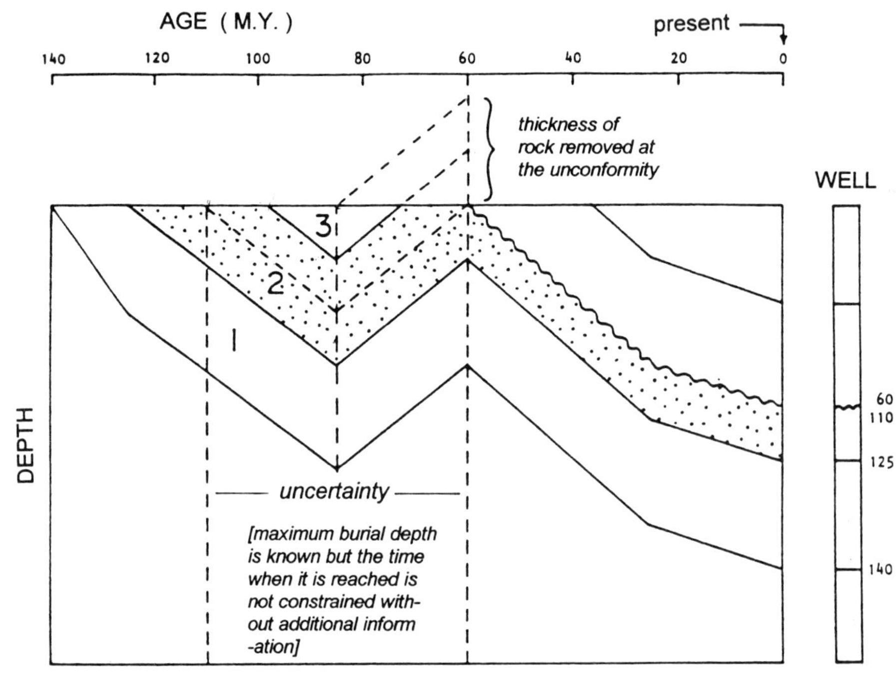

Figure 3.14. Burial history reconstruction for a section containing an erosional unconformity where the amount of rock removed is known. In this example the uncomformity separates rocks with ages of 60 Ma (above) and 110 Ma (below).

removed by erosion is known. In this example Unit 3 reached its maximum thickness at 95 Ma but was completely lost by erosion by 72.5 Ma. Unit 2 lost approximately half its thickness, and all rocks younger than 110 Ma are removed before reburial started at 60 Ma. Thus, the unconformity penetrated in the well separates rocks with ages of 60 Ma (above) and 110 Ma (below). Note that the age at the unconformity (110 Ma) is not the age when uplift started or the uplift age. While maximum depth of burial can usually be obtained (see below), the timing of the maximum depth of burial (shown at 85 Ma in the figure) is often difficult to constrain in practice. It is usually difficult to decide whether uplift was slow or rapid, but sometimes fission track information provides a useful control.

Each unconformity in a well implies a period of uplift and erosion, and in areas with multiple unconformities the burial history diagram must reflect these events. Figure 3.15 shows a burial history diagram that includes 6 unconformities (van Wijhe, 1987). Uplift at *A* resulted in the loss of 1000 m of section and brought the top of source rock 1 to the surface. Uplift *B* also corresponds to a major uplift, but *C* through *F* are relatively minor periods of uplift.

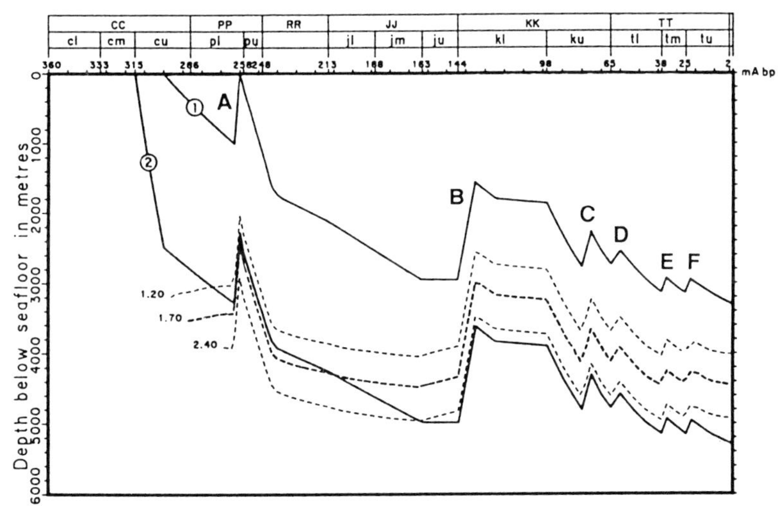

Figure 3.15. Burial history graph of a well just north of the Broad Fourteens basin, North Sea, showing (1) the burial history of the top of the Upper Carboniferous source rock interval, and (2) the burial history of the base of the Upper Carboniferous source rock interval. Calculated isovitrinite reflectance lines are superimposed (Van Wijhe, 1987). Periods of uplift at times A-F all correspond to some erosional loss at the surface and will be represented by unconformities in the well.

3.6.2. Reconstruction

Sediment lost: In practice, the commercial computer models reconstruct the burial histories for sections with unconformities, but it is necessary to provide information on ages and rock losses. In areas with many wells and good geological control, the standard geological methods provide a way of estimating the amount of rock lost at an unconformity. When only one well is available the problem is more difficult, but in principle any non-reversible depth-related parameter can be used to determine the amount of uplift. Careful reconstruction of unconformities that occur in the deeper and hotter parts of the burial history is much more critical than for those occurring shallow (Waples et al, 1992b).

Organic materials change irreversibly with depth, and vitrinite, coal, and biomarkers, have all been used in reconstructing sections with unconformities. Vitrinite reflectance profiles commonly show abrupt increases in reflectivity at unconformities, and Dow (1977) suggested that if the trend below the unconformity is extrapolated to the value at the unconformity, then the vertical distance corresponds to the amount of rock lost by erosion (Figures 3.16 and 3.17). This technique requires good vitrinite reflectance data and can be applied only when there has been limited reburial, because significant reburial raises temperature and smoothes out the differences. The surface corresponding to an

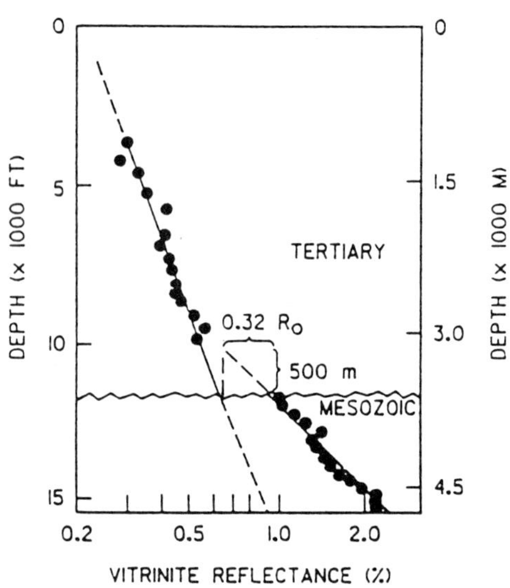

Figure 3.16. Graphical solution proposed by Dow (1977) for using vitrinite reflectance for estimating the amount of uplift and erosion at an unconformity.

unconformity was once probably subaerially exposed, and R_o values right at, or just below, the erosional surface may show anomalous values (generally high) due to oxidation. For the special case where the present day surface is erosional, vitrinite reflectance values can be extrapolated to a typical surface value (Figure 3.18). Most researchers use 0.2% R_o as the surface value, but Link and Bustin (1989) extrapolated to a surface value of 0.15% R_o in their study in northern Canada, where they showed that erosional loss varied from 0.7 to 4.7 km across the area. In western Wyoming a major basement uplift during the Late Cretaceous led to considerable thinning of Upper Cretaceous rocks (Warner and Royse, 1987). Vitrinite reflectance values below the later Hogsback (Darby) are about 0.5%, and extrapolating the R_o - depth trend to a surface value of 0.2% R_o gives the thickness of eroded rock as 7200 ft (2196 m) (Figure 3.19). This value was used by Warner and Royse (1987) in reconstructing the burial history for the area (Figure 3.20). To the south at Whitney Canyon, R_o profiles show an erosional loss of more than 10,000 ft (3050 m).

Permian rocks outcrop at the surface in the southern Junggar basin of China. Here Carroll et al (1992) used regional geology, surface R_o values, and subsurface R_o profiles to calculate that 5.25 km of Triassic and Jurassic rocks had been removed by erosion. In the central Brooks Range of Alaska discontinuities in vitrinite reflectance values show that there has been 13 km or more of uplift (Johnsson et al, 1993). The presence of metamorphic rocks at the surface in many areas suggests that even greater amounts of unroofing can occur. For additional discussions on using R_o values to constrain erosional loss, see Middleton and Schmidt (1982), Armagnac et al (1989), Katz et al (1988), and Bordenave (1993). Goodarzi et al (1992) have combined graptolite reflectance information with vitrinite reflectance data to calculate erosional loss of section in the Queen Elizabeth Islands, Arctic Canada.

Coalification is a depth-related, non-reversible process (Hacquebard, 1977) and has been used to estimate the amount of overburden removed. Nurkowski (1984) measured calorific value, equilibrium moisture contents, and ash contents of near-surface coals in Alberta, western Canada, and concluded that between 900 and 1900 m of sediment had been removed, with the greatest amount of removal occurring in a west-southwesterly direction. Majorowicz et al (1990) have also used coalification data to estimate erosional losses in the southern Alberta Plains.

Many of the organic compounds in sedimentary rocks are not thermally stable, and their structures change irreversibly in response to rising temperatures. Biomarker transformations have been especially well studied (see Chapter 5) and used to recognize the reduction in reaction rates induced by the decreased temperatures due to loss of section. The fit between theoretical sterane and hopane isomerization, and sterane

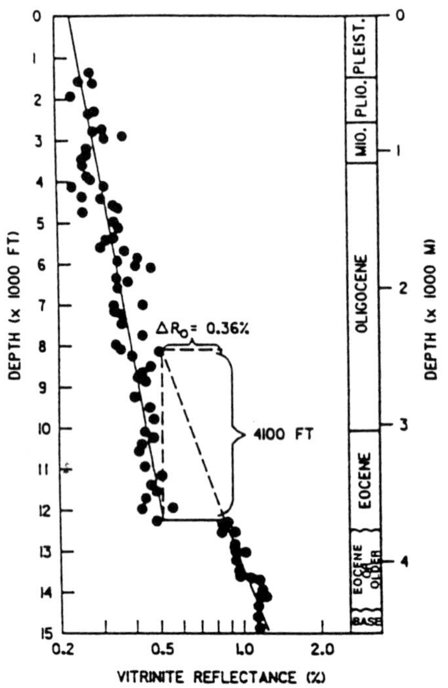

Figure 3.17. Use of vitrinite reflectance data to estimate erosional loss of section at an unconformity (Katz et al, 1988).

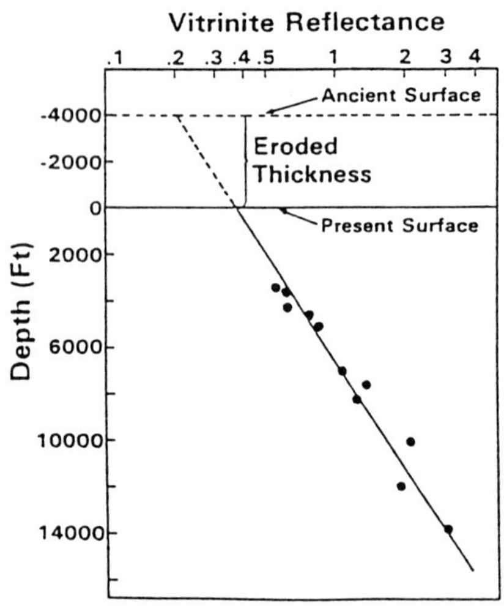

Figure 3.18. Method for using vitrinite reflectance data for estimating eroded thickness when the present day surface is erosional.

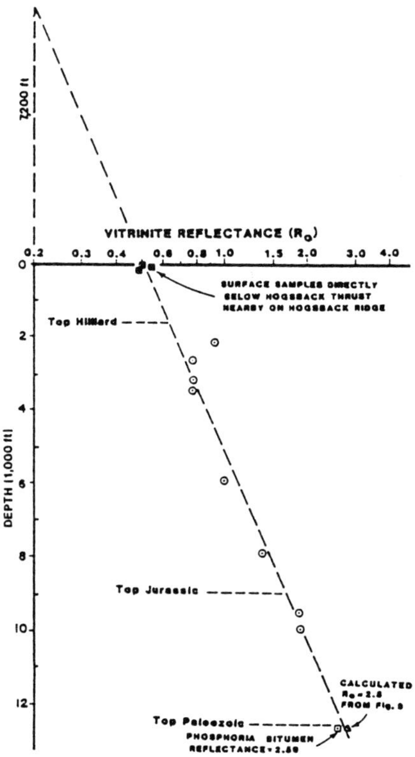

Figure 3.19. Vitrinite reflectance values versus depth in the Mobil 22-19G Tip Top well, western Wyoming showing that about 7,200 ft (2190 m) of rock has been eroded at the well site (Warner and Royse, 1987).

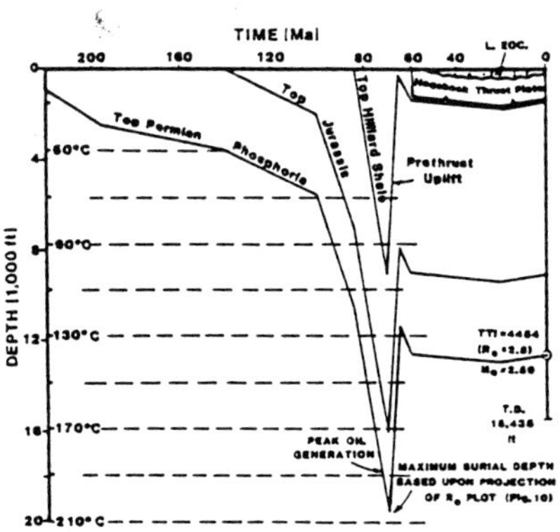

Figure 3.20. Burial history diagram for the Mobil 22-19G Tip Top well, western Wyoming. Approximately 7200 ft (2190 m) of Late Cretaceous strata was removed by uplift and erosion during a 5 m.y. period prior to Hogsback (Darby) thrust emplacement (Warner and Royse, 1987).

aromatization curves with the corresponding measured ratios led Mackenzie and McKenzie (1983) to suggest greater uplifts in the Paris Basin and North Sea than had previously been proposed. They also applied the method to the structural inversion in the Lower Saxony Basin. The method has the added advantage that the isomerization and aromatization reactions are frozen in by uplift and consequent cooling, and this provides an absolute time for the uplift. Walters and Kotra (1990) used data from the same biomarker isomerizations and aromatization reactions in the Newark Basin rift valley sequence and concluded that about 2400 m of rock had been removed by erosion starting at approximately 50 Ma.

Shale compaction during burial leads to steadily decreasing porosities and as a consequence of the diminished quantity of low-density water, shale bulk densities increase. The present-day bulk density should reflect the maximum depth of burial, so that values could be used to estimate the amount of uplift and the thickness of eroded sedimentary rock. This method has been used by Bond (1984) for the San Juan basin, where a plot of bulk density versus depth was constructed for the Cretaceous shales on the north flank of the basin and then compared with the standard density versus depth plot. The comparison showed that an adjustment of somewhere between 5500 and 7500 ft (1678 and 2288 m) was needed, with the best fit corresponding to 6500 ft (1983 m) of uplift.

Changing the bulk density of rocks also changes their seismic interval velocities. Since compaction is not reversible, uplifted rocks have an interval velocity that is higher than would be anticipated for the present depth. Marie (1975) first proposed this method and applied it to inversion of the Bunter Shale Formation in the North Sea basin by comparing measured interval velocities with a standard curve developed for interval velocity versus depth. The same method was used by Glennie and Boegner (1981) to estimate the amount of inversion in the Sole Pit basin area of the southern North Sea, and Cope (1986) later found that it gave results generally comparable to those obtained from vitrinite reflectance data for the area. A subsequent detailed study by Bulat and Stoker (1987) (based on data from 300 wells and eight stratigraphic units ranging in age from Carboniferous to Tertiary) showed that for most of the southern North Sea the median uplift values are consistent with the known geologic history. They found that establishing the absolute magnitude of uplift from interval velocities was problematic due to other geologic factors. Strictly, the method can only be applied to areas with uniform lithology, similar pressure histories, and the same pore fluid composition. Interval seismic velocity analysis has also been used offshore Norway in the Barents Sea. Theis et al (1993) combined R_o data from individual wells with seismic data for interpolation between wells. In the studied area uplifts varied from 0 to >3200 m over distances of about 500 km.

Sonic transit times (obtained from electric logs) are a function of porosity, and values decrease with increasing compaction and reflect depth of burial. A plot of the log of the interval transit time, *t*, (usually in msec/ft) versus depth gives a straight line (Magara, 1976), but to use this for estimating loss of section it is necessary to know surface values. Magara, who first suggested this method, derived a surface value of 200 $msec.ft^{-1}$ for the Western Canadian basin. In this case, that corresponds to an uplift of about 4500 ft (1373 m) (Figure 3.21). Connolly (1989) also used shale transit times in western Canada and combined sonic log data with vitrinite reflectance data to define a smooth westward increase in the amount of erosion in northwest Alberta, with losses ranging from 800 m to 2400 m along a trend towards the Rocky Mountains. Sweeney et al (1986) used a similar approach to calculate the amount of overburden removed in the Uinta basin, Utah. They estimated a normal compaction trend (above the overpressured zone) with *t* decreasing at 0.000096 $msec.ft^{-1}.ft^{-1}$, which is within the range obtained by Magara (1976) for the Cretaceous shales of Western Canada where values lie between 0.000085 and 0.000147 $msec. ft^{-1}.ft^{-1}$. Overburden removed was generally in the range from 5000 to 6600 ft (1525 to 2013 m) with an estimated uncertainty of $\pm$1000 ft ($\pm$305 m). Goy (1979) used this method to demonstrate more than 1 km of uplift in the Paris Basin, and Conrad and Griffith (1983) used it in the Amadeus basin of central Australia. In the Bighorn basin of Wyoming, Heasler and Kharitonova (1996) developed an improved exponential model of compaction and used it with sonic log transit time data for 29 wells. Calculated erosional loses varied from 0 to 5600 ft (0 - 1700 m) with an average of 2500 ft (800 m). Feinstein et al (1996) combined organic maturity data with sonic porosities to evaluate both erosional loses and paleotemperatures in the Mackenzie Plain, Canada. Sonic transit times cannot be used to calculate erosional loss when shales are overpressured and undercompacted, because under these conditions shale composition is no longer a simple function of burial depth. In practice it can often be difficult to recognize situations where shales have been overpressured in the past but are now normally pressured.

Sandstone porosity generally shows a non-reversible decrease with depth of burial, but regional trends are commonly obscured by variable clay contents, secondary cementation, abnormal pressures, and other factors that produce a wide scatter in porosity values (see discussion on *Compaction* above). In the Dandaragan Trough of the Perth Basin, western Australia, Thomas (1984) avoided these problems by plotting

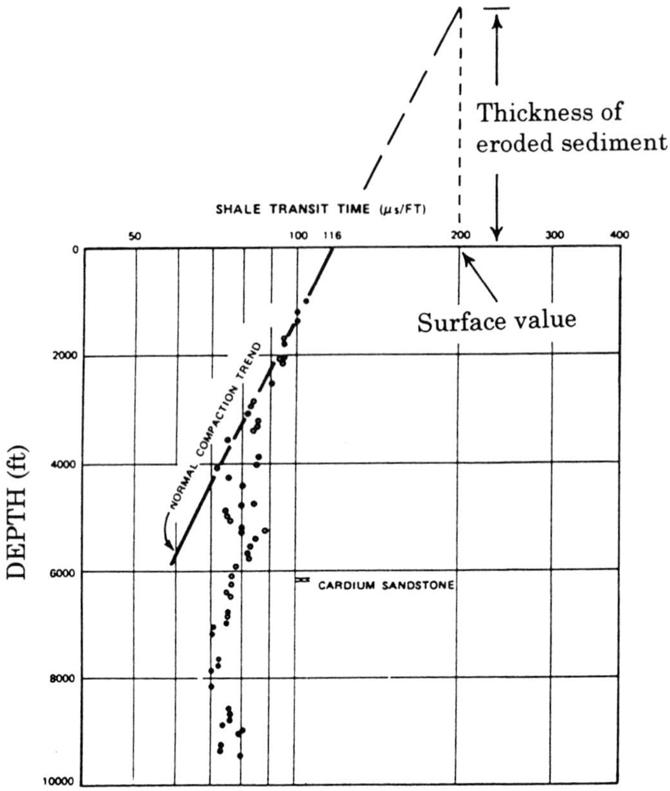

Figure 3.21. Shale transit time-depth plot for the Pacific Amoco Ricinus 16-29-34-8-W5 well, western Canada (Magara, 1976). Shale transit times for the normal compaction trend have been extrapolated to a typical surface value to provide an estimate of the thickness of eroded sediment. Notice that below about 4500 ft abnormal pressures become common and these transit time values should not be used in the extrapolation.

porosity values (obtained from sonic logs) for the best 6 m of every 30 m of section. Porosity data for the relatively uniform Yarragadee Formation led to values for uplift that range from more than 3500 m in the east to less than 2000 m in the west. Thomas (1984) claimed that this approach provided maximum values (corresponding to the cleanest, best-sorted sands that lacked secondary cementation) and avoided any sonic "shale" effects.

Temperatures derived from fluid inclusions in cements can be combined with geothermal gradients to give depths at the time of cement formation. Comparison with present-day gradients and depths of burial then provides an estimate of the amount of uplift. McClimans and Videtich (1989) used this method in the south of England and calculated that uplifts varied from 150 ft (46 m) in the east to 4500 ft (1373 m) in the west. In northeast Utah, Narr and Currie (1982) used fluid inclusions to calculate erosional loss that ranged from 1112 to 9482 ft (339 to 2892 m), but found that the maximum thickness removed (9482 ft, 2892 m) was considerably more that the value of 5988 ft (1826 m) estimated from the stratigraphic record. There was insufficient information to decide whether this discrepancy was due to errors in interpreting fluid inclusion data, or in reconstructing the stratigraphic record. Sarwar and Friedman (1994) combined fluid inclusion information with clay diagenesis to estimate that the erosional loss in the Adirondacks of New York was as much as 8000 m.

The mass-238 isotope of uranium usually decays by alpha-particle emission, but can also decay by fission (see Chapter 5). In the latter case damaged tracks form in the host minerals. However, at elevated temperatures the damage is repaired by annealing and the tracks become shorter and eventually disappear. Uplift and erosion that lead to the formation of unconformities cause cooling, and the temperature dependence of fission-track annealing can be used to establish the amount of uplift. The method is usually based on measurements made for apatite and the details are discussed later (Chapter 5). In the Taranaki Basin of central New Zealand, Kamp and Green (1990) used apatite fission track annealing (AFTA) to show that in the Fresne well there had been 3.0 $\pm$ 0.3 km of uplift and that it had occurred at a rate of 0.75 mm/yr. Farther east in the Surville and North Tasman wells there has been less uplift and estimates were $\leq$2.0 $\pm$ 0.5 km and 1.10 $\pm$ 0.3 km, respectively. Permian limestone forms the surface in the Grand Canyon area of the Colorado Plateau, western U.S.A., and Dumitru et al (1994) used AFTA data to show that the area cooled from about 90 to 100 ^{o}C in the Late Cretaceous (75 Ma), corresponding to a post-Laramide-age removal of about 1.3 - 4.5 km of section. This is consistent with the 2.5 - 3.5 km of Mesozoic section still present to the north in Utah, and shows that it previously extended over a much greater area of the Colorado Plateau. AFTA methods were also employed to show that approximately 2.0 - 2.5 km of uplift occurred in the Pine Point area of the Northwest Territories, Canada (Arne, 1991). O'Sullivan et al (1993) used AFTA techniques in the northeastern Brooks Range, Arctic National Wildlife Refuge, Alaska. In this complex area, with duplexing of thrust sheets, they were able to demonstrate four major phases of cooling associated with uplift and erosion. Minimum erosion rates were estimated to be 0.4 - 0.7 mm/m.y., if the geothermal gradient was 30 ^{o}C/km, and 0.6 - 1.0 mm/m.y., if the geothermal gradient was 20 ^{o}C/km. Other studies using fission track studies include those by Omar et al (1989, 1994), Fitzgerald (1994), and Grist et al (1995).

Schwarz et al (1985) suggested a method based on remnant magnetization, and Pantano and Lerche (1989) have developed a method that uses inversion of vitrinite reflectance data and sterane isomerization data. A somewhat different approach was suggested by Zhao and Lerche (1993) and instead of relying on a single parameter, they used a range of geological inputs in a procedure called thermal indicator tomography. An application to the Piceance Basin, Colorado, showed a loss of 4000 ft (1220 m) of section.

Ages: Ages immediately above and below an unconformity are almost always derived from paleontological information. In fact, unconformities are often first recognized from breaks in the expected fossil sequence.

Timing of maximum depth of burial: It requires a lot of geological control to constrain the time when maximum depth of burial was reached. Apatite fission track annealing may provide a useful method and Gleadow et al (1984) used it to work out the uplift history of the Antarctic Mountains. They concluded that there had been 4 km of uplift with a minimum average rate of 55 m/m.y. in the Cenozoic, but a lower rate of 15 m/m.y. between 154 ± 7 and 66 ± 4 Ma. Fitzgerald and Stump (1991) also found AFTA data to be useful in constraining uplift rates in Antarctica. Cerveny et al (1988) have used it for the Himalayas, and Roberts and Burbank (1993) for the Tetons.

Brown (1991) proposed an apatite fission-track stratigraphic technique to quantify tectonic uplift and accompanying erosion. (The method is analogous to that used for tectonic subsidence by Streckler et al, 1988). Fission-track methods can be applied in areas where only crystalline basement remains and none of the sedimentary section is preserved.

3.7. ROTATION OF BEDS (INCLUDING FOLDING)

Previous discussions have referred to the thickness of individual horizontal units, but sediments are frequently not horizontal. For inclined beds the relevant dimension on a burial diagram is not the true vertical thickness (TVT), but rather the vertical distance between the top and bottom of the rock. This would, of course, be greater than the rock thickness when the rock is dipping. In calculating sediment accumulation rates it is important to use true thicknesses and not drilled thicknesses.

If the dip of a rock unit changes, its vertical dimension changes, and the lines representing the top and bottom of the unit on the burial diagram diverge. In the extreme case of a bed that is rotated into a vertical position,

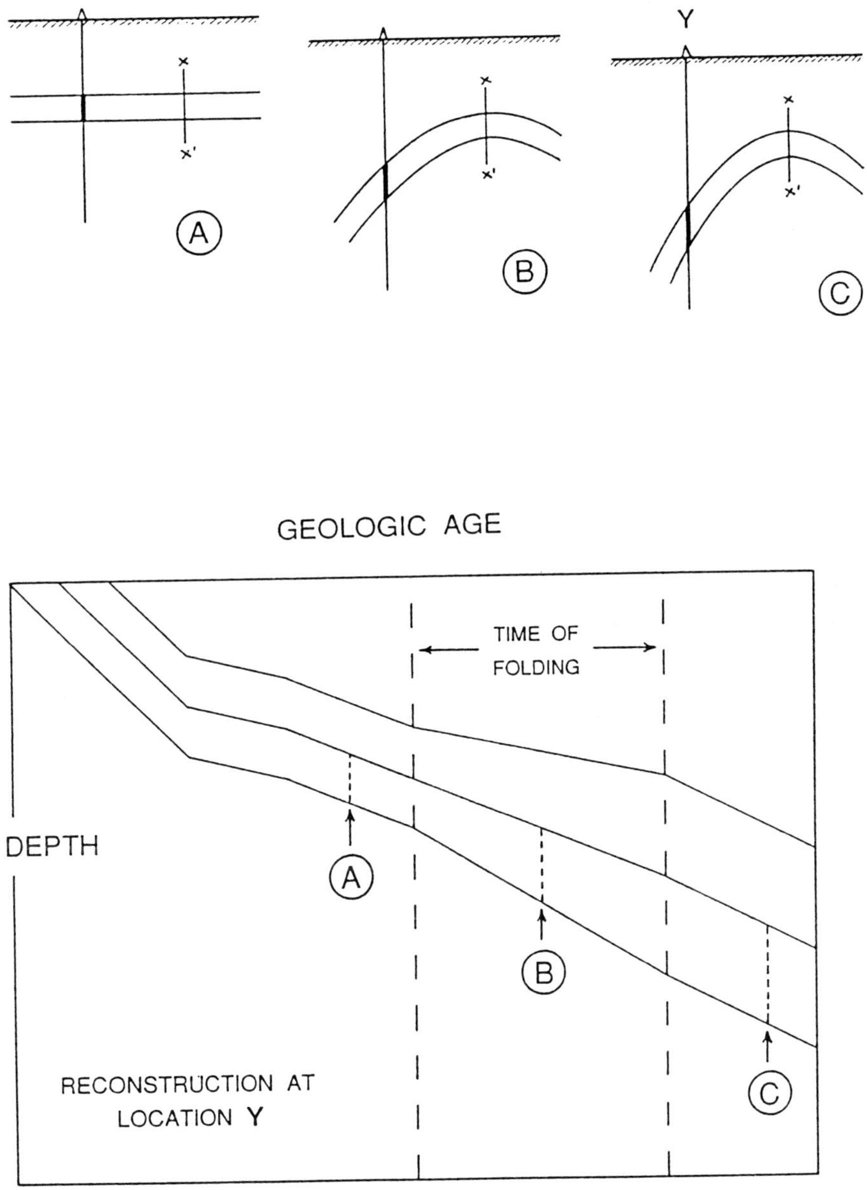

Figure 3.22. Three stages (A, B, C) for a rock unit being folded showing changes in cross section (top) and burial history (bottom). Lines on the burial history plot diverge during the time of folding.

the drilled thickness would be infinite. Figure 3.22 illustrates the case where an anticline has formed by folding and the well of interest, Y, is drilled out on one limb. Initially this rock unit (at the well location) is horizontal and has a vertical dimension equal to the stratigraphic thickness, but during the time of fold development thickness steadily increases. At the location of the well drilled on the crest of the anticline the thickness does not change through time. In practice, a whole package of strata is likely to be folded, producing significant thickening of the entire section.

Failure to recognize a dipping unit would lead to incorrect thickness values being used in a burial history reconstruction. This may happen in a rank wildcat well (for which no dip meter data are available), or in areas with limited well control. When a presumed vertical well has in fact deviated, the calculated depth profiles can be too deep, and the error would be consistent throughout the time span of the reconstruction, except for the initial time interval (Figure 3.23). Well deviation is more likely, and more complicated, when the beds are dipping steeply because the drill bit tends to "walk into the beds" and drill across them. The drilled distance from the top to bottom of a bed may actually be less than in a vertical hole drilled at the same location.

The presence of an angular unconformity shows that the beds below the unconformity were tilted before they were eroded. A schematic burial history diagram that gives the development of an angular unconformity in terms of the four stages of initial sediment accumulation, tilting, uplift/erosion, and final reburial is given in Figure 3.24. Sediments accumulated for the 70 m.y. from 250 to 180 Ma, were then rotated (folded)

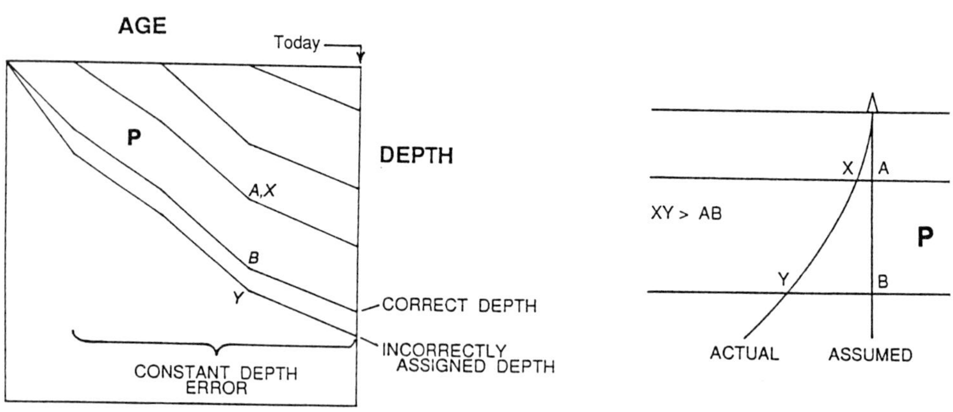

Figure 3.23. Burial history diagram showing burial profiles for the correct thickness of Unit P and an incorrectly assigned (and too great) vertical thickness.

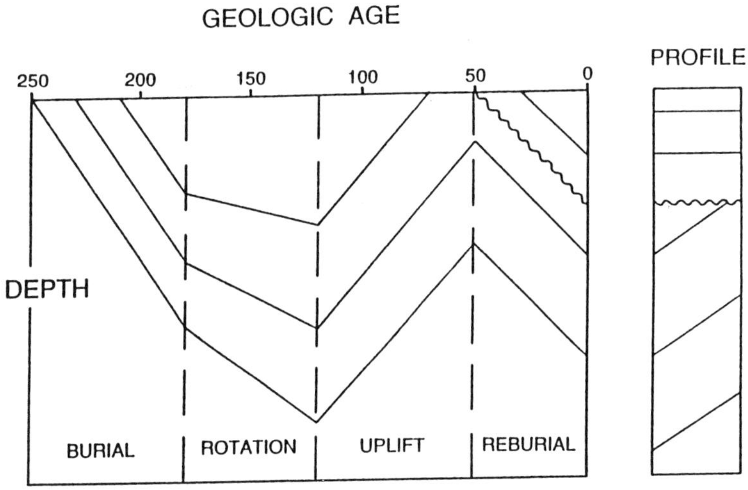

Figure 3.24. Burial history diagram for a well location that cuts across an angular unconformity. The reconstruction considers the four stages of sediment accumulation, rotation, uplift and erosion, and final reburial. The right-hand profile shows the rock sequence as it is today.

from 180 to 120 Ma, and subsequently uplifted. Erosion stopped and reburial occurred from 50 Ma to the present. The sediments deposited over the last 50 m.y. are horizontal, and are separated from the underlying dipping beds by the unconformity.

3.8. FAULTING

Wells drilled across faults may encounter repeated sections, truncated sections, or sections with formations missing. The latter situation is used for the illustration given in Figure 3.25, where rocks on both sides of the fault are buried at uniform rates, but the rate is higher in the west than it is to the east. A growth fault might provide an example of this type of behavior. The burial curves for formations penetrated by Well A (drilled on the downthrown side of the fault) are steeper than those developed for Well B, because in Well B they have not been buried as deeply in the same amount of time. Well X penetrates the fault at 10,000 ft (3050 m) and shows burial curves like those in Well A above the fault, and

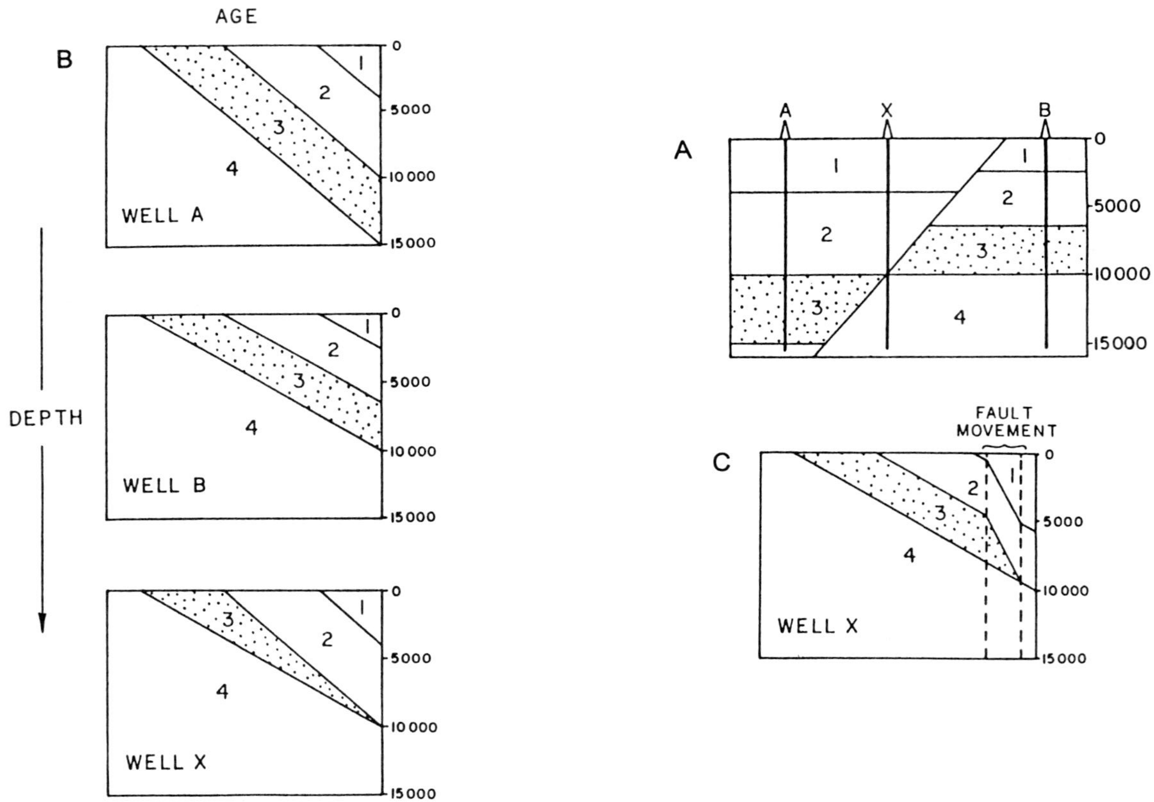

Figure 3.25. (A) Geological cross section for a faulted area where Well X has crossed the fault but Wells A and B have not. (B) Burial histories for locations A, B, and X. (c) Burial history at the location of Well X showing fault movement occurring over a limited period of time.

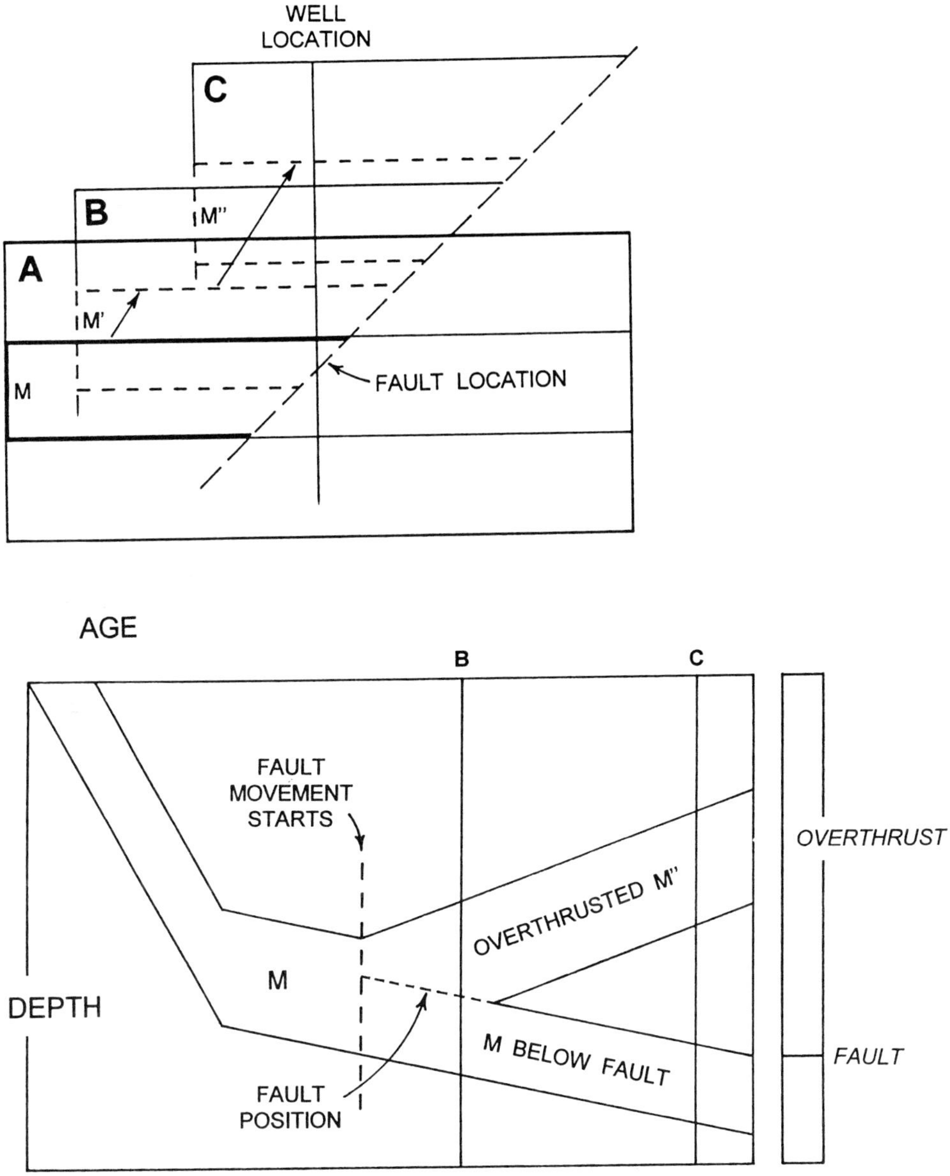

Figure 3.26. (Top) Geological cross section of an area with an overthrust fault showing the initial position (A), and subsequent movement to B and C. (Bottom) The corresponding burial history diagram shows an initial thickening at the well location at time B, and then a separation of Unit M into two parts as a repeated section is drilled at time C.

burial curves like those in Well B below the fault. In Part C of the figure the burial curves are shown for the case where faulting occurs over a limited time interval. Overthrusting can lead first to a local thickening of section, and then to repeated sections. Unit M in Figure 3.26 is cut by a thrust fault at the well location being reconstructed. As the overthrusted sheet moves from position A to B, and then to C, the amount of Unit M drilled first increases (B), but then separates into two separate parts (C). Examples involving thrust faulting are given in Section 10.6.

3.9. INTRUSIONS AND DIAPIRS

Sedimentary sections can be intruded by mobile igneous rock materials that are emplaced rapidly and at high temperatures. Such intrusions modify the burial histories of rocks in the area, as well as disturbing the local thermal regime. Temperature effects are discussed in Section 5.6. A horizontal intrusive (sill) must create space for itself, and this requires a pressure that is close to lithostatic and capable of lifting the overlying rock column. Intrusion would be followed by an isostatic readjustment, due to the greater density of the igneous body, and this leads to a somewhat greater burial depth. The amount of adjustment for a single sill is relatively minor and can usually be ignored.

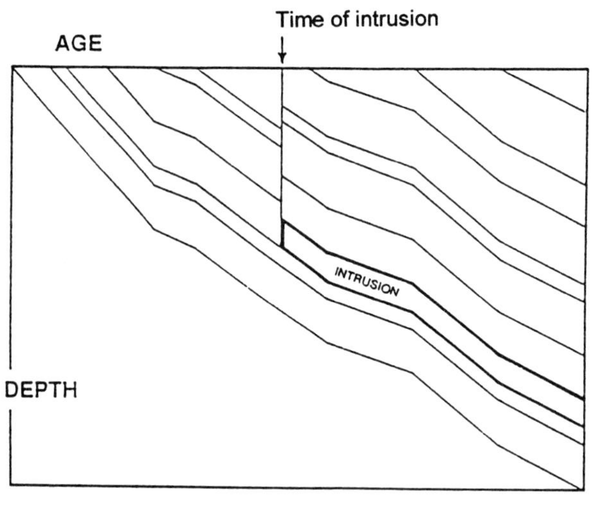

Figure 3.27. A schematic burial history diagram in which an igneous sill is intruded into a sedimentary section.

Burial history diagrams show igneous intrusives appearing as an instantaneous thickening with no subsequent changes in thickness through time. Figure 3.27 gives an example of a burial history diagram in which the intrusive lifts the overlying sediments. Wang et al (1989) have constructed a burial diagram in which the intrusive buries underlying rocks more deeply without changing the depths of the overlying sedimentary layers. Extrusive igneous rocks can be treated just like other layers that are added to the sedimentary pile (Johnson et al, 1993), except that their accumulation rates (m/m.y.) are much higher.

Salt may intrude sedimentary sections, and in diapirs can develop thicknesses in excess of 6000 m (Halbouty, 1979). Diapirism is accompanied by withdrawal of adjacent salt and development of a rim syncline. These effects are discussed in more detail in Section 10.7. Igneous intrusions reach their maximum thickness almost instantaneously and then remain at constant thickness. In contrast, salt may continue to thicken over long periods, and can even show periods with no change before thickening resumes. It is also possible for the salt to get thinner through time, for example in the vicinity of the rim syncline adjacent to a salt diapir. Figure 3.28 shows a burial diagram for a section containing a salt layer that developed into a diapir. Salt accumulated from 100 to 95 Ma and was then buried until diapirism started at 60 Ma ago and lasted for 40 m.y. During this time the diapir developed at location B and a rim syncline at A, and burial histories for both locations are combined in Figure 3.28.

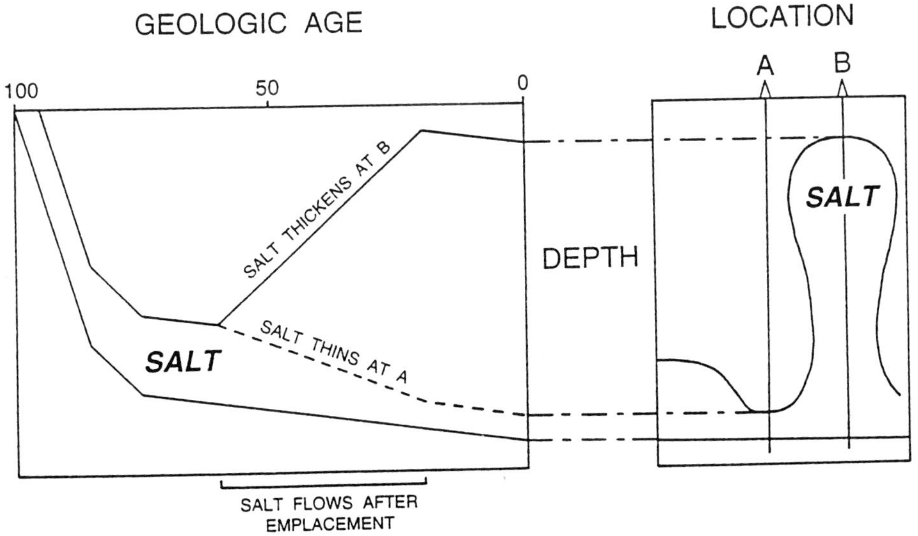

Figure 3.28. Burial history diagram for a section containing a salt layer that develops into a diapir with an associated rim syncline.

For simplicity overlying beds are not shown, but during diapir formation additional sediment would accumulate in the column overlying the rim syncline. Salt thinning can also be produced by salt dissolution (Anderson and Knapp, 1993).

3.10. WATER DEPTH (PALEOBATHYMETRY)

Most rocks are deposited in aqueous environments, and in the majority of cases these are marine. In constructing burial history diagrams, it is generally more appropriate to use Mean Sea Level (MSL) as the reference for depth rather than the top of the sediment pile. Water can then be treated like a "pseudo rock" with appropriate values for thickness and thermal conductivity. The so-called "Vail Curve" (see for example Haq et al (1989)) gives relative changes in global sea level based on onlap of coastal deposits in marine sequences. However, local sea level may differ significantly from the sequence of global cycles because of purely local geologic processes. There are few unambiguous techniques for establishing paleo water depth, but in general the problem gets more difficult as water depth increases. It is usually fairly easy to identify very shallow water and intertidal zones, which have features such as desiccation cracks. Also, the rapid physico-chemical changes in shallow waters produce an environment occupied by many organisms that dig deep burrows for protection. In the subtidal zones, conditions fluctuate less and in general the number and diversity of organisms decreases towards deeper water. *Foraminifera* can be useful as paleobathymetry indicators (Bandy and Arnal, 1969), but fine resolution is often not possible (Boltovskoy and Wright (1976): Figure 3.29). Gernant and Kessling (1966) showed that there were systematic trends in *Foraminifera* distributions in the middle part of the Oligocene Frio formation of Texas. Increasing water depth led to an increase in (1) total species, (2) calcareous forms, (3) planktonic/benthic ratio, (4) internal complexity of arenaceous species, and (5) the size and shell thickness of calcareous species. Faunal dominance, arenaceous forms, and the presence of chitin linings in calcareous species all decreased.

Although the presence of photosynthetic organisms helps define the photic zone, this may not always relate simply to depth because light penetration is affected by water turbidity. Within the photic zone, Golubic et al (1975) have suggested that the distribution of various species of microborers in carbonates is a potential depth indicator. They based their proposal on studies of contemporary species. The distribution of sighted ostracodes in chalks has been used to infer the depth of the photic zone,

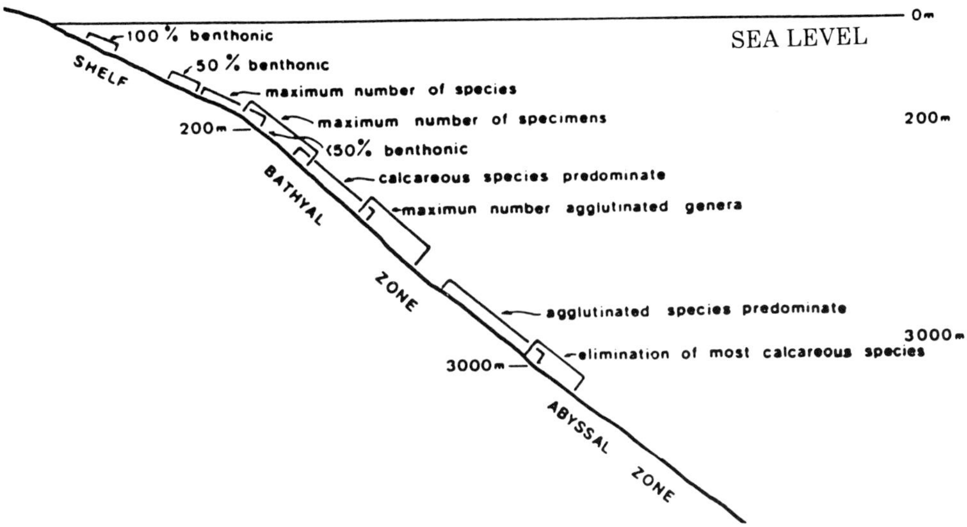

Figure 3.29. Summary of changing benthic foraminiferal characteristics with water depth (Boltovskoy and Wright, 1976).

and by making suitable assumptions about turbidity can be used to estimate paleobathymetry (Kontrovitz and Myers, 1988; Puckett, 1991). The presence of algae also generally indicates the photic zone, and this includes coral reefs.

The absence of carbonate fossils in deep-water sediments is due in part to the position of the carbonate compensation depth. This can only be used as a general guide because the variation of its depth with time and temperature, and its geographic persistence, are not well known and are difficult to predict from geochemical models. A further complication arises because carbonate material may be preserved in deep waters if there is very rapid sedimentation that protects the deposited material. A turbidity current that carries coccolith tests into deep water would provide a possible setting where this could happen.

The character of the sediments themselves often provides a good guide to depositional environments and water depths (Figure 3.30), but unfortunately most preserved sedimentary rocks were deposited in epicontinental seas and on continental shelves, where it is especially difficult to constrain water depth, particularly for level-bottom sediments.

Establishing the elevation of non-marine depositional environments is difficult. Palynological studies may help but have large uncertainties.

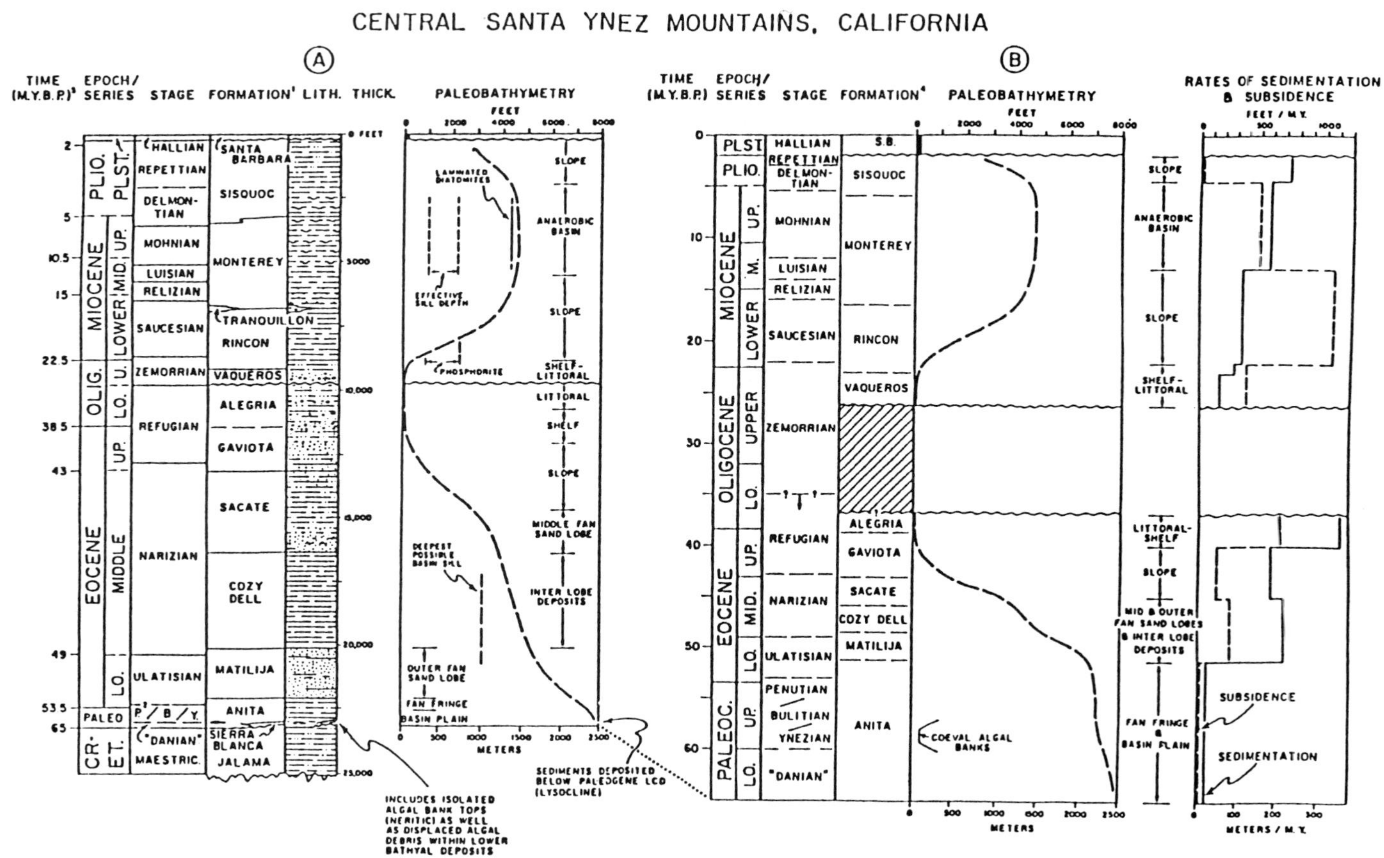

Figure 3.30. Changing bathymetry inferred from data for depositional environment (Ingle,1980).

Other approaches used to estimate paleoelevations have included (1) inferred paleoslope on river systems, and (2) estimated distance from coast line. In general, for thick non-marine sequences to be preserved in the geologic record they must have formed fairly close to sea level.

3.11. EXCEPTIONS TO SITE-SPECIFIC RECONSTRUCTIONS

Depth-age burial profiles are almost always constructed for a single geographic location, but occasionally it can be useful to relax this restriction. Some situations are listed below:

(1) Burial histories for a given rock unit at various locations can be superimposed on a single diagram to facilitate comparisons. For example, lines showing source rock behavior at various locations across a basin, from its center to the flanks, could be displayed on one diagram. In the Williston Basin of the north central U.S. several organic rich intervals are potential source rocks (Burrus et al, 1996), and Figure 3.31 shows burial histories for one of them, the Bakken Shale, at various locations (Webster, 1984). In the basin center it is now mature, but it is still immature on the flanks. Other examples of this type have already been given in Figure 3.27 (where salt behavior at two different locations was compared on a single diagram), and in Figure 3.15 (showing multiple unconformities and two source rocks).

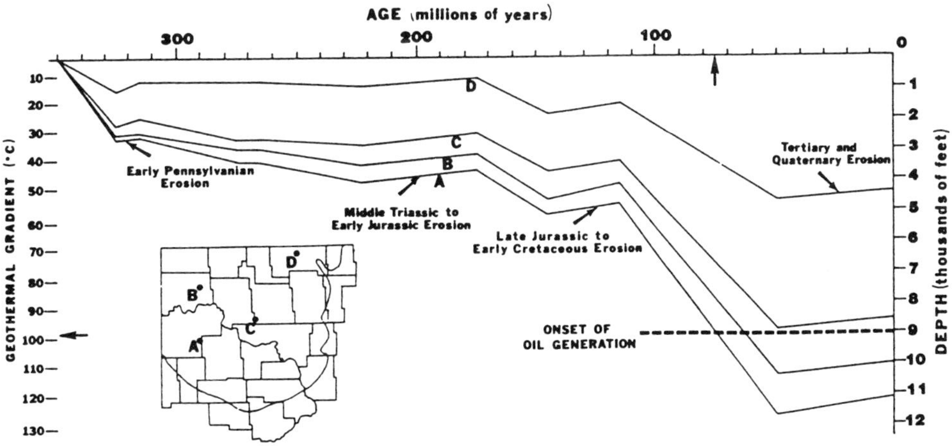

Figure 3.31. Burial history profiles for the Bakken Formation in wells A, B, C, and D at 4 different locations in the Williston Basin, north central U.S. (inset map) (Webster, 1984).

(2) If there is considerable lateral movement, it is sometimes helpful to follow a single element of rock through its burial history. For example, in a subduction complex it may be more useful to follow the progress of one part of the downgoing plate even through it may move hundreds of kilometers laterally. In some cases this can also be true in overthrust terrains where substantial lateral movements are involved.

(3) One of the major uses of thermal modeling is to calculate the time of petroleum generation, but after generation migration may move the oil to a reservoir that is a long way from the source rock, both laterally and/or vertically. The subsequent thermal history of the oil in the reservoir may be very different from that of the source rock. Calculation of oil stability (and its survival or cracking to gas) depends on the thermal history of the reservoir. In cases where long-distance migration is involved, it is often appropriate to break the burial diagram into two parts and track the source rock until generation/migration occurs and then follow the reservoir rock. In the case of some 2-D models, or the more complex 3-D models, this is handled by the model itself (see Chapter 9).

3.12. TIMING OF STRUCTURE DEVELOPMENT

Burial history diagrams are developed for a single location and show the depth of any given formation at the selected location at any point in time. If reconstructions are made for a number of different geographic positions, then at any given geologic time the depths of a formation can be obtained for each position and contoured to give a structure map (Wavrek and Barker, 1988).

The principle is illustrated by Figure 3.32, which shows a simple schematic example that includes four wells (A, B, C, and D) drilled across a present day anticline. In this case a burial history plot has been constructed for each location and combined to give a block diagram. This shows that the anticlinal structure developed late. Although the location of a mature source rock under the anticline looks ideal, a combination of the burial history reconstructions with maturity modeling for the source rock leads to the conclusion that the source rock generated oil before the anticline developed. The trapping configuration was too late to trap crude oil and would only be a target for late-stage gas.

Structure reconstructions through time can be made manually by combining data from many one-dimensional burial history plots. For example, McLimans and Videtich (1989) used data from 25 locations in the south of England to construct simple structural maps and then used these to help establish the maturity history of the Lower Jurassic. 2-D and 3-D computer models are capable of calculating and displaying cross sections or contour surfaces, respectively. These are discussed later in Chapter 9.

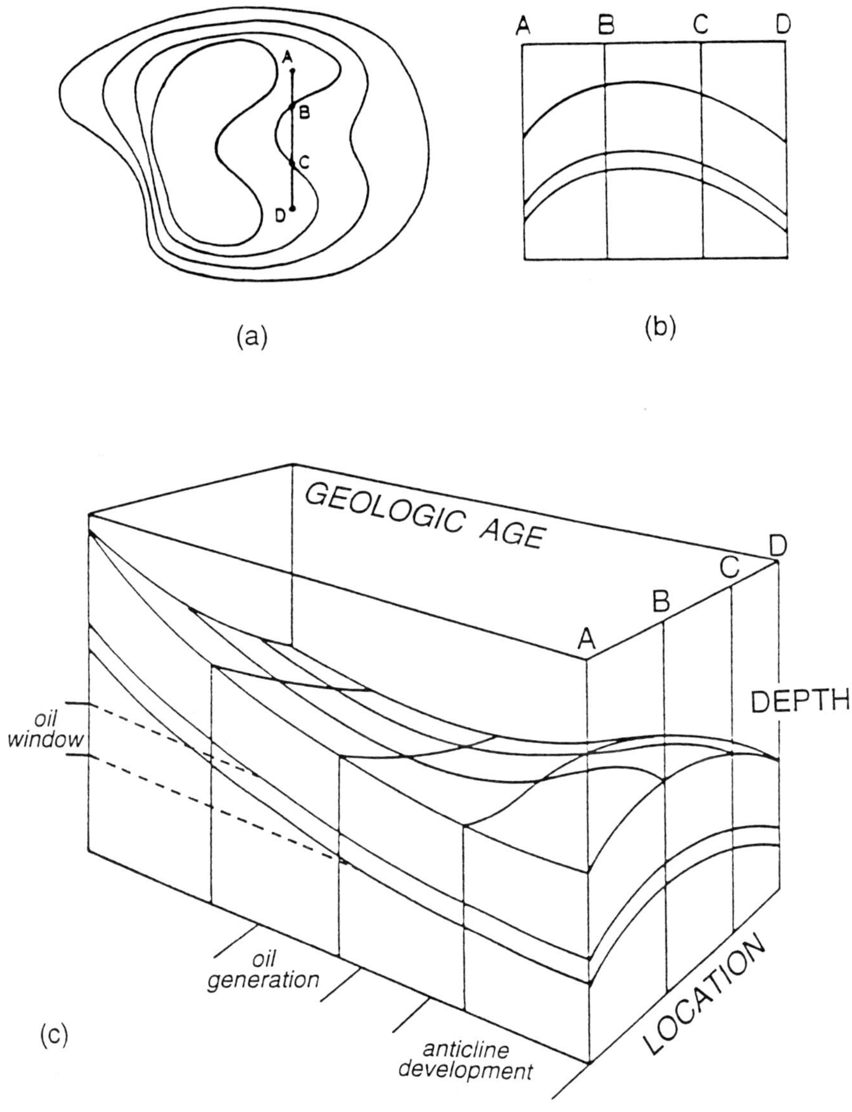

Figure 3.32. Schematic example showing (a) a simple basin with structural nosing, (b) the cross section established by drilling Wells A, B, C,and D, (c) a depth-location-age diagram showing how structure develops through time.

3.13. GEOLOGIC AGES

3.13.1. Radioactive Decay

Construction of burial diagrams and their use in calculating organic maturity requires absolute, not relative, ages. The evolution of life forms through geologic time has provided a sequence of changing fossils that can be used to establish a relative time scale as one fossil type succeeds another. Based on biostratigraphic information of this sort a worldwide sequence of time intervals has been named to give the geologic time scale (e.g., Harland et al 1982 and 1990; see also Appendix III). Index fossils, or assemblages of fossils, are especially valuable since they were present over limited but well defined time spans and have widespread geographic occurrence. In practice, most petroleum geologists use paleontological data to establish the time period, and then rely on published dates to establish the absolute age (in millions of years).

The discovery of radioactive decay, and the realization that decay rates were independent of the chemical environment (mineral, rock) or physical conditions (pressure, temperature) lead to the development of "*decay clocks*" that are capable of giving absolute geologic ages (Faure, 1986; Parrish and Roddick, 1985; Geyh and Schleicher, 1990; Dickin, 1995). Radioactive parent atoms decay to daughter products with first order kinetics so that:

$$N = N_o e^{-\lambda t} \quad (3.8)$$

where λ is the decay constant, t is time, N_o is the initial number of radioactive atoms, and N is the number of radioactive atoms remaining. The half life is the time required for half of the radioactive parent atoms to decay, and the decay constant is related to the half life, $t_{1/2}$, by $t_{1/2} = \ln 2/\lambda = 0.693/\lambda$. The most widely used methods of geochronology are outlined below.

3.13.2. Potassium-Argon Dating

Potassium is made up mostly of the mass 41 and 39 isotopes but also includes about 0.11 percent of potassium-40. Potassium-40 is radioactive and has a branched decay that leads to both argon-40 and calcium-40. The potassium-argon decay scheme has a half life of 1.3 b.y. and is the process used in dating minerals. It is assumed that initially the rock or mineral had no argon, but that as the potassium atoms decay the concentration of argon steadily increases. From a knowledge of the potassium content, argon content, and decay parameters an age can be calculated.

The neutral radiogenic argon atom that replaces a potassium ion in the mineral lattice can be lost from the lattice to give an apparent age that is too low. Rate of loss depends on:

(1) *temperature*: at elevated temperatures diffusion rates increase, and if the temperature is high enough, all the radiogenic ^{40}Ar is lost, so that metamorphism resets the age to zero. Mild heating may give partial argon loss that leads to meaningless intermediate ages. A potassium-argon "age" actually dates the time when argon loss by diffusion becomes negligible, and this happens when the temperature falls below the "*closure temperature*", or "*blocking temperature*". Figure 3.33 shows schematically how different ages may be obtained from two igneous rocks that were intruded at the same time, but cooled at different rates. Mineral A has a high closure temperature, and becomes a closed system for argon soon after intrusion when temperatures are still high. The corresponding age of "a" is close to the time of intrusion. Mineral B has a much lower closure temperature, so that the intrusion takes longer to cool to a sufficiently low temperature for argon to be retained, and the age obtained for fast cooling is *b*. Slower cooling has a greater effect on age for Mineral B and reduces the measured age to *b'*. Actual values for apparent mineral ages adjacent

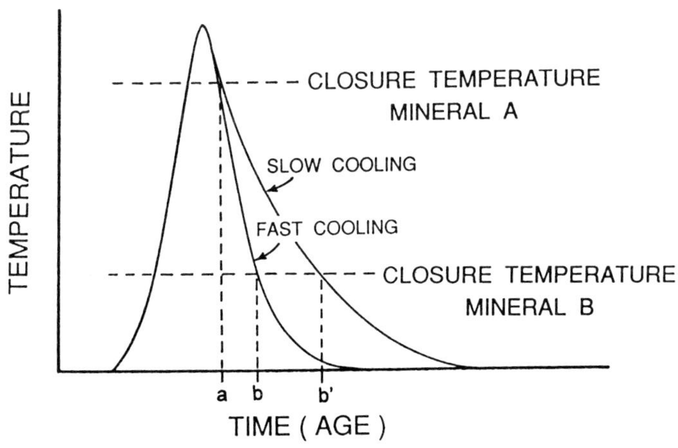

Figure 3.33. A potassium argon "date" records the time when the temperature falls below the "closure temperature" (also called the "blocking temperature"). For slow cooling conditions this can give a significantly younger age than for rapid cooling.

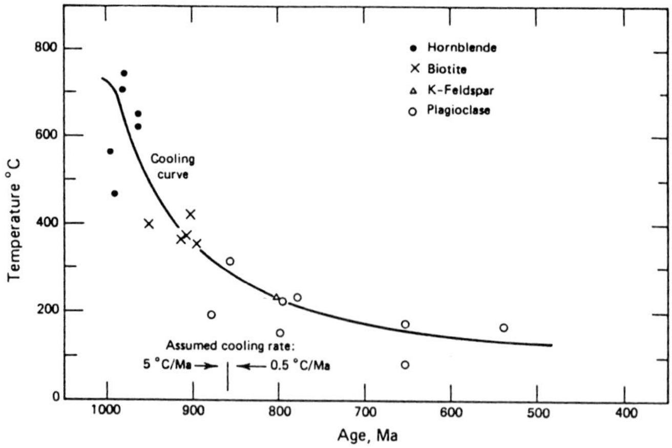

Figure 3.34. Cooling curve for the Haliburton Highlands, Grenville structural province, Ontario, derived by determining both a date and a closure temperature (Faure, 1986).

to an intrusive are given in Figure 3.34 and show higher closure temperatures for hornblende and lower values for the feldspars.

(2) *the mineral*: generally biotites and amphiboles give reliable ages but igneous feldspars do not. However, potassium feldspar overgrowths that develop in sedimentary rocks are typically very pure and some have a sanidine-like, monoclinic structure that appears to be highly retentive for radiogenic argon. Glauconites are unreliable if buried.

Dating sedimentary rocks is difficult. Many minerals contain a core of detrital material (with an age controlled by the parent rock), and an authigenic overgrowth that reflects the time of growth in the sedimentary environment (Glasmann et al, 1989b). Dating of the grain gives an intermediate age with a value that depends on the relative contributions of the core and the overgrowth. For example, Girard et al (1988) separated detrital feldspar from Lower Cretaceous arkosic sandstones in Angola and used K-Ar dating to show that the overgrowths developed at 98 $\pm$ 16 Ma, while the age of the detrital cores was 516 $\pm$ 26 Ma. Clearly, a *"whole grain"* age would give some geologically meaningless intermediate age.

Illite grains of various sizes were separated from core samples and dated by Glasmann et al (1989a) as part of a study of hydrocarbon entrapment in the Huldra and Vaslefrikk reservoirs of the North Sea. Large illite grains (>0.5 mm) were generally contaminated and had Paleozoic detrital cores that approached the age of 432 $\pm$ 10 Ma obtained for muscovites, and these were consistent with a Caledonian provenance. Fine-grained material (<0.1 mm) gave the most reliable (i.e., youngest) ages ranging from 98 - 38 Ma. Because illite growth appeared to be the last diagenetic event prior to hydrocarbon migration into the reservoir, the ages suggest a prolonged period of illite formation, with the youngest age corresponding to the time when illite formation was terminated by oil accumulation. The combination of these two events provides a useful check on the validity of geochemical basin-history models. The nearby Veslefrikk field appears to have filled 5 - 7 m.y. after Huldra. Liewig et al (1987) also used K-Ar dating for illites in the Brent sandstone of the North Sea. Their study combined this data with Rb-Sr dating and fluid inclusion thermometry and suggested that reservoir filling occurred about 40 Ma. Diagenetic illites in the Lower Brent Group sands were dated by Hamilton et al (1987) using the K-Ar method, and in addition, they used rubidium-strontium dating to obtain provenance ages for detrital muscovites. Barnes et al (1992) dated illites in the St. Peters Sandstone and combined ages with a burial history curve to show that illitization occurred at a depth of 2.7 - 3.4 km, while Ziegler et al (1994) combined K-Ar ages for illites with burial history to establish temperatures at the time of illite formation.

As temperature rises with increasing depth of burial illite grains and other fine-grained minerals lose argon and measured ages decrease with depth. Larger grains often contain an (older) detrital core, or grew over an extended period of time, and both of these effects lead to older ages. The trends are summarized in Figure 3.35 and an example from Perry (1974) is given in Figure 3.36. In the Ordos basin, China, Zhao et al (1996) found that the <0.2 mm clay mineral fraction gave the same K-Ar ages over a considerable depth range and interpreted this as showing the time of a short-duration, high-temperature event. Illite dates may also be complicated by the formation of illite from the conversion of smectite with increasing depth when potassium ions are present.

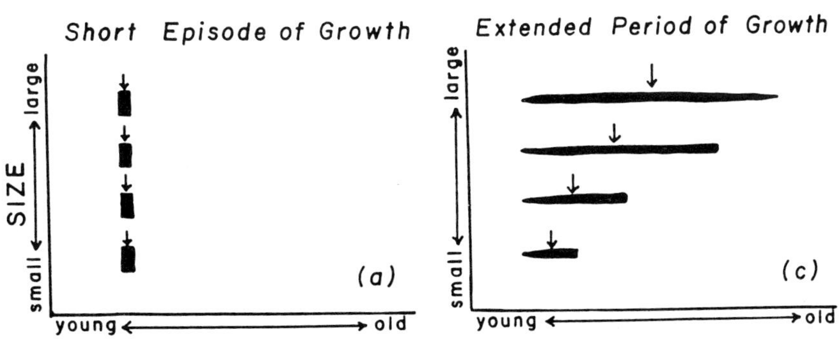

Figure 3.35. Idealized illite age spectra showing relationship between mean age of illite and size of separated fractions for two hypothetical cases. The arrows indicate the mean age of each fraction (Lee et al, 1989).

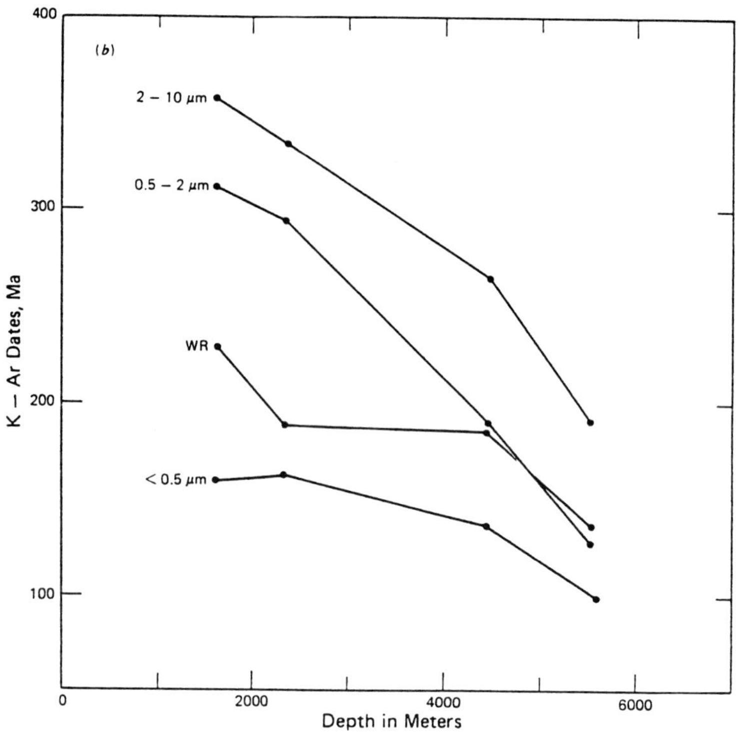

Figure 3.36. Decrease of apparent potassium-argon ages of whole-rock samples and grain-size fractions from the same shale. The lowering of the K-Ar dates is caused by the loss of radiogenic Ar-40 from detrital potassium feldspar and mica as a result of burial metamorphism. WR is "whole rock". (From Perry, 1974).

In the classical K-Ar method, potassium and radiogenic ^{40}Ar are measured on different fragments of the same sample. They can be measured on exactly the same piece of sample by using fast neutron irradiation to convert the ^{39}K to ^{39}Ar, and then measuring the ^{39}Ar and ^{40}Ar in a single mass-spectrometric analysis (McDougall and Harrison, 1988). Using this approach, it is possible to calculate apparent ages for different parts of a single grain. This is done by stepwise heating, so that argon released from different locations in the grain can be separately analyzed isotopically. At first argon is released only from the most easily depleted sites, which are the ones that the argon would be lost from most easily in nature. Apparent ages would be low. At higher temperatures argon would be lost from the more retentive interior sites, and true ages would be obtained, so that with this technique it is often possible to get reliable ages from partially metamorphosed samples. Argon-39 techniques are difficult to apply to clastic rocks because the diagenetic overgrowths and the detrital cores of potassium feldspar grains release their argon over similar temperature ranges (Girard and Onstett, 1991). However, the ability to work with the same small samples for both potassium and argon is a major advantage and has been refined by using laser heating of small areas.

3.13.3. Rubidium-Strontium Methods

The rubidium-87 isotope is radioactive and decays to strontium-87 with a half life of 4.9×10^{10} years. In principle this could be used like potassium-argon dating, but in practice it is complicated because rocks already contain some ^{87}Sr. The effect of this so-called "*common strontium*" can be circumvented by analyzing several different minerals with different ratios of (total Rb/total Sr). A plot of $^{87}Rb/^{86}Sr$ vs $^{87}Sr/^{86}Sr$ for three different minerals from a single rock is given in Figure 3.37. At the time the rock is formed the $^{87}Sr/^{86}Sr$ *"initial ratio"* is "IR", but the ratio steadily increases through time as additional radiogenic ^{87}Sr is added from the decay process. If a mineral contains relatively little rubidium (MIN 1), then in a given time the ^{87}Sr does not increase as much as for a rubidium-rich mineral (MIN 3). The line defined by the values for the different

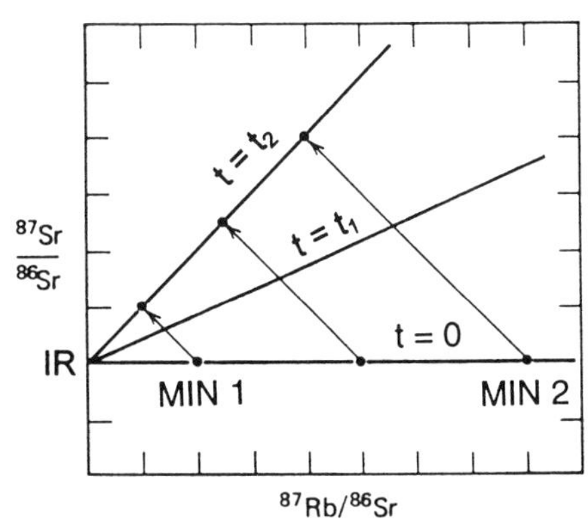

Figure 3.37. Rubidium-strontium isochron diagram showing how the increase in the Sr-87/Sr-86 ratio depends on the rubidium content of the different minerals in the rock (MIN1, MIN2). "IR" is the so-called initial ratio, and t is time ($t_2 > t_1$).

minerals will always be a straight line, and this is called an "*isochron*". Its slope is proportional to the age of the rock, and its intercept on the vertical axis will be the strontium initial ratio.

Examples of rubidium-strontium isochron plots for clay mineral separates are given in Figures 3.38 and 3.39. Ravenhurst et al (1986) separated the clay-sized fraction from a 10 m section of the Horton sandstone and analyzed five samples with varying rubidium contents that gave an isochron corresponding to an age of 300 $\pm$ 6 Ma (Figure 3.38). The situation presented by Morton (1985) for the Woodford shale of west Texas (Figure 3.39) is more complex. The coarser clay-mineral fraction contained the 2M polymorph and gave an age of 539 $\pm$ 1 8 Ma. This is apparently the age of the source region for the detrital material eroded into the Woodford shale. In contrast, the finer fraction (<0.2 mm, 1Md polymorph) gave an age of 301 $\pm$ 4 Ma, which dates an episode of diagenesis.

The $^{87}Sr/^{86}Sr$ ratio for seawater has varied through time and the changes are well documented (Hess et al, 1986). This provides a way of dating marine sediments by comparing the $^{87}Sr/^{86}Sr$ values obtained from carbonate or phosphatic fossil materials with the curve. Rundberg and Smalley (1989) have used this for dating Cenozoic sediments in the northern North Sea and claim a precision of $\pm$ 0.5 - 3 m.y.

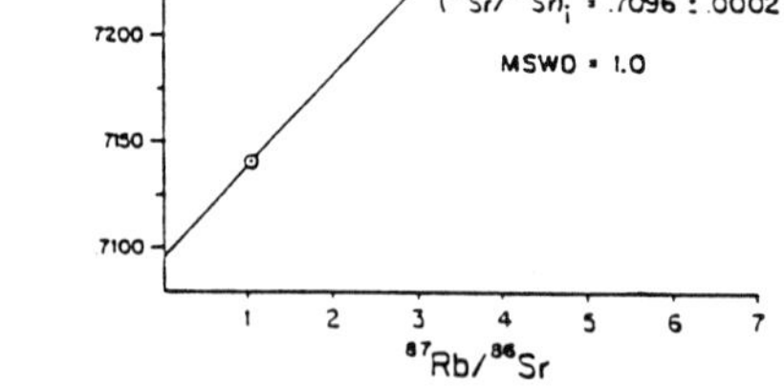

Figure 3.38. Rubidium-strontium isochron derived from clay separates. These were obtained from a 10 m stratigraphic interval of the Horton sandstone, 7 km NE of Truro, Nova Scotia (Ravenhurst et al, 1986).

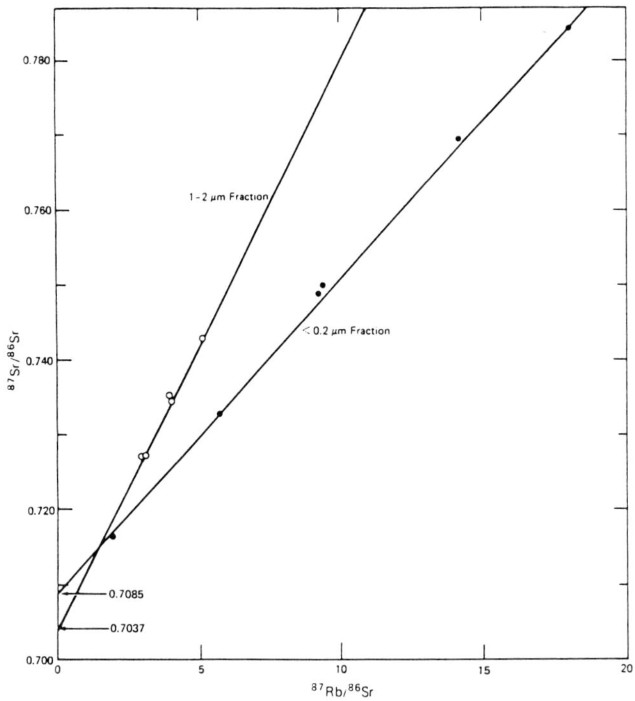

Figure 3.39. Rubidium-strontium isochron plots for two different grain-size fractions from the Woodford Shale, west Texas. The coarser fraction has an age of 539 $\pm$ 18 Ma, and the finer-grained fraction 301 $\pm$ 4 Ma corresponding to the ages of the source region and a diagenetic event, respectively (Morton, 1985).

3.13.4. Other Dating Techniques

Lead Methods: All three naturally occurring isotopes of uranium are radioactive and in each case a series of unstable intermediate isotopes decays to a final stable isotope of lead. The decay of thorium-232 also ends in a stable lead isotope (^{208}Pb). Although several parent-daughter pairs can be used for dating, the ages obtained by the different methods are often not the same and are called "*discordant*". An alternative way of using the data is to plot ($^{206}Pb/^{238}U$) vs ($^{207}Pb/^{235}U$) to give what is known as a "*concordia plot*" where the curve corresponds to the same ("*concordant*") ages for the two parent/daughter decay schemes. Deviations from the curve can be interpreted in terms of various geologic processes involving lead loss from the system. Usually, the lead methods are more useful for older samples, and have been less important for dating Phanerozoic samples and sediments in general.

Fission Tracks: Uranium can decay by fission, giving two heavy fragments that recoil from each other and produce considerable lattice damage in the host mineral. The number of damage tracks ("*fission tracks*") is proportional to the amount of uranium and the mineral age (Wagner andvan den Haute, 1992). This method is discussed later under Fission Track Annealing Methods (Section 5.3.3).

Carbon-14: In addition to two stable isotopes, carbon has a radioactive isotope of mass 14. Carbon-14 is produced in the upper atmosphere by the interaction of cosmic rays with nitrogen-14. It is ultimately oxidized to $^{14}CO_2$ and incorporated into the biosphere where it comes into equilibrium with the other carbon isotopes. Even thought it decays with a half life of 5730 years, it is constantly being replenished, so that the relative amount of ^{14}C remains essentially constant.

The exchange of ^{14}C between an organism and the rest of the carbon cycle continues until the organism dies. At that time the remains of the organism form a carbon reservoir which is effectively isolated, and the amount of ^{14}C it contains slowly decreases due to radioactive decay. If the decay rate and the original $^{14}C/^{12}C$ ratio are known, this process can be used to date the organic matter. Although many corrections are necessary, the carbon-14 dating method is useful back to about 50,000 years b.p. It could be used, for example, to establish the age of organic matter in recent sediments, or to decide whether an accumulation of hydrocarbons in recent sediments is indigenous or has migrated in from some deeper and older source (that now contains no ^{14}C). It is also useful for establishing recent sedimentation rates.

Magnetic Reversals ("magneticochronology"): The Earth's magnetic field has periodically reversed direction. These reversals appear to have occurred very rapidly and to have been synchronous globally. The sequence of normal and reversed fields is quite well known, and key reversals are well dated (e.g., Berggren et al, 1985; Hailwood, 1989), allowing the pattern of reversals to be used as a dating tool.

CHAPTER 4

PRESENT - DAY TEMPERATURES

4.1. THERMAL REGIMES

4.1.1. Introduction

Temperature is a critically important factor in modeling the generation of petroleum and a temperature range of only 100 °C is enough to cover the entire age range of known oils. If the rate of petroleum generation roughly doubles for every 10 °C rise in temperature, then a 100 °C temperature rise increases the rate by a factor of 2^{10} or 1024. This is about the same as the range in age from the youngest oils (say, 1 m.y.) to the oldest Precambrian oils that are >1,000 m.y. Temperature is also important in controlling inorganic reactions, such as the dewatering of clays and the mineral transformations that can create or destroy porosity. It also has an important role in some of the processes that have a marked effect on petroleum composition. Biodegradation of crude oils is limited to temperatures below about 70 °C, because the bacteria that consume oils cannot tolerate higher temperatures. At higher temperatures, oils are thermally degraded to give lighter oils, then condensates, and finally dry gas. Although we are mainly interested in present day temperatures corresponding to drillable depths of about 30,000 ft (9.2 km), both the source rocks and the reservoir may have been buried deeper in the past. At 30,000 ft (9.2 km) and an average geothermal gradient, temperature will be roughly 270 °C, and even for higher geothermal gradients temperatures are unlikely to exceed 375 °C. 50,000 ft (15.3 km) of burial covers virtually all rocks that will be drilled, but deeper temperatures can have an important impact on thermal regimes. Detailed temperature information can be useful in understanding tectonic evolution of basins, and as an exploration tool.

The earths thermal energy is either residual from the process of planetary accretion, or comes as a consequence of radioactive decay. Heat is transferred by conduction, convection, and radiation, but radiation is not important for the earth systems we will be considering. Convection makes an important contribution to Mantle processes, but is not generally important in the crust except locally where water flow rate is high, such as in recharge areas. Most heat transfer in the crust occurs by conduction, where energy is moved by transfer of kinetic energy through intermolecular collisions. Heat generated by radioactive decay comes mainly from the elements potassium, uranium and thorium, though the quantities of their radioactive isotopes have diminished through time, and are now about half what they were 3 b.y. ago. Their heat production currently accounts for approximately 80% of the heat radiated from the Earth, and the remaining 20% is due to cooling (Turcotte, 1980). The heat is being transferred from the earths hot interior to its cool surface where it is radiated into space. This implies that temperature should increase with subsurface depth and, apart from some nearsurface perturbations, this is what is generally observed in deep wells and mines.

On a closely sampled temperature profile, the local temperature gradient, dT/dz, is given by Fourier's law if heat transfer is by conduction; radiation and convection are negligible; and steady state conditions prevail. The law takes the form:

$$Q = -K \, . \, dT/dz \tag{4.1}$$

where Q is heat flow in the direction of z, T is temperature, and K is the thermal conductivity. When heat flow is constant with depth, then over some depth range ($D_2 - D_1$) the corresponding temperature range of ($T_2 - T_1$) leads to:

$$Q = -K \, . \, (T_2 - T_1) \, / \, (D_2 - D_1) \tag{4.2}$$

where T_1 is the temperature measured at D_1. $(T_2 - T_1) / (D_2 - D_1)$ is the temperature increase per unit depth, and is called the "*geothermal gradient*". It follows that:

HEAT FLOW = THERMAL CONDUCTIVITY OF ROCK x GEOTHERMAL GRADIENT.

In modeling thermal regimes it is important to have an appreciation for the significance of each of these three terms. They are discussed individually below.

4.1.2. Heat Flow

Heat flowing out of the Earth can be measured at the surface, and values are reported either as mW/m^2 or as $mcal/cm^2.sec$. A heat flow of 10^{-6} $cal/cm^2.sec$ (equal to 41.84 mW/m^2) is known as the "*heat flow unit*" or "*hfu*". Values for heat flows in different geologic settings are summarized in Figure 4.1, and values for specific areas are given in Table 4.1. In general heat flow is higher in geologically active areas such as active ocean ridges, rifts, and back-arc basins. Old shield areas, old oceanic crust, and some fore-arc basins are at the low end of the range with heat flows below 50 mW/m^2. The worldwide average is close to 63 mW/m^2 (1.5 hfu), and Allen and Allen (1990) have expressed this in more familiar terms as being equivalent to the heat from a 100 Watt light bulb spread over the area of a tennis court. Basement heat flow can show significant variations even over a limited area, and Thorne and Watts (1989) contoured values for the North Sea where they found that differences of up to 0.4 hfu occurred in as little as 50 km.

Table 4.1. Selected heat flow values (Data assembled by Platte River Associates).

Geologic Province/Specific Example	HFU	mW/m^2
General Range	0.5 - 9.6	20 - 400
Average Worldwide	1.5	63
Average Oceanic	1.6	67
Average Continental	1.4	58
Eastern USA	1.36	57
England/Wales	1.41	59
Great Plains, Western USA	1.5 - 2.5	62 - 104
Young Extensional Basins (Active)	>1.9	>80
Older Extensional Basins	1.0 - 1.5	40 - 60
Pannonian Basin, E-Cent. Europe	2.0 - 2.5	85 - 105
Pattani Trough, Gulf of Thailand	1.9 - 2.4	80 - 100
Gulf of Leon Margin, W. Mediterr.	1.6 - 2.1	65 - 90
Gulf of Suez, Egypt	1.5 - 2.0	60 - 84
Viking Graben, North Sea	1.3 - 2.0	56 - 82
Uinta Basin, Utah	1.0 - 1.6	40 - 65
North Biscay Margin, East Atlantic	0.8 - 1.0	36 - 43
Old Shield Areas	<1.0	<40
Balkan-Ukrainian Shield	0.7 - 1.0	30 - 40
Canadian Shield	0.8	34
West Australia	0.9	39

Heat reaching the earth's surface is due in part to subcrustal heat flow out of the cooling earth, but also has a contribution from heat produced by the radioactive decay of potassium, uranium, and thorium. These elements concentrate in the crust (where there can be significant variations among different rock types). Table 4.2 presents data assembled by Ungerer et al (1990) showing total surface heat flow in different provinces compared with the contribution from crustal radiogenic heat.

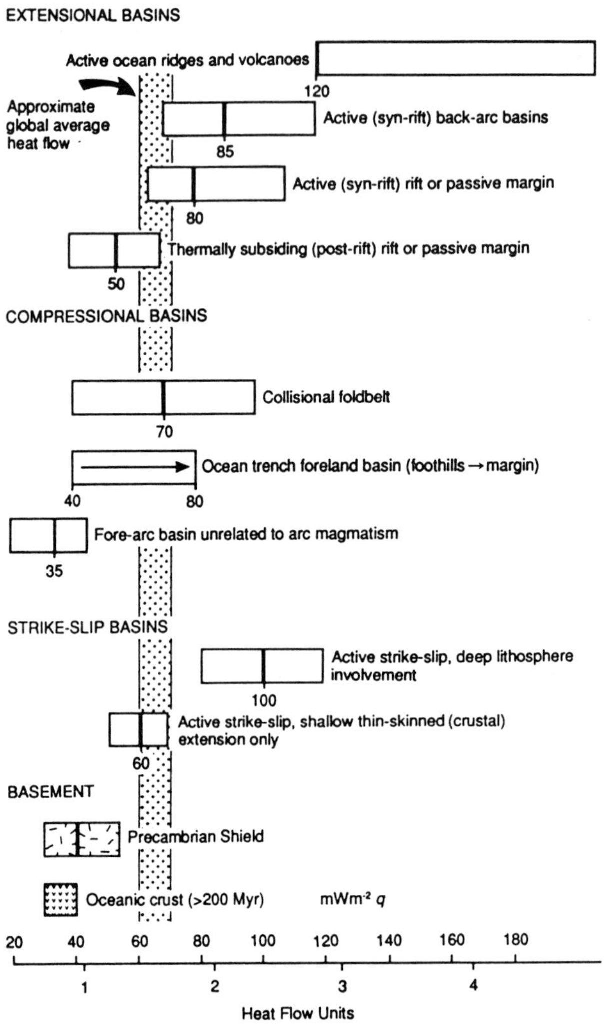

Figure 4.1: Summary of the typical heat flows associated with sedimentary basins of various types (from Allen and Allen, 1990).

Table 4.2. Variability of surface and reduced heat flow (in mW/m^2) for various terrestrial provinces. Data assembled by Ungerer et al (1990). The reduced heat flow approximates subcrustal heat flow at the base of the crust and is obtained by subtracting estimated radiogenic heat production of the crust.

PROVINCE	TOTAL SURFACE HEAT FLOW	REDUCED HEAT FLOW	REDUCED/ TOTAL, %
Sierra Nevada	37 ± 13	18 ± 3	49
Eastern USA	57 ± 17	33 ± 4	58
England, Wales	59 ± 23	23 ± 3	39
Australian Shield:			
Central	83 ± 21	27 ± 6	32
West	39 ± 8	26 ±6	67
Canada:			
Superior Province	34 ± 8	21 ± 1	62
Churchill Province	44 ± 8	37	84
Baltic Shield	36 ± 8	22 ± 6	61
Ukranian Shield	37	25	67
Indian:			
Archean	49 ± 7	33	67
Proterozoic	76 ± 3	54	71
France Hercynian	68 ± 20	27	38

The contributions to surface heat flow, Q, are related by:

$$Q = A D + Q_r \tag{4.3}$$

where A is radiogenic heat production, D is the radioactive crust thickness, and Q_r is the reduced heat flow.

Anomalous, non-equilibrium heat flow values can be caused by seasonal or longer climatic variations, such as the withdrawal of glaciation. However, much bigger variations occur when the basic assumption of heat transfer by conduction breaks down. This happens in areas with active subsurface groundwater flow (Lam and Jones, 1984; Majorowicz and Jessop, 1981b). In general heat flow values are probably no better than ± 10%. Values obtained within 5 - 10 km of hot springs or other surface hydrothermal manifestations are suspect from a regional point of view (Sass et al, 1989).

It follows from Equation 4.2 that surface heat flow values can be combined with thermal conductivities for the rocks to provide a value for present day geothermal gradients. This provides a useful way of constraining paleogeothermal gradients and is discussed more fully in Chapter 5.

4.1.3. Thermal Conductivities

Heat is transferred through the earth mainly by conduction (since fluid movement at depth is usually slow), and for a given heat flow it is the thermal conductivity of the rock that controls geothermal gradient, and hence temperature at any depth. Selected values for thermal conductivity of some of the rocks, minerals, and fluids important in basin modeling are listed in Table 4.3. The range for rocks (coal to evaporites) is more than an order of magnitude, though for the more common sedimentary materials it is somewhat smaller. The range of values shown for most rocks is depth-dependent, and reflects the influence of variable porosity and changing mineralogy (Somerton, 1992). In general, the higher thermal conductivities are for evaporites, some dolomites, and sands, while low values are observed for coal, shale, limestone, oil, gas, and water.

A rock is a porous aggregate with a thermal conductivity that depends on porosity, mineralogy, and temperature, as well as on the size, shape and arrangement of the grains. The pore space in the rock is filled with water, oil, or gas, and all these fluids have low thermal conductivities. As compaction or cementation decreases porosity, the relative amount of fluid is reduced, and the thermal conductivity of the rock rises. For any specific value of porosity (ϕ) the thermal conductivity of the rock (K_r) can be calculated using the following relationship:

$$K_r = (K_w/K_m)^{\phi} \times K_m = K_w^{\phi} \times K_m^{1-\phi} \qquad (4.4)$$

where K_w and K_m are the thermal conductivities of the water and mineral grains, respectively. Equation 4.4 is valid only if the K_m/K_w ratio is not too large (Ungerer et al, 1990), but has been shown to give an adequate description for clays (Morin and Silva, 1984). Figure 4.2 shows how the thermal conductivity of a sandstone steadily decreases as the porosity (and water content) increases. Lee and Bethke (1994) used the equation:

$$K = (5.53 - 4.4\phi) \times 10^{-3} \qquad (4.5)$$

Table 4.3
Selected thermal conductivities (Data from Clark, 1966; Horai, 1971; Gretener, 1981; Brigaud et al, 1990)

MATERIAL	W/m/°K	ROCK	W/m/°K
Quartz	7.8	Sandstone	1.9 - 7.4
Feldspar	2.3	Quartzite	3.7 - 7.5
Calcite	3.4	Clay	1.8
Halite	5.9	Mudstone	1.5 - 2.3
Anhydrite	5.1	Shale	0.6 - 2.9
Gypsum	1.1	Chalk	2.9
Analcime	1.3	Marl	0.9 - 2.8
Mica	2.3	Limestone	1.9 - 3.0
Siderite/ankerite	3.0	Dolomite	1.9 - 5.4
Pyrite	19.2	Granite	2.2 - 3.4
Rutile	5.1	Rhyolite	2.5
Ice	2.1	Diabase	1.9 - 2.2
Water	0.6	Dunite	4.0 - 5.6
Oil	0.2	Basalt	1.5
Gas	0.1		

to relate thermal conductivity (in cal/cm.sec °C) to porosity, ϕ. Because compaction decreases porosity with depth this implies that thermal conductivity increases with depth, and as a consequence, for constant heat flow the geothermal gradient must also vary with depth.

Values for thermal conductivities of minerals are lower at higher temperatures, but the values for high thermal conductivities decrease faster, and at depth there is less contrast in the values for different minerals. The spread in thermal conductivities at 200 °C is roughly half of what it is at the surface. Quartz has the highest temperature coefficient of thermal conductivity of all of the common rock-forming minerals, and the quartz content of a rock has a major influence on changing thermal conductivity with depth (Robertson, 1979; McKenna et al, 1996). Clay-rich shales on the other hand, have only a weak depth dependence for thermal conductivity (Blackwell and Steele, 1989). It appears that changes in their thermal conductivity caused by loss of water during compaction are offset by changing orientations of the clay platelets. With increasing depth the random orientation typical of shallow depths is replaced by a more oriented packing in which most of the particles become sub-horizontal. Since clay minerals have good thermal conductivity along the grains, but poor conductivity across them, this reduces the thermal conductivity for the shale and offsets the increases due to the loss of low thermal conductivity compaction water. In the Paris basin Demongodin et al (1993) found from *in situ* measurements using high resolution temperature logs that the vertical thermal conductivities of shaley formations were significantly lower than predicted by an isotropic model. They interpreted this as being due to anisotropy in thermal conductivity, and this was consistent with their observation that the effect increased with depth. Champion et al (1984) used the equation:

$$K_{m,T} = K_{m,20} \cdot [293/(273 + T)] \qquad (4.6)$$

to relate the matrix conductivity at 20 °C ($K_{m,20}$) to the matrix conductivity at T °C ($K_{m,T}$). For a geothermal gradient of 25 °C/km a sample with a matrix conductivity of 3.0 W/m/K at 20 °C has a thermal conductivity that varies from 3.1 W/m/K at the surface (10 °C) to 2.2 W/m/K at a depth of 5 km where the temperature is 135 °C. The TEMISPACK finite difference 2-D model (Doligez et al, 1987) uses a temperature dependency that includes porosity (Burrus et al , 1991), although the temperature dependence for the

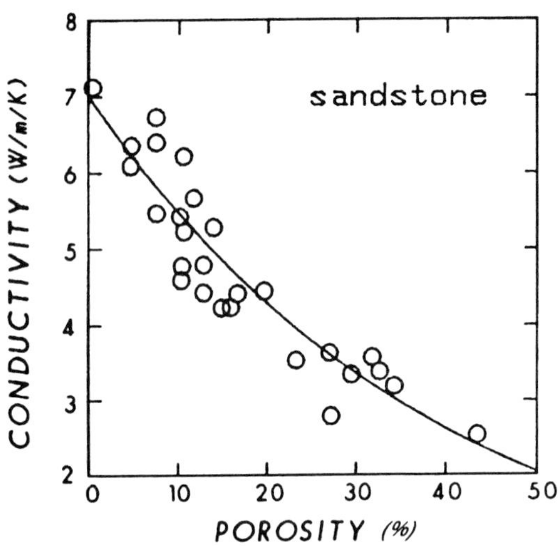

Figure 4.2: Effect of porosity on measured thermal conductivity for some clean, water saturated sandstones. The solid line was calculated using Equation 4.3. (from Brigaud et al, 1990).

thermal conductivity of water in the pores is relatively small. It has the form:

$$K = [K_w^{\phi} . K_m^{1-\phi}] / [1 + a(T - T_o)] \quad (4.7)$$

where K is the conductivity at temperature T and a is a constant.

In situ measurements of thermal conductivity are time-consuming, expensive, and not always reliable (Beck et al, 1971; Brigaud et al, 1990), and log-derived values are still tentative (Vacquier et al, 1988). Laboratory determinations are generally acceptable (except for some very porous or friable materials). The thermal conductivity of soft sediments or friable rocks are best measured with line source, time-transient methods (Vacquier, 1985; Sass et al, 1984), but a divided bar apparatus is used for rocks that can be machined into suitable cylindrical samples. Recently other, more convenient, transient methods have been developed (Middleton, 1993). Drill cuttings provide adequate materials for laboratory determinations, and for a porous monomineralic rock Equation 4.4 can be used to obtain bulk thermal conductivity. When several minerals are present more elaborate relationships are needed, and the simplest of these uses the geometric formula:

$$K_r = (K_1)^{p1} \times (K_2)^{p2} \times \ldots\ldots \times (K_i)^{pi} \quad (4.8)$$

where K_i and pi are the thermal conductivity and volume fraction of the i th component, respectively. This gives good agreement with measured bulk-rock conductivities and is generally within $\pm$ 10%. The other commonly used method is the Maxwell Method described by Gallagher (1987) and Beck et al (1971). Both Sass et al (1971) and Gallagher (1987) found that the Maxwell method slightly underestimates the values, while the geometric method slightly overestimates. An average of the two gives the best values. Although thermal conductivity can be estimated from mineral composition plus porosity (Beck, 1976; Zimmerman, 1989), it is always preferable to measure values directly.

In situ porosity and lithology data from electric logs can be combined with laboratory calibrated thermal conductivities for the pertinent minerals to establish *in situ* thermal conductivity. For three Uinta Basin wells Brigaud et al (1990) found good agreement between their *in situ* values and those determined by laboratory measurements on 565 samples. However, transferring laboratory-determined values to the natural system is not always easy, and Brigaud et al (1990) have listed the following problems: (1) a large number of samples must be measured to characterize adequately the stratigraphic section, (2) the rock matrix conductivity, most commonly determined in laboratory measurements, must still be modified to apply to *in situ* conditions, and (3) some lithologies, in particular shales, may be altered during drilling and sampling so that laboratory measurements can contain systematic errors.

4.1.4. Geothermal Gradients

Values for geothermal gradients vary from about 10 °C/km up to 50 °C/km or higher, with average values being roughly 25 °C/km. High average geothermal gradients are associated with geologically active areas, such as rifts, while low values are common in shield areas, subduction complexes, and active delta systems. Local variations can be significant, and Figure 4.3 shows that in Oklahoma the gradients vary from 2.3 °F/100 ft to 1.3 °F/100 ft over a distance of less than 100 miles. For this range of gradients the corresponding temperatures at 10,000 ft (3050 m) would differ by 100 °F. Petroleum generation rate roughly doubles every 18 °F (10 °C) and 100 °F corresponds to a difference in rate of generation by a factor of approximately 2^5, so that at a depth of 10,000 ft (3050 m) petroleum generation would be more than 30 times faster in the hotter area. The range of geothermal gradients in the Norwegian North Sea is not so extreme (Baird, 1986), but still varies from 1.7 °F/100 ft to 2.4 °F/100 ft in only 40 miles. Western Canadian basin gradients in southern Alberta range from a low of 18 °C/km in the west to a high of more than 32 °C/km, with as little as 50 miles separating the high and low values (Majorowicz et al, 1990). At a subsurface depth of 3 km temperatures would be 69 °C and 105 °C (assuming 15 °C at the surface). This difference of 36 °C corresponds to a difference in rate of petroleum generation of approximately $2^{3.6}$, or a factor of 12. In the northern part of Alberta, geothermal gradients can be even higher (Lam and Jones, 1984), but in the west central Peace River Arch they range form a low of 25 °C/km to a high of more than 35 °C/km (Bachu and Cao, 1992). Geothermal gradients in the Rhine Graben and Indonesian basins extend from 30 °C/km to 70 °C/km, while in the Gippsland basin (southeast Australia) and the Niger Delta gradients range from 20 to 40 °C/km and 17 to 27 °C/km, respectively (Perrodon, 1983). In the Llanos sub-Andean basin of Colombia the cooling effect of flowing water gives very variable gradients leading to differences in geothermal gradients of 25 °C/km over less than 50 km (Bachu et al, 1995). The overall trend is from 50 °C/km in the east to 20 °C/km in the west.

Terrestrial heat flow is constrained by the boundary conditions at the surface and "deep". This lower limit is normally put at the Moho and assigned a temperature of 1330 °C, but it can vary locally during the

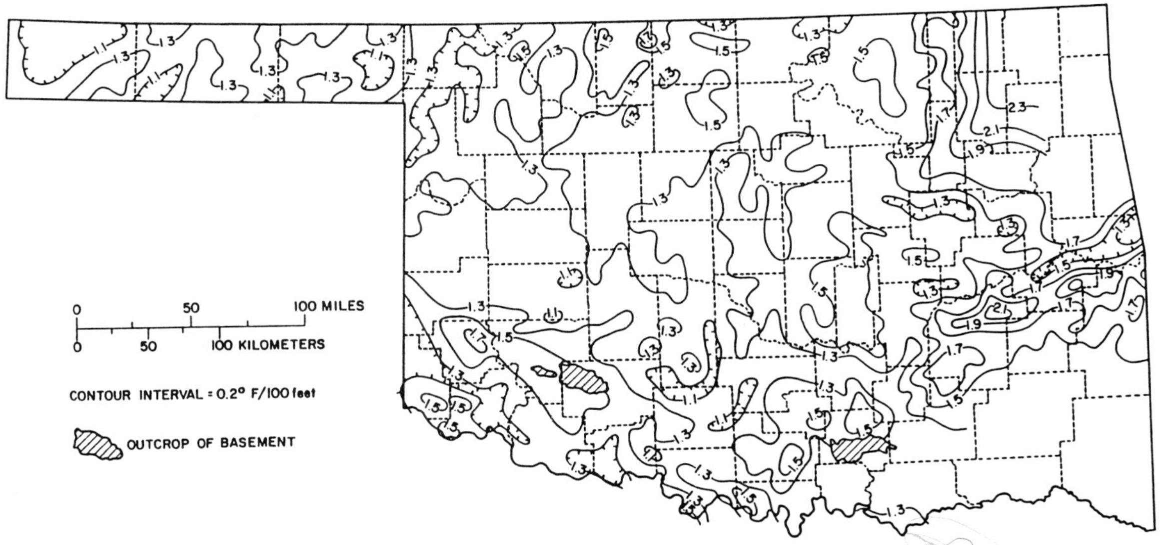

Figure 4.3: Geothermal gradient map of Oklahoma, U.S.A. (Harrison et al, 1983). The figure shows that gradients can vary from 2.3 °F/100 ft to 1.3 °F/100 ft over a distance of less than 100 miles. For this range of gradients the corresponding temperatures at 10,000 ft (3050 m) would differ by 100 °F. Since petroleum generation rate roughly doubles every 18 °F (10 °C) this represents a difference in rate of generation by a factor of approximately 2^5, so that petroleum generation would be more than 30 times faster at a depth of 10,000 ft (3050 m) in the hotter area.

development of crustal spreading centers or volcanoes. The surface boundary is usually close to 15 ± 5 °C and is discussed in more detail later. For a fixed heat flow, geothermal gradient is inversely related to thermal conductivity (Equation 4.1), and highly conductive rocks give low gradients, while poor conductors give high gradients (Figure 4.4).

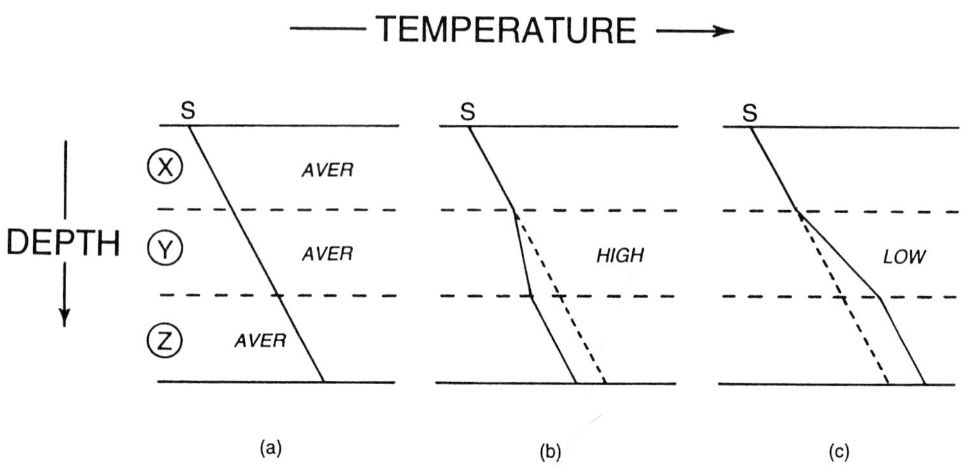

Figure 4.4. Variation of temperature gradients with thermal conductivity. Layers X and Y have average thermal conductivities. The value for Layer Y is (a) average, (b) high, and (c) low. S is surface temperature.

Figure 4.4 shows schematically a three layer section in which the top and bottom layers (X and Z) always have the same average thermal conductivity. If the middle unit also has an average thermal conductivity, then the geothermal gradient will be the same all through the section (Figure 4.4a). When the central unit has a higher than average thermal conductivity it conducts heat more easily, and as a consequence the contrast in temperatures between the top and bottom is reduced and the geothermal gradient falls to a lower value. However, the heat flow remains the same throughout the entire section and the thermal conductivity of Layers X and Z are unchanged, so that the top and bottom geothermal gradients are the same as in Figure 4.4a, and the temperature profile of Figure 4.4b develops. Conversely, if the central unit, Y, has a lower than average thermal conductivity, it will generate a higher geothermal gradient and the overall profile of Figure 4.4c will form when steady state conditions are reached.

Salt has a very high thermal conductivity and evaporites in a section produce marked changes in geothermal gradients. The Upper Permian section in the Delaware basin of Texas and New Mexico includes evaporites in the Salado and Castile formations, and has an average geothermal gradient of only 1.15 °F/100 ft (21 °C/km). In contrast, Hill (1984) found that the average of 37 gradients below the Upper Permian was higher with a value of 1.54 °F/100 ft (28 °C/km). Burrus and Audebert (1990) showed that below the salt in the Gulf of Lions, France, the geothermal gradient was higher, but the temperature was 20 °C lower than expected in the absence of the salt. Their temperature profile is shown in Figure 4.5 and closely resembles the schematic profile of Figure 4.4b. A very similar

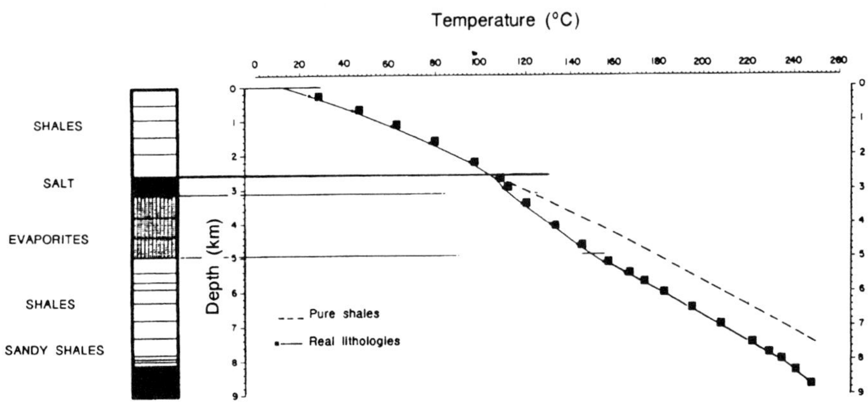

Figure 4.5. Temperature profile through a section containing an evaporite (from Burrus and Audebert, 1990).

profile was reported by Hite et al (1984) in southeast Utah. Here the gradient through the evaporite-rich Paradox Formation is much less than gradients above or below, and temperatures in the underlying rocks are depressed. The effect of an insulating layer is seen in the southern North Sea where Oxburgh and Andrews-Speed (1981) report that the deep high temperatures are due to a thick insulating cover of Tertiary sediments (equivalent to Layer Y in Figure 4.4c). Abnormally-pressured intervals often have low thermal conductivities because they are undercompacted, with correspondingly higher water contents. This leads to a higher geothermal gradient (Figure 4.6). In the Delaware basin of west Texas, for example, the gradient is 25.1 °C/km through the overpressured zone compared with the average geothermal gradient of 21.0 °C/km (Luo et al, 1994). Coals, with their very low thermal conductivities, have a considerable blanketing effect that produces higher temperatures in thedeeper rocks (Pollack and Cercone, 1994). If the coals are shallow there can be significant effects on the thermal regime for all the deeper

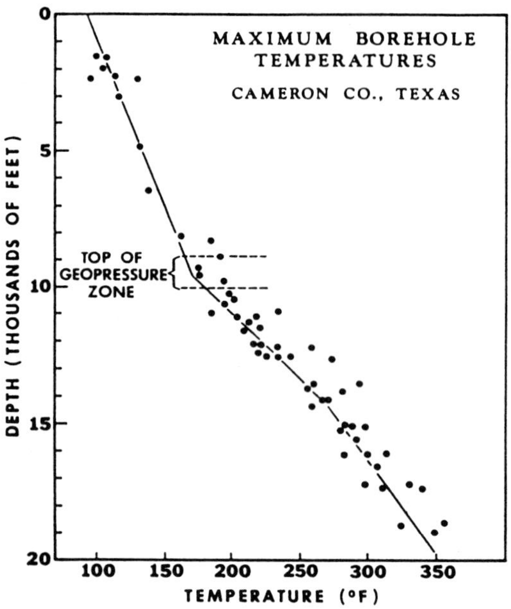

Figure 4.6. Maximum temperatures recorded in boreholes in Cameron County, Texas, showing effect of abnormal pressure on geothermal gradient (from Jones, 1975).

sediments. Cercone et al (1996) measured the thermal conductivities of coals from the Appalachian basin and obtained values of 0.2 ± 0.15 Wm^{-1} K^{-1}. They used these low values to explain anomalous thermal maturity in the basin, where the elevated temperatures below the coal leads to higher values in the maturity indicators (such as R_o), even after the coals are removed by erosion. This is illustrated schematically in Figure 4.7

An ideal geothermal gradient could be calculated from the simple expression:

(bottom hole temperature - surface temperature)/(total depth - elevation of kelly bushing).

In practice this is complicated by near-surface problems (discussed below), and by the difficulty of getting bottom hole equilibrium temperatures. Additional problems occur in situations where hole deviation is not well constrained, since it can be difficult to obtain a true vertical depth. Also, on some offshore wells the kelly bushing may be as much as 30 m above mean sea level and appropriate corrections must be made.

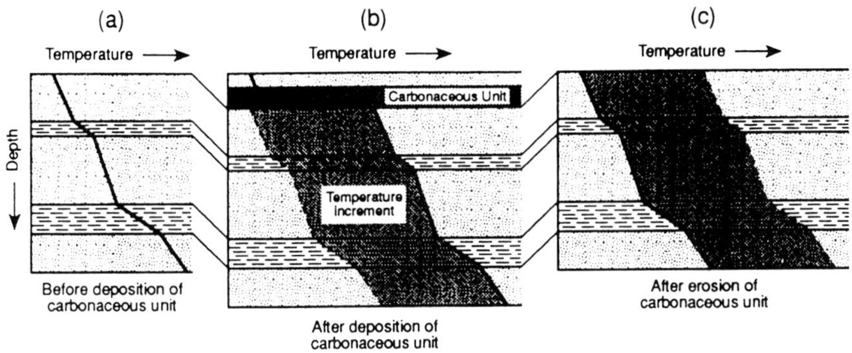

Figure 4.7. Schematic representation of the insulating effect caused by a carbonaceous or organic-rich unit deposited above a normal stratigraphic section. The large "anomalous" temperature increase can be caused by a relatively thin low thermal conductivity layer (Cercone et al, 1996).

Real geothermal gradients are generally non-linear and this poses problems in extrapolating beyond total depth (TD) in a well. Breaks in gradients are caused by variations in thermal conductivities for the rock units being drilled, and the actual temperatures at TD will depend on whether drilling stops at the top or bottom of an interval with a higher or lower than average thermal conductivity. This can be illustrated by reference to Figure 4.8. When conductivities of Layers X and Z are average and Layer Y is above average, and drilling stops near the top of Layer Y, then the stabilized bottom hole temperature (BHT) would be slightly lower than expected from an average geothermal gradient. If drilling stops near the bottom of Layer Y the geothermal gradient calculated from the surface temperature and a stabilized BHT (CE) will be considerably lower than gradient BF. Similarly, if the thermal conductivity of Layer Y is high, extrapolated geothermal gradients from the top (A) and bottom (D) of Y give quite different apparent geothermal gradients. Because geothermal gradients are very commonly obtained from a single BHT and an average surface temperature the gradient can be too high or too low, and clearly will give erroneous values when extrapolated deeper. Blackwell and Steele (1989) have given an extreme example of this potential problem for the USGS Watson No. 1 well, near Big Springs, eastern Kansas. The temperature profile is reproduced here as Figure 4.9 and shows multiple

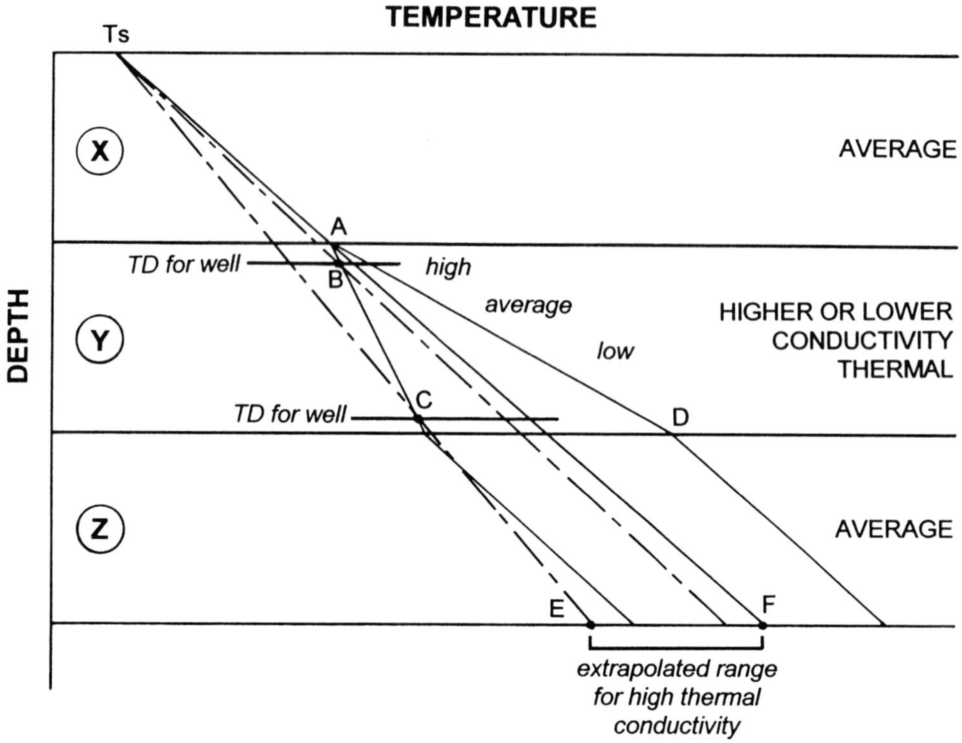

Figure 4.8. Schematic temperature versus depth profiles for a 3-layer system illustrating situations where Layers X and Z always have an average thermal conductivity, but where the conductivity of Layer Y may be higher or lower than average. Dashed lines show possible extrapolated geothermal gradients when drilling stops at different depths in Layer Y. Ts is the surface temperature.

segments with the different geothermal gradients corresponding to the various lithologies. If drilling had stopped at 220 m, and a stabilized BHT obtained, then a straight line trend to an average surface temperature would have given a geothermal gradient of 44.7 °C/km. Drilling to 500 m and extrapolating that value of stabilized BHT to the surface temperature would have a produced a much lower value for the geothermal gradient of 31.0 °C/km. Extrapolating from a single, local geothermal gradient to the regional value should be done with care.

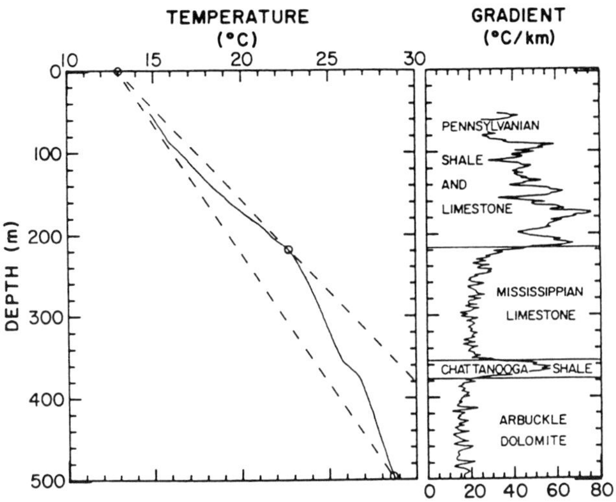

Figure 4.9. Temperature-depth and temperature gradient-depth plots for USGS Watson No. 1 well in eastern Kansas near Big Springs (SE1/4SW1/4SE1/4 sec. 18 T. 18S, R. 23E). The generalized stratigraphy encountered in the hole is shown. Dashed lines show the results that would be obtained by calculating an average gradient for the well using the mean annual surface temperature and a BHT from 220 m depth or from 500 m depth. (From Blackwell and Steele, 1989).

4.2. PRESENT - DAY TEMPERATURES

4.2.1. Introduction

The present day thermal regime in any area is a final product of the variety of geological processes that have operated. Depending on the specific details, it may or may not be helpful in constraining past thermal development. However, the present represents the only opportunity to actually measure temperatures at a specific time in the development of the basin. In spite of this significance the temperatures usually obtained from wells are notoriously unreliable.

Temperatures are commonly recorded by maximum reading thermometers attached to electric logging sondes, and bottom hole temperatures recorded at total depth are reported on the log headings. The process of drilling the well involves circulating drilling mud that was initially at surface temperature, so the hole is cooled and BHTs are always low, often by 10's of degrees (Figure 4.10). In the Wattenberg field of the Denver basin, for example, Meyer and McGee (1985) reported temperatures that were between 11 and 61 °C too low. Temperatures measured in the drilling mud are also usually too low, but Drill Stem Tests (DSTs), or Repeat Formation Testers (RFTs), may give reliable temperatures if there is enough fluid flow, but in general they are also somewhat low (Baird, 1986). Expanding gas entering the hole during the DST may lead to lowered temperatures.

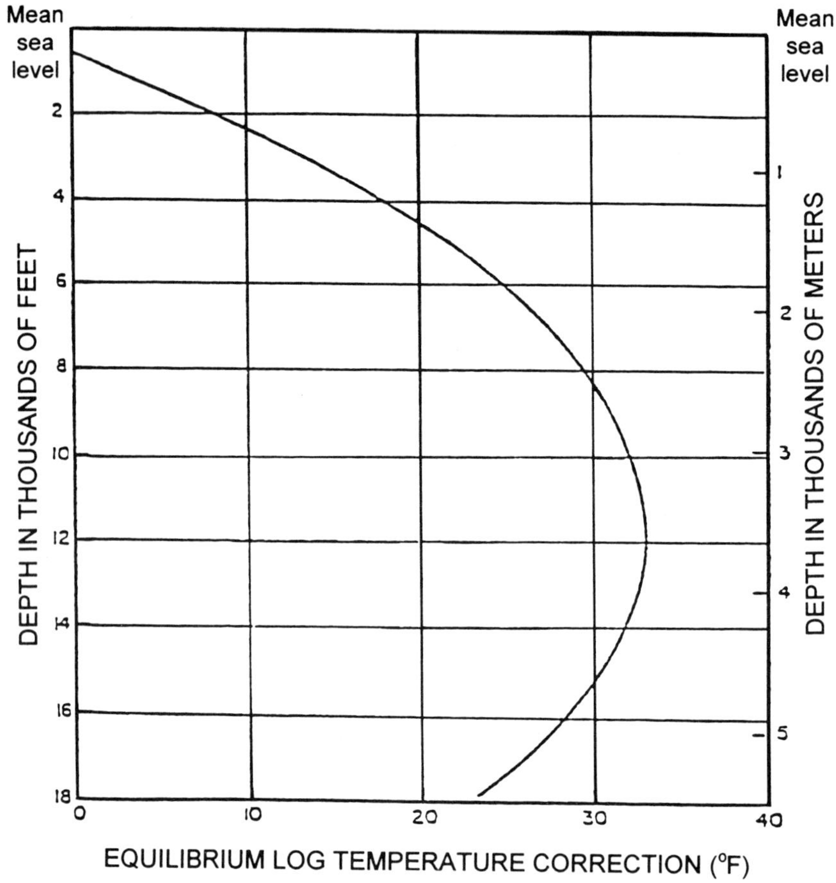

Figure 4.10. Empirical downhole temperature correction curve from Kehle (1971) and Jones (1975) for Gulf of Mexico wells. The temperature correction is to be added to the reported BHT.

Logging operators occasionally calculate temperatures reported on the log heading from TD and geothermal gradient. For example, in the Michigan basin values are sometimes obtained using 0.56 °C/100 ft plus a surface temperature of 15 °C, and if the reported values match these values exactly they are almost certainly not valid subsurface temperatures (Vugrinovich, 1989).

A true formation temperature (also called an *"equilibrium temperature"*, *"virgin temperature"*, or *"stabilized temperature"*) could be obtained by letting the hole return to thermal equilibrium with its surroundings and warm up to its original temperature. In practice this takes far too long to be a practical approach because equilibration time appears to be roughly twice the drilling time (for holes less than about 3000 m deep). Logging old holes that have cooled for a year or more can be a useful way of getting valid temperature profiles, but again is rarely practical because they are normally cemented when they are abandoned. Various empirical relationships have been used to estimate temperature corrections, but none is wholly satisfactory since the reduction in temperature caused by drilling depends on depth, drilling rate, and the thermal properties of the rock and mud. In the simplest case measured temperatures are assumed to be 10 - 15% too low. In the North Sea, Thorne and Watts (1989) increased borehole temperatures by 15% to "approximately counteract the cooling due to mud circulation". This approximation is appropriate for a mean circulation time of 10 hours (Carstens and Finstad, 1981). More elaborate empirical correction curves have been proposed by Harrison et al (1983) for Oklahoma, and by Jones (1975) for use along the coast of the Gulf of Mexico (Figure 4.10). It is important to remember that these curves are obtained from a generalized trend through data points that show considerable scatter (Cheung, 1978).

It may be possible to use a few carefully-documented wells in an area to develop a local correction scheme that can then be used to correct other wells. Willett and Chapman (1987) used this approach in their study of the Uinta Basin, and reported that while data were available for 2000 wells only 95 of these had multiple temperature measurements that would permit extrapolation to equilibrium conditions (see below). The high quality data were used to develop the expression:

$$T_c = 6.93 - 1.67z^2 + 0.101z^3 + 0.0026z^4 \qquad (4.9)$$

which related the temperature correction, T_c, to the depth, z. This relationship was then applied to wells that lacked detailed temperature information to get a better estimate of their stabilized bottom hole temperatures. When data for cessation of mud circulation were available, a more elaborate expression could be used. Even a very limited number of high quality temperature determinations in an area can be extremely helpful.

4.2.2. Extrapolations and pseudo-Horner Plots

When mud circulation stops (and it may continue after drilling has stopped) the rocks cooled by the circulating mud start to warm up. If several logging runs are made, each one will measure the temperature at a different length of time after circulation ceased, and successive measurements will give steadily increasing temperatures as the hole is warming back towards thermal equilibrium. In some cases it may be possible to extrapolate the temperature trend to give the stabilized bottom hole temperature. Although this can be done visually (Figure 4.11), several mathematical procedures have been developed for extrapolating temperatures to infinite time.

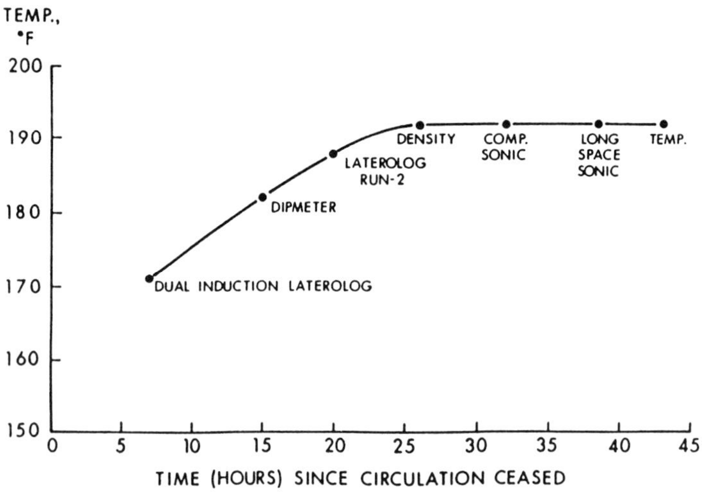

Figure 4.11. Schematic example showing how temperatures measured on successive electric log runs can be extrapolated to infinite time to give a corrected BHT.

The most widely used method for extrapolation involves the so-called *"pseudo-Horner Plot"*. This procedure was originally developed by Horner (1951) for extrapolating pressure data, but was later extended to temperature data by Dowdle and Cobb (1975), Fertl and Wichmann (1977), Fertl (1985), and others. The basic equation is:

$$t_m = t_{bh} - C \log [(T + \Delta T) / \Delta T] \quad (4.10)$$

where t_{bh} is the stabilized bottom hole temperature, t_m is the temperature measured by the logging tool thermometer at some time ΔT after mud circulation stopped, T is the mud circulation time since drilling stopped, and C is a constant. As ΔT increases indefinitely the term $(T + \Delta T)/\Delta T$ decreases towards a limit of 1.0, and a plot of $\log [(T + \Delta T)/\Delta T]$, or sometimes $(T + \Delta T)/\Delta T$, against temperature can be used to establish temperature after thermal re-equilibration. An example is given in Figure 4.12. This technique has been used in many areas around the world (e.g.,

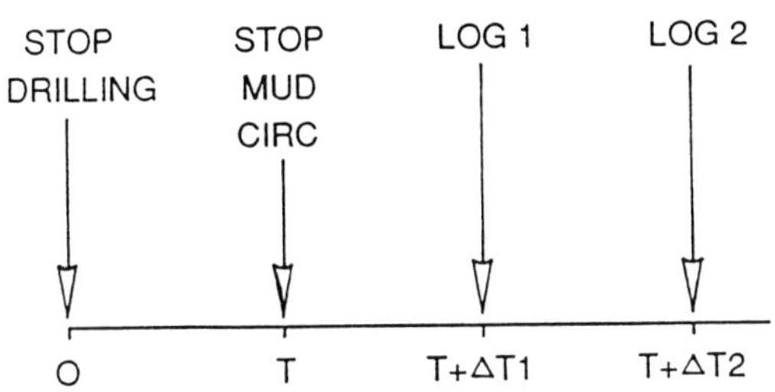

CIRCULATION DATA:

T	ΔT	(T + ΔT)/ΔT	log(T + ΔT)/ΔT	t_m (°C)
1.5	6.08	1.25	0.0957	111.0
	9.38	1.16	0.0644	116.0
	12.42	1.12	0.0492	119.5

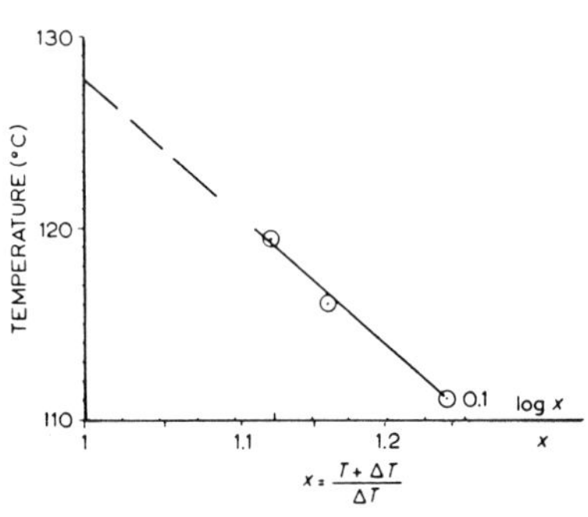

Figure 4.12. Example showing the procedures used in constructing a Horner plot (data are taken from Chapman, 1983).

in the N. Sea by Carstens and Finstad, 1981; in the Utah-Wyoming thrust belt by Deming and Chapman, 1988; in Australia by Russell and Baillie, 1989; and in the Delaware basin, Texas, by Luo et al, 1994).

The Horner-plot procedure only gives the temperature at the bottom of the hole and cannot be used to give temperatures at shallower depths, or to give temperature profiles. The only way to get this type of data is to measure extrapolated BHTs at various stages of drilling, i.e., during temporary cessation of drilling at intermediate depths (see, for example, Figure 4.5). Although the Horner plot method is simple and widely used, Beck and Shen (1989) have compared it with more rigorous methods and concluded that the Horner method underestimates temperatures somewhat. Dowdle and Cobb (1975) had also recognized that the Horner plot underestimates temperature, and they suggested that the theoretical line on a Horner plot actually had curvature and is not a true straight line. Other alternatives to Horner extrapolations have been discussed by Russell and Baillie (1989). Parasnis (1971) suggested a graphical method for extrapolating temperatures to infinite time as a way of calculation equilibrium temperatures in wells, and Russell and Baillie (1989) have used it in exploration wells offshore Tasmania. In the Niger delta Nwachukwu (1976) applied a modified Lachenbruch-Brewer equation to calculate true formation temperatures, and concluded that equilibrium temperatures were reached about 48 hours after cessation of mud circulation. Figure 4.13 shows his data for the first 26 hours together with the final equilibrium temperature. In contrast, Oxburgh and Andrews-Speed (1981) claimed that it took "a number of years" for a deep hole to return to thermal equilibrium. A recent review of thermal extrapolations has been given by Hermanrud and Shen (1989), and Carstens and Finstad (1981) have discussed the sensitivity of the Horner method to various sources of error.

4.2.3. Gas Hydrate Stabilities

Geothermal gradients can be especially difficult to estimate in rank frontier areas offshore where there is little, if any, well control. Regional setting can be a useful guide, and in many cases this is all that is available. In some offshore areas there are strong seismic reflectors that often parallel the bottom topography and are known as "*bottom simulating reflectors*" (*BSRs*). These are due to the presence of gas hydrates in the sediments (Kvenvolden et al, 1993; Kvenvolden, 1993) and can be used to infer geothermal gradients (Shipley et al, 1979; Yamano et al, 1982). Gas hydrates (also called "*clathrates*") form when the cage-like structure of ice traps methane or other small molecules within the cavities of the crystal lattice (Makagon, 1981; Sloan, 1990; Sloan et al, 1994). Trapping is purely

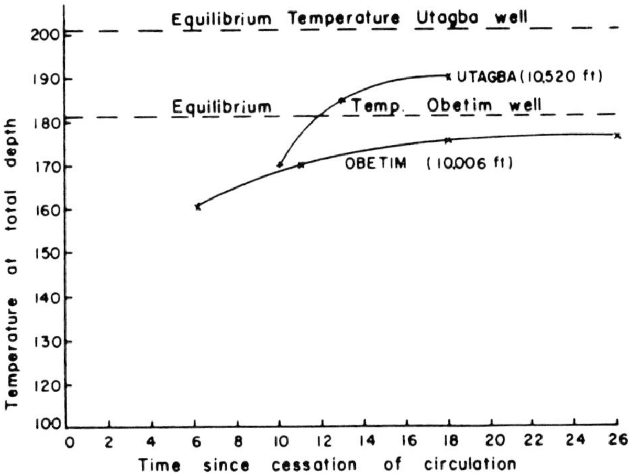

Figure 4.13. Temperature-buildup curve for Obetim and Utagba wells and calculated equilibrium temperatures. Temperatures are in °F and time is in hours. (From Nwachukwu, 1976).

physical with no chemical bonding, but the presence of the guest molecules stabilizes the host ice structure so that it remains the stable phase at higher temperatures. The phase diagrams for hydrates of methane and other gases are well known, and show that the gas hydrates can survive to temperatures of 15 °C or more as gas pressure increases. In the Arctic, gas hydrates form in water depths greater than 335 m, but in the subtropics higher pressures are needed and the depths exceed 600 m.

A BSR corresponds to the depth and temperature in the sediments where gas hydrates are no longer thermally stable and break down to give gas and water. At the base of the hydrate stability zone where there is free gas, a strong reflector (with a negative polarity kick) develops, but within the hydrate itself there are very weak internal reflections because the acoustic impedance values for the gas hydrates are very close to those for recent sediments (Figures 4.14 and 4.15). Because the gas in recent sediments is mainly biogenic (Kvenvolden, 1995), and therefore essentially pure methane (see Chapter 1), the relevant phase diagram for the gas hydrate stability is the methane-seawater diagram. This provides the temperature at the base of the BSR. Depth can be found from the seismic two-way travel time, and a combination of this depth with temperature inferred from hydrate stability gives the present day geothermal gradient. In a few cases higher hydrocarbons with a presumed thermogenic origin have been reported in gas hydrates (Brooks et al, 1984, 1986).

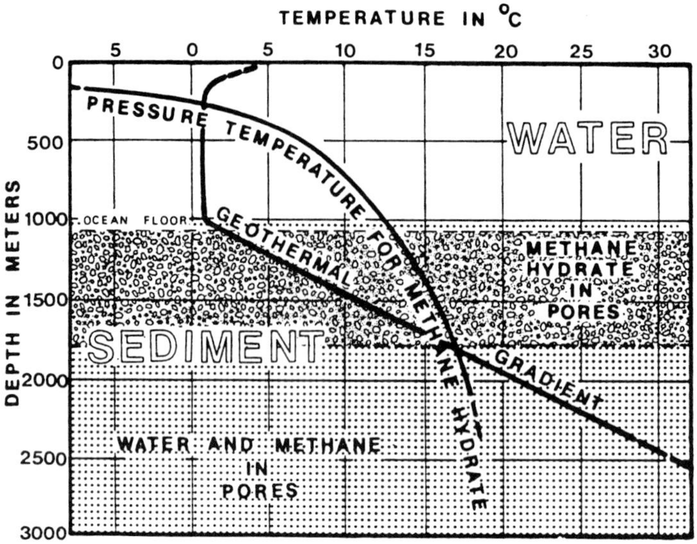

Figure 4.14. Diagrammatic illustration of the zone of gas hydrate formation and its relationship to depth and geothermal gradient (Hedberg, 1980).

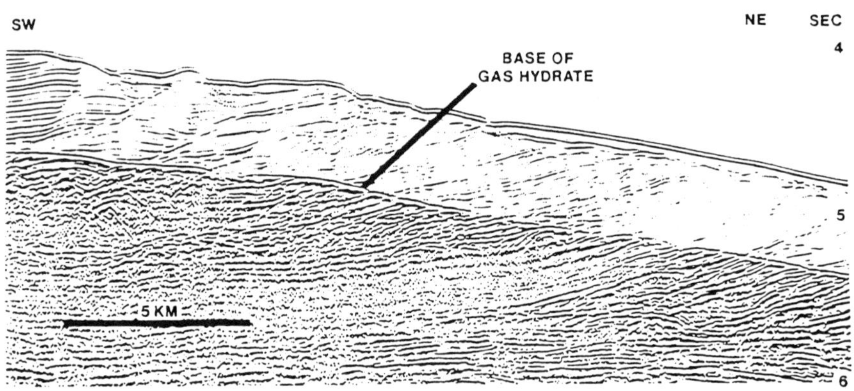

Figure 4.15. A 12-fold multichannel reflection profile from crest and eastern flank of Blake Outer Ridge showing the Base of Gas Hydrate as a bottom simulating reflector (BSR) (Shipley et al, 1979).

What is measured by this method is actually the geothermal gradient in the gas hydrate zone. Stoll and Bryan (1979) showed that gas hydrate formation lowers the thermal conductivity of the sediments by about 30%, so it is necessary to correct for this change. It is also important to do the measurements and calculations for flat areas because topographic highs have larger surface areas that lead to faster cooling and produce low apparent gradients. Diapirs can also disturb local geothermal gradients, and this is reflected in variations in the depth of the base of the hydrates. Over salt diapirs the BSRs become shallower corresponding to an elevated geothermal gradient, but the reverse is true over shale diapirs where gradients are lower and the BSR is deeper (MacLeod, 1982). Finally, it is important to be sure that reflections are in fact due to changes in hydrate stability, and are not produced by some other diagenetic change (Hein et al, 1978). In a seismic section given by Carlson et al (1985) the BSR produced by the gas hydrate occurs at a 2-way travel time of 3.5 sec, whereas the diagenetic change BSR is fifty percent deeper at 5.2 sec 2-way travel time.

In areas of permafrost shallow geothermal gradients may be very low. On the Siberian platform temperatures do not usually exceed 20 °C down to 2000 m (corresponding to a geothermal gradient of less than 10 °C/km). At shallower depths temperature is commonly not more than 3 - 4 °C (Anon, 1987).

4.2.4. Chemistry of Formation Waters

Na-K-Ca geothermometers: Any temperature-dependent process has the potential for constraining geologic temperatures, either today or in the past. Cation exchange among subsurface waters, feldspars, and feldspar-alteration products is controlled by temperature (especially above 100 °C), and the Na/K and Na/Ca ratios have been related to temperature by Fournier and Truesdell (1973). While this approach has been used successfully in geothermal areas it needs to be modified for sedimentary basins which have higher salinities and pressures but generally lower temperatures (Land and McPherson, 1992). Kharaka and Mariner (1989) have used the composition of subsurface waters to estimate temperatures in the range 30-200 °C, and they also introduced a magnesium-lithium chemical geothermometer that gives improved results over the magnesium corrected Na-K-Ca geothermometer. They cautioned that waters taken from gas wells may not be representative of the waters at depth.

Silica geothermometers: Silica geothermometers are generally applicable to waters from sedimentary basins with temperatures higher than about 70 °C (Kharaka et al, 1977). The silica content of groundwater has been related to heat flow by Swanberg and Morgan (1979, 1980) using the relationship:

$$T_{SiO2} = mQ + b \qquad (4.11)$$

where T_{SiO2} is the temperature derived from the silica geothermometer, Q is the regional heat flow, and m and b are constants with values of 670 $\pm$ 67 $^{o}Cm^{2}W^{-1}$ and 13.2 $\pm$ 5.1 ^{o}C, respectively. Swanberg and Morgan (in Sass et al, 1989) give a heat flow map of the United States plotted from data obtained in this way. 200,000 silica-heat flow determinations were employed, and in general heat flow contours are similar to those based on more conventional methods. All of the chemical geothermometers based on the concentration of cations in subsurface waters are empirical, and have to be calibrated using data from well-documented areas.

4.3. DRILLING-INDUCED COMPLICATIONS

4.3.1. Abnormal Pressures

We have already discussed the effect that drilling has on cooling the bottom of a well, but other thermal changes can also be induced by drilling procedures. Drilling rate may be higher in abnormally pressured intervals because the rocks often have higher porosities. High porosities due to undercompaction are a consequence of pressure developed during disequilibrium compaction, and are less likely to occur if high pressures are due to organic maturation. When penetration rate increases there is less time for thermal equilibration and the mud temperatures are somewhat lower and rock temperatures higher. The opposite occurs when drilling rate is reduced. Wells are usually cased before drilling into an abnormal pressure interval, and this requires a reduction in drill bit size which can lead to changes in drilling rates and cause changes in apparent temperatures.

Undercompacted rocks have higher porosities, and the increased water content reduces their thermal conductivities and leads to higher geothermal gradients through the overpressured interval. If drilling stops in the abnormally pressured zone, or immediately below it, the situation is analagous to that shown in Figure 4.8, and the geothermal gradient will appear anomalously high.

4.3.2. Cement

When casing has been recently cemented there will be anomalously high temperatures caused by exothermic reactions as the cement solidifies. The higher temperatures may be detected by continuous reading thermometers (but these are not reliable for estimating geothermal gradients).

4.4. GEOLOGIC COMPLICATIONS

The methods outlined above for estimating temperature give a single value for a single location, usually at TD, for the well being studied. It is then assumed that (1) temperatures can be extrapolated (or interpolated) vertically and horizontally, (2) temperatures are controlled by conduction, and (3) the system is thermally at equilibrium. There are areas where all of these individual assumptions are not justified.

4.4.1. Waterflow

The assumption of conductive heat transfer may break down near surface where gravity driven groundwater movement leads to forced convective heat transfer. This results in lowered temperatures and geothermal gradients beneath regional groundwater recharge areas, undisturbed temperatures and gradients beneath areas of lateral groundwater flow, and increased temperatures and gradients beneath regional groundwater discharge areas (Vugrinovich, 1989) (Figure 4.16). The western Canadian basin has been particularly well studied (Hitchon, 1984; Majorowicz et al, 1985), and here groundwater motion, and hence geothermal gradient, is related to topography. In the Prairies basin of western Canada, Jones and Majorowicz (1987) used a data base of 65,000 bottom hole temperatures to show that temperatures were elevated in areas of discharge, but depressed in areas of recharge. Similar studies have been reported for the Michigan basin (Vugrinovich, 1989), the Canadian part of the Williston basin (Majorowicz et al, 1986), and in overthrust/foreland basin systems (see Section 10.6.1). A study of water recharge into the Rhine graben led Person and Garven (1992) to conclude that "a regional topography-driven groundwater flow system was established that shifted the depth to the oil window between recharge and discharge areas by more than 1 km". They also found that an upward water flow of 1 mm/year moved the top of the oil window hundreds of meters. In the East Midlands basin of the United Kingdom, Wilson and

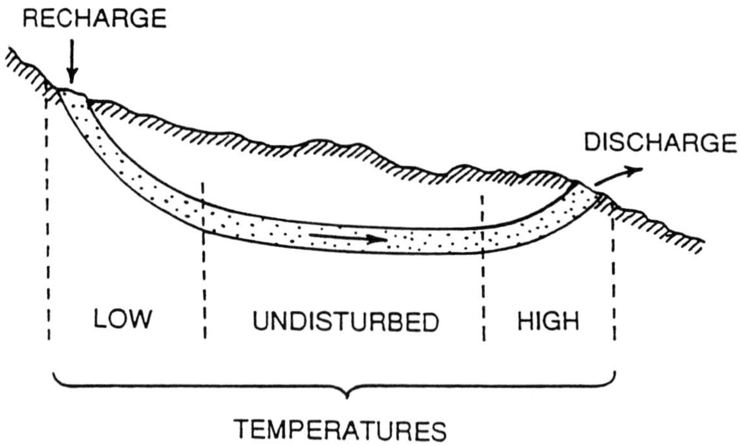

Figure 4.16. Schematic figure showing cooling in the area of recharge, and warming in the area of discharge.

Luheshi (1987) used a 2-D model of water flow developed by Smith and Chapman (1983) to explain the Eakring temperature anomaly. Although the predicted trends were generally correct, they felt that a more comprehensive 3-D model would be needed to fully understand the thermal regime in the area. Wood and Hewett (1982) have argued that the formation of spontaneous eddy currents, with velocities of the order of 1 m/year, is common in sandstones, even for average geothermal gradients. However, their modeling used a rather high permeability of 10 darcys. Recharge may be active over long distances (Bethke and Marshak, 1990) and can often be recognized from (1) salinity patterns, (2) tilted oil water contacts, (3) sloping potentiometric surfaces, and (4) the presence of biodegraded oils.

4.4.2. Lateral Heat Flow

One dimensional thermal models generally assume vertical heat flow, but around salt domes there may be a significant lateral component (see Section 10.7.3). This situation also occurs on a larger scale when the top of the asthenosphere rises to shallower depths during crustal thinning and extension. When lateral transfer of heat is important it is usually called "*thermal refraction*". Heat may be redistributed by convection (Hulen et al, 1993) and leads to geothermal gradients that are very low because temperatures are evened out by the circulating fluids. Convection seems more likely to occur locally than to affect a whole basin (Ludvigsen et al, 1993).

4.4.3. Faults

Faults can be seals or conduits. When they are conduits they provide a pathway for cold surface waters to penetrate into the subsurface, or for hot waters to move up from depth. On a regional scale they can play an important role (see Section 4.4.1, above), but they can also produce very localized temperature anomalies. Growth faults are a characteristic feature of the progradational margin of the Gulf of Mexico, and commonly have associated high temperatures (Figure 4.17). Data for onshore south Texas from Bodnar and Sharp (1988) are plotted in Figure 4.18 and show considerable variations in the depths of individual isotherms. Gradients calculated from the 350 °F isotherm vary from as high as 2.4 °F/100 ft to as low as 1.75 °F/100 ft. Temperature anomalies are often greatest close to the faults (see, for example, the fourth growth fault from the east in Figure 4.17), and this is usually the location of petroleum reservoirs that are either sealed against faults or in rollover anticlines. Average geothermal gradients in south Texas are in the 20 - 30 °C/km range, but values of 55 - 100 °C/km have been recorded along some faults (Cooper, 1978). In the northern North Sea, Carstens and Finstad (1981) have used the distribution of faults associated with rifting to explain the regional pattern of geothermal gradients.

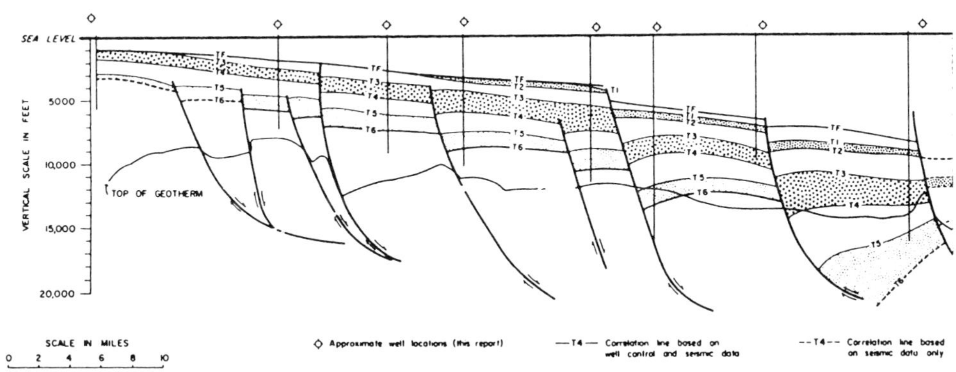

Figure 4.17. Approximate East-West section parallel to the U.S. - Mexican border (Rio Grande River) with a superimposed isotherm (Bebout and Agagu, 1980).

The relative movement of one side of a fault against the other may produce frictional heating. However, published evidence for frictional heating along faults is equivocal. While many studies have presented evidence for local heating, heat flow distribution across faults (Brune et al,

1969), and vitrinite reflectance across thrust faults (Bustin, 1983; Underwood et al, 1988) suggest that frictional heating is not a major heat source. Interpretations are complicated because faults can act as conduits for flow of hot waters, and when shearing occurs the small sized particles formed are more easily oxidized, and in many cases this leads to anomalously high values for vitrinite reflectance. It seems likely that thermal effects, if any, are limited to a very small area immediately adjacent to the faults, and that they are not significant in causing enhanced maturation of large volumes of source rock.

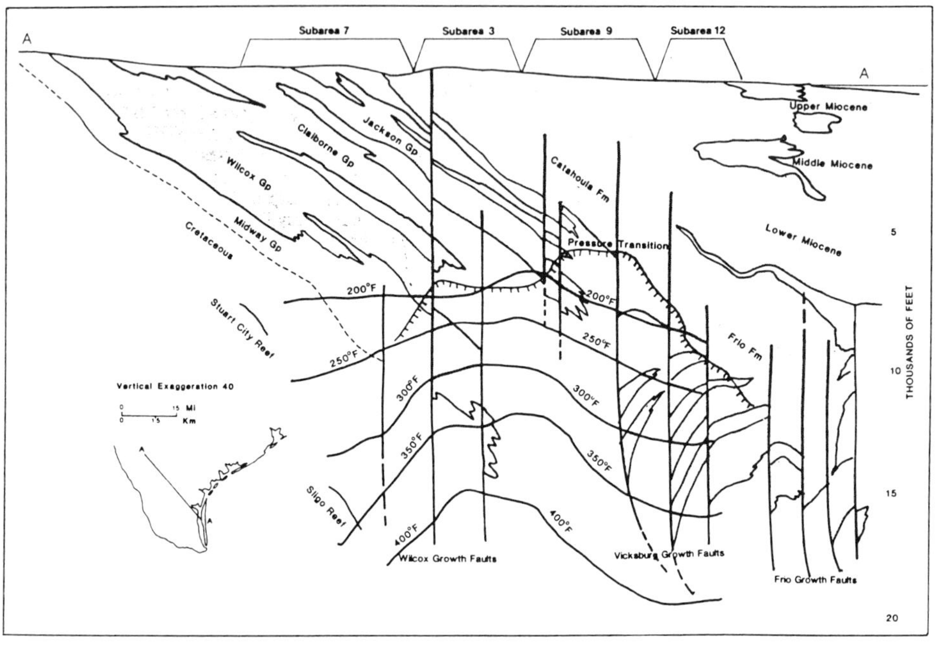

Figure 4.18. Generalized cross section of south Texas Gulf Coast with isotherms and the top of the geopressure transition zone shown (Bodnar and Sharp, 1988).

4.4.4. Lack of Thermal Equilibrium

Not all environments are at thermal equilibrium. This is especially true near-surface where there can be rapid fluctuations in response to changes at the surface (Cathles and Nunns, 1991). These changes include not only diurnal and seasonal fluctuations, but also larger events such as glaciation (either being applied or removed) (Johansen et al, 1996). Response to these changes depends on the magnitude of the change and the specific heat of the rock units. The development of extensive permafrost may also disturb the local thermal regime (Nixon, 1986).

When cool near-surface sediments are being carried into the subsurface rapidly they may not have sufficient time to warm up to thermal equilibrium. This happens with the rapid sediment accumulation that is characteristic of most of the large Tertiary delta systems. It also happens in subduction complexes where the cold downgoing plate is rapidly subducted. Both environments typically have below average geothermal gradients. Conversely, in areas of rapid uplift and erosion, near-surface temperatures may be anomalously high.

4.4.5. Salt Domes

Salt has a very high thermal conductivity and in salt diapirs heat is conducted from the deep subsurface up into shallower levels. Here it may lead to higher temperatures immediately above the diapirs (Figure 4.19), but lateral heat conduction, or heat refraction, is also significant (Vizgirda et al, 1985; O'Brien and Lerche, 1984; Evans, 1977; Rashid, 1978; Rashid and McAlary, 1977) (Figure 4.20) (see also Section 10.7.3). Thermal refraction of significant quantities of heat away from salt diapirs and out into the surrounding sediments can lead to isotherms that become subvertical near the diapirs (Anderson et al, 1991). The high temperatures adjacent to diapirs can drive convection and further disturb the local thermal regime (Figure 4.20).

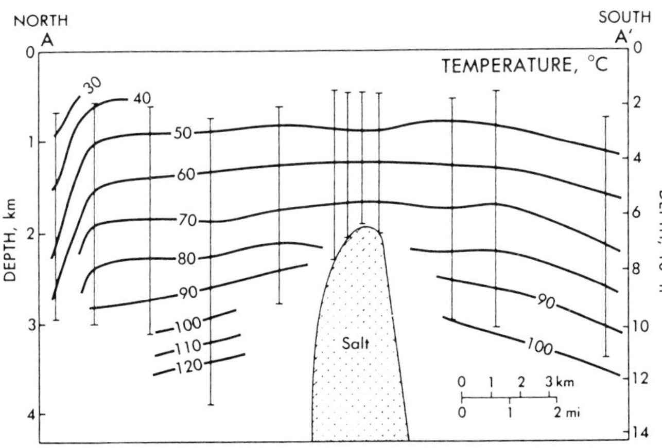

Figure 4.19. Cross section showing estimated position of isotherms in the vicinity of the Welsh dome (Bennett and Hanor, 1987).

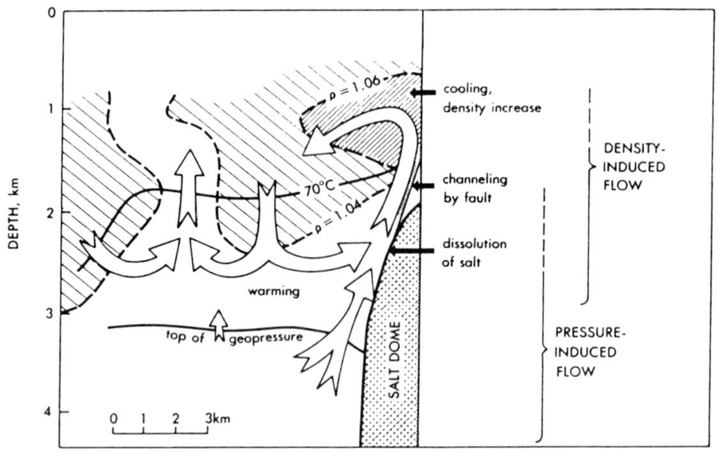

Figure 4.20. Possible pattern of large-scale solute transport on the north flank of the Welsh dome (Bennett and Hanor, 1987) implying a redistribution of heat and altered positions of the local isotherms.

4.4.6. Unrecognized Igneous Intrusions

Silicate melts are intruded at high temperatures and can impact thermal indicators such as vitrinite reflectance (Horvath et al, 1986; Jaeger, 1964; Kettel, 1983; Wang et al, 1989; Summer and Verosub, 1989). If the igneous intrusion was recent, temperatures may still be above the local equilibrium values. When drilling stops short of penetrating the intrusive, and its presence is not recognized, then the high current temperature could be extrapolated to surface temperature to give an unrealistically high present day geothermal gradient.

4.4.7. Temperature Anomalies Over Oil Fields

Well coverage is usually greater over oil and gas fields, but temperature data from these wells can be misleading because many oil and gas accumulations have positive temperature anomalies associated with them (Roberts, 1981; Zielinski, 1985; Meyer and McGee, 1985; McGee et al, 1989). These locally high temperatures are usually attributed to the lateral and vertical migration of hot fluids that lead to "focusing" over structures. However, it is worth noting that many seals (but especially overpressured shales) are good insulators with low thermal conductivities.

This is the situation shown schematically in Figures 4.4c and 4.6 where the temperatures below the poor conductors are higher than in situations where they are absent.

4.4.8. Anthropogenic Complications

Recent changes at or near the surface may lead to non-equilibrium near-surface temperature regimes. Human activities that impact temperatures include deforestation, filling of large lakes, tunneling, and paving.

CHAPTER 5

PALEOTEMPERATURES

5.1. REPRESENTING TEMPERATURES

At any given geographic location heat flow and geothermal gradient may have varied considerably over geologic time, and the present day values could be very different from those that occurred in the past. Temperature variations have a major role in influencing the progress of petroleum generation, and it is important to be able to constrain paleotemperature history as accurately as possible. Although geologic setting can provide some generalized estimates of changing temperature through time, more detailed values require other techniques, and in the last few years a wide range of temperature dependent processes have been evaluated for their potential in providing reliable paleotemperatures. Before these can be discussed in detail it is necessary to consider the way in which temperature variations through time are displayed graphically. Typically this is done by overlaying the temperature isotherms on a normal burial history diagram with a linear depth scale.

Figure 5.1 shows the simplest case where a linear geothermal gradient does not change over the time period considered, and in this example remains constant at 25 °C/km from 300 Ma to the present. If surface temperature remains unchanged at 20 °C then a subsurface temperature of 70 °C will always be 2 km below the surface. Similarly all the other isotherms on the diagram will be horizontal and at constant depth, and there will be uniform vertical spacing between the isotherms.

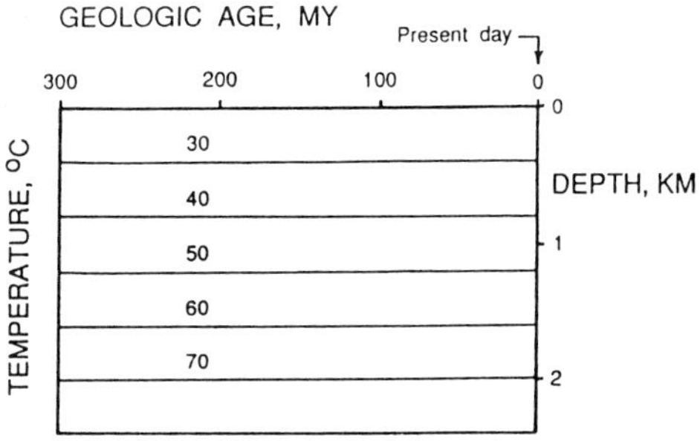

Figure 5.1. Isotherms plotted on a depth-geologic age diagram for the last 300 m.y. corresponding to a constant geothermal gradient of 25 °C/km.

In nature few gradients, if any, are linear because different rock types have different thermal conductivities (see Equation 4.2). When a non-linear gradient is superimposed on a burial diagram the spacing of the isotherms at any given time will be variable, with higher gradients producing closer spacings. Figure 5.2 is an example of a situation where the gradient has been 35 °C/km down to 3 km, but only 20 °C/km below that. A geothermal gradient may be linear at any given time, but the absolute value of the gradient may change through time. Figures 5.3a and b show the two cases where an area is cooling down or warming up. In Figure 5.3a the present day gradient is 20 °C/km, but 150 Ma it was 25 °C/km and 200 Ma it was 40 °C/km. In all cases the gradients are linear.

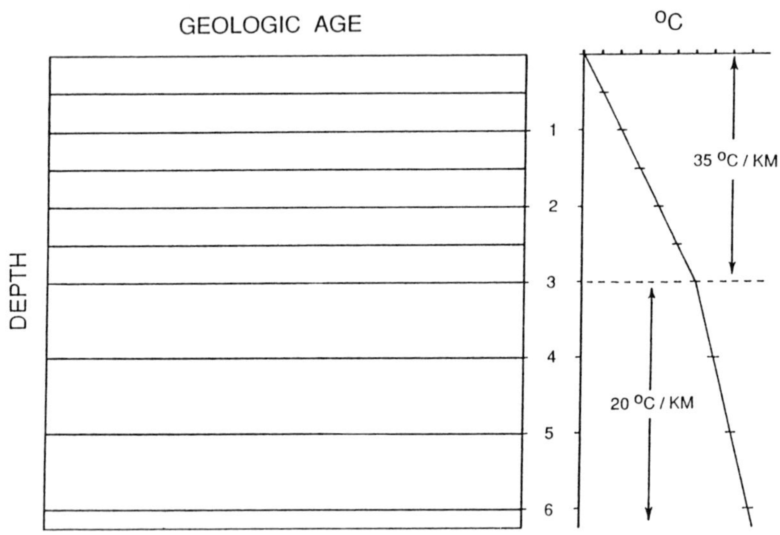

Figure 5.2. Isotherms for a non-linear geothermal gradient that remains unchanged through time. In this example the geothermal gradient is 35 °C/km down to 3 km, but decreases to 20 °C/km at greater depths.

The conductivity of individual units in the sedimentary rock pile is modified through time by compaction and cementation. Larger changes will occur if shales become overpressured and remain undercompacted, and abnormally pressured intervals show lower thermal conductivities and therefore produce higher gradients. Abnormal pressures can be a transient phenomenon with development and dissipation of the high pressures occurring over a relatively short time period.

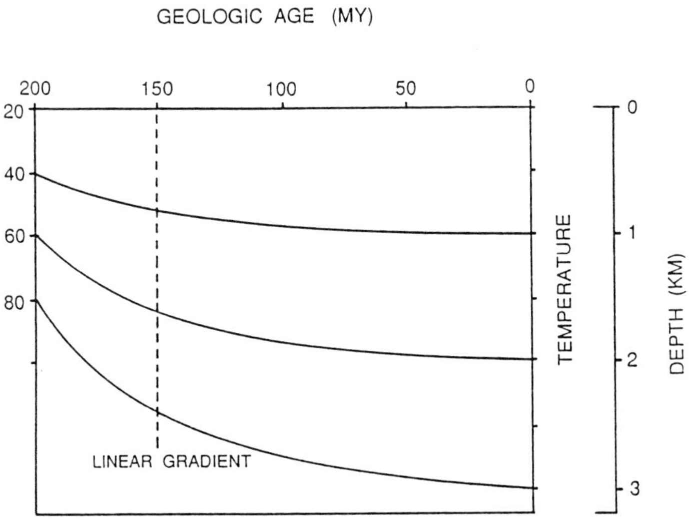

Figure 5.3a. Isotherms for a geothermal gradient that decreases through time.

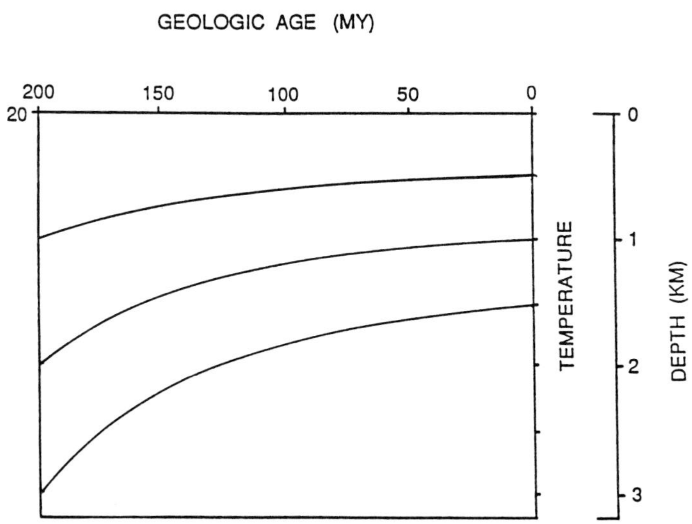

Figure 5.3b. Isotherms for a geothermal gradient that increases through time.

Real world geothermal gradients may change considerably through time, but can often be treated as combinations of the simple profiles given above. In the San Juan basin of New Mexico, for example, geothermal gradients were initially relatively low, but then increased to a maximum at 37 Ma before decreasing to their present value (Figure 5.4). Many other examples are given in the following pages. Although the temperature may be falling smoothly through time in some area of interest, most commercial thermal modeling programs are not capable of handling the temperature data in this way. They require a given gradient at a given time (or calculate it from heat flow at a given time), and curves like those given in Figure 5.3a must be broken down into a limited number of straight line segments. The actual number allowed will depend on the program, but it is worth remembering that the deepest and hottest part of the burial history is usually most important in controlling oil generation/survival, and so this part of the temperature regime should be approximated by the largest number of straight line segments.

5.2. GEOLOGIC SETTINGS

5.2.1. Introduction

For a given geothermal gradient, a value for burial rate implies a heating rate. Since geothermal gradients range from 25 °C/km to 45 °C/km, and burial rates from 10 m/m.y. to 500 m/m.y., heating rates are in the general range from 0.25 °C/m.y. to 22.5 °C/m.y. (Table 5.1). Most rates are between 0.3 and 15 °C/m.y.

Heat flow and temperature gradients can vary widely through time in different geologic environments. For most settings enough areas have been studied in detail that the general trends for heat flow and temperature are well known (Klemme, 1975). Data of this sort provide a useful initial guide in evaluating previously unexplored basins.

5.2.2. Geologic Settings

Rifts/Passive Margins: Rifting is often initiated over a local hot-spot where heat flow and geothermal gradients are high (>50 °C/km). With continued extension of the rift, and the evolution of a passive margin, geothermal gradients fall to average values through time (Royden et al, 1980). Isotherms reflecting these changes have been published for the Reconcavo basin, Brazil (Cohen, 1985) (Figure 5.5), and for the Aaiun

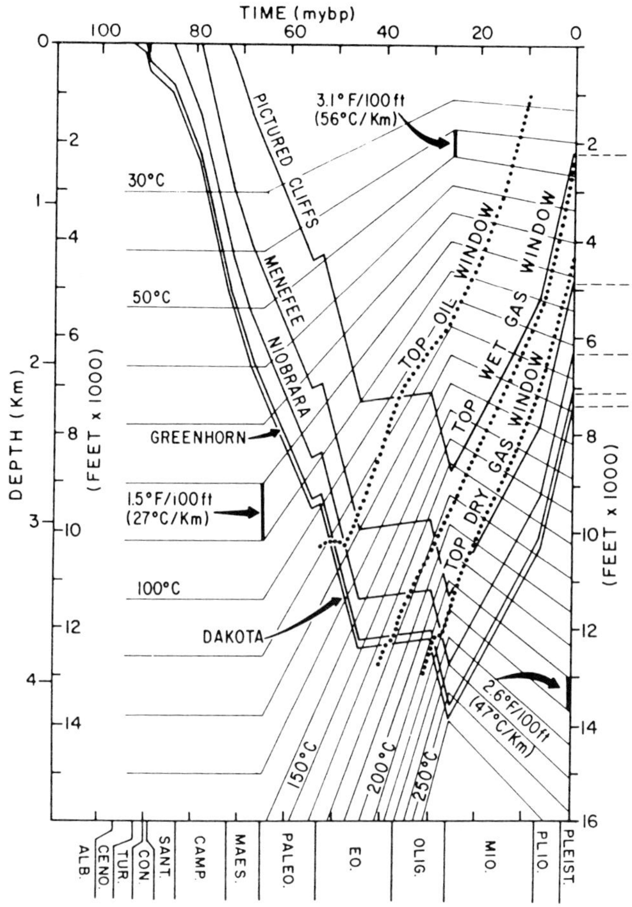

Figure 5.4. Burial history diagram for rocks in the San Juan basin, New Mexico, with superimposed temperature isotherms. The geothermal gradient was initially low, but warmed up to a maximum at 37 Ma and then cooled to its present value (from Bond, 1984).

basin, west Africa (Duncan and Roman, 1986). Teichmuller and Teichmuller (1986) related changing geothermal gradients in the middle Upper Rhine graben to an Eocene rifting event followed by cooling and onset of recent new rifting and spreading (Figure 5.6). Their estimates of paleogeothermal gradients were based on vitrinite reflectance data and coalification gradients obtained from deep boreholes.

Tertiary Deltas: Deltas show the opposite trend to passive margins. Delta piles are built as rivers dump large quantities of cold materials producing a thick sediment accumulation that is out of thermal equilibrium and leads to depressed temperatures. As the delta overruns the first depocenter, or switches to a different depocenter, the deltaic sediments warm up to the regional average (see Section 10.5.2).

Table 5.1
Heating rates in °C/m.y for various combinations of geothermal gradient and burial rate.

BURIAL RATE m/m.y.	GEOTHERMAL GRADIENT 25 °C/km	35 °C/km	45 °C/km
10	0.25	0.35	0.45
50	1.25	1.75	2.25
100	2.50	3.50	4.50
250	6.25	8.65	11.25
500	12.50	17.50	22.50

Interior Basins: Intracratonic basins are formed by a variety of different mechanisms, but the structurally simple basins often show relatively uncomplicated thermal histories, with geothermal gradients that do not seem to have changed much through time. However, these older basins are generally no longer compacting and water penetrating on the basin flanks may modify the thermal regime, as shown by Willett and Chapman, (1987) for the Uinta basin. Other examples are given in Section 10.2.

Overthrusts: Overthrusts can show very complex temperature histories because one thermal regime is superimposed on top of another. The actual temperatures that develop will depend on the relative rates of thrusting and uplift versus erosion. Under some circumstances a reversal in temperatures with depth is possible. Specific examples are discussed in Section 10.6.1.

Combinations: At any given location, a variety of different geologic processes may operate over time and often the different thermal regimes must be combined. For example, if a delta progrades into an opening rift the steadily cooling system may thermally "undershoot" in the area of the delta before recovering to the regional average. Complex thermal changes can develop in other geologic settings formed as a result of multiple processes, such as a rifted margin overridden by overthrusting (e.g., Arkoma basin of Oklahoma and Arkansas). It is often helpful to indicate the sequence of geologic processes on the burial history diagram.

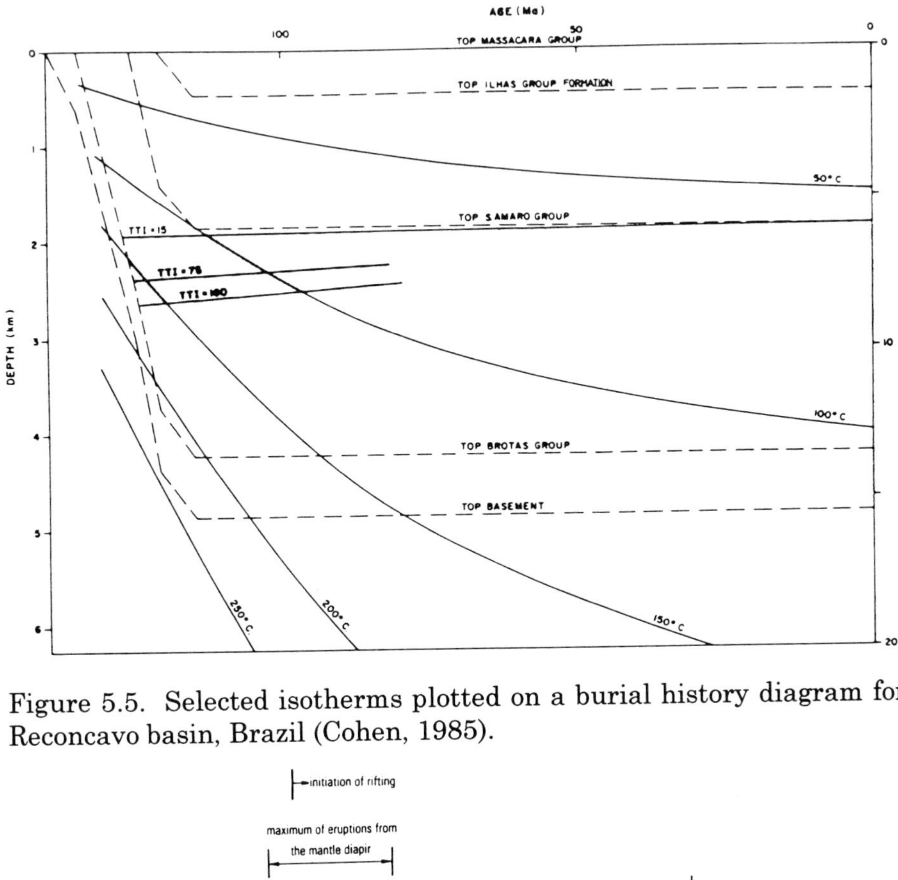

Figure 5.5. Selected isotherms plotted on a burial history diagram for the Reconcavo basin, Brazil (Cohen, 1985).

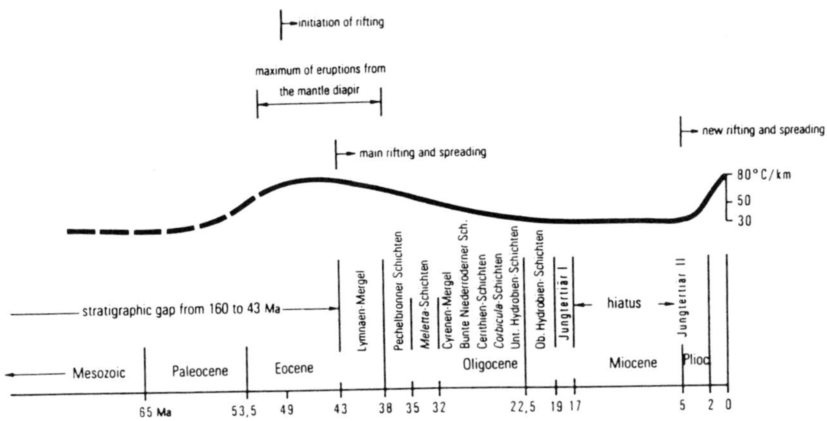

Figure 5.6. Variation of geothermal gradients during the development of the middle Upper Rhine Graben, deduced from results of coalification studies (from Teichmuller and Teichmuller, 1986).

5.3. DIRECT MEASUREMENTS

5.3.1. Fluid Inclusions

Introduction: Fluid inclusions provide a useful and important way of deriving paleotemperatures. These inclusions are typically microscopic defects that develop in crystals during or after crystal growth and trap small volumes of the fluids present in the vicinity of the crystal. They can provide unique information about the composition and physical conditions of the environment where the host mineral grew (Hollister and Crawford, 1981; Roedder, 1984; Shepherd et al, 1985; Mullis, 1987; Burrus, 1988; McClimans, 1987; Touray, 1994; Goldstein and Reynolds, 1994). Burley et al (1989) give examples of color photomicrographs of fluid inclusions.

Information derived from fluid inclusions relates to the time of inclusion development, and it is critically important to relate this to the burial history of the rock unit (Bodnar, 1994). For example, in a sandstone the grains give information about conditions in some precursor granitic melt, but only the later overgrowths can provide information about the sedimentary environment in which the sandstone evolved. In practice, it is overgrowths and cements, especially fracture-filling cements, that offer the best material for fluid inclusion measurements (Figure 5.7). Quartz and calcite are most commonly used, but anhydrite, albitized feldspars, barite, and ankerite are occasionally useful. In carbonate systems calcite and dolomite cements can be used (Montanez, 1994; Qing and Mountjoy, 1994).

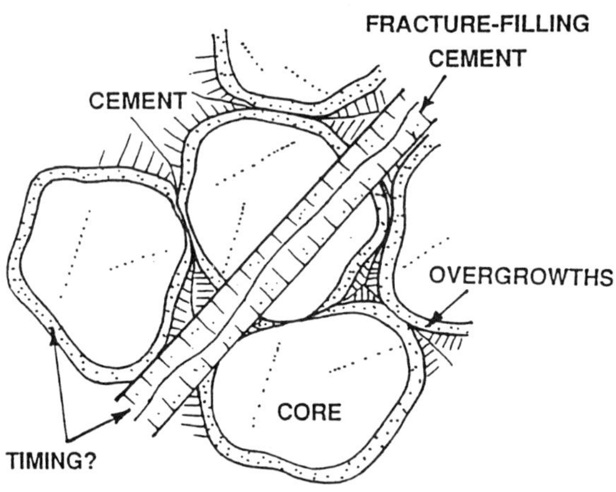

Figure 5.7. Locations of fluid inclusions in fracture-filling cements and overgrowths. Inclusions in the cores of the quartz grains have not sampled the sedimentary environment.

Principle: The principle used in determining paleotemperatures can best be described by reference to a fluid inclusion formed by trapping a homogeneous aqueous phase of low salinity. At room temperature such inclusions generally contain a vapor bubble. The pressure-temperature diagram for the aqueous solution (Figure 5.8) shows that at low pressure but high temperature there is vapor (V), but that at higher pressures most of the diagram is in the liquid field (L). Lines of equal density ("*isochores*") have been added to the figure. In addition, selected geothermal gradients can be superimposed because, for a normally pressured system, P and T are related through the geothermal gradient. During crystal growth an inclusion forms at the P, T values corresponding to point B in Figure 5.8. Since a fixed mass of material is now trapped in a constant volume the density is fixed, and on cooling P and T will be related by the relevant isochore (in this example 0.943 gm/cm^3). This isochore eventually intersects the Liquid-Vapor phase boundary, and it is at this point that the vapor phase (bubble) appears, and subsequently increases in size as the host mineral is cooled to room temperature. Laboratory heating can reverse this sequence of events, and the temperature at which the bubble disappears can be measured to give the "*homogenization temperature*", T_h. These microthermometry techniques are normally used to examine inclusions in doubly polished "thick" thin-sections that are roughly twice as thick as normal thin sections. Thinner or thicker sections can be used, but it is important that unnecessary heating is avoided during cutting and polishing. The measured T_h value defines the isochore (for any given composition), but does not constrain the position along it. The value for the actual growth temperature will depend on pressure and requires a so-called "*pressure correction*" (Potter, 1977). Estimating this correction is one of the main problems in using fluid inclusions to establish paleotemperatures. In the absence of some independent measure of pressure it can sometimes be estimated from the geothermal gradient and an assumed hydrostatic (or possibly lithostatic) pressure regime.

Composition: The composition of the system must be known in order to select the correct phase diagram for interpreting T_h in terms of crystallization temperature. Salinity is frequently an adequate indicator of composition, and even though other ions may be present, it is commonly expressed as an equivalent weight percent of NaCl. The salinity of the liquid can be obtained by measuring the freezing point depression for the inclusion fluids by freezing the sample in liquid nitrogen, letting it warm up, and noting the temperature when the last ice crystal melts ("T_m"). At higher salinities the isochores on the P-T diagram are not as steep, but it usually takes about 3-4 wt% difference to change the slope of the isochores

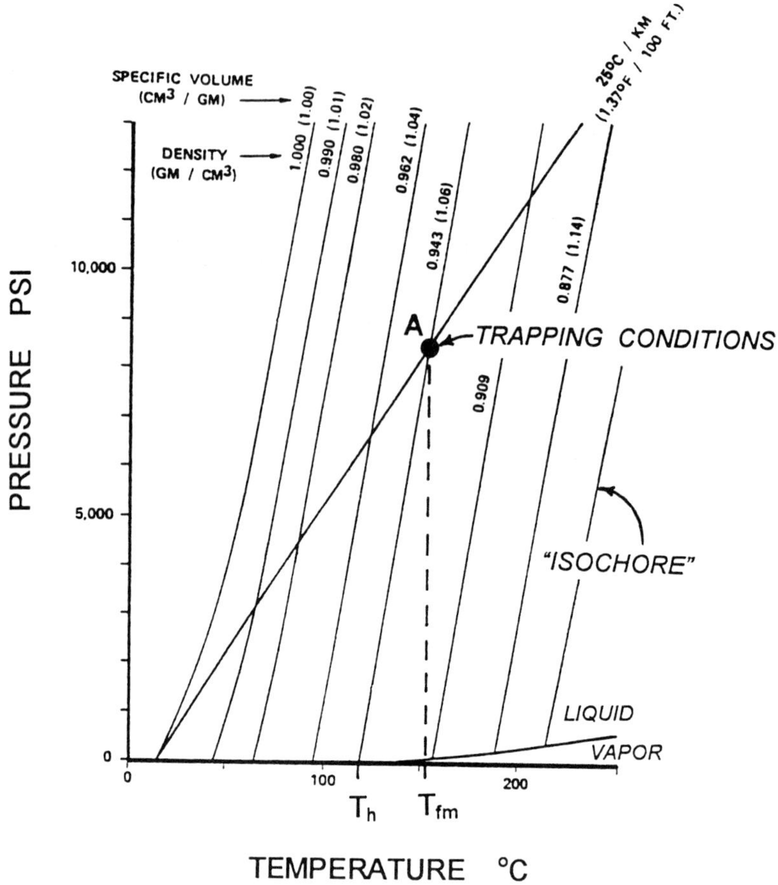

Figure 5.8. Pressure-temperature-density (or specific volume) diagram for water. A geothermal gradient of 25 °C/km for hydrostatically pressured fluids is superimposed. An inclusion that forms at Point A will have a formation temperature of T_{fm} and this will be higher than the homogenization temperature of T_h.

enough to affect temperature interpretations. More detailed analysis of fluid inclusion fluids can be made, either by extracting the fluids and analyzing them (using Atomic Adsorption spectrophotometry (AA) or other suitable technique), or by *in situ* analysis (such as laser Raman spectroscopy).

The gas content of inclusions can be important because it raises the bubble point curve (i.e., the V/L boundary) to higher pressures as well as changing the slopes of the isochores. Methane is an especially significant component (Hanor, 1980; Pagel et al, 1986). Figure 5.9 shows that when homogenization temperature (Th) is measured for a gas-bearing inclusion the incorrectly assumed isochore (through C, using the aqueous solution vapor-liquid boundary) leads to a trapping temperature that is too high. The correct isochore is defined by the line passing through B on the bubble point (L/V) curve for the gas-containing system (Walgenwitz et al, 1990). A further complication in methane-bearing fluid inclusions is the formation of clathrates that give melting points above 0 °C and confuse salinity determinations.

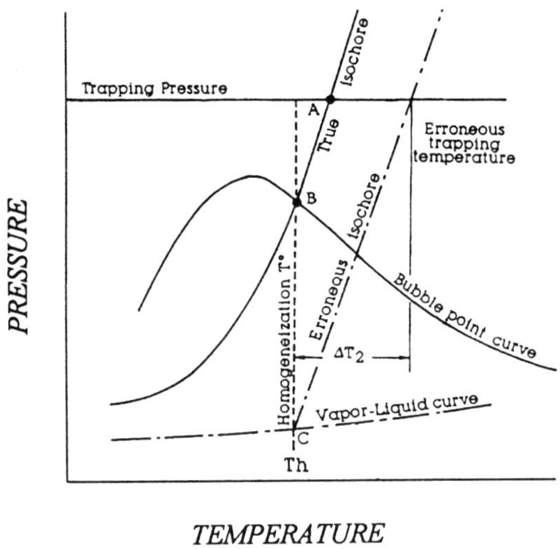

Figure 5.9. Pressure -temperature diagram for an aqueous system ("Vapor-Liquid curve") and methane-bearing system ("Bubble point curve") showing the potential effect of dissolved methane on the estimation of trapping temperature (Walgenwitz et al, 1990).

Oil-bearing inclusions: In many petroliferous areas inclusions have trapped migrating oil. When oil-filled inclusions and aqueous inclusions coexist in the host mineral, and both inclusion types are distributed along the same growth zone, they are generally considered syngenetic. Small oil-bearing inclusions may be difficult to distinguish from aqueous inclusions, but can often be recognized by their fluorescence under ultraviolet light. Oil is much more compressible than water and as a consequence its PVT diagram is very different with a bubble point curve that occurs at higher pressures. The isochores are much flatter and pressure rises more slowly with increasing temperature, but the actual values will depend on the composition of the particular oil and will have to be determined separately for each case. Often data for oils produced in an area can be used to select the appropriate PVT diagram, but they do need to be from the same depth (i.e., same maturity), and losses of methane and light ends during retrieval can be a problem. There have been attempts to analyze the oils trapped in the inclusions (see Burruss, 1989).

In samples containing syngenetic oil and aqueous inclusions the difference in the two phase diagrams provides a potential method for deriving both paleotemperatures and paleopressures. This can be illustrated by reference to Figure 5.10 where isochores for aqueous and oil-bearing inclusions are superimposed on a single PVT diagram. If the aqueous inclusions have T_h of 126 °C they will define the isochore labeled "A2", and the host mineral must have crystallized at some P-T conditions lying along this line. If cogenetic oil-bearing inclusions have T_h of 105 °C they define isochore "C2", and constrain inclusion formation to some P-T pair that lies along this line. But both types of inclusions developed at the same temperature and pressure and the only point that is on both isochores is where the two isochores cross, which is at point B' (Figure 5.10). This corresponds to a growth temperature of 162 °C and a growth pressure of 38 MPa. In practice measured homogenization temperatures span a range of temperatures and Figure 5.11 shows data from Narr and Burruss (1984) for oil-bearing and aqueous inclusions in the Little Knife field, North Dakota, U.S.A. The maximum and minimum values for these aqueous inclusions were used to define the isochores A1 and A2 of Figure 5.10, while corresponding values for the oil-bearing inclusions defined isochores C1 and C2. This leads to the conclusion that fluid inclusion development occurred within the P-T field outlined by the heavy lines of Figure 5.10. Walgenwitz et al (1990) were able to use this method with coexisting oil and aqueous inclusions in authigenic feldspars from the Pinda reservoir, offshore Angola basin, to estimate both pressures and temperatures at the time of mineral growth. Samples from a depth of 3340 m had oil bearing inclusions with T_h of 110 $\pm$ 6 °C, and aqueous inclusions with T_h of 150 $\pm$ 11 °C. Decrepitation data (assuming 600 bars rupture pressure in feldspars) were used to constrain the oil isochore and a range of plausible trapping temperatures between 191 and 198 °C calculated. Corresponding pressures were slightly above hydrostatic but close to present-day values.

Complications: Fluid inclusions present a potentially powerful method for constraining paleotemperatures, but there are several possible sources of error that have to be considered. If the host mineral is buried to greater depths after the inclusion formed, then the very high pressures generated

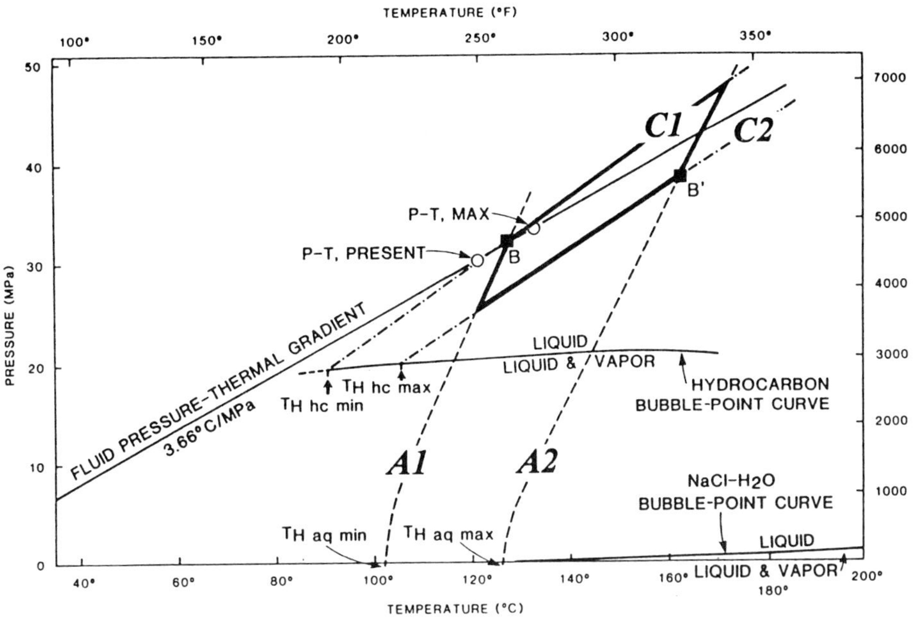

Figure 5.10. Pressure-volume-temperature (PVT) diagrams for saline solution (25 wt.% NaCl) and for Little Knife hydrocarbons (Narr and Burruss, 1984) with selected isochores superimposed (A1, A2 for the aqueous solution; C1, C2 for the crude oil).

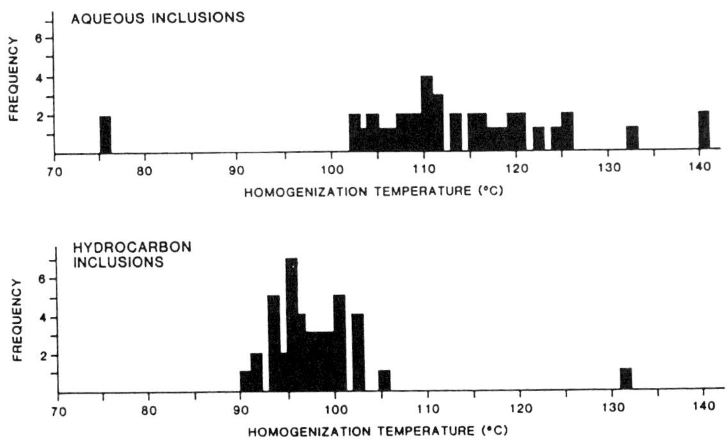

Figure 5.11. Homogenization temperatures reported by Narr and Burruss (1984) for aqueous and hydrocarbon fluid inclusions in the Little Knife field, North Dakota.

within the inclusions can lead to stretching and an increase in inclusion volume (Bodnar and Bethke, 1984; Burruss, 1987; Prezbindowski and Larese, 1987; Barker and Goldstein, 1990; Prezbindowski and Tapp, 1991; Touray, 1994). This reduces density, and the PT values correspond to an isochore that has a higher temperature at any given value of pressure (B to C, or D to E in Figure 5.12). Subsequent cooling leads to bubble formation at a higher temperature and a correspondingly higher value of Th. These trends have been duplicated in laboratory experiments (Figure 5.13).

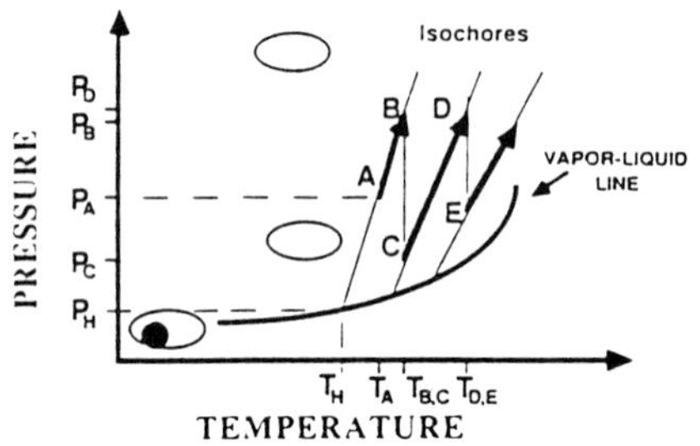

Figure 5.12. Generalized PVT diagram showing how stretching of inclusions decreases their density and moves the corresponding PT from one isochore to another (Prezbindowski and Larese, 1987).

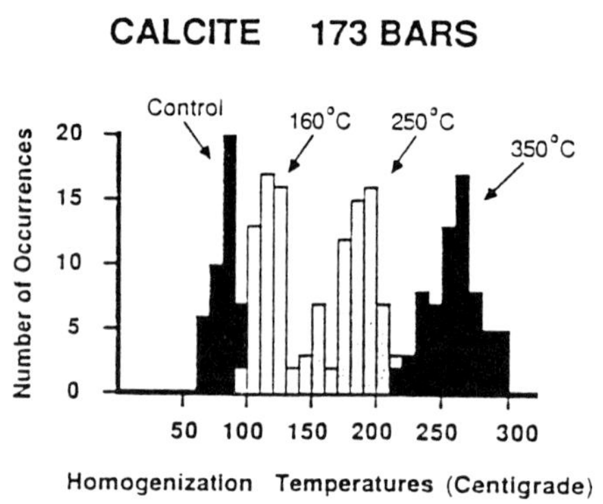

Figure 5.13. Measured homogenization temperatures for fluid inclusions in calcites that were heated to various temperatures (as indicated) under a confining pressure of 173 bars (Prezbindowski and Larese, 1987).

Stretching depends on inclusion size and composition, and on host mineral hardness. Smaller inclusions are less affected by stretching than large ones, and angular inclusions stretch more than those that are near-spherical. Inclusions in soft minerals (halite, gypsum, anhydrite, barite) are most likely to stretch, hard minerals least likely (quartz). Fluid composition has an effect on stretching because pressure rises more slowly with temperature in an oil-bearing inclusion. Stretching leads to inferred temperatures that are too high. When temperatures deduced from fluid inclusions are compared with other sources of temperature data the inclusion values are commonly higher than expected. This lead Pagel et al (1986) (in Gabon) to suggest "hydrothermal fluid circulation". In W. Canada Tilley et al (1989) proposed "... hot fluids migrated updip... to produce unexpectedly high geothermal gradients", while in southern England McClimans and Videtich (1989) found "...temperatures greater than can be accounted for by present surface temperature, geothermal gradient and depth". Burley et al (1989) (Tartan, N. Sea) reported that "...these temperatures are interpreted to indicate hot migrating fluids...". In an ideal situation stretching would reset the temperature to the value corresponding to the maximum depth of burial (Barker and Goldstein, 1990). This is most likely to occur for large, aqueous inclusions in a soft mineral, but cannot be relied on as temperatures derived from T_h are almost always intermediate. Fortunately, it seems to take about 200 MPa (equivalent to roughly 200 °C of overheating) to stretch inclusions in quartz (Bodnar et al, 1989), and they appear to give reliable data in most cases (Emery and Robinson, 1993).

In using fluid inclusions to obtain paleotemperatures it is important to consider the following:

(1) Paleotemperature data should be derived from inclusions that formed at the time the host mineral was crystallizing, but in many cases later rehealed cracks give "*secondary inclusions*" that formed at some unknown time after crystal growth. Distinguishing among primary and various generations of secondary inclusions may be staightforward, but sometimes can be very difficult (see Roedder (1984) for distinctions).

(2) Necking down and pinching off after separation of a vapor phase can give two quite different inclusions with different filling temperatures (e.g., Jensenius, 1987).

(3) Cement formation in the vadose zone may trap random proportions of air and water and give anomalously high temperatures. Often these inclusions are accompanied by a large number of single phase inclusions.

(4) Although fluid inclusions are very common they may be absent in minerals or cements of interest. In many cases, particularly quartz overgrowths, they can be very small and difficult to work with. Improving analytical equipment may help solve this problem (Wilkinson, 1994).

5.3.2. Vitrinite Reflectance

Introduction: Organic materials are sensitive indicators of past thermal history (Cooper, 1977), and vitrinite reflectance, TAI, coal rank, and biomarkers have all been used. Vitrinite alters non-reversibly (like other organic matter types) in response to rising temperature and passing time because the reactions are kinetically controlled. The changes that occur in both chemical composition and physical characteristics have been discussed in Chapter 1, and it was pointed out there that vitrinite reflectance is the most widely used parameter for establishing organic maturity. Reflectivity values for vitrinite can also be used to infer paleotemperatures and past geothermal gradients because at higher geothermal gradients R_o values increase more rapidly with depth (Robert, 1980; Russell and Baillie, 1989). An example of the changing gradient was included in Dow (1977) (Figure 5.14) where the steeper trend below the Jurassic-Triassic unconformity shows that the geothermal gradient at that time was higher than it was later in the Mesozoic. A similar example was given by Bordenave (1993)

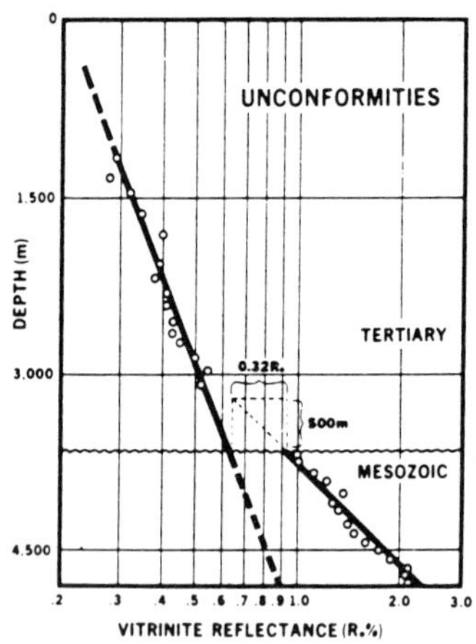

Figure 5.14. Vitrinite reflectance profile for an Indonesian well with an erosional unconformity (Dow, 1977). The slope of the R_o/depth curve for the Mesozoic (below the unconformity) is greater than in the Tertiary indicating a higher geothermal gradient.

for a rift setting. It showed an unconformity with R_o gradients below it corresponding to high heat flows in the rift stage, and lower gradients above the unconformity consistent with todays lower geothermal gradient. Tang and Lerche (1993) reported a vitrinite reflectance gradient of 0.25 $\%R_o$/km in the Tertiary section of the Parsons F-09 well in the Mackenzie Delta, but only 0.09 $\%R_o$/km in the Cretaceous section. Inversion of this data (Lerche, 1990) was used to obtain the geothermal gradients. Based on a maturation study in northern Canada, Link and Bustin (1989) suggested that vitrinite reflectance was related to geothermal gradient as follows:

Peel Plateaux	0.32 log%R_o/km	40-45 ^oC/km
Southern Mackenzie Delta	0.23 log%R_o/km	30-35 ^oC/km
Eagle Plain	0.18 - 0.29 log%R_o/km	20-35 ^oC/km

In the Ordos basin, China, Zhao et al (1996) related a profile of 0.36 %R_o/km to a paleogradient of 57 ^oC/km. When using these data it is important to remember that R_o values are also affected by time, and Dow (1978) published profiles from the U.S. Gulf Coast that varied from 0.065 %R_o/km for the Plio-Pleistocene to 0.375 %R_o/km for the Cretaceous.

Calibration: If vitrinite values are to be interpreted in terms of actual temperatures, then calibration is needed. It appears that temperature has the dominant influence in increasing R_o values (especially above 130 ^oC), and the effect of heating time is usually less significant than the "noise" in the experimental determination of R_o values (Barker, 1983; Price, 1983; Kalkreuth and McMechan, 1984; Barker, 1989). Since R_o is not reversible, reflectance values act like a maximum reading thermometer. Barker and Pawlewicz (1986) have developed an empirical calibration based on data from more than 35 different geologic systems with known temperatures and heating times ranging from 0.01 m.y. to >30 m.y. Their correlation data leads to the following equation:

$$\ln R_m = 0.0078\, T_{max} - 1.2 \qquad (5.1)$$

based on a least squares regression for over 600 values. T_{max} is the temperature exposure of the vitrinite and is not the same as the pyrolysis T_{max}. R_m is the mean vitrinite reflectance. The data suggest that about 70% (r^2=0.7) of the variability in the vitrinite reflectance values can be explained by considering temperature alone. However, it is important to note the wide range of values – at an R_m of 1.0 the corresponding temperature ranges from 60 ^oC to nearly 250 ^oC.

Equation 5.1 can be written as:

$$T_{max} = 128 \ln R_m + 154 \qquad (5.2)$$

which is slightly different from the equation proposed by Barker (1988), which is

$$T_{max} = 104 \ln R_m + 148. \tag{5.3}$$

The difference in T_{max} is 6 °C at R_m = 1.0, and 10 °C at R_m = 1.5. Equation 5.3 tends to underestimate temperature at low vitrinite reflectance values (<0.5%), and to overestimate it at high values (>2%) (Johnsson et al, 1993).

Laboratory calibrations using bomb experiments have been carried out by Bostick and Pawlewicz (1984) and the results are shown in Figure 5.15. Vitrinite reflectivities have also been calibrated against temperature by using T_h values from fluid inclusions in calcites that are thought to be at their maximum temperatures (Barker and Goldstein, 1990) (Figure 5.16). Some of the scatter in the data was due to the effects of geologic age, especially at low temperatures. Russell and Bone (1989) and Aizawa (1990) have also used fluid inclusion homogenization temperatures to calibrate vitrinite reflectance values. Aizawa obtained the relationship:

$$R_{max} = 0.352 \exp 0.00976 \times T \tag{5.4}$$

where T is the minimum value of the homogenization temperature. The relationship is applicable to Tertiary sedimentary basins.

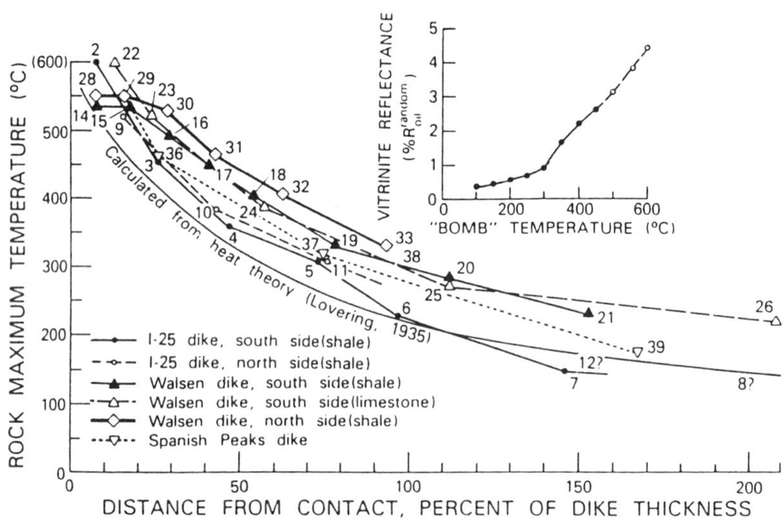

Figure 5.15. Vitrinite reflectance related to temperature by laboratory heating ("Bomb" temperature, Inset). These data have been used to estimate maximum temperatures at various distances from igneous dikes. The temperatures show good general agreement with temperatures calculated theoretically based on an intrusion temperature of 1230 °C and a country rock temperature of 30 °C (Bostick and Pawlewicz, 1984).

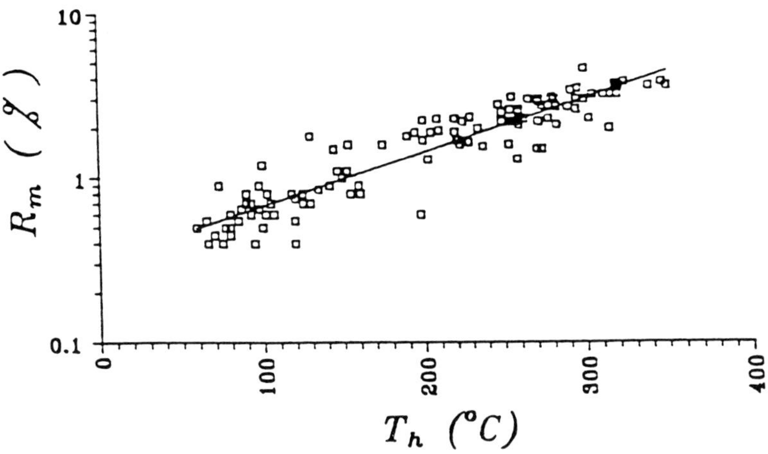

Figure 5.16. T_h of aqueous inclusions in calcite versus vitrinite reflectance (R_m). (From Barker and Goldstein, 1991)

Complications: Some of the difficulties encountered in using vitrinite reflectance as a maturity indicator were discussed in Chapter 1. These included faulting, unconformities, igneous intrusions, bitumen-impregnated vitrinites, and cavings. In addition the following potential problems need to be considered:

(1) There are many experimental problems in generating high quality reflectance data and it is important that the selection of vitrinite particles is made by an experienced operator (see Section 1.9.4). Poor polishing that leaves a pitted or scratched surface leads to unreliable results. Most measurements are made on kerogen separates, but it is possible to make determinations from polished rock samples. While these provide a smaller number of vitrinite particles to measure, their relationship to the rock matrix is less ambiguous and the data better (Barker, 1996).

(2) Vitrinite reflectance profiles do not always increase smoothly with increasing depth, even when unconformities, igneous intrusions, and recycled organic matter are absent. Law et al (1989) described "kinky" profiles from a variety of U.S. and Canadian basins where breaks in slope do not necessarily correspond to changes in lithology or changes in present day geothermal gradients. They argued that these profiles were due to perturbations of thermal gradients caused by the same processes that lead to abnormal and/or subnormal subsurface pressures.

(3) Localized thermal anomalies, such as igneous intrusions, may have modified the regional R_o trends. If well control is limited, and the

intrusion is not actually recognized or drilled into, the disturbed R_o profile may be mistaken as being typical of the region.

5.3.3. Fission Track Annealing Methods

When uranium-238 in a mineral decays by fission, two heavy particles are produced and they recoil in opposite directions with considerable energy (~200 MeV). This produces intense linear damage zones as the highly charged particles interact with the ions that form the mineral lattice (Figure 5.17). Damage *"tracks"* are approximately 14.0 - 15.6 μm long and tens of angstroms in diameter and are visible with an electron microscope. Suitable etchants will preferentially remove material from the damaged area and enlarge the tracks so that they can be seen with an optical microscope. Examples of photomicrographs are given by Gleadow et al (1983, 1986) and Green et al (1986). Because the number of tracks generated depends on the age of the mineral and the amount of uranium (which can be measured), fission track studies were used first for dating minerals (Fleisher, 1975: Wagner and van den Haute, 1992). However, it soon became clear that minerals exposed to elevated temperatures gave unreliable ages, because at the higher temperatures displaced ions in the damaged part of the mineral lattice (the track) diffuse

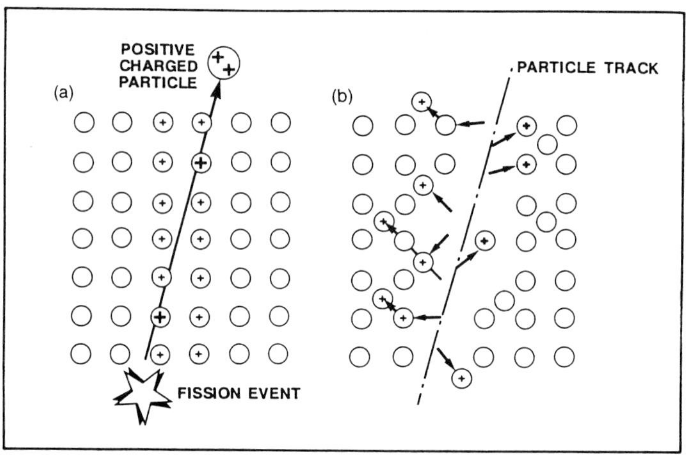

Figure 5.17. Illustration of the way uranium fission produces tracks in a crystalline solid. As a result of the fission event a positively charged particle passes through the lattice causing localized ionization (A). The newly charged ions repel one another and produce a line defect along the path of the charged particle (B) (after Naeser, 1979).

back towards their original lattice positions, and steadily reheal the damage in a process known as "*track annealing* " or "*track fading*". Given sufficiently high temperatures and enough time, the lattice damage will be completely annealed and no tracks will remain. Apparent ages will drop to zero (Figure 5.18). It is this temperature-dependent rehealing process that leads to the possibility of developing a geothermometer (Naeser, 1993). As lattice damage is thermally annealed track lengths are steadily reduced, and temperature history is deduced from the track length distributions.

Laboratory studies, supported by studies in nature, show that annealing rates have a different temperature dependence in different minerals. Tracks in apatite can be completely removed in 1 hour at 370 °C, while in nature temperatures over the range from 20-150 °C seem appropriate for ages of 1-100 m.y. An apatite sample exposed to temperatures of 110 °C for 10 m.y. will lose all its tracks by annealing, and O'Sullivan et al (1993) used the Burnham and Sweeney (1989) model to

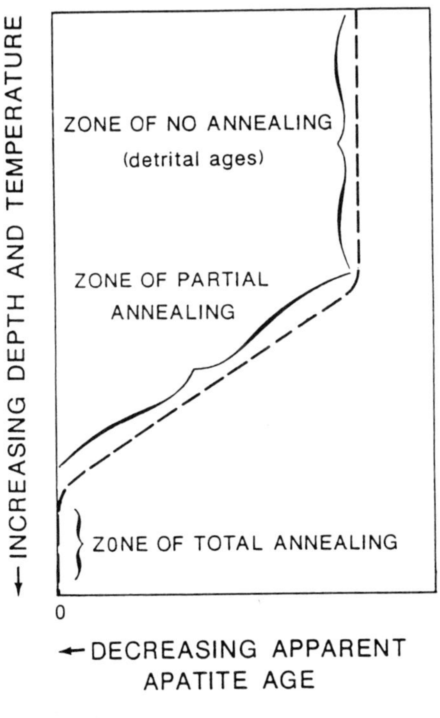

Figure 5.18. Increased amount of track annealing as temperature rises due to increased depth of burial. Apparent ages decrease to zero as tracks are removed. (Naeser, 1990).

calculate the corresponding R_0 values. They derived a value of 0.7%. Thus apatite annealing spans much of the temperature range of interest in petroleum generation studies, and for this reason apatite is the mineral most commonly used. Occasionally supporting data from zircon and sphene can be helpful because they anneal at somewhat higher temperatures. For other minerals annealing temperatures are too high for use in organic maturation studies. Since tracks are introduced continuously throughout the life of the sample, the oldest ones will have been influenced by the overall thermal history, while those formed more recently will only record the thermal effects from the time they formed up to the present. Thus fission track annealing offers the potential for inferring not just a temperature at a specific time, but the entire thermal history through the life of the studied mineral. Apatite Fission Track Analysis ("*AFTA*") has become an important tool for documenting past thermal histories (Gleadow et al, 1986; Green et al, 1989) and has been applied in more than 40 basins worldwide (see reference list in Naeser, 1993).

Detrital apatites for use in thermal studies of sedimentary sections are usually obtained from sands. The quantity of sandstone needed varies considerably, and for poorly sorted, immature, arkosic sandstones, where apatite concentrations are relatively high, as little as 50-100 g is usually enough. In contrast, it may be necessary to collect 2 kg or more if the sandstones are clean and well sorted. Samples can be composited as long as the depth range does not cover a temperature increment greater than 4 °C (which corresponds to 130 m for a geothermal gradient of 30 °C/km). 6 to 8 samples evenly spaced down a 3 - 4 km well are usually sufficient, but there is no point in collecting samples at temperatures above 120 °C. Fine to coarse sand (100-200 μm) is needed and Green et al (1989) have found that more than 80% of sandstones contain enough apatite for thermal studies.

An apatite concentrate of better than 80% is needed for track studies and is obtained using heavy liquid and magnetic separation techniques. The grains are mounted in epoxy, polished, etched, and counted. Because thermal interpretations depend on determining track lengths, only confined tracks can be used, and those truncated by the mineral surface have to be disregarded. If possible 100 confined tracks are counted in each of 20 grains. The uranium concentration is determined by irradiating the sample with thermal neutrons to induce additional track formation. For sedimentary rock samples an external detector mode has to be used to monitor the neutron flux, and a piece of muscovite or plastic is irradiated with the sample (Naeser, 1986; Gleadow, 1981).

Figure 5.19 shows how track length distributions reflect different thermal histories. In the case of Burial History A, apatites are not heated

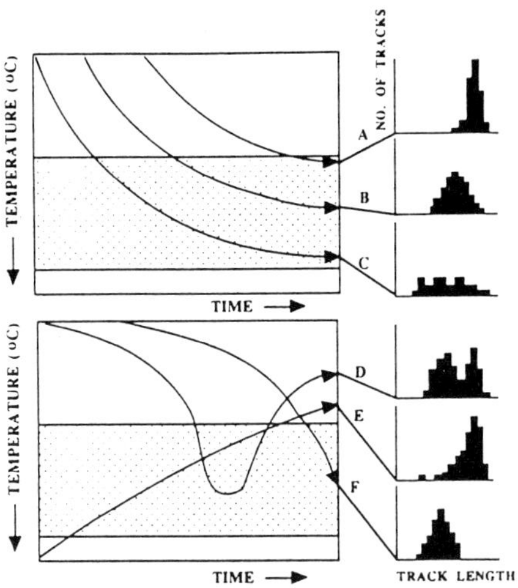

Figure 5.19. Temperature-time paths and resulting apatite track-length distributions for rocks that have experienced different thermal histories.
Top: Track distributions produced as a result of simple progressive burial of apatites to different levels in the zone of annealing.
Bottom: Track distributions for more complex burial histories including a past thermal event (D), slow cooling (E), and recent heating (F).
(From Gleadow et al, 1983).

enough for annealing to be significant, and there is a single length population centered around the unannealed track length of about 15 μm. For Burial History B, the apatites enter the temperature range for annealing earlier, resulting in many shortened tracks, but those tracks produced by recent uranium decays have had less time to shorten by partial annealing, and remain close to their original length. This results in a broader distribution of track lengths. The distribution is even broader for Burial History C where higher temperatures have been reached (due to greater burial) and older tracks have been heated for a longer time. In the more complex burial history represented by D, older tracks have been partially annealed, but the younger tracks formed after cooling have not been shortened at all, resulting in a bimodal track length distribution. Track length distributions shown for E, and F have been produced by slow cooling on uplift, and a recent heating event, respectively.

AFTA has wide applicability and can be used for most rock types. Suitable sample materials include cores and cuttings from boreholes, as well as outcrops and basement. It can be applied to red beds where organic maturity indicators are not present (e.g., Carter et al, 1995), but generally the AFTA techniques cannot be used with carbonates because they lack apatites. Kamp and Green (1990) used AFTA techniques to provide information on both the thermal and tectonic history of the southern part of the Taranaki basin, which is located between the North and South Islands of New Zealand. They collected 4 - 8 1 kg sand samples at evenly spaced depths in each of four wells. This involved compositing over ~50 m intervals. Apatites were separated, mounted, polished, and etched using standard procedures, and they attempted to measure one hundred confined, horizontal tracks in each apatite grain where possible. The AFTA data showed that only one of the four wells studied, Kupe, was still at its maximum temperature. Fast Pliocene-Holocene burial produced a very rapid rise in temperature, and fission track length, which takes about 10 m.y. to adjust, had not completely annealed to reflect present day temperatures. Uplift and erosion caused cooling in the other three wells –

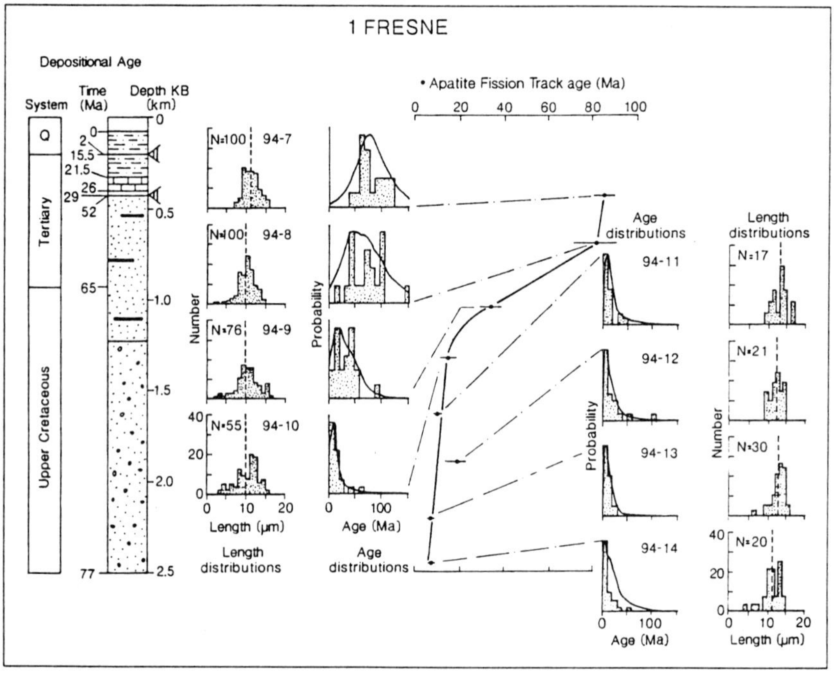

Figure 5.20. Apatite single-grain fission track age and track length distributions of samples from the 1 Fresne well, Taranaki basin, New Zealand (Kamp and Green, 1990). Error bars are ± one standard deviation of mean or pooled (as appropriate) apatite fission track ages. Track length distributions are normalized to 100 tracks.

by 80 °C in the case of Fresne, 40 °C in Surville, and 25 °C in N. Tasman corresponding to the removal of 3.0 $\pm$ 0.3 km, 2.0 $\pm$ 0.5 km, and 1.0 $\pm$ 0.3 km, respectively. AFTA showed that at the location of the Fresne well petroleum generation predated the formation of potential trapping structures, but in the Kupe area the traps were already developed when generation started. This area was considered to be much more prospective. Fission track and related data for the Fresne well are shown in Figure 5.20.

Other thermal studies using AFTA techniques have included those by Gleadow and Duddy (1984), Gleadow et al (1984), Moore et al (1986), Johnsson (1986), Green (1989), Hill and Gleadow (1989), Naeser et al (1990), Fitzgerald and Stump (1991), Roden (1991), Brown (1991), McCulloh (1993), Burtner et al (1994), and Burtner and Nigrini (1994), and Armstrong et al (1996).

Several factors can complicate the use of apatite fission track methods and need to be carefully evaluated. In detrital volcanic grains there may be multiple populations of tracks, and these must be distinguished so that only those relevant to the sedimentary environment are used. The presence of inherited tracks can be a major limitation in determining a unique thermal history for sedimentary sequences. Also, there are compositional variations in apatites, and tracks in chlorapatite appear to be more resistant to thermal annealing than those in fluorapatite and may anneal at temperatures as much as 30 °C higher (Green et al, 1986; Naeser, 1993). Hydroxyapatite resembles fluorapatite in its annealing characteristics. In their study of the thermal evolution of the Idaho-Wyoming-Utah overthrust, Burtner et al (1994) used compositional data obtained from electron microprobe analyses to divide their apatites into two groups with different chlorine contents. One had >0.5% Cl, and the other <0.5% Cl. By treating these "F-ap" and "F-Cl-OH-ap" groups separately rather than as a pooled set they improved their confidence in the thermal model developed for the area.

5.3.4. Biomarker Transformations

Introduction: The bitumen in rock extracts and oils contain a wide variety of compounds including the complex but characteristic molecules derived directly from organisms. Originally called chemical fossils, they are now generally known as biological markers or biomarkers (Mackenzie 1984a; Philp and Lewis, 1987; Waples and Machihari, 1991; Peters and Moldowan, 1992; Moldowan et al, 1992). As part of their biochemical processes organisms synthesize compounds with very specific structures. The ring numbering nomenclature and the conventions used to specify the 3-dimensional architecture of the biomarkers were described in Chapter 1 (see Figures 1.37 and 1.38). Many of the biomarkers have variable, but limited, thermal stabilities and change in structure and composition as they are buried with the sediments. These changes can be used to indicate maturity levels (Chapter 1; Figures 1.39, 1.40, and 1.41; Table 1.4), but can also be interpreted more specifically in terms of past thermal history.

Reactions: Although a wide variety of biomarker rearrangements, and degradations occur in nature very few satisfy the rigorous requirements for use as a thermal indicator. An ideal maturity parameter based on compounds *A* and *B* would derive from a reaction of the form:

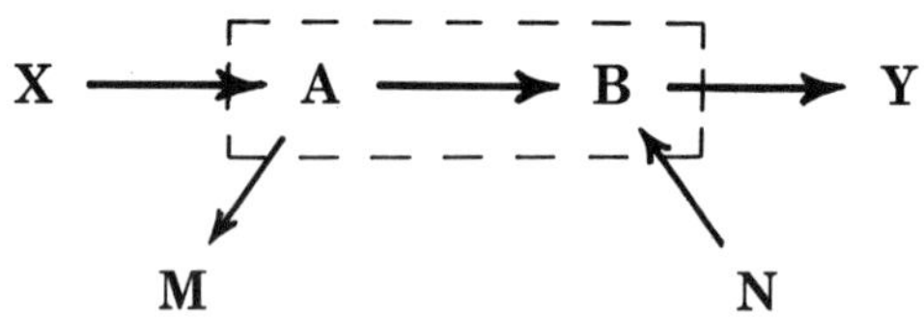

(5.5)

where *A* and *B* are single compounds. *A* should have no currently active precursor source (no formation from *X*); *A* should have no fate other than transformation to *B* (no conversion to *M*); *B* should have no source other than *A* (no formation from *N*); *B* should be stable (no conversion to *Y*); the reaction should not proceed to completion in less than the full range of maturity of interest (see Curiale et al, 1989). The additional side reactions do not pose an intrinsic problem, and the mathematical treatment would be straightforward. The main problem is one of documentation and availability of appropriate E and A values for each of the additional reactions. Analytical methods are well established and relative abundances of biomarkers in oils and rock extracts are routinely determined by GCMS. About 10-100 g of rock is needed to provide sufficient solvent extract for the analytical procedures.

The only three biomarker reactions that have been used so far involve sterane and hopane isomerization and a sterane aromatization:

(1) *C_{29}-sterane isomerization at C-20*: The orientation of groups around the C-20 carbon in the C_{29}-sterane leads to two possible configurations called R and S (see Chapter 1). The biochemically-produced sterols that are precursors to the steranes have only the R configuration, and this is inherited by the steranes. With increasing temperature the S configuration (which is of comparable thermal stability to the R configuration) is formed, and the ratio 20S/(20R + 20S), starts at zero and rises to an equilibrium value of 0.54.

(2) *C_{32}-homohopane isomerization at C-22*: This reaction is analagous to the sterane isomerization at C-20. Homohopanes start with R stereochemistry, but hydrogen exchange at the C-22 carbon leads to the

formation of increasing amounts of the S stereoisomer. The ratio 22S/(22R + 22S) rises from an initial value close to zero to an equilibrium value of 0.61 (Seifert and Moldowan, 1980).

(3) *Aromatization of C-ring monoaromatic sterane*: The conversion of steranes with one aromatic ring to those in which all 3 six-membered rings are aromatic is different from the isomerization reactions ((1) and (2)) in that it is a chemical reaction in which loss of hydrogen converts the monoaromatic steranes (MAS) completely to triaromatic steranes (TAS). The ratio TAS/(MAS + TAS) goes from 0 to 1.0. However, the reaction may be more complex than a simple reactant-product relationship because at higher maturity levels the value often drops below 1.0. Beaumont et al (1985) recommended caution in using data from aromatization reactions where the ratio is not between 0.2 and 0.8.

All three reactions appear to be first order and, like any other reaction, reaction rates and temperature dependency are controlled by the relevant E and A values. Initially E and A values were deduced solely from geologic studies (Mackenzie and McKenzie, 1983), but the introduction of hydrous pyrolysis provided a technique for measuring values in the laboratory (Ritter et al, 1993) (although there are still some uncertainties about the possible change of reaction mechanism at the higher temperatures). Figure 2.10 of Chapter 2 is an example of using the Arrhenius equation plot (ln k versus 1/T) to obtain E and A values for the North Sea and the Pannonian basin. The ratios of product/(reactant + product) for all three reactions can be determined to $\pm$ 0.04 (Mackenzie and McKenzie, 1983; Beaumont et al, 1985), and Suzuki has shown that this corresponds to $\pm$ 5 °C.

Applications: The use of biomarker reactions to establish paleo-temperatures was first suggested by Mackenzie and McKenzie (1983). They proposed using the three reactions discussed above, and showed that by combining them they could get information on both temperature and timing. They applied the techniques to extensional basins where temperature histories can be stated as a simple function of β, the McKenzie stretching factor. The activation energies they obtained for the two isomerization reactions were the same (91 kJ/mol), but the A values differed (Table 5.2). The monoaromatic sterane aromatization had a much higher E (200 kJ/mol) and in old basins, such as the North Sea, both isomerization reactions are almost complete before appreciable aromatization has occurred, whereas in young basins, such as the Pannonian Basin in Hungary, aromatization is almost complete before appreciable isomerization of the steranes has occurred. The progress of pairs of reactions can be plotted conveniently on diagrams that show sterane isomerization versus hopane isomerization (Figure 5.21); or sterane isomerization versus sterane aromatization (Figure 5.22); or hopane

Table 5.2
Estimates of kinetic parameters for biomarker reactions (Mackenzie and McKenzie, 1983).

REACTION	E (kJ/mol)	A (sec^{-1})	λ
Sterane isomerization			
at C-20	91	k_2 0.007	
		k_1 0.006	1.174
Hopane isomerization			
at C-22	91	k_2 0.025	
		k_1 0.016	1.564
Steroid hydrocarbon			
aromatization	200	k_2 1.8×10^{14}	

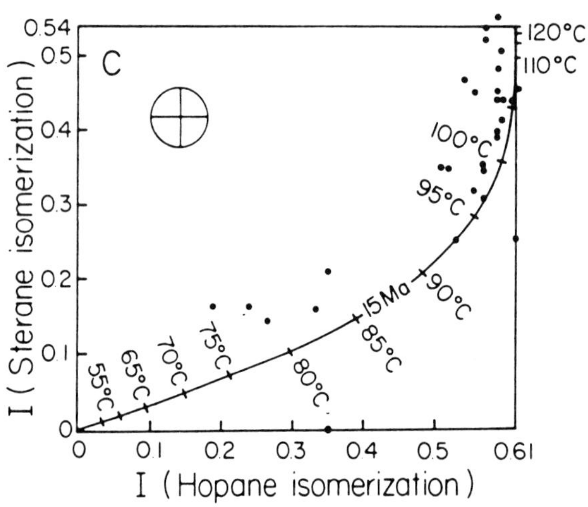

Figure 5.21. Predicted trend for sterane and hopane isomerizations using the selected thermal history (solid line) compared with some observed values from the Alberta foreland basin, western Canada (Beaumont et al, 1985).

isomerization versus sterane aromatization, and are called I-I, A-I, and A-I diagrams, respectively.

The ideal progress across the diagrams can be calculated for any heating model because this specifies the particular combinations of time and temperature. Actual data points can then be compared with the modeled conditions. Diagrams with a family of curves for different sets of conditions (ages, say) are commonly used (Figure 5.23). This is often a convenient way of evaluating the effects of uplifts. Rullkotter and Marzi (1989) extended the kinetic data that Mackenzie and McKenzie obtained geologically for the North Sea by adding hydrous pyrolysis values (Figure 2.10). The revised values for E's and A's were then used to evaluate the paleothermal regime in the Michigan basin, U.S.A. Beaumont et al (1985) used methods very similar to those of Mackenzie and McKenzie (1983) in their study of the Alberta foreland basin, western Canada, and the two A-I diagrams are shown in Figure 5.23. Based on this data they concluded that the distribution of paleogeothermal gradients has remained essentially unchanged from the Cretaceous to the present, with values decreasing systematically away from the foothills of the Rocky Mountains. They attributed this to recharge of meteoric water on the uplifted margin, and a consequently depressed thermal regime. After deep circulation and heating, the water then discharges and raises the geothermal gradient in that area. They showed that the temperature distribution through time was consistent with their lithospheric flexural model for the basin.

The study of the northern Newark basin of the northeast U.S. by Walters and Kotra (1990) provides a good example of the way biomarker data can be used to help decide among alternative models in a basin where paleoheat flows, timing of uplift, and the thickness of sediments removed by erosion are poorly constrained. The approach used was similar to that suggested by Mackenzie and McKenzie (1983), and relied on sterane and hopane isomerizations and sterane aromatization to limit possible thermal histories. They plotted calculated biomarker trends as a function of depth for a variety of burial histories to develop a family of lines that could be compared with measured biomarker ratios. By varying the thicknesses of sediments lost during the main erosional event they concluded that the best fit between experimental data and the theoretical curves corresponded to 2400 - 2800 m of rock eroded. The modified burial history was combined with fluid inclusion temperature data to confirm their preferred geothermal gradient of 25 ^{o}C/km.

A simplified model using only sterane epimerizations was proposed by Suzuki (1990) for use in basins where variations in both the burial rates and geothermal gradients are low. He applied it in the Niigata basin, Japan. Figure 5.24 shows ideal trends in sterane epimerization ratios (20S/(20R + 20S)) for heating rates ranging from 0.1 to 1000 ^{o}C/m.y.

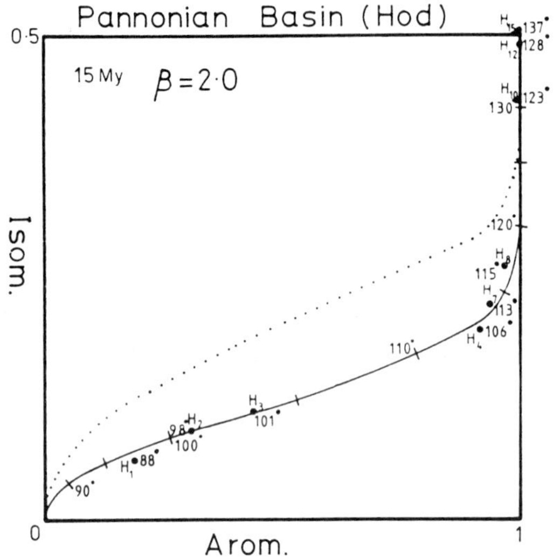

Figure 5.22. Predicted trend for sterane isomerization and monoaromatic sterane aromatization in the Hod-1 well, Pannonian basin, Hungary (solid line). Present day temperatures are shown, and measured biomarker ratios are given ("H"). (Data from Sajgo and Lefler, 1986).

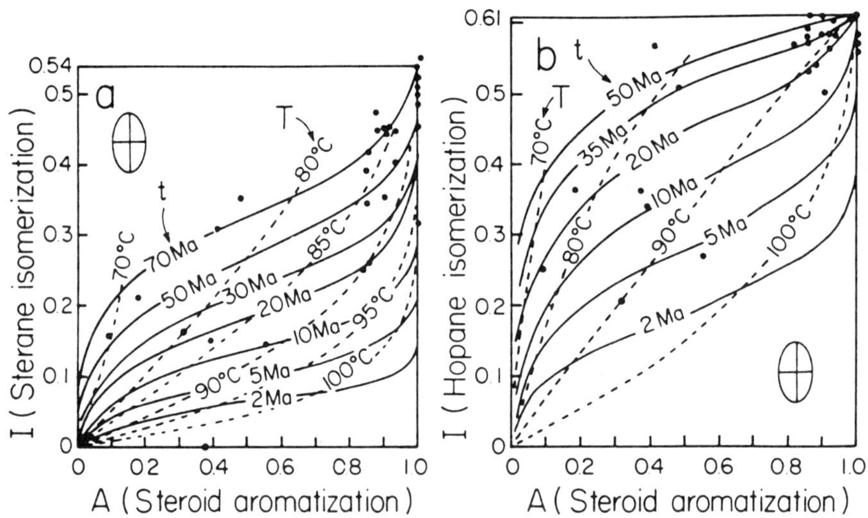

Figure 5.23. Families of calculated curves for sterane isomerization/ monoaromatic sterane aromatization (a), and hopane isomerization/ monoaromatic sterane aromatization (b) for selected temperatures and geologic ages. Rock data obtained in the Alberta forehand basin are superimposed (Beaumont et al, 1985).

calculated using E and A values from Alexander et al (1986). Ratios measured in the Yasuda R-7 well are consistent with a heating rate of about 40 °C/m.y. For these very young rocks (with a maximum temperature that was reached in the Quaternary) a value of 0.1 for 20S/(20R + 20S) corresponds to a maximum temperature of about 100 °C, and a value of 0.2 to about 115 °C.

As part of their study of paleoheat flux in the North Sea, Armagnac et al (1988) used both biomarker transformations and vitrinite reflectance. Based on data presented for a single well, together with data from more than 40 proprietary well data sets, they recommended activation energies of 50 kJ/mol and 20 kJ/mol for the sterane and hopane isomerizations,

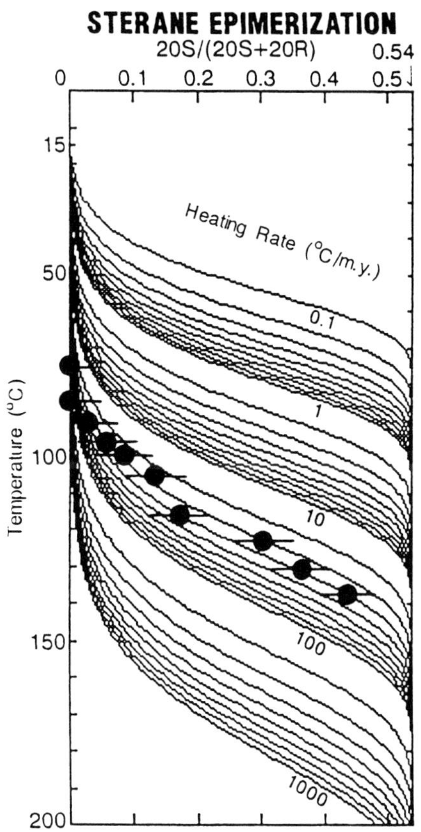

Figure 5.24. Graph showing the relationship between the degree of sterane isomerization and maximum temperature at various heating rates (Suzuki, 1990).

respectively. No A values were obtained by their techniques. Their E values are considerably different from those used by Mackenzie and McKenzie (1983) and may have been influenced by the fact that the isomerizations did not reach equilibrium, with sterane ratios having a maximum value of only 0.3.

Other studies using biomarker ratios to estimate paleotemperatures have included those by Suzuki (1984); Hong et al (1986) in the Linyi Basin, China; Mackenzie et al (1985) for the Nova Scotia continental margin, and Issler and Beaumont (1987) for the Labrador and West Greenland continental margins; Mackenzie et al (1988) in NW Germany; and Marzi et al (1990). Hermansen (1993) combined biomarker data with R_o data, and Ritter et al (1993) combined it with NMR aromatization parameters.

Limitations/problems: Biomarker methods are applicable only up to about $R_o = 1.0$ because at higher maturity levels reactions are complete and no further changes occur. Also, above this maturation level the biomarkers are reduced to very low concentrations or are absent. Even at low maturity levels some rocks may not have sufficient quantities of biomarkers for accurate analyses. It is important to establish that the biomarkers are indigenous to the sediments being extracted and have not been depleted by migration out, or contaminated by migration in. The best samples for biomarker analysis are those taken from low permeability, organic-rich cores (Beaumont et al, 1985). Overpressured shales are less likely to be contaminated by material moving in. RockEval pyrolysis often provides a useful screening technique and gives an anomalously high S1 (bitumen peak) in contaminated samples. Note that (1) drilling muds frequently contain organic additives and oil-based muds are commonly used in some areas, (2) mineral environment (catalysis) may effect rates of transformation (Strachan et al, 1989), and (3) in outcrop samples weathering can change biomarker ratios (Clayton and King, 1987).

Biomarker kinetic data may be unreliable or not available. Peters and Moldowan (1991) and Wavrek (1992) have claimed that the sterane isomerization is more complex than generally assumed. Abbott et al (1990) have questioned the basic assumption that the C_{29}-sterane isomerization at C-20 is in fact a single, thermally-controlled reaction, and have presented data which suggest that two steps are involved. First the sterane is stripped from the kerogen with a rather "immature" S/R ratio (X to A, Equation 5.5), and then is later matured as part of the bitumen. Both steps are first order and Abbott et al (1990) have derived appropriate kinetic values.

5.3.5. Clay Transformations

Recent sediments often contain unstable detrital phases, and these convert to more stable materials as temperature rises with increasing depth of burial (Weaver, 1989). Some clay transformations are especially important because they occur in the temperature range where petroleum is being generated, and since clays are a major constituent of shales, paleotemperature methods based on their conversions should have wide applicability. Of particular interest is the conversion of smectite, first to mixed layer smectite/illite, and finally to illite. This has been described by Perry and Hower (1970, 1972), Hower (1976), and many others. This conversion also releases water and may be significant in generating abnormal subsurface pressures (Burst, 1969; Bruce, 1984).

The precise nature of the material identified as mixed layer illite/smectite ("*I/S*") by XRD, and the details of the mechanisms that convert it ultimately to illite, are still imperfectly known (Essene and Peacor, 1995). Even evidence on the role of time is conflicting. Weaver (1979) concluded that time had no influence. Based on studies of young sediments from the Colorado River delta in southern California, Jennings and Thompson (1986) suggested that the reaction is only kinetically controlled up to 175 °C, and that above this it is controlled solely by temperature.

The progress of the smectite-to-illite conversion is irreversible and acts like a maximum reading thermometer. The conversion has been empirically related to temperature, and Hoffman and Hower (1979) were among the first to suggest that clay minerals could serve as geothermometers (Figure 5.25). Now the most widely used reaction is the reduction in the number of smectite (expandable) layers in the I/S structure and the parallel changes in degree of ordering. Although detailed kinetic models are available to describe the reactions (Pytte and Reynolds, 1989), the simplified scheme introduced by Hoffman and Hower is often used. This shows that the major change from random orientation (R = 0) to short range order (R = 1) occurs in the temperature range from 100 - 110 °C for Tertiary to Cretaceous age sediments. This roughly overlaps the range of petroleum generation (Pollastro and Schmoker, 1989). For the short heating times typical of hydrothermal environments and very young basins (Jennings and Thompson, 1986; Ramsayer and Boles, 1986), the reduced time seems to require somewhat higher temperatures. Figure 5.26 shows changes in the much younger Colorado delta sediments lagging behind those for Gulf of Mexico mid-Tertiary clays at any given temperature. Above about 175 - 180 °C the conversion from R = 1 to R = >3 (long range order) is similar in both age ranges. There is evidence for a dual, serial reaction mechanism for the conversion of smectite to illite. Vasseur and Velde (1993) have studied samples with widely ranging time-temperature

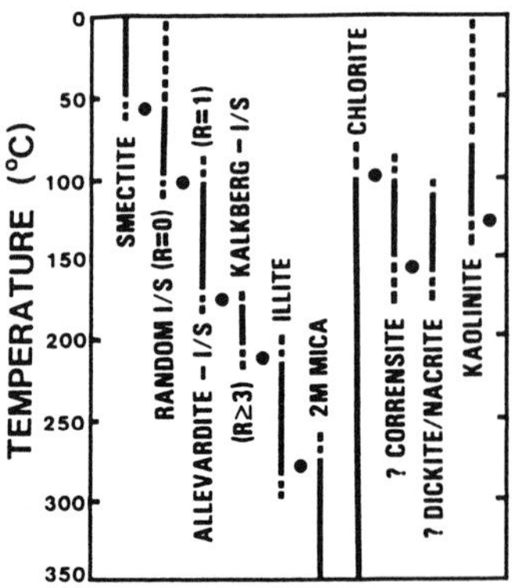

Figure 5.25. Ranges of temperatures for the stabilities of common clay minerals and the ordering of mixed-layer illite/smectite (I/S) in diagenetic and low-grade metamorphic settings (Hoffman and Hower (1979), modified by Pollastro (1990)). Solid circles indicate approximate phase-limiting or phase-transitional temperatures.

histories and treated them as the products of a two-stage reaction with a first step that leads to the formation and growth of a disordered I/S structure (R=0), and a second step where the smectite layers disappear as the ordered form develops (R = 1). The first step terminates at about 50% smectite. Both steps were considered to be first order, and Vasseur and Velde (1993) computed E and A values for each of them.

The increase in order from random I/S (R = 0) to the ordered form (R = 1) often occurs over a very short depth interval and can be used to constrain the 100 - 110 °C isotherm. Commonly in exploration the variation of I/S order, or percent illite layers in the mixed layer I/S, is determined as a function of depth for the vertical section penetrated by the wells, and then used to establish temperature. In the Mancos shale of the southern Rocky Mountains and the Colorado Plateau, Nadeau and Reynolds (1981) used this approach to show that I/S ordering correlated well with coal rank, while Pollastro and Scholle (1986) have successfully used the method to study clays in chalks. Figure 5.27 shows data generated by Pollastro and Barker (1986) for the Wagon Wheel well, Pinedale Anticline, Wyoming. The present day gradient obtained from temperature measurements in the well is parallel to the geothermal gradient derived from both vitrinite reflectance and I/S (which show excellent agreement). However, the present temperatures are lower than the paleotemperatures, and this implies considerable uplift and cooling. In the western Canadian basin of Alberta, Connolly (1989) used the smectite-illite conversion as a way of estimating maximum temperatures, and

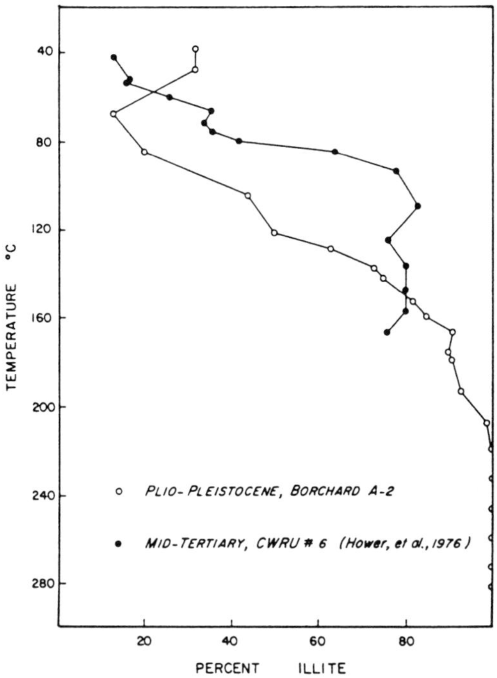

Figure 5.26. Comparison of percent illite layers in the <0.1 mm fraction of samples from the Plio-Pleistocene Borchard A-2 wells with the Mid-Tertiary CWRU No. 6 well (Howard et al, 1976). Data are from Jennings and Thompson (1986) and are plotted against logged well temperatures.

compared this information with the temperatures derived from vitrinite reflectance. Although values derived by both techniques were generally consistent, when there were differences the temperatures derived from vitrinite reflectance were lower by as much as 10 °C

Clay transformation studies have been carried out in the North Sea and have included those by Pearson and Small (1988) in the Viking Graben, and Glassman et al (1989b) in the Bergen High. In the Viking Graben ordering of illite occurs at a mean temperature of 93 °C and a mean vitrinite reflectance of 0.64%. On the Bergen High the current bore-hole temperatures in the upper part of the wells show reasonably agreement with the temperature gradient based on the temperature of disappearance of smectite (70 °C), the first occurrence of R = 1 I/S (100 °C), and the first occurrence of R = 3 I/S (180 °C). Calculated and measured gradients do not agree as well for the deeper part of the section.

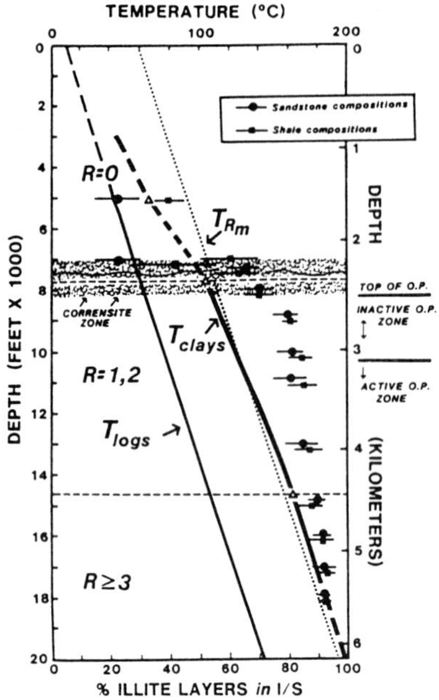

Figure 5.27. Percent illite layers and ordering (R) of mixed-layer illite/smectite (I/S) in sandstone and shale versus depth to sample in the El Paso Natural Gas, Wagon Wheel No. 1 well, Pinedale area, Wyoming. The present day temperature curve determined from logs (T_{logs}); the maximum burial temperature curves determined from I/S ordering (T_{clays}); and vitrinite reflectance (T_{Rm}) profiles are superimposed. Zones of active and inactive overpressuring (O.P.) are shown. (From Pollastro, 1990).

The conversion of smectite to illite requires potassium and it is usually assumed that this is made available from detrital orthoclase. In the Gulf Coast the abundance of potassium feldspar decreases with depth and it is absent below 3500 m. The rate of increase of illite layers in the I/S is greatest over the depth range in which the feldspar disappears (Hower et al, 1976).

Macchi et al (1990) studied clays in the Morecombe Bay gas field, Irish Sea, and suggested that Gulf of Mexico data/models for progressive illitization may not be applicable in other areas. Anjos (1986) reached a similar conclusion in the Campos basin of Brazil and attributed it to the presence of iron-rich smectites. Although temperature seems to have the dominant role in smectite conversions, geologic age, pore fluid composition, and the composition of the mineral reactants many also be involved. Elliot et al (1991) studied the I/S conversions in the Denver basin and concluded that although smectite converted to I/S mainly in response to increasing temperature that accompanied increasing depth of burial, there were local factors such as solution chemistry, mineralogy of adjacent shales, and hydrologic and tectonic factors.

Illite crystallinity (defined in terms of the width at half height of the 10 angstrom illite XRD peak) and illite polytype distribution have been used to help reconstruct paleotemperatures (Guthrie et al, 1986; Kisch, 1987; Yonkee et al, 1989; Wilcoxon et al, 1990).

Practical considerations: Short range variations in I/S ratios are not uncommon and it is important to sample a range of lithologies. Cores are the preferred samples because they eliminate the possibility of contamination by cavings (Freed and Peacor, 1989), but cuttings and outcrop samples have been used successfully. XRD is used to charaterize oriented, ethylene glycol-saturated specimens. Ordering in I/S is determined by peak positions in the low angle region, and composition estimated from the angular position of the I/S diffraction peak near 17° 2θ. While the <2 μm fraction is commonly analyzed better results are obtained from the 0.5 μm or less fraction (Pollastro, 1990). In general temperature determination works best for rocks with ages between 10 and 350 Ma and is less predictable for very young and very old rocks (Pollastro and Schmoker, 1989).

5.3.6. Miscellaneous Methods

Electron Spin Resonance (ESR): The electron spin resonance technique has not been widely used, but has potential for application to all types of organic matter (Marchand and Conard, 1980). This technique determines the concentration of free electrons in the organic matter that were produced as *"free radicals"* when chemical bonds were thermally broken. While small free radicals quickly react, larger molecules (particularly those containing fused aromatic rings) stabilize a free electron by spreading it over the whole molecule (*"delocalization"*) and it can survive. The higher the temperature experienced by the organic matter, the more bonds that were broken, and the greater the number of free electrons in the organic matter. Stabilization of electrons depends on organic matter type, and the technique requires calibration for each type of organic matter. Few examples have been published since Pusey's original paper (Pusey, 1973), but esr has been applied to the northern North Sea basin by Cooper et al (1975), who inferred that maximum paleotemperatures were reached in the Tertiary.

$^{40}Ar/^{39}Ar$ Methods: When the temperature of a mineral rises any argon that has been generated from the decay of potassium-40 begins to diffuse out of the host mineral lattice, and the rate of this diffusional loss is temperature dependent. Harrison and Be (1983) and Harrison and Burke (1988) have used this to develop a technique of *"thermochronology"* for microcline. Figure 5.28 shows that the most easily depleted argon sites have lost radiogenic argon (and so give low apparent ages), while argon is still trapped in the most retentive sites. The sample from 6200 m shows a much greater loss of labile argon than the shallower sample from 3725 m.
Oxygen isotope methods: Because isotopes of an element have slightly different physical properties the distribution of the ^{16}O and ^{18}O isotopes among coexisting oxygen-bearing phases is temperature dependent. The deviation from a uniform composition (i.e., the difference between the

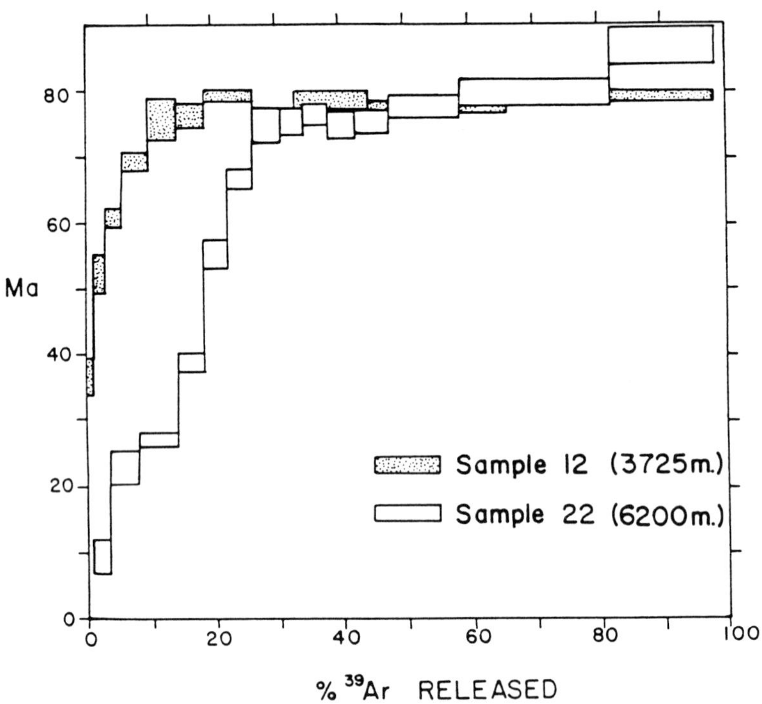

Figure 5.28. Distribution of apparent $^{40}Ar/^{39}Ar$ ages within two crystals of detrital microcline from deep drill cores taken in the southern San Joaquin Valley, California. The sample from a depth of 6.2 km shows evidence of a recent thermal episode that has caused loss of radiogenic ^{40}Ar (Harrison and Burke, 1988).

$^{18}O/^{16}O$ ratios for two phases) is known as "*fractionation*". For a pair of minerals, A and B, in isotopic equilibrium, a "*Fractionation Factor*" can be defined by the expression:

$$\alpha_{AB} = [(H/L)_A] / [(H/L)_B] \quad (5.6)$$

where α_{AB} is the fractionation factor between the minerals A and B, and L and H are the light and heavy isotopes, respectively. With increasing equilibration temperature the value of α decreases towards 1.00, and it is this temperature dependence that offers the possibility of using a pair of oxygen-bearing minerals as a geothermometer (Faure, 1986). Fractionation of oxygen isotopes between a mineral and its aqueous growth medium is temperature dependent and can be expressed by a relationship of the form:

$$10^3 \ln \alpha_{\text{min-water}} = A \times 10^6 T^{-2} + B \quad (5.7)$$

where α is the fractionation, A and B are constants, and T is in Kelvins. For the dolomite-water system Land (1985) showed that the temperature dependence of the fractionation α is given by:

$$10^3 \ln \alpha_{\text{dolo-water}} = 2.78 \times 10^6 T^{-2} + 0.91 \quad (5.8)$$

and Friedman and O'Neil (1977) found that for the calcite-water system the relationship was:

$$10^3 \ln \alpha_{\text{calcite-water}} = 2.78 \times 10^6 T^{-2} - 2.89. \quad (5.9)$$

Data are also available for quartz, siderite, chlorite, ankerite, kaolinite and phosphates. (Some other mineral pairs are discussed later under "Paleosurface Temperatures", Section 5.5.5.) Since the fractionation α is a difference between the isotope values of the mineral and water, the growth temperature of the mineral cannot be obtained without knowing the $\delta^{18}O$ for the water. Most studies using oxygen isotopes have considered several mineral-water pairs and have often combined the information with fluid inclusion homogenization temperatures to try and develop a consistent temperature picture (Kupecz and Land, 1991; Qing and Mountjoy, 1992; Martin et al, 1994)

In order to use the partitioning of stable isotopes in quartz-illite (or I/S) to establish paleotemperatures in shales (Weaver et al, 1984) it must be possible to separate at least two coexisting minerals from the shale. Also, the magnitude of the isotopic fractionation between the minerals, and the manner in which it varies with temperature, must be known. The minerals must have reached isotopic equilibrium with one another at the maximum temperature to which the sample was exposed and they must not have undergone subsequent (retrograde) isotopic exchange (Eslinger et al, 1979). The accuracy of the quartz-illite/smectite geothermometer is probably no better than ± 15 °C when the equilibration temperature is 40 °C, and drops to ± 30 °C when the equilibration temperatures rises to 150 °C. Over the range from 40 - 150 °C an error in smectite expandability of 10% would lead to an error in temperature of about 10 °C (Eslinger et al, 1979).

Silica transitions: Opal-A occurs as unaltered biogenic siliceous remains of diatoms, radiolarians, and sponge spicules. It is X-ray amorphous and metastable, and with the increasing temperature that accompanies burial diagenesis it transforms to stable microcrystalline quartz. This transition proceeds via an intermediate metastable stage called opal-CT that is disordered α-cristobalite and α-tridymite (Mitsui and Taguchi, 1977; Calvert, 1983; Laschet, 1984; Keller and Isaacs, 1985; Iijima, 1988). The principle d-spacing in the opal-CT decreases until quartz forms, and these values have been calibrated as a function of temperature for the Miocene Monterey Formation California by Keller and Isaacs (1985). They found that the temperature of transition from opal-A to quartz varied with the percent of biogenic silica in the rocks and their diagram is reproduced here as Figure 5.29. In addition, Kastner et al (1977) have found that the rate of conversion of opal-A to opal-CT is much higher in carbonates than in clay-rich sediments. Pisciotto (1981) related opal-quartz changes to burial histories for various locations in the Monterey Formation of California.

5.4. CALCULATION FROM THERMAL CONDUCTIVITY AND HEAT FLOW

The methods discussed so far have all relied on measuring some present day value of a temperature-dependent process to provide a paleotemperature at a particular time in the past. An alternative approach to the problem of constraining past temperatures is to combine heat flow with values for thermal conductivity to calculate a geothermal gradient and hence temperature at any depth. The relationship between these terms is given by Equation 4.1 and has been discussed in Section 4.1.1.

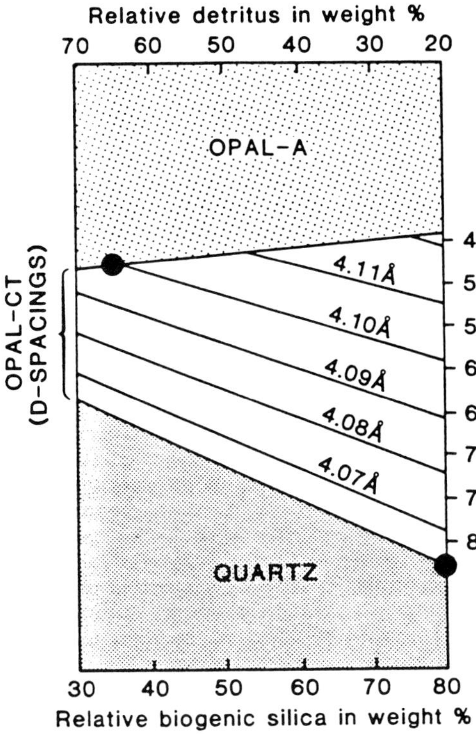

Figure 5.29. A preliminary temperature scale for silica diagenesis in the Monterey Formation derived by Keller and Isaacs (1985). The combined effects of time and temperature are shown. Temperatures of two points of silica-phase transformation (black dots) in the Point Conception COST well were used to calibrate temperature for this empirical silica-phase zonation.

Heat flow data can be measured at the surface today or past values can be inferred from a knowledge of basin evolution (e.g., Jarvis and McKenzie, 1980). Paleogeothermal gradients for the area of the Tartan Field in the North Sea were estimated by Burley et al (1989) using the method described by Jarvis and McKenzie (1980). A thermal conductivity of 2 W/m.°C was used since this is appropriate for a sandstone with 20 - 30% porosity. Calculations for an extension model that involved stretching over 60 m.y., and a stretching factor (β) of 2, lead to a peak geothermal gradient of 45 °C/km during crustal extension. This was followed by a gradual decrease to 25 °C/km, which is somewhat lower than the measured value of 32 °C/km. An alternative approach using fluid inclusion-derived temperatures for quartz cements deposited simultaneously on high and low blocks gave a geothermal gradient of 34 °C/km.

Inferred paleoheat flow depends on the thermal model used to explain basin development. For example, Nunn et al (1984) attribute the formation of the Michigan basin, USA, to a thermal anomaly that developed about 450 Ma, and then decayed. Initially heat flow was approximately 1.8 hfu but steadily decreased to the present value of 1.1 hfu. This model allowed them to calculate heat flow through time and hence to establish temperature at any depth. They consider the temperature distribution in the subsiding basin to be made up of two parts. One is due to the postulated thermal anomaly and is called "excess temperature", while the other is the equilibrium temperature after the lithosphere has cooled.

Roufosse (1987) used detailed tectonic histories to calculate heat flow variations through time for the western Barents Sea. Projected values ranged from 1.0 to about 1.7 hfu and showed three pulses corresponding to the Hercynian/Uralian, Kimmerian, and Laramide orogenies. Other examples are included in Chapter 10. Cao and Lerche (1988) have developed a method based on inversion of vitrinite reflectance data to provide paleoheat flow values. Using this technique for the Navarin Basin COST No. 1 well in the Bering Sea showed a generally warming trend from 1 hfu at 70 Ma to 1.75 hfu at present. In contrast the rift setting of the Canning Basin, Australia has a present day heat flow of 1.4 hfu compared with a calculated value of 4.0 hfu at 375 Ma. This high paleoheat flow is consistent with rifting.

Use of heat flow and thermal conductivity to calculate temperature gradients assumes that heat is transferred only by conduction and that the system is in thermal equilibrium. Locally both of these assumptions may not be valid. Figure 5.30 shows a schematic illustration of the changing thermal regime caused by an instantaneous deposition of a thick sediment layer (Vik and Hermanrud, 1993). After sediment deposition the system readjusts towards a new thermal equilibrium by heating, not only the recently deposited sediments but also the whole sediment pile. The effect is to produce a reduced heat flow until the new equilibrium is established. Present day nearsurface values may be perturbed by seasonal (or longer) climatic fluctuations, and, locally, by man-made structures such as tunnels.

It is relatively easy to put thermal conductivity values for each rock layer into a model as subsequent sedimentary layers are added. Commonly default values for thermal conductivity are used, but current values can be measured directly (Brigaud et al, 1990) (see Section 4.1.3). However, thermal conductivity depends on lithology, and for any given rock type will vary with depth as compaction changes the water/matrix ratio. This implies a steadily changing geothermal gradient with depth for a

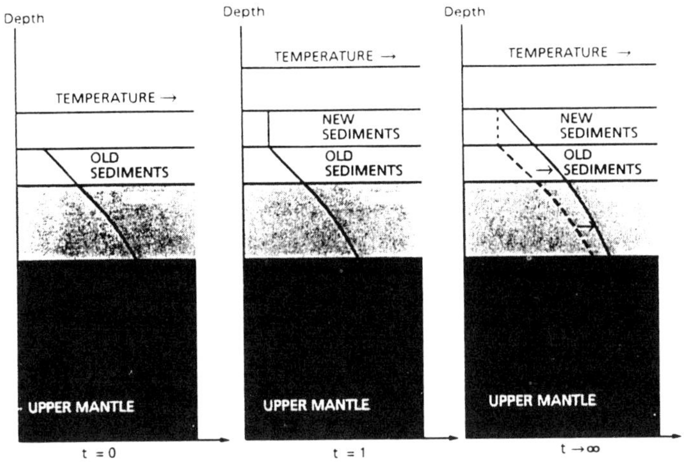

Figure 5.30. Schematic illustration of the changing thermal regime caused by an instantaneous deposition of a thick sedimentary layer (Vik and Hermanrud, 1993).

compacting rock. In addition, a variety of empirical relationships have also been developed, such as the one by Evans (1977) for the North Sea which relates thermal conductivity (K) to porosity (%), sonic velocity (V, m/sec) and bulk density (d, g/ml^3):

$$K = -0.0499 - 0.160V + 3.60d - 5.50. \tag{5.9}$$

In making these calculations it is important to include heat generated within the rocks by the processes of radioactive decay. The amount of heat is lithology dependent so that quantities will vary as the sediment pile accumulates (or is eroded) (Rybach and Cermak, 1982; Deming and Chapman, 1989). In the Provencal Basin radiogenic heat accounts for 33 mW/m^2 of the total heat flow of 67 mW/m^2 (Burrus and Audebert, 1990). The basin is in the western Mediterranean and forms part of the southern end of the West European rift system. In the prerift sediments the Upper Crust radiogenic heat production contributes 2 mW/m^3, and the Lower Crust 0.9 mW/m^3. With 20 km and 15 km of equivalent radiogenic thickness for the Upper and Lower Crust, respectively, Burrus and Audebert (1990) used data from Ryback and Cermak (1982) to calculate radioactive heat generation and its contribution to the overall heat flow. Keen and Lewis (1982) measured the concentration of potassium, uranium and thorium, the heat producing radioactive elements, in sediments from 14 deep wells on the eastern Canadian margin. Calculated heat production ranged from 0.3 to 1.8 mW/m^3. This increased heat flow by 15% and temperatures were raised by as much as 17 °C. The impact of this additional heat on petroleum generation was evaluated and showed that generation would occur earlier and/or shallower.

When using heat flow to calculate geothermal gradient, the rock types in the sedimentary column must be known, but in situations with extensive uplift and erosion there may be no remaining record of the rock types removed. Erosion removes rocks of higher porosity, so that the remaining sediments have a lower average porosity and higher average thermal conductivity, which for a constant heat flow leads to a lower calculated geothermal gradient.

5.5. PALEOSURFACE TEMPERATURES

Present day geothermal gradients are commonly calculated using bottom hole temperature together with the surface temperature. Although temperatures at the surface can fluctuate widely throughout the year, these variations are rapidly damped with depth, and by about a few hundred meters are effectively controlled by mean surface temperatures. In reconstructing paleogeothermal gradients it is necessary to estimate mean surface temperatures and often this is not easy to do. Climate and proximity to the equator provide a first estimate, while rock types, fossil assemblages, and isotopic ratios can all provide useful constraints.

5.5.1. Latitude Variations

Climate reconstructions throughout the geologic record (Habicht, 1979; Crowley and North, 1991; Frakes et al, 1992) provide an overall framework for estimating surface temperatures. When coupled with paleolatitude based on plate tectonic reconstructions, this can give a first approximation to surface temperature. For example, in the last 125 m.y. India has moved from near the South Pole, across the equator, to its present position, producing fairly predictable changes in average surface temperatures. Using climate trends with plate reconstructions for specific locations is complicated by uncertainties about year-round climate compared with purely seasonal fluctuations, which can be quite substantial. In general climate reconstructions get less reliable as time before present increases.

It appears that some time periods were significantly warmer than at present, and the mid-Cretaceous (covering 90-120 Ma) is a time when there may have been no ice and temperatures were warmer than now. This conclusion is based on evidence from coral reef distributions, the locations of both invertebrate and vertebrate fossils, and oxygen isotope data. Also, the paleotemperatures deduced from oxygen isotope data given in Figure 5.31 show that in the early Eocene (50-55 Ma) temperatures reached their highest Cenozoic values. This is consistent with the formation of lateritic soil horizons that extend to 45° N and S (Frakes et al, 1992). Their presence implies warm temperatures with seasonal rainfall. Paleosols have also been a useful source of information on paleoclimate (Mack and Jones, 1994; Mack et al, 1994).

Figure 5.32 summarizes the distribution of temperatures with latitude for glacial and interglacial conditions as well as for the present day. It shows that equatorial temperatures do not change much, while in contrast temperatures at high latitudes show sizable variations that can be significant in thermal modeling.

5.5.2. Geographic Setting

While most sediments are deposited in marine environments, some important source rocks have developed in a terrestrial setting, especially in lakes. Temperatures on land may be very different from those in the marine environment at the same latitude, although it is often assumed that only rocks deposited relatively close to sea level are likely to be preserved in the geologic record. In some situations the pollen record provides information about local vegetation, and this can be useful in putting limits on possible temperatures. Mid-Eocene sediments in Antarctica, for example, show a beech forest with ferns, consistent with a temperate climate. Snail shells, periglacial features, oxygen and hydrogen isotope ratios in lake sediments, soil carbonates, cave deposits, tree rings, and ground water have all been used to constrain paleotemperatures, especially those in the recent past (see Stute et al (1992) for selected references). The temperature dependence of the solubility of noble gases in water (particularly argon, krypton and xenon) has been developed into a useful technique for establishing paleotemperatures in the last 0.5 m.y., or so. The water equilibrates with gases in the atmosphere in the unsaturated soil zone. This water then becomes isolated from the atmosphere as it moves through the aquifer and retains the surface temperature information. The principle of the method has been known for 20 years and Stute et al (1992) have recently applied it to the Carrizo aquifer in Texas.

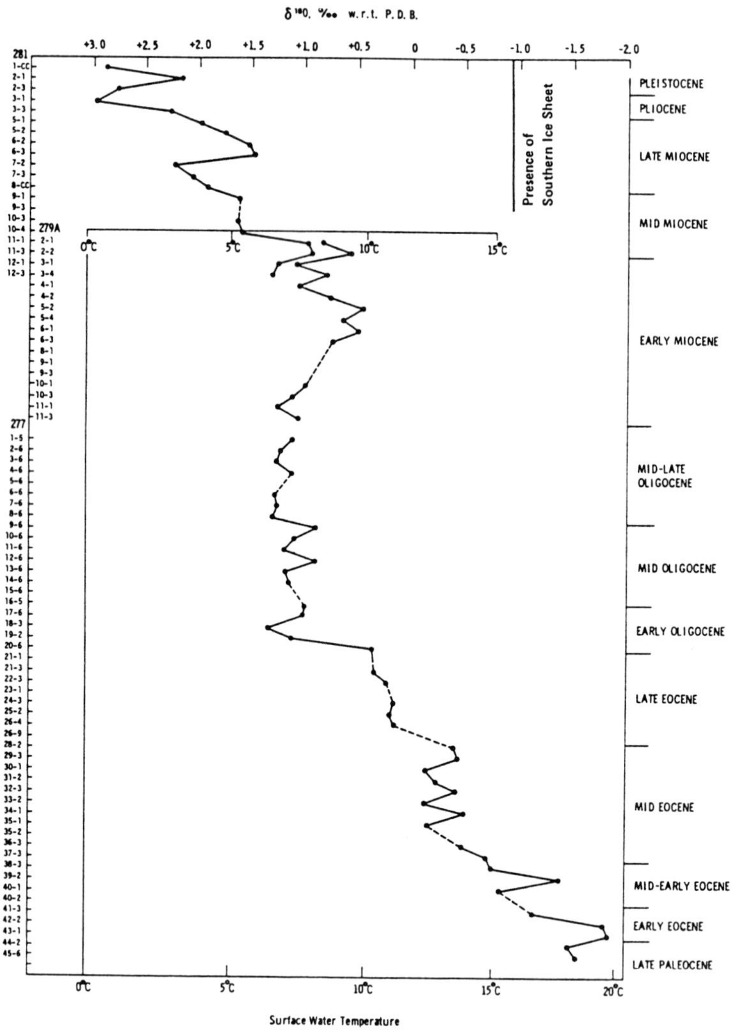

Figure 5.31. Surface ocean water temperatures for the Cenozoic derived from oxygen isotopes for samples from the southern hemisphere (Habicht, 1979)

They found temperatures as low as 15.4 °C compared with a present day temperature of 21.8 °C.

Even in a marine setting there can be major temperature differences between coastal shallow waters and deep marine environments. The latter commonly have temperatures near 4 °C, whereas shallow water more closely reflects local climate. In near-polar regions water temperatures seem to show less variation with depth. In their reconstruction of paleosurface temperatures in the Oseberg area of the North Sea, for example, Dahl et al (1987) used a steadily decreasing temperature (from 20 °C to 10 °C) as water depth increased through time from 20 m to 300 m.

5.5.3. Rock Type

In some cases the nature of the rocks themselves can be useful indicators of climatic conditions. Coral reefs only grow where the water temperature is above 25.5 °C (Figure 5.33), and their presence in a geologic section limits estimates of surface temperature. Similarly, the presence of laterites or bauxites indicates a tropical to subtropical weathering regime. At the other end of the scale, tillites and glacial striations show the influence of ice (Hambrey and Harland, 1981), and correspondingly low temperatures. Schmoker (1986) has used the presence of shallow-water carbonates, together with evidence for a lower paleolatitude, to conclude that Paleozoic temperatures in the Anadarko basin of western Oklahoma

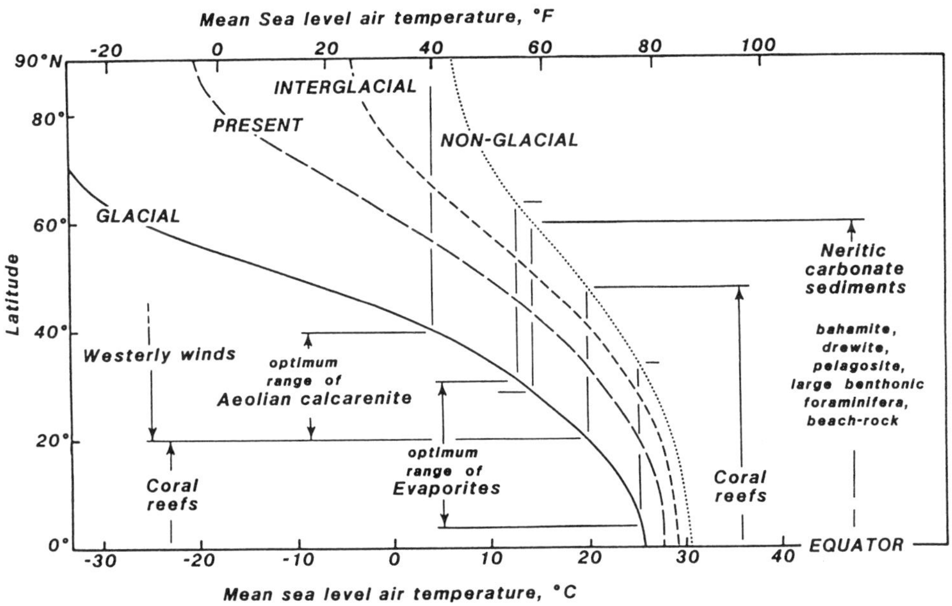

Figure 5.32. Surface mean temperatures as a function of latitude for the present, glacial, non-glacial, and interglacial periods (from Habicht, 1979).

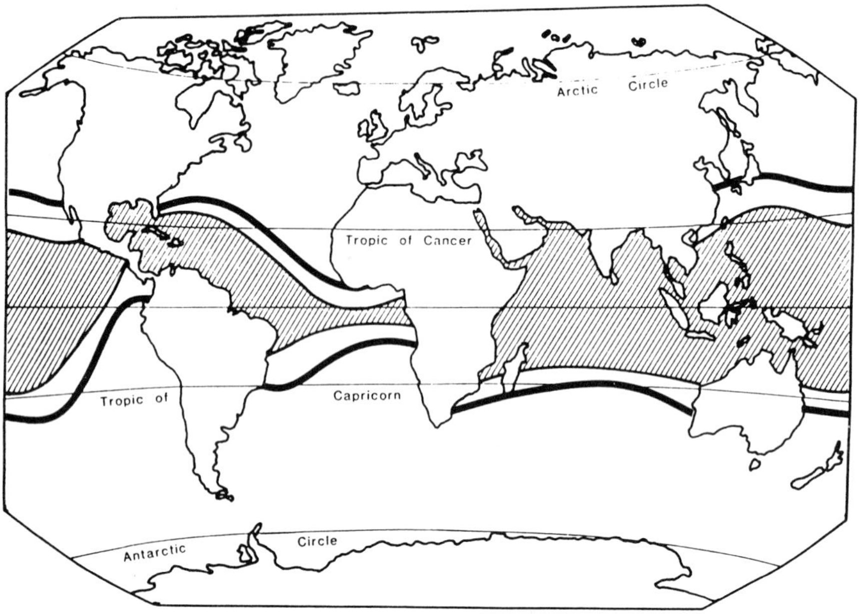

Figure 5.33. Present world distribution of coral reefs (from Habicht, 1979).

were about 10 °C higher than present day temperatures. In his thermal reconstruction, surface temperatures in the area were assumed to decline from 27 °C to a present day value of 16 °C (Figure 5.34). England and Bustin (1986) also modeled the western Canadian basin with a decreasing surface temperature since the Oligocene.

Evaporites form where there is a moisture deficit, and this is generally assumed to be in hot, arid, subtropical areas at a latitude of approximately 30° (Drewry et al, 1974; Frakes, 1979). However, Hardie (1991) has stressed the importance of tectonic setting and suggested a much wider range for evaporite formation, and Gordon (1975) has accumulated data to show that evaporites usually form within about 50° of the equator. While evaporites can be used as a general guide to paleoclimate, they are not infallable indicators of high temperatures and only indicate a moisture deficit.

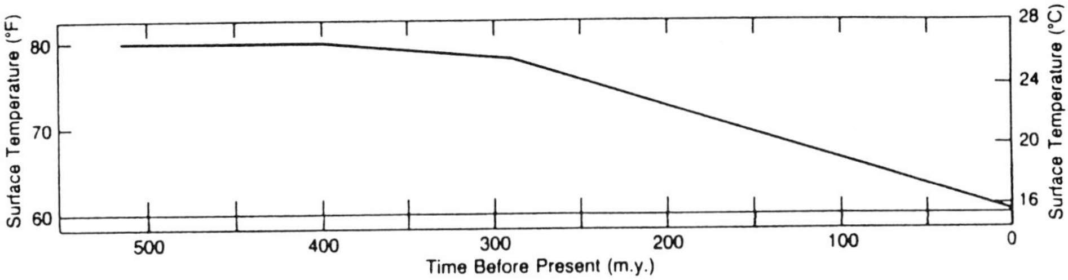

Figure 5.34. Inferred mean surface temperatures in western Oklahoma for the last 500 m.y. (Schmoker, 1986).

5.5.4. Fossil Indicators

Organisms respond to changing environmental conditions including temperature. For example, in cold climates animal size tends to be larger than in warm climates ("*Bergmann's Rule*"), and Strauch (1968) has reported variations in the size of adult *Hiatella arctica* that appear to be temperature related, and concluded that there has been a decrease in temperature since the mid-Cenozoic. Morphology may also be influenced by temperature, and Reynolds and Thunell (1986) related morphologically distinct forms of *Turborotalia pachyderma* to paleoclimate change and seasonal changes in water temperature. In using the presence of specific fossil forms to show temperatures the temperature tolerance has to be deduced from a study of contemporary forms, and this comparison gets less reliable as older fossils are used.

Coiling direction of planktonic foraminiferas has been used to help establish seawater paleo-temperatures (Bandy, 1960). Ericson (1959) and Be and Hamlin (1967) reported that *Turborotalia pachyderma* were sinistrally coiled in water temperatures below 7.2 °C, and Green (1960) found that 98% of Arctic populations were sinistrally coiled. However, Cifelli (1971) claimed that the temperature boundary separating dextral and sinistral populations was not clear. Jenkins (1967, 1968) has inferred alternating colder and warmer water temperatures off New Zealand based on ten alternating sinistral and dextral populations of *T. pachyderma,* and Belford (1989) has reported apparent rapid temperature fluctuations for offshore Tasmania.

5.5.5. Isotopic Methods

The isotopic methods discussed above for estimating paleo-temperatures can also be used to provide paleosurface temperatures. The fractionation of oxygen isotopes between calcite and water is especially important. Isotopic fractionations are temperature dependent (Faure, 1986) so that, for example, carbonate shells growing in isotopic equilibrium with sea water will have oxygen isotope values that are related to temperature by an expression of the form:

$$T\,^{\circ}C = A - B(del_c - del_w) + C(del_c - del_w)^2 \qquad (5.10)$$

where del_c and del_w are the oxygen isotope values ($^{o}/_{oo}$) for the calcite and water respectively. Craig (1965) originally determined the constants to be A = 16.9, B = 4.2, and C = 0.13, but slightly different values have been suggested by O'Neil et al (1969), Shackleton (1974), and Anderson and Arthur (1983).

Miller et al (1987) used the oxygen isotope ratios for *foraminifera* to document steadily decreasing ocean temperatures from 70 Ma to the present, and found synchronous changes in the benthic and planktonic forms. They attributed the changes to ice free conditions from the Paleocene to the Eocene, with subsequent periodic ice growth and decay from then up to the present. Oxygen isotope measurements on planktonic *foraminifera* were also used by Savin et al (1985) to derive surface and near-surface temperatures for the Miocene ocean. Kennett and Barker (1990) have used this technique for latest Cretaceous to Cenozoic sediments in the Weddell Sea, and in Ross Island oxygen isotope data (Pirrie and Marshall, 1990) showed that polar regions were cool in the Late Cretaceous relative to lower latitudes with an inferred trend of ~0.3 °C per degree of latitude. Ditchfield and Marshall (1989) obtained a mean temperature of 25 °C (high 26.5 °C) for U. Cretaceous chalks, and found that banding corresponded to an annual range of about 4.5 °C. Feary et al (1991) used the oxygen isotope composition of *foraminifera* tests to establish the Cenozoic paleotemperature trend for northwest Australia. They were able to relate this to carbonate buildups in warm periods and complete lack of carbonates in cool periods, such as the Oligocene. Estimated minimum temperature for carbonates was 16.5 °C, recognizing that temperatures can probably be estimated to only $\pm$ 1.5 - 2 °C (Savin, 1977).

In applying oxygen isotopes for the determination of near-surface temperatures it is necessary to know the oxygen isotope value for the water (del_w of Equation 5.10). This can be difficult to estimate because it varies with salinity and the quantity of water tied up in polar ice. One possible solution is to use benthic *foraminifera* that grow in deep water where they are assumed to have formed at constant temperature, and so can be used to estimate $\delta^{18}O$ for the water. Any biogenic carbonate must be in isotopic equilibrium with water, and if it is not the oxygen isotope values will not be solely a function of temperature. For example, Erez (1978) showed that marine organisms that derive nutrients from symbiotic algae, which perform isotopic partitioning, may provide misleading isotope information and erroneous paleotemperatures. Mineralogic effects can be important, especially the difference between an initial precipitation as aragonite rather than calcite. For aragonite, Horibe and Oba (1972) determined the A, B, and C constants of Equation 5.10 to be 13.85, 4.54, and 0.04, respectively. Also, diagenesis may erase or modify the original isotopic ratios producing anomalous temperature estimates (Killingsley, 1983). In rocks older than Jurassic there is very little calcium carbonate shell material suitable for $\delta^{18}O$ measurements.

Paleotemperatures can also be derived from water and biogenic silica (opal, chert), or from water and biogenic phosphate (apatite). Unfortunately it does not seem possible to combine two water-mineral pairs and eliminate the need to know the isotopic composition of the water (Faure, 1986).

5.6. COMPLICATING FACTORS

Determinations of paleotemperatures in the subsurface are subject to a wide range of uncertainties, because many factors that perturb temperature leave little evidence in the rocks. Abnormal pressures due to disequilibrium compaction may cause undercompaction and significantly change thermal conductivities, but then decay until the rocks today are normally pressured. This is less likely to be a problem when overpressuring has been caused by organic maturation. Also, abnormal pressures that approach lithostatic values can induce hydraulic fracturing and produce short-lived bursts of hot fluids. Deep aquifers that were active in the past may no longer be flowing.

Faults that acted as conduits for hot (or cold) water flow in the past may now be effective seals (Majorowicz and Jessop, 1981; Bethke and Marshak, 1990). Royden (1991) has suggested that in pre-Tertiary time hot springs (>100 °C), rising through near-vertical fractures, may have locally annealed fission tracks in apatites from parts of W. Virginia and Virginia. Vitrinite reflectance, CAI, and detailed geological investigations of the Valley and Ridge Province of Virginia led Lewis and Hower (1990) to conclude that warm fluids expelled during Alleghenian tectonic events could have played at least as important a role as that of combined stratigraphic and tectonic burial. Other studies of the effects of local increases in the thermal maturity of sediments due to hydrothermal fluids have included those by Barker (1983), Law et al (1986), Ilchik et al (1986), and Walters and Kotra (1990). In the Tartan field of the North Sea authigenic quartz cements deposited in fractures give fluid inclusion homogenization temperatures that are 15-30 °C higher than those obtained from contemporaneous cements away from the fracture (Burley et al, 1989). Temperatures show temporal fluctuations (Figure 5.35) that seem to indicate a pulsed flow through the fractures. A past role for convective heat flow was invoked by Law et al (1989) to make calculated and observed "kinky" vitrinite reflectance profiles match. An unrealistically high apparent thermal conductivity of 12 W/m°C would have been required in some parts of the section.

Dewatering of clays produces considerable amounts of water (equal to roughly 50% of the volume of the final dehydrated clay) and this can act as a heat transfer agent as it moves towards the surface (deCaritat, 1989). Tillman and Barnes (1983) suggested that in the northern Appalachian Basin geothermal gradients were raised as high as 40 °C/km before cooling to a more average value of 25 °C/km. Smith and Chapman (1983) have provided a useful review of the thermal effects of subsurface fluid flow, and aspects of topographically-driven recharge are discussed in Section 10.6.1.

Igneous intrusions may produce very intense local heating, but the extremely high temperatures last for only a short time. Intrusions are emplaced rapidly at high temperatures, usually in the range 600 - 1200 °C, with acidic rocks at the low end of the range and basic magmas at the high end. Cooling is rapid, and generally takes less than about a million years

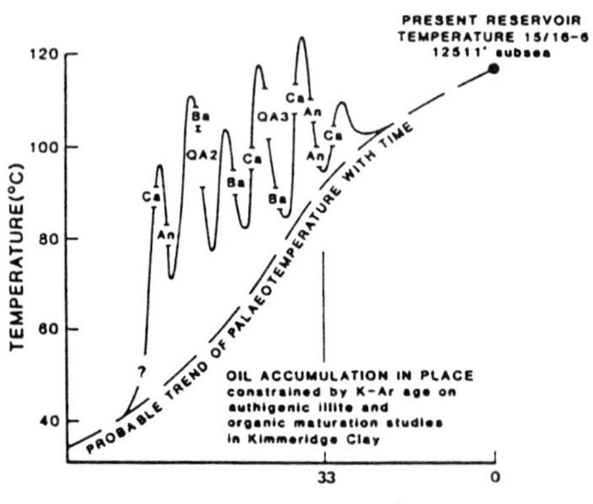

Figure 5.35. Time-temperature trend for authigenic minerals in the Tartan downthrown block, North Sea (Burley et al, 1989).

(Jaeger, 1968). Quantifying the impact of this temperature perturbation can be difficult, but the thermal effects of dike and sill intrusion are usually quite limited. As a general guide intrusions do not affect sediments more than 2D from the dike (or sill), where D is the width of the dike (Dow, 1977), although the effects can be considerably less (Barker and Bone, 1995). In the Newark basin, Walters and Kotra (1990) used biomarker ratios and I-I and A-I plots to show that a short, approximately 5 m.y. long, thermal pulse pushed temperature gradients to about 75 °C/km. They also considered the effects of flood basalts and concluded that their accumulation as a series of individual flows had little direct effect on the underlying sedimentary rocks. While a single intrusive may have only a minor effect on petroleum generation, it can have a major impact on local vitrinite reflectance values. These may be very misleading, especially if drilling is stopped short of the actual sill.

In areas of rapid sedimentation, or rapid uplift/erosion, thermal equilibrium may not be reached and organic maturation values will appear anomalous. This explanation was proposed by England and Bustin (1986) for the western Canadian sedimentary basin, and by Cazier et al (1995) in the Llanos foreland basin of Columbia. Sediment accumulation rates of up to 113 m/m.y. led Link (1988) to a similar conclusion for the Eagle Plain of northern Canada. This is consistent with the 1-D finite element theoretical model of deBremaecker (1983) that showed temperature distribution will be out of equilibrium if sediment accumulation rates exceed approximately 100 m/m.y.

CHAPTER 6

TIME / TEMPERATURE COMBINATIONS AND THE LOPATIN METHOD

6.1. EARLY MODELS

Bergius in 1913 appears to have been the first person to recognize the combined effects of time and temperature in maturing organic matter. Like many of the early investigators, he was interested in coal, in part because of its economic importance and in part because coalification was better understood at that time. Cane (1950) studied the algal coal called torbanite and obtained a global activation energy of 48.5 kcal/mol, but Pitt (1961), also working with coals, was the first to suggest that maturation is most appropriately described by a set of parallel, first order reactions. The first model for the generation of petroleum from kerogen was not introduced until 1967 (by Louis and Tissot), and was subsequently described by Tissot (1969) and Tissot and Pelet (1971). Good historical reviews that cover the early attempts at modeling are given by Wright (1980), Waples (1984), Lewan (1992), and Hermanrud (1993).

Many of the early models made the simplifying assumption that time had a minor role and could be ignored. This, of course, implies that the oil generating threshold can be given solely in terms of temperature. Usually the values were put at about 60 - 65 °C, but with a range of around 50 - 90 °C, regardless of age. Some authors (e.g., Pusey, 1973) did specify an age range where the temperature value was applicable. However, Philippi (1965) reported the onset of petroleum generation at much higher temperatures, roughly 115 - 140 °C, in some very young Californian basins, and many other exceptions have been reported (see Section 1.5.4). It appears that in oil and gas generation age (time) has a significant effect in modifying the role of temperature.

A widely-quoted nomograph that relates coal rank, maximum temperature, and heating time (equated to age) was published by Karweil in 1956 and is reproduced here as Figure 6.1. Geologic age is plotted against temperature, and a series of curves used to give the equivalent organic maturity for any time/temperature combination. Since the diagram was developed for coals, rank indicators such as volatile content are included, but the equivalent vitrinite reflectance values have generally been used in extending applications to oil and gas generation. Figure 6.1 shows that for a temperature of 160 °C and an age of 30 m.y. (point A), the corresponding R_o is 2.0%. If heating temperature is reduced, say to 100 °C, then following the 30 m.y. time line leads to point B where the corresponding R_o value is now lower and close to 0.75%. When temperature remains unchanged at 160 °C, but time is reduced to 10 m.y. (point C), the equivalent R_o falls to 0.93%. Bostick (1973, 1979) has given some examples of applying the Karweil approach to petroleum source rocks. Experience suggests that the Karweil nomograph may overestimate the role of time.

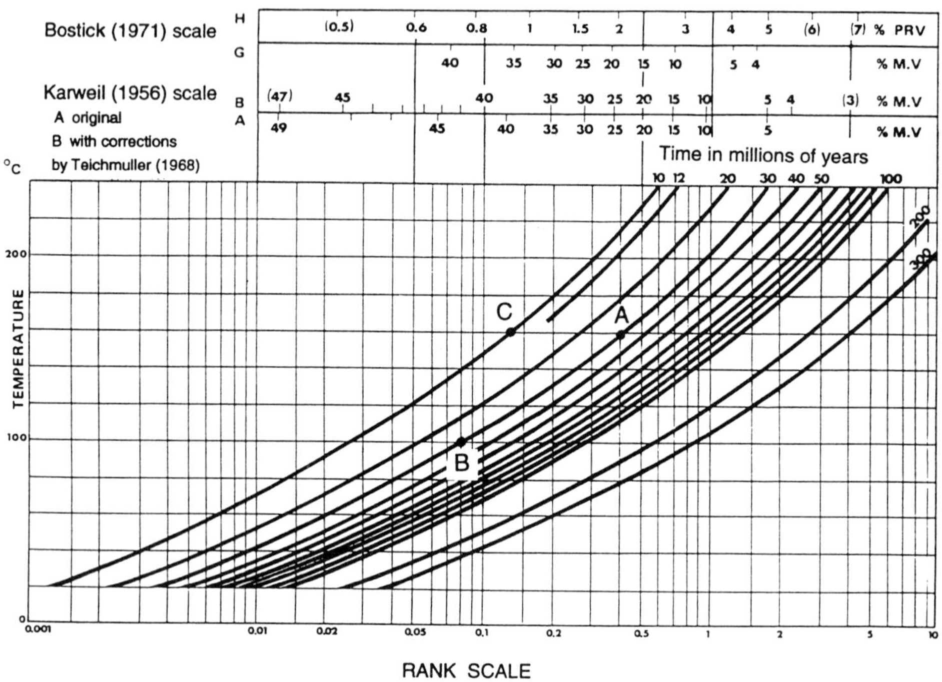

Figure 6.1. Time-temperature nomograph based on Karweil (1956) with additions by Robert (1988). The figure was originally developed for coals and gives rank (top) as vitrinite reflectance (PRV) or percent volatile matter (MV).

The importance of time in the reactions that generate petroleum was recognized by Connan (1974), and his empirical study equated formation age to "cooking time" and maximum temperature to reaction temperature. However, a rock may spend a considerable time at low temperature, where reaction rates are negligible, and Lopatin (1969) seems to have been the first to introduce the idea of an *"effective heating time"*, which he defined to be the time spent above 100 °C. Hood et al (1975) considered the effective heating time to be the time spend within 15 °C of the maximum, and they combined this time estimate with the maximum temperature to obtain a *"Level of Organic Metamorphism"* (LOM) (Figure 6.2). On this scale the onset of generation occurs at a LOM of 8 - 9, and destruction of oil is at 11 - 12. The values are roughly equivalent to a reaction rate that doubles every 10 °C, but implies rather low global activation energies of 18 - 33 kcal/mol over the LOM range of 9 - 16. Shiboaka and Bennett (1977) also considered time in setting up a time-temperature-R_o plot, and Pigott (1985) developed an empirical maturation method.

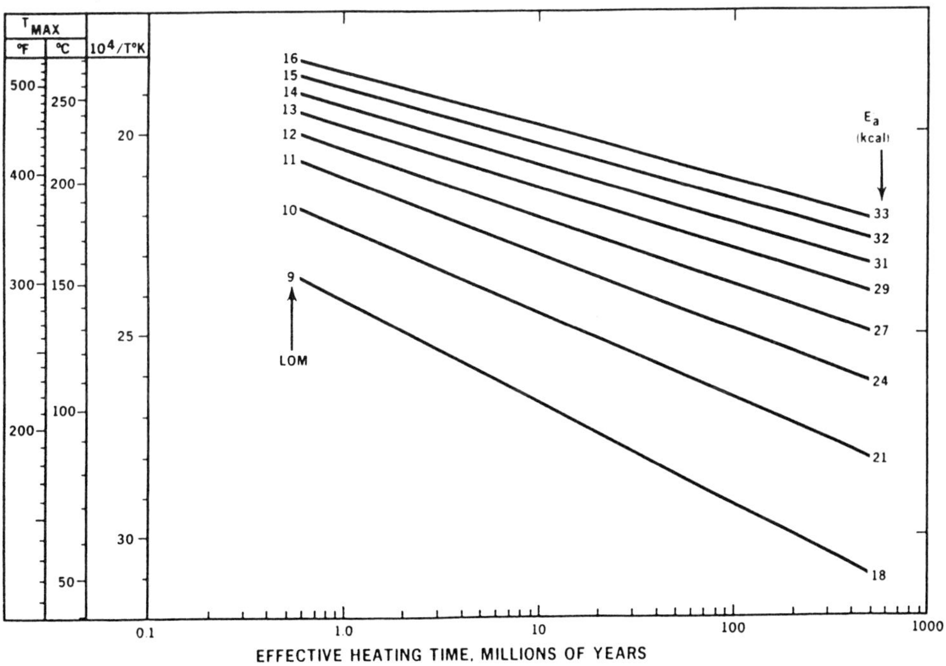

Figure 6.2. Diagram used by Hood et al (1975) to calculate level of organic metamorphism (LOM). It is based on maximum heating temperature (vertical scale), and a horizontal scale of effective heating time (defined as the time spent within 15 °C of the maximum temperature).

6.2. THE LOPATIN METHOD AND CALCULATION OF TTI

The thermal models currently being used fall into two broad catagories. The simplest ones rely on keeping track of some time-temperature integrator through time, and then comparing the values with calibration data obtained from well-documented basins. They are being steadily replaced by the more rigorous models based on first order kinetics, and these are described in Chapter 7.

The most widely used calibrated method for establishing the depth to the onset of petroleum generation (as well as determining depth limitations on oil and condensate occurrence) was proposed by Lopatin in 1971. His method is based on the assumption that the rate of the petroleum generation reaction doubles with every 10 °C increase in temperature. The Lopatin method has been described in English by Waples (1980) who also improved it by using more extensive and better calibration data. Lopatin considered the increase in maturation that occurred in each 10 °C temperature increment throughout the burial history, and selected 100 - 110 °C for his reference interval. Since reaction rate doubles for every 10 °C rise in temperature, when the temperature interval is 10 °C above the reference, the reaction (i.e., generation) will go twice as fast, while in the temperature interval 10 °C lower (i.e., 90 - 100 °C) it will proceed only half as fast as in the reference interval. Thus a temperature factor δ can be defined for each interval n, such that $\delta = r^n$, where $r = 2$. A list of temperature factors (δ) for each of the temperature increments (n) is given in Table 6.1.

The maturity increase in any 10 °C temperature interval will be directly proportional to the amount of time spent in that interval but will vary as the n th power of the temperature, thus:

$$\Delta Maturity = (\Delta t_i)(r^{n_i}) \tag{6.1}$$

where *ΔMaturity* is the incremental increase in maturity in the i th temperature interval. Total maturity achieved by a given rock unit is the sum of the increments in maturation attained in each of the individual 10 °C intervals. Lopatin coined the term *"Time Temperature Index"* (TTI) for the maturity, it can be given either for the incremental values, or as a cumulative TTI.

The method proposed by Lopatin requires a burial depth-versus-age profile, together with an estimate of the geothermal gradient through time. No geochemical data are needed (they enter the model indirectly through the calibration data). The first step in using Lopatin's method is to plot the

Table 6.1
Temperature factors (δ) used by Lopatin for each 10 °C temperature increment.

TEMPERATURE INTERVAL, °C	δ	
20 - 30	2^{-8}	1/256
30 - 40	2^{-7}	1/128
40 - 50	2^{-6}	1/64
50 - 60	2^{-5}	1/32
60 - 70	2^{-4}	1/16
70 - 80	2^{-3}	1/8
80 - 90	2^{-2}	1/4
90 - 100	2^{-1}	1/2
100 - 110	2^{0}	1
110 - 120	2^{1}	2
120 - 130	2^{2}	4
130 - 140	2^{3}	8

burial profile and then superimpose on it a temperature grid for every 10 °C. An example is shown in Figure 6.3 for a Jurassic rock that was deposited at 150 Ma and buried steadily until 100 Ma, when the burial rate decreased for 50 m.y. before increasing again. The rock is now 5 km deep. For a geothermal gradient of 25 °C/km the 10 °C isotemperature lines can be superimposed on the depth profile as shown. From the figure we find that the rock entered the 20 - 30 °C interval at 150 Ma and left it at 139 Ma, so that 11 m.y. were spent at temperatures between 20 and 30 °C. For this interval n is -8, and for r = 2 the increase in maturity is given by:

$$(\Delta Ti)(r^{ni}) \quad = \quad 11 \times 2^{-8} \quad = \quad 11 / 256 \quad = \quad 0.04. \qquad (6.2)$$

Similarly, the rock spends 10 m.y. between 30 and 40 °C, and the increment in maturity is 10 x 2^{-7}, or 0.08. The cumulative increase is 0.12. This process of calculating the maturity increments for each 10 °C temperature increase can be continued until the present day depth is

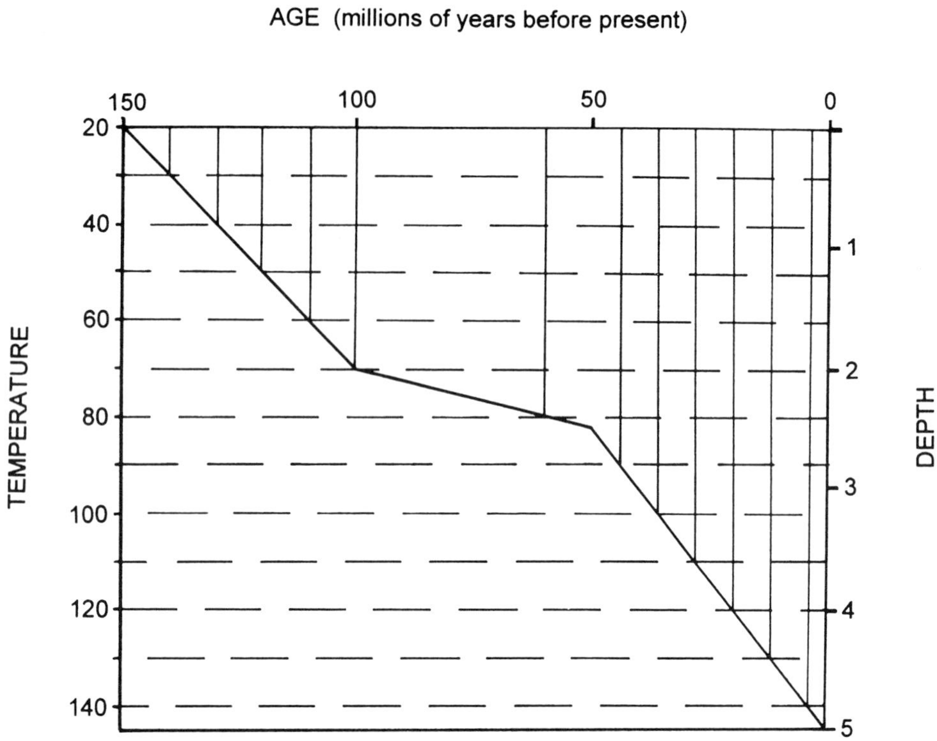

Figure 6.3. Burial history for a Jurassic source rock (150 m.y. old) that is currently 5 km deep in an area where the geothermal gradient has remained unchanged at 25 °C/km.

reached, in this case 5 km. Table 6.2 shows the values obtained for each increment in TTI and the cumulative TTI. Note that the sum of the time intervals will equal the age of the rock. Cumulative TTI has no intrinsic meaning and can be interpreted in terms of organic maturity only by reference to calibration data obtained from well-studied basins. Calibration data from Waples (1980) are reproduced here as Table 6.3, and show that the rock is still in the oil window at present because the value for cumulative TTI has not exceeded the value of 160 which defines the base of the oil window. But when did it begin generating? Again, referring to the calibration data shows that generation starts at a cumulative TTI of 15, and for the example being considered this was reached when the temperature got close to 100 °C, corresponding to 36 Ma (Figure 6.3).

Table 6.2
Calculated incremental and cumulative time-temperature indexes (TTIs) for the burial history and geothermal gradients shown in Figure 6.3.

TEMPERATURE INTERVAL °C	2^r	TIME SPENT IN TEMPERATURE INTERVAL IN M.Y.	INCREMENT IN TTI	CUMULATIVE TTI
20 - 30	2^{-8}	11	0.04	0.04
30 - 40	2^{-7}	10	0.08	0.12
40 - 50	2^{-6}	9	0.14	0.26
50 - 60	2^{-5}	11	0.34	0.60
60 - 70	2^{-4}	9	0.56	1.16
70 - 80	2^{-3}	40	5.00	6.61
80 - 90	2^{-2}	16	4.00	10.16
90 - 100	2^{-1}	8	4.00	14.16
100 - 110	2^{0}	8	8.00	22.16
110 - 120	2^{1}	7	14.00	36.16
120 - 130	2^{2}	8	32.00	68.16
130 - 140	2^{3}	9	72.00	140.16

Table 6.3
Correlation of TTI values with other widely used maturity indicators and their relationships to important stages in the generation and preservation of oil and gas (from Waples, 1980).

STAGE	TTI	R_o	TAI
Onset of oil generation	15	0.65	2.65
Peak oil generation	75	1.00	2.9
End of oil generation	160	1.30	3.2
Upper TTI limit for occurrence of oil with <40° API	~500	1.75	3.6
Upper TTI limit for occurrence of oil with <50° API	~1000	2.0	3.7
Upper limit for occurrence of wet gas	~1500	2.2	3.75
Last known occurrence of dry gas	~65,000	4.8	>4.0

In order to use the Lopatin method with published calibration data it is necessary to:
(1) express temperature in degrees Centigrade
(2) express age in millions of years
(3) use the 100 - 110 °C interval as the reference ($\delta = 2^0$).

Any rock element is followed through its 10 °C temperature increments regardless of whether it is being buried or uplifted. On uplift temperature factors fall rapidly – for example, from 100 to 70 °C the rate of generation falls by 2^3, almost an order of magnitude.

The example used to illustrate the principle of Lopatin's method was geologically very simple, and involved only steady burial with a uniform and unchanging geothermal gradient. Real situations are usually much more complex, and frequently involve changing geothermal gradients through time, and periods of uplift as well as burial. The Lopatin method still sums the individual TTIs for each 10 °C interval regardless of whether the rocks are being buried or uplifted, or whether isotherms are constant or changing. The method can be applied to any geological model, regardless of its complexity, but a "well-thought out, detailed reconstruction will obviously yield more reliable results than one which is largely based on guess-work. In some cases, however, even a very crude or approximate model may be able to answer important questions about hydrocarbon generation or preservation" (Waples, 1980).

6.3. CALIBRATION

The Lopatin method is empirical, and a given cumulative TTI can only be related to maturation status by calibration. This commonly involves equating cumulative TTI values to equivalent vitrinite reflectance values or other maturity indicator (e.g., TAI) (Table 6.3).

Lopatin used 10 °C increments ($\Delta T = 10$) and r^n with $r = 2$. There is no theoretical reason why ΔT or r should not be changed, but in practice ΔT is almost always put equal to 10 °C. Waples (1980) showed that it is not possible to select a value for r from purely theoretical considerations, and he used vitrinite reflectance (and equivalent TAI data) to calibrate the value of r. This calibration was based on 402 samples from 31 worldwide wells that spanned an age range from Paleozoic to Quaternary, and a vitrinite reflectance range from 0.4 to 6%. His plot of TTI vs R_o is shown together with the calculated curve obtained by using $r = 2$ (Figure 6.4). Values of r from 1.0 to 10.0 were tried, and while there is generally good

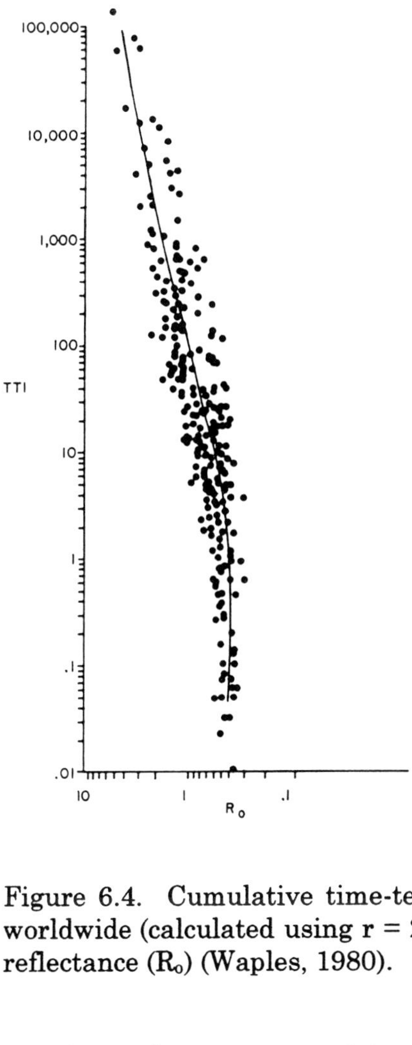

Figure 6.4. Cumulative time-temperature indexes for 31 wells in basins worldwide (calculated using r = 2 in Equation 6.1) plotted against vitrinite reflectance (R_o) (Waples, 1980).

correlation from 1.6 to 2.5, 2.0 provides the best fit. The scatter around the r = 2.0 line may be due to experimental errors in the R_o and TAI data; uncertainties in the geologic models for past thermal histories, especially paleotemperatures; or variations in organic matter type. Use of a constant value of r over a wide temperature range implies that activation energy increases, with global values that would be 12 kcal/mol at 25 °C, 20 kcal/mol at 100 °C, and 26 kcal/mol at 160 °C. Other authors have chosen to use different values for r. Kettel (1983) (based on only four wells) used 1.6, and Pitt (1986) used 2.3.

In attempting to overcome the inherent limitations of the Lopatin method Galitsyn (1973), Lopatin and Bostick (1973) and Lopatin (1976) suggested decreasing the value of r with increasing temperature. Lopatin

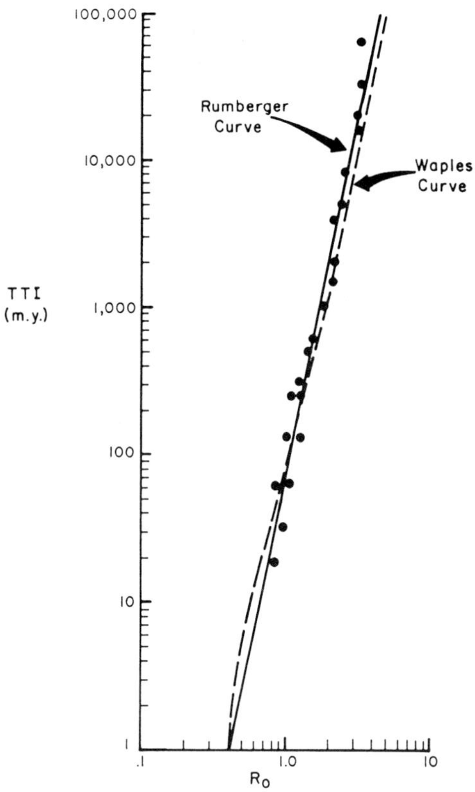

Figure 6.5. Time-temperature indexes (TTIs) calculated for the Shell Rumberger 5 well, Beckham County, Anadarko basin, Oklahoma (Hood et al, 1975), compared with the Waples calibration data (Waples, 1980).

(1976) also suggested increasing ΔT from 10 °C to 15 °C, and then to 30 °C. Gretener and Curtis (1982) modified the Lopatin scale by referencing it to a 10 °C, unburied state. There have been few recent attempts to further refine the Lopatin method because commercial modeling software that uses the more rigorous kinetic procedures is now readily available (see Chapter 7).

The TTI - R_o calibration of Waples (1980) is still widely used but consistently gives R_o values that are significantly higher than those calculated using Arrhenius-based methods, especially at higher temperatures (Morrow and Issler, 1993). This is true even when values are corrected for compaction effects (Dykstra, 1987). The most appropriate calibration data will be those obtained in a geologic setting similar to the one being modeled. For example, Bond (1984) used the Rumberger well in the Anadarko basin to provide a local calibration, but in this case the TTI - R_o trend is quite close to the Waples trend (Figure 6.5). For rifted/passive margins the TTI - R_o calibrations obtained by Issler (1984) from the eastern Canadian margin are often used successfully, and they seem to work well for other young, high heat flow areas. Issler and Snowdon (1990) concluded that the major problem in using Lopatin's method was due to the use of Waples calibration data. They claimed that in many cases it is necessary to assume paleothermal gradients significantly lower than present values in order to make calculated and observed values agree. The Issler calibration "appears to give good results for continuously subsiding basins with heating rates of the order 1 - 10 °C/m.y." (Issler, 1984; Horvath et al, 1988). Figure 6.6 shows a comparison of the data obtained by Horvath et al (1986) for the Pannonian basin with the calibration curves of both Issler and Waples. Clearly, the Issler curve is a better fit in this young, high heat flow basin.

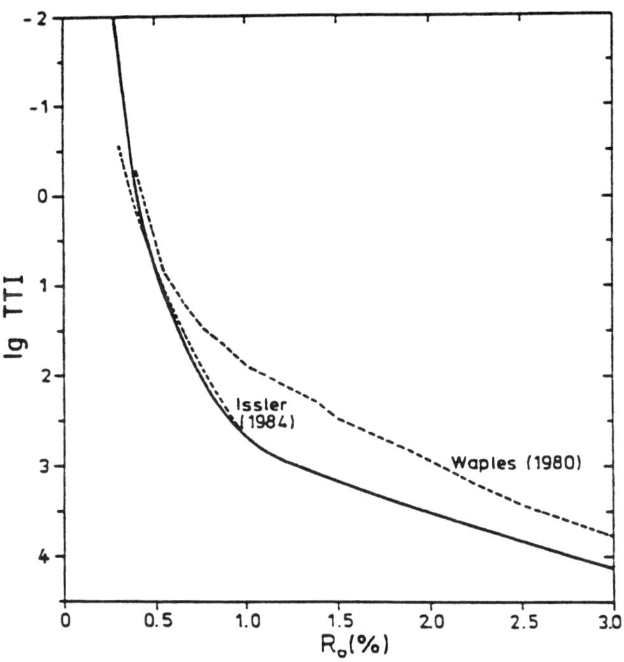

Figure 6.6. Relationship between calculated TTI and vitrinite reflectance in the Pannonian Basin obtained using the calibration data of Waples (1980), Issler (1984), and Horvath et al (1985) (solid line). (From Horvath et al, 1985).

In some cases it is possible to use data from one or more well-documented wells to generate a local calibration curve (e.g., Schmoker, 1986). This approach would be useful, for example, when a well drilled on a high block finds oil but only penetrates immature source rock. In this situation it is important to establish the depth to mature source rock, and local calibration based on the available wells increases confidence in the extrapolation to greater depths in the area.

6.4. OIL WINDOW

Organic matter maturation increases along a burial time-temperature track, and for the Lopatin method comparison with calibration data shows when it reaches a value corresponding to the onset of petroleum generation. As maturation continues it leads through peak generation to the end of the oil window. Calculation for the onset of petroleum generation, and the position of the floor for oil, can be made for the top, bottom, or middle of any unit in the reconstructed burial history. This provides a way of establishing the position of the oil window through time (Waples, 1980).

In the simplest case where all rock units are buried at the same uniform rate, in an area with a constant geothermal gradient, they will all become mature at the same depth, because by then they will all have experienced the same combination of time and temperature. In this case both the top and bottom of the oil window will be represented by horizontal lines (Figure 6.7). Most basins are not this simple, and the depth of the oil window will vary through time. It is common for sedimentary basins to fill rapidly at first, but then more slowly with time, so that in general the burial history profiles show an initial rapid increase in burial depth, followed by a decreasing sediment accumulation rate. The effect of this on the position of the oil window is shown in Figure 6.8 where the oil window rises to shallower depth through time. With decreasing burial rate the onset of generation becomes steadily shallower as the rocks get younger. Even while rocks remain at a constant depth and temperature they continue to mature (in the same way that a cake in an oven continues to cook even though the oven temperature is not rising). As a consequence the oil window tends to move to shallower depths through time.

In situations where sediment burial rate increases rapidly, the oil window will be tilted towards greater depths. Skjervoy and Sylta (1993) provide an example from the southwest Horda Platform, N. Sea, where the

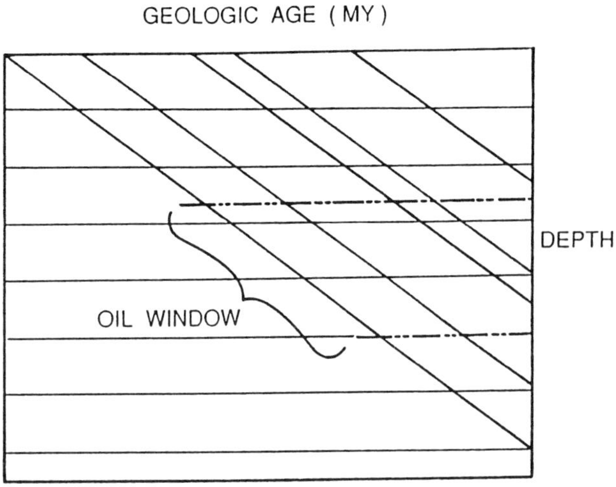

Figure 6.7. The depths of the top and bottom of the oil window remain constant for uniform burial rates and a constant geothermal gradient.

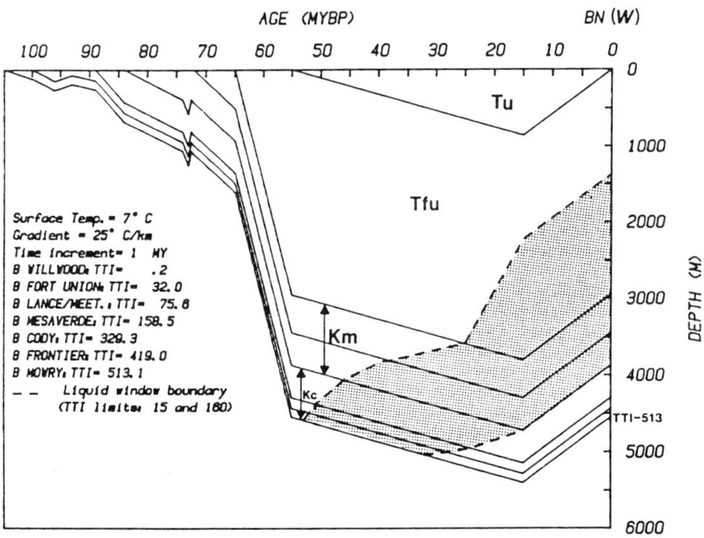

Figure 6.8. Time-temperature reconstruction for a location in the Big Horn basin, Montana where geothermal gradient has remained constant at 25 °C/km and surface temperature is unchanging at 7 °C. The oil window moves to shallower depths from 55 Ma to the present in response to decreased burial rates and final uplift. (From Hagen and Surdam, 1984.)

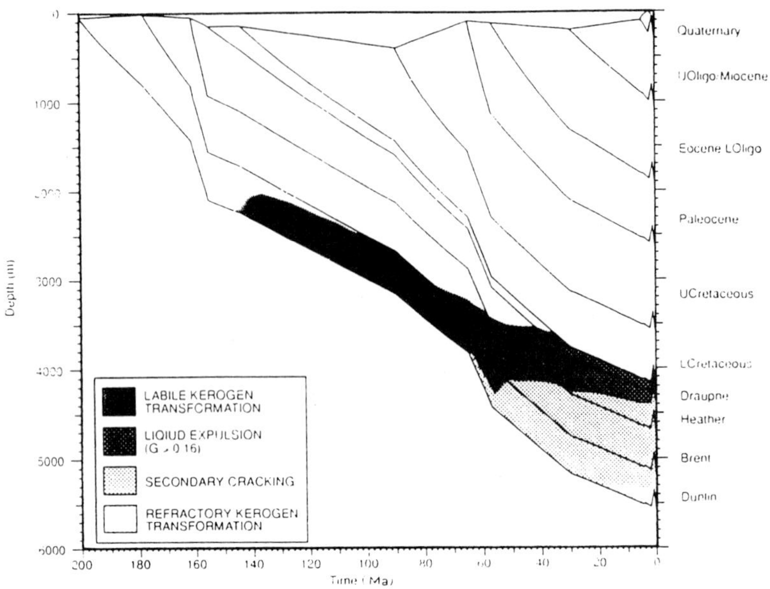

Figure 6.9. Oil window that steadily increases in depth through time for a location on the Horda Platform, North Sea. (From Skjervoy and Sylta, 1993.)

top of the oil window was initially 2000 m deep but has moved steadily deeper through time to a depth exceeding 4000 m (Figure 6.9). Most real burial histories show time segments with varying rates of burial combined with periods of uplift, so that the depth of the oil window may vary irregularly through time.

These calculations of varying maturity with geologic age can be made for all the rocks in the burial reconstruction, but only those units with high organic contents will have the potential to be effective source rocks. Combining information on organic content with the position of the oil window shows when source rocks are actively generating and expelling oil in the modeled area (see Section 8.3). These calculations are, of course, not restricted to the Lopatin method and maturity can be calculated from first order kinetics (Chapter 7) and superimposed on burial history plots. It is important to recognize that the position of the oil window only shows maturation status, and on uplift this is not simply related to generationrate, which steadily decreases. Vitrinite reflectance values for a rock now at the surface may show that it is in the oil window, but it is not currently an effective source rock because the rate of generation and expulsion is essentially zero.

6.5. LIMITATIONS OF THE LOPATIN METHOD

The Lopatin method has the advantage of simplicity and in spite of serious defects (Waples et al, 1992b) is still often used (e.g., Lee and Bethke, 1994; Shdidi et al, 1995; Parnell and Carey, 1995; Zhao et al, 1996). Its use is often justified in cases were there is only a limited geological framework, or inadequate geochemical control does not merit use of more elaborate models (e.g., Graham et al, 1990; Newell et al, 1993). It is most reliable at lower maturity levels, and with care can be used to indicate onset of oil generation. The simplifying assumption that reaction rate doubles for every 10 °C begins to break down at higher temperatures, so that the Lopatin method only provides a rough guide to the depth and timing of the oil floor and the onset of gas generation. Since it is a calibrated method it is best used under conditions similar to those used for calibration. In general it does not work well for the extremes of sedimentation rate, and results can be unreliable for young basins with high geothermal gradients, or old basins with low geothermal gradients. In establishing source rock maturity the Lopatin method appears to overestimate the role of time (Bustin, 1989) and underestimate the effect of temperature (Quigley and Mackenzie, 1988).

The Lopatin method works best for Type II, oil-prone kerogen. This is the major kerogen type in many of the World's prolific source rocks (including Middle East and Venezuelan carbonates, the Kimmeridge Clay of the North Sea, and the Woodford and Bakken shales of the U.S. mid-continent), so the Lopatin method has wide potential applicability. However, it does not easily cope with other types of organic matter, and there is no way to deal with mixtures of different kerogen types. For Type I kerogen the simplest way is to work with equivalent vitrinite reflectance values, rather than directly with cumulative TTI. Onset of generation will then be at an equivalent R_o = 0.7%, rather than R_o = 0.5%. This implies that for a section with multiple organic-rich beds, some dominated by Type I kerogen and others by Type II kerogen, there will be two different oil windows. It is important to note that vitrinite is a Type III, wood-derived kerogen, and so does not behave in exactly the same way as algal Type II kerogens that generate most of the oil. The difference in maturity at any given depth will depend on the particular time-temperature history they have been exposed to.

CHAPTER 7

TISSOT-ESPITALIE TYPE KINETIC MODELS

7.1. INTRODUCTION

The Lopatin method has been very important in drawing attention to thermal modeling as a way of calculating organic matter maturity. It provides a simple method that does not need organic geochemical insights, or require major computing power, and is easily accessible to exploration geologists. However, the method does have serious limitations because the basic assumption that reaction rate doubles for every 10 °C increase in temperature is not valid over most of the temperature ranges considered for oil and gas generation and preservation. Also, the Lopatin method is not applicable to all types of organic matter. More rigorous kinetic models have been developed to circumvent these limitations.

The thermal breakdown of the large complex kerogen "molecule", that occurs in the process of petroleum generation, involves breaking off smaller units by the cleavage of a variety of different bonds. Bonds break roughly in the order of their bond energies with the weaker bonds rupturing at lower temperatures. Bond cleavages that have to be considered include C – C, aromatic C = C, C – N, C – S, C – O, and C = O, among others, but the situation is further complicated because bond energies vary with the nature of the surrounding groups. For example, the strength of a C – C bond depends on its position in a chain, and it is easier to cleave a very long chain off a large ring complex than to break off a one-carbon group (i.e., methyl), which is why gas is generated at greater depths, corresponding to the higher temperatures that provide more thermal energy. Ideally a mathematical model would treat petroleum generation as a series of parallel first order kinetic reactions, each with its own appropriate E and A values, and then consider subsequent reactions of the products in a series of sequential reactions. Clearly, the rigorous treatment of this system is a daunting prospect, and models use a variety of simplifying assumptions. Different models are distinguished mainly by the nature of the assumptions they make.

7.2. MODELS

The first model of petroleum generation that explicitly accounted for both time and temperature through first order kinetics was proposed by Tissot in 1969, and later discussed more fully by Tissot and Espitalie (1975). It has since been further developed, refined, and applied by geoscientists at the French Petroleum Institute (IFP) (see for example, Tissot et al, 1987).

The model treats petroleum generation as a 2-step process:

KEROGEN $\Rightarrow$ OIL $\Rightarrow$ GAS.

The thermal breakdown of the kerogen is described by 6 independent first order reactions, and the subsequent cracking of oil to gas by one, giving a total of 7 values for activation energies, and 7 values for frequency factors (Figure 7.1). In addition, each of the kerogen components has a different potential to generate oil, and the oil can convert to gas, so that the additional 7 values leads to a total of 21 parameters that have to be considered by the algorithm.

An alternative model developed at the Lawrence Livermore Laboratory (LLNL) in California, also considers multiple parallel reactions, but includes the possibility of generating gas directly from the maturing kerogen, rather than just as a product of cracking oil (Sweeney et al, 1987; Burnham and Sweeney, 1990) (Figure 7.2). This model provided the option

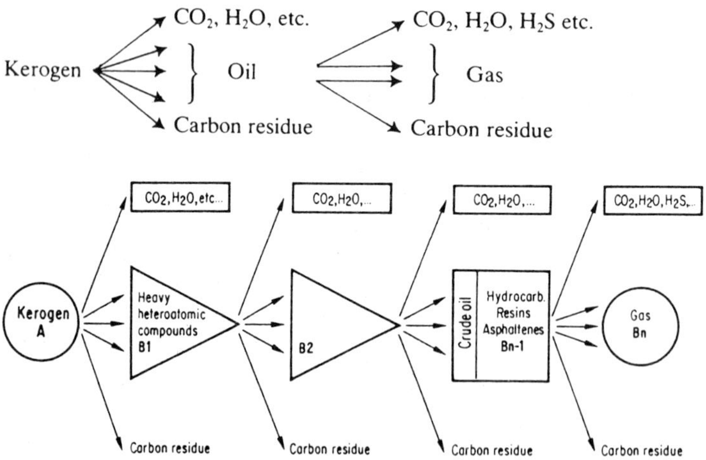

Figure 7.1. General scheme of kerogen degradation used by Tissot and Espitalie (1975) in their kinetic model.

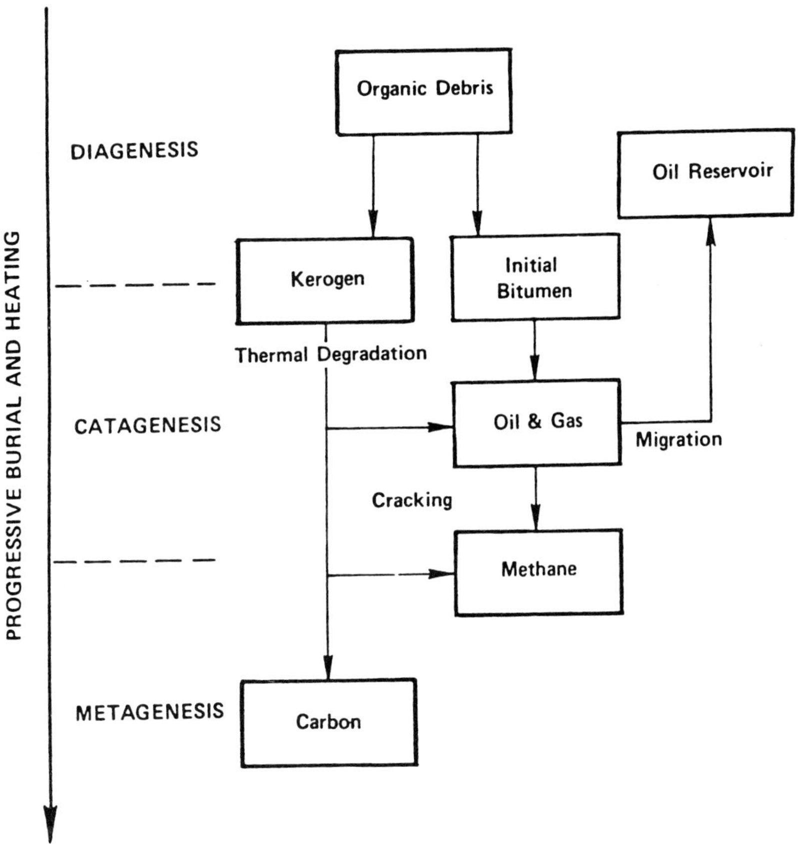

Figure 7.2. Lawrence Livermore (LLNL) kinetic model (Sweeney et al, 1987).

of specifying a suite of activation energies that cover a Gaussian distribution of values (Braun and Burnham, 1987), but assumed only a single value for A. Both the IFP and the LLNL models have been continuously developed over the last few years to provide more rigorous mathematical treatments, and other groups have also built similar models that use parallel reactions with first order kinetics.

The kinetic scheme that describes thermal maturation of organic matter is combined with burial history (Chapter 3), and thermal history (Chapter 5), to model the evolution of thermal maturity through time for a given area. A summary of typical input data and results for a classical 1-D model is given in Figure 7.3.

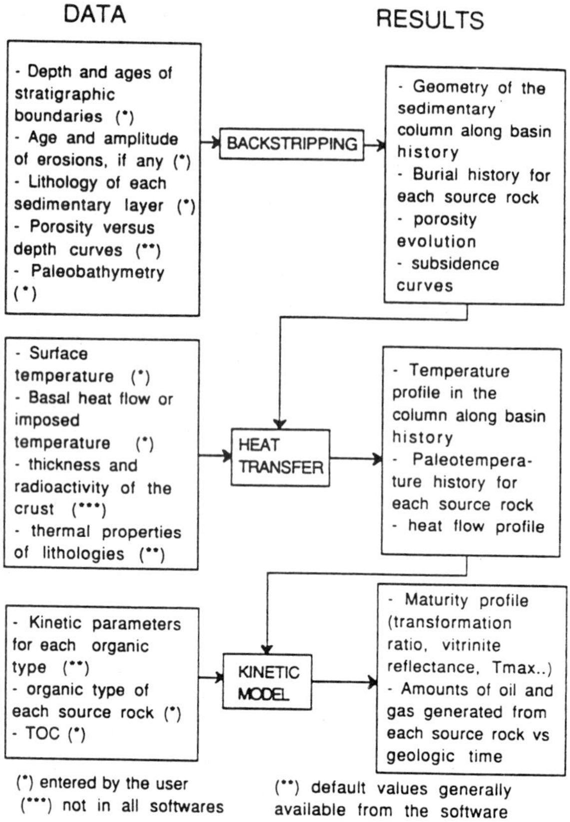

Figure 7.3. A generalized scheme showing the combination of burial reconstruction, heat transfer, and petroleum generation in a 1-D thermal model (from Ungerer, 1993).

7.3. SOURCES OF E AND A DATA

7.3.1. Introduction

Kinetic models require as input the kinetic parameters for each of the reactions modeled, or at least global values appropriate for the organic matter in the rock being modeled. Initially many people used single E, A pairs (e.g. Wood, 1988; Barker, 1989; Mudford and Best, 1989; Mudford, 1990). This is most likely to be valid for Type I kerogen because it can be closely approximated by a single E value (Burnham et al, 1987; Ungerer and Pelet, 1987). Sweeney et al (1987) have used 52.4 kcal/mol for the

Type I Green River kerogen and showed that the calculated maturity agrees well with the reported geology.

In general global values seem to be very low compared with laboratory-determined values for bond cleavage. The Lopatin method's basic assumption of reaction rate doubling every 10 °C corresponds to only 20 kcal/mol at 100 °C whereas C – C bond cleavage is very much higher, generally of the order of 79 kcal/mol. The reason for this apparent dilemma was discussed in Section 2.2, where it was shown that a set of parallel first order reactions leads to a combined, "global", value that is very much lower than any single reaction being considered. It was also pointed out in Chapter 2 that E and log A appear to be linearly related. Sometimes it is possible to deduce A and E values from geological studies (e.g., Dahl and Augustson, 1993), but the main problem involves establishing a reliable thermal history, and this can introduce considerable uncertainty. It is now more common to obtain kinetic parameters from laboratory pyrolysis, even though there are concerns that the short term, high temperature laboratory experiments may involve different reaction mechanisms, making extrapolation to geological times and temperatures questionable (e.g., Snowdon, 1979). Michelsen and Khorasani (1995) have pointed out that the temperature range for laboratory pyrolysis corresponds to a reaction rate where there is relatively little change with temperature, while in contrast, for the natural temperature range there is a higher rate of change. Small laboratory errors can extrapolate to large errors in the geologic setting. In spite of these potential problems, pyrolysis data are commonly used, and sufficient data have been accumulated that it is now possible to assign default values to the major kerogen types.

7.3.2. Pyrolysis

Activation energies and frequency factors can show a wide range of values for different organic matter types, and ideally should be determined separately for each type of organic matter in the study area. This is done most conveniently by pyrolysis, because this laboratory procedure combines the effects of time and temperature. The shape and position of the pyrolysis Peak 2 (S2) (see Section 1.9.4) reflects the distribution of E and A values. While it is possible to fit a pyrolysis peak with a single A and E pair to give global values, it is preferable to adopt a suite of values. Usually a dozen pairs (more or less) are used, and the fit is optimized by running pyrolysis at 3 or 4 widely different heating rates (Figure 7.4). These usually range from a fraction of a °C/min up to 50 °C/min, or more. High heating rates may lead to poor temperature control and Espitalie et al (1993) recommend six heating rates over the range from 2 °C/min to 30 or

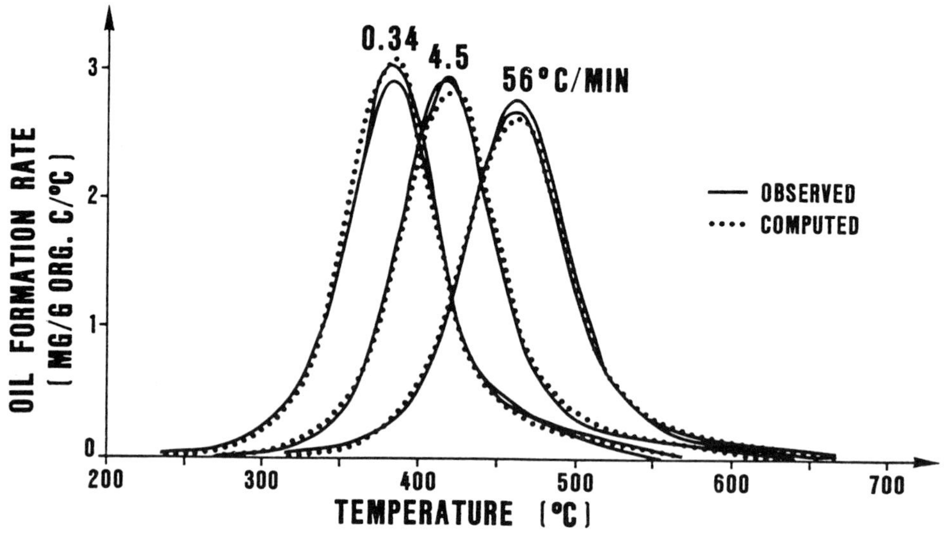

Figure 7.4. Pyrolysis response at three different heating rates. Experimental data are shown as solid lines and these give some indication of the reproducibility of the method. Computed data (dotted lines) are obtained from a best fit of E and A data.

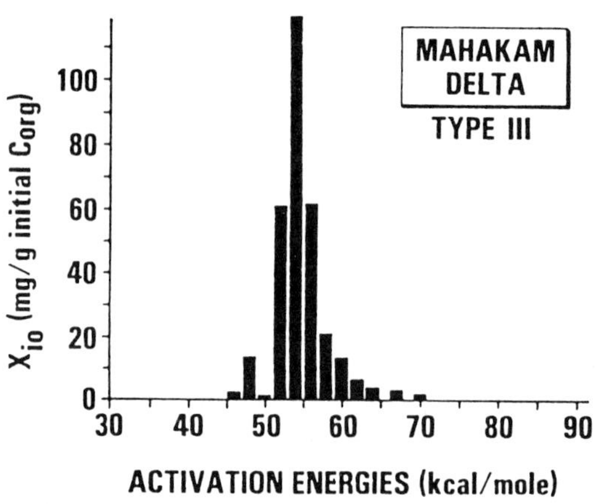

Figure 7.5. Typical representation of the suite of activation energies describing the thermal breakdown of a Type III kerogen, in this case from the Mahakam Delta, Indonesia. Activation energies are given in 2 kcal/mol increments.

40 °C/min. Ungerer and Pelet (1987) used 0.34, 4.5, and 56 °C/min and found that optimizing the fit of pyrolysis data with a suite of E values required the higher and lower heating rates to differ by at least two orders of magnitude. They claimed that A values have a less dramatic role on the cracking rates for unimolecular (first order) reactions over the normally observed range from 10^{14} to 10^{18} sec^{-1}. Each individual bond in the kerogen has its own environment, so that bond energies form a continuum. Digitizing this is arbitrary and values are normally given in 1 or 2 kcal/mole increments over the range from 40 to 60 kcal/mole (e.g., Figure 7.5). Replicate analyses by Burnham (1994) led to a standard deviation of about 1 kcal/mol (with a corresponding two-fold change in A), but Burnham et al (1987) have shown that small temperature fluctuations in the pyrolysis apparatus lead to large variations in the E and A values determined. Examples are given in the figures. The best match between the pyrolysis curve and a suite of activation energies is generally obtained by minimizing an error function, which is the sum of squared differences between the experimental rates of release and those calculated from the kinetic model (Ungerer, 1990). Several programs for fitting kinetic data have been published (e.g., KINETIC, OPTIM). Frequency factors are commonly held constant, though this seems to be done more to simplify the mathematics than out of any detailed insights into the kinetic processes. Matching can also be simplified by assuming a Gaussian distribution for the E values and a constant value for A (Braun and Burnham, 1987; Mackenzie and Quigley, 1988), but detailed studies have shown that in general (and especially for Type II kerogens) the suite of E's is not symmetrical around the central value.

A useful alternative mathematical procedure was developed by Freeman and Carroll (1958). This only requires a single heating rate and makes no assumptions about the order of reaction. Delvaux et al (1990) described a modified version of their procedure that used a 25 °C/min heating rate with previously-extracted samples. The method described by Klomp and Wright (1990) also determines activation energies without prior assumptions of reaction order or frequency factors. Activation energies are determined as a function of kerogen conversion.

Experimental uncertainties during laboratory pyrolysis lead to errors in the derived values for E and A suites (Jarvie, 1991; Espitalie et al, 1993). Although Lakshmanan et al (1991) have claimed that laboratory measurements are not sufficiently precise to give kinetic parameters that can be extrapolated to geologic processes, this appears to be a minority viewpoint. Nielsen and Dahl (1991) considered how well kinetic parameters could be determined from RockEval pyrolysis data and what

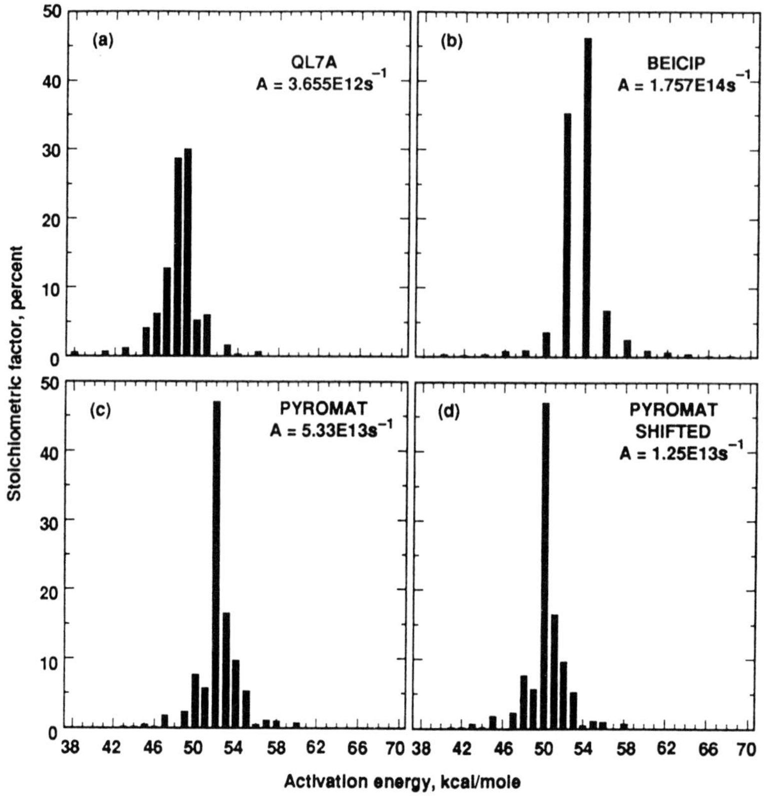

Figure 7.6a. Suites of activation energies and stoichiometric factors obtained for the La Luna source rock, Venezuela using (a) Rock-Eval kinetics, (b) BEICIP kinetics, (c) Pyromat kinetics, and (d) shifted Pyromat kinetics (Sweeney et al, 1990).

impact these errors have on geological predictions of generation depth and timing. Reproducibility of T_{max} to within 1 °C lead to negligible errors in the hydrocarbon generation history. When T_{max} was determined to within 2 °C, then for a geothermal gradient of 35 °C/km and a heating rate of 2 °C/m.y. peak generation occurred at 134 $\pm$ 4 °C, corresponding to a depth uncertainty of $\pm$ 114 m. In a similar study for isothermal hydrous pyrolysis, Nielsen and Barth (1991) found that the experimental errors lead to an uncertainty of about $\pm$ 400 m for the depth of the top of the oil window when geothermal gradient was 30 °C/km and the heating rate was 2 °C/m.y.

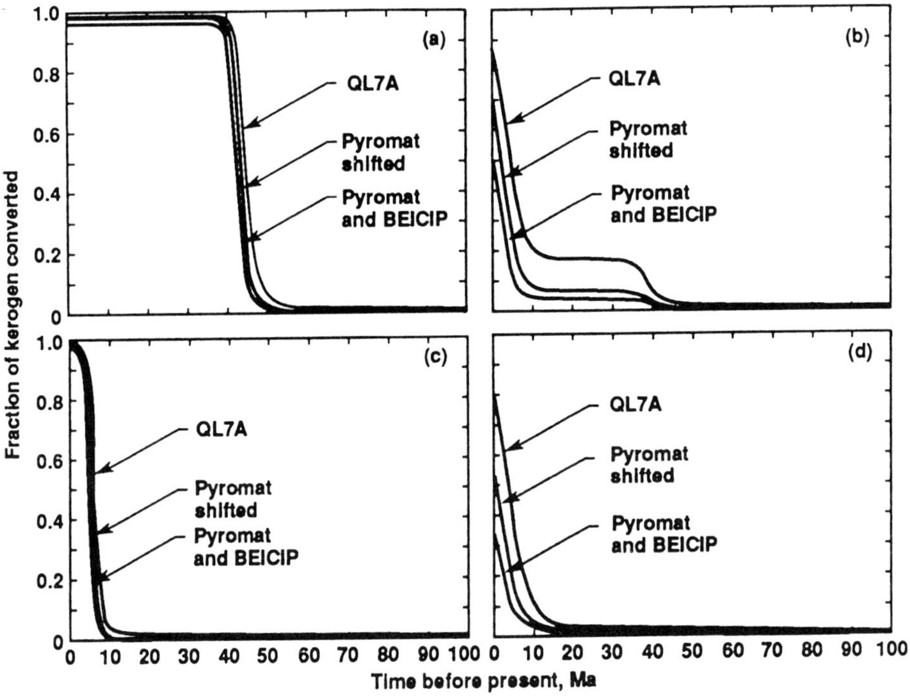

Figure 7.6b. Calculated fraction of kerogen converted to oil as a function of time for the four sets of activation energies given in Figure 7.6a and four different burial histories. The widest spread was found for an area where petroleum generation started about 45 Ma, but was subsequently suspended due to uplift 35 Ma, and then resumed about 15 Ma (Sweeney et al (1990).

Comparisons of some of the more widely used methods for obtaining suites of activation energies from laboratory pyrolysis have been published by Burnham et al (1987, 1988) and Sweeney et al (1990). Figure 7.6 shows the various distributions obtained by Sweeney et al (1990) for the Cretaceous La Luna carbonate source rock of western Venezuela. The significance of these differences for the calculated timing of petroleum generation is also shown for several areas with different thermal histories. In most cases the agreement is adequate, but it is poorest for an area where petroleum generation started about 45 Ma, but was suspended due to uplift 35 Ma, and then resumed about 15 Ma (Sweeney et al, 1990). All activation energy data sets lead to the conclusion that the La Luna is still generating in this area, and this is consistent with the observed abnormal pressures.

The determination of kinetic parameters from pyrolysis experiments assumes that the mechanism for hydrocarbon formation (and the nature of the compounds actually formed) are the same at laboratory heating rates of 0.1 to 50 °C/min as natural heating rates of approximately 10^{-11} °C/min (= 5.3 °C/m.y.). Also, because of the nature of the time/temperature trade off, laboratory heating experiments subject organic materials to much higher temperatures (~500 °C), where reaction mechanisms may be different (i.e., different E, A) than those occurring at roughly the 100 °C required for generation in nature (Snowdon, 1979; Shaefer, 1990). Both whole rock samples and separated kerogens have been used for the pyrolytic determination of kinetic parameters. Separation of kerogen involves an additional, time-consuming step (Durand and Nicaise, 1980), but avoids problems due to an overlapping bitumen response from Peak 1 (S1). However, kerogen separation rarely approaches 100% recovery of the organic matter, and there is always some uncertainty about just how representative the recovered material is. In general, low organic carbon rocks give low pyrolysis yields compared with isolated kerogen. While pyrolysis of whole rock samples is faster and more convenient, the mineral matter may adsorb heavy ends and lead to thermal cracking reactions (Espitalie et al, 1984) in a so-called *"matrix effect"*. Dembicki (1992) claimed that isolated kerogens may not provide realistic samples for kinetic parameter determination, and could give non-representative values. His conclusion was based on pyrolysis studies using separated kerogen and various kerogen/mineral mixtures. In each case E and A values were calculated by fitting the pyrolysis curve using the KINETIC program of Braun and Burnham (1987). At low organic/matrix ratios hydrocarbons were apparently adsorbed by quartz, calcite, and dolomite and released at higher temperatures, leading to higher apparent activation energies (see Chapter 2, Figure 2.4) (Figure 7.7). In contrast, kaolinite and bentonite appeared to catalyze the thermal breakdown of the kerogen, and E values for these systems were lower than for the isolated kerogen. Except for kaolinite, E value suites for the kerogen/matrix combinations approached values for the separated kerogen at high organic values corresponding to roughly 3 - 4 wt% TOC. Dembicki (1992) stressed the importance of using representative samples for pyrolysis, and not working with the organic-rich ones because they are easier to analyze. Pelet (Pelet and Dembicki, 1994) criticized Dembicki's conclusion, and by using his data to simulate a simple

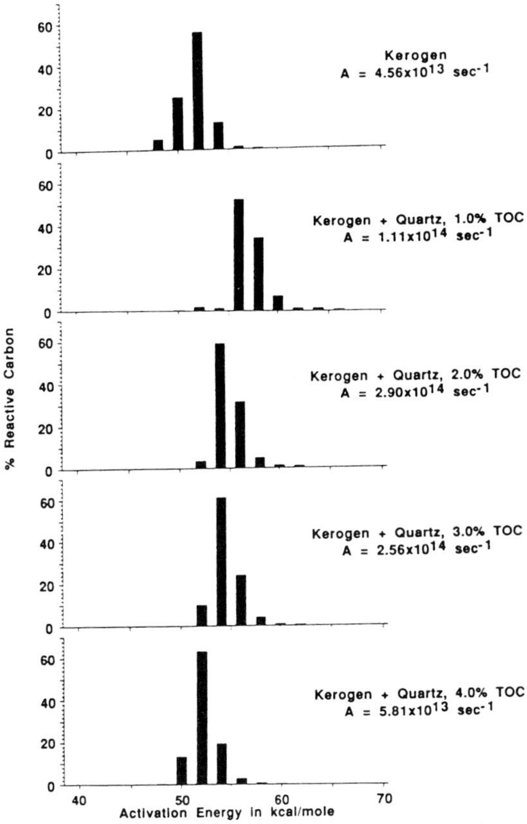

Figure 7.7. Comparison of activation energy suites obtained by pyrolysis of separated kerogen, and various kerogen-quartz mixtures (Dembicki, 1992).

geological model concluded that the results obtained were geochemically unlikely. However, it seems that considerations of matrix effects and mineral catalysis during pyrolysis become less significant as organic carbon content rises, and above a couple of percent can probably be ignored. Quigley and Mackenzie (1988), for example, found no significant matrix problems. An additional problem has been reported for samples that contain salt because this produces high apparent values for T_{max}, S2, and HI (Issler and Snowdon, 1990).

Standard pyrolysis methods heat small samples in flowing helium at pressures close to 1 atmosphere, but pyrolysis in an open flowing system allows the volatile products to escape, and so does not simulate real world conditions as well as closed system pyrolysis. This is especially true for reactions that involve cracking of mobile components (Ungerer, 1990). In open systems, heavy ends are not mobilized, and values obtained for total extractables are low. The rapid removal of oil-like materials from the heated zone also prevents cracking to gas, and these *"secondary reactions"* are not simulated. In spite of all these potential difficulties dry pyrolysis is widely used to obtain activation energies, and the values obtained appear to work quite well.

Subsurface conditions involve the presence of water and higher pressures, and are better simulated by hydrous pyrolysis (Lewan et al, 1979; Lewan, 1993; Ritter et al., 1993; Barth and Nielsen, 1993). These techniques need much larger samples, require several days for each experiment, and need a large number of experiments, and so they are less commonly used. For example, Castelli et al (1990) pyrolyzed 20 g samples of crushed and extracted rock for 4 - 336 hours at 260 - 330 °C and 15 - 17 MPa. In addition detailed analyses of the products were needed before kinetic parameters could be obtained. Horsfield et al (1993) compared closed and open system pyrolysis methods and concluded that they are closely similar, and either experimental procedure can be used to generate meaningful data.

In order to get kinetic parameters from organic matter by pyrolysis it is essential to have immature samples. This can present difficulties in areas with limited well control, because the wells may only have penetrated the source rock where it is mature, and in some areas there may be no immature source rocks at all. In mature samples, part of the generating ability has been used up, and the remaining organic matter will have a distribution of activation energies that is quite different from the immature starting material. Specifically, it will show diminished amounts of the components with low E values (Figure 7.8). Schaefer et al (1990) have given an example of activation energy distributions determined at various maturity levels for Posidonia Shale samples from northern Germany. For situations where samples of different maturities are available from a single rock unit with uniform organic matter type, Nakayama and Cao (1993) have developed a method based on HI data obtained from RockEval pyrolysis. It is also important to recognize the possibility of lateral variations in organic matter type and to determine E and A values for multiple, representative samples with good areal distribution across the study area. The development of some reservoirs is controlled by the location of paleotopographic highs, and often these can be areas where conditions have been more oxidizing, and where terrestrial organic matter may have been relatively more abundant. Here organic matter may appear gas prone (not only because it is partly terrestrial but also because it is oxidized), while the laterally equivalent rocks contain

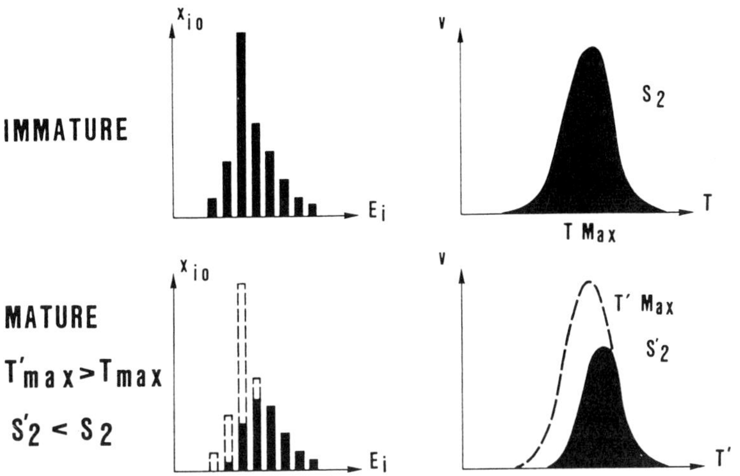

Figure 7.8. Changing activation energy suites corresponding to organic materials of different maturities. As maturity increases low activation energy values become less prominent. (From Tissot et al, 1987).

unoxidized oil-prone organic matter. It is important to consider these lateral facies changes in selecting appropriate immature, time equivalent samples.

7.3.3. Default Values

The increasing amount of kinetic data being generated for the various kerogen types shows an overall consistency, and it is possible to assign generalized E and A suites (*"default values"*) for each of the 3 major kerogen types, as well as for the sulfur-rich Type II ("II-S") (Hunt et al, 1991). Even in areas with good control, it is often simpler to use these default values than to obtain laboratory-derived numbers (Table 7.1; Figure 7.9). This is particularly true in preliminary thermal models. Most commercial modeling programs provide default values for each of the organic matter types, so it becomes a problem of establishing which of the organic matter types is present. When samples are available this is done routinely from RockEval HI/OI pyrolysis data or from van Krevelen diagrams, but both of these techniques give results that are the weighted average of all the organic material in the samples. It is possible to have a mixture of Type I and Type III kerogens that plots right on the Type II track of the van Krevelen or HI/OI diagrams (Figure 7.10). In the basins of the Central African rift subsystem Genik (1993) found that large amounts

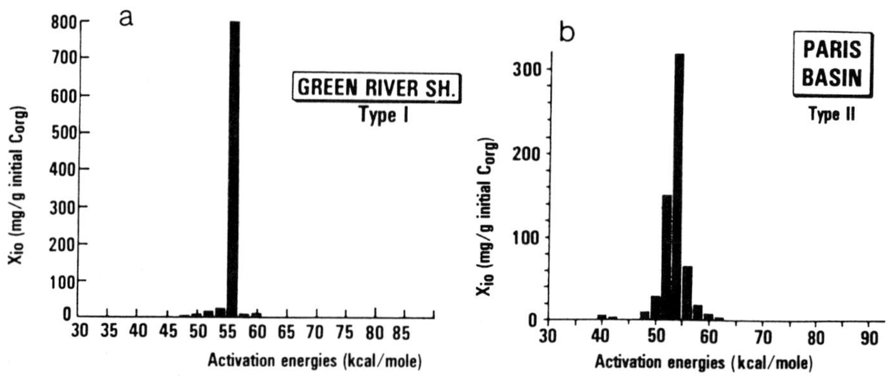

Figure 7.9. Typical activation energy suites for Types I and II kerogens (Type III is shown in Figure 7.5) (Tissot et al , 1987).

of kerogen that appeared to be Type II, based on Hydrogen Index values in the 300 to 600 range, were in fact mixtures of Type I and III kerogens. At high maturity levels all three tracks on a van Krevelen type diagram converge (Figure 7.10), and it is difficult to make distinctions among organic matter types and to deduce which path they followed across the diagram. For very mature samples this makes it difficult to decide what type of organic material was originally present. In addition to the chemical methods, microscopic examination of separated kerogen provides important supporting information for establishing organic matter type, but again for very mature samples making distinctions can become difficult or, if the organic matter is blackened, impossible.

Modeling undrilled areas frequently involves extrapolating long distances from well control. In such cases it is important to consider possible regional changes in organic matter type. For example, in the Norwegian sector of the North Sea, Baird (1986) showed a systematic distribution of kerogen types in the Kimmeridge Clay. Type I kerogens occurred mainly along the Central graben, with Type III being the dominant type in platform areas to the east. Type II occupied the intermediate area.

Dixon et al (1992) have compared the Type III default values used for the commercial MATOIL program (BEICIP, France) with those obtained by laboratory pyrolysis for Type III kerogens in the Beaufort/Mackenzie area of northwest Canada. Modeling with the default values gives onset of generation at 7 Ma, and the measured values give 17 Ma. The Beaufort/Mackenzie kerogens appear to be somewhat unusual in having a greater than normal potential for low activation energy reactions. In cases

Table 7.1
Selected suites of default values for the various kerogen types from different research groups (Data assembled by Platte River Associates, 1993).

LLNL Kerogens

Type I

Cracking P-Primary S-Secondary	Fract. (0-1)	Activation Energy (kcal/mole)	Arrhenius Constant (1/my)	Reaction Constants
P	0.07	49.0	1.6e+27	primary oil = 650
P	0.90	53.0	1.6e+27	primary gas = 70
P	0.03	54.0	1.6e+27	
S	1.00	54	3.2e+25	2ndary gas yield=50%

Type II

Cracking P-Primary S-Secondary	Fract. (0-1)	Activation Energy (kcal/mole)	Arrhenius Constant (1/my)	Reaction Constants
P	0.05	49.0	9.5e+26	primary oil = 350
P	0.20	50.0	9.5e+26	primary gas = 65
P	0.50	51.0	9.5e+26	
P	0.20	52.0	9.5e+26	
P	0.05	53.0	9.5e+26	
S	1.00	54	3.2e+25	2ndary gas yield=50%

Type III

P-Primary S-Secondary	Fract. (0-1)	Activation Energy	Arrhenius Constant	Reaction Constants
P	0.04	48.0	5.1e+26	primary oil = 50
P	0.14	50.0	5.1e+26	primary gas = 110
P	0.32	52.0	5.1e+26	
P	0.17	54.0	5.1e+26	
P	0.13	56.0	5.1e+26	
P	0.10	60.0	5.1e+26	
P	0.07	64.0	5.1e+26	
P	0.03	68.0	5.1e+26	
S	1.00	54	3.2e+25	2ndary gas yield=50%

Tissot & Espitalie′ Kerogens

Type I

Cracking P-Primary	Fract. (0-1)	Activation Energy (kcal/mole)	Arrhenius Constant (1/my)	Reaction Constants
P	0.027	10	4.750e+05	primary oil = 895
P	0.071	30	3.040e+16	primary gas = 0
P	0.152	50	2.280e+25	2ndary gas yield=0%
P	0.170	60	3.980e+30	
P	0.388	70	4.470e+32	
P	0.192	80	1.100e+34	

Type II

Cracking P-Primary	Fract. (0-1)	Activation Energy (kcal/mole)	Arrhenius Constant (1/my)	Reaction Constants
P	0.032	10	1.270e+05	primary oil = 695
P	0.049	30	7.470e+16	primary gas = 0
P	0.361	50	1.480e+27	2ndary gas yield=0%
P	0.219	60	5.520e+29	
P	0.167	70	2.040e+35	
P	0.172	80	3.800e+35	

Type III

Cracking P-Primary	Fract. (0-1)	Activation Energy (kcal/mole)	Arrhenius Constant (1/my)	Reaction Constants
P	0.073	10	5.200e+03	primary oil = 315
P	0.168	30	4.200e+16	primary gas = 0
P	0.228	50	4.330e+25	2ndary gas yield=0%
P	0.289	60	1.970e+32	
P	0.156	70	1.200e+33	
P	0.086	80	7.560e+31	

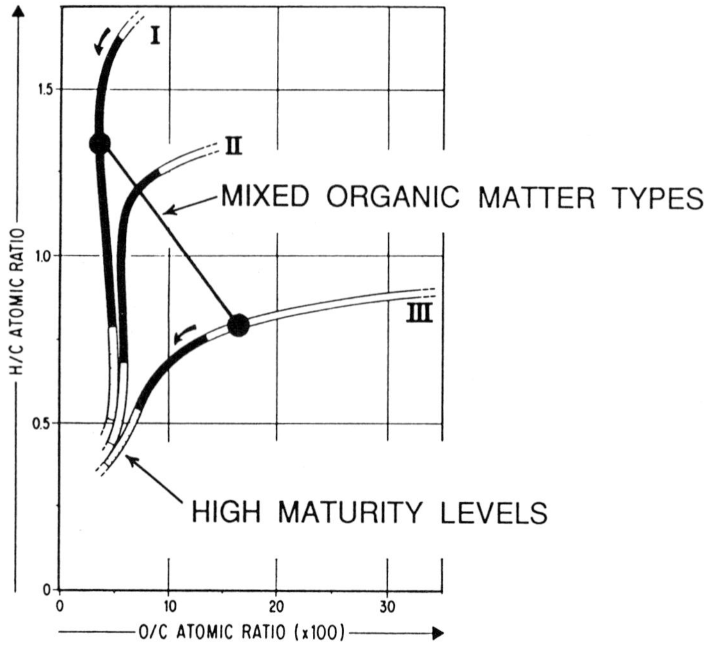

Figure 7.10. Van Krevelen diagram showing how a mixture of Type I and Type III kerogens could plot on the Type II track. High maturity samples are difficult to assign to a particular track.

where the potential source rock contains a mixture of different kerogens, it is possible to combine the various activation energy suites in the proportions of the different organic matter types and use this in the modeling. Most commercial models offer this option.

7.3.4. Outcrop Samples

In the very early stages of exploration when there are few, if any, wells, there may be no subsurface samples. Often the only materials available are from outcrops. These may show the environment of deposition for important rock units, and give an indication of marine, deltaic, carbonate reef, lacustrine, or fluvial character that can then be used to select appropriate default values for the likely organic materials. Analysis of outcrop samples can give direct information about the organic materials, but these data need to be treated with caution because of the possible effects of weathering and oxidation (Leythaeuser, 1973; Clayton and King, 1987). Whenever possible, road cuts, water wells, and quarries should be used to get out of the zone of weathering. In temperate regions or high latitudes this is not usually difficult, but it may be impossible in the tropics where weathering can be a severe problem, and the effects may penetrate several hundred meters. Oxidation of pyrite to sulfate can be used to monitor the extent of weathering (Lo and Cardott, 1995). All of the effects of weathering lead to pessimistic evaluations.

Bacterial degradation of the kerogen exposed in an outcrop is rapid, and results in decreased hydrogen content as the aerobic bacteria consume the H-rich alkane side chains. In addition, direct oxidation leads to an increased oxygen content in the kerogen, and the net effect of these two processes acting together is to move a point on a van Krevelen diagram towards the gas-prone track (Figure 7.11). As a consequence, an outcrop sample may be characterized as a Type III kerogen and the corresponding E and A values used in modeling deep burial, when in fact the unit actually contained Type II Kerogen. Oxidation of outcrop organic matter generally leads to values of total organic carbon that are too low (because of loss of material by oxidation) (Clayton and Swetland, 1978), and also gives vitrinite reflectance values that are anomalously high.

7.3.5. Oil-Source Rock Correlation

Early stages of exploration commonly target high blocks, and often oil samples are available before the corresponding source rocks have been drilled. Also, in areas where there has been long distance migration, oil samples may be available long before source rock samples are obtained. In many cases details of oil composition can be used to deduce the nature of the source rock (e.g., Alexander et al, 1992), and this information helps constrain the modeling procedure. For example, a waxy, high pour point, low sulfur crude oil would suggest a source rock formed in an environment dominated by terrestrial organic matter, such as a lake, or a deltaic, near-shore setting. High pristane/phytane ratios and light carbon isotope values would support this, as would the $C_{27\text{-}29\text{-}30}$-sterane distribution discussed in Serction 1.11.3. Comparison of carbon isotope values for saturate and aromatic fractions of oils is especially useful in distinguishing between terrestrial- and marine-sourced oils (Sofer, 1984), and Killops et al (1994) have described an example from the Taranaki basin, New Zealand. A wide range of biomarkers and biomarker ratios has been proposed for relating oils to specific types of organic matter in source rocks, and Table 7.2 gives some commonly used compounds that help to show the nature of the source rock. Some, such as gammacerane, suggest an environment – in this case a high salinity sabkha. Others, such as oleanane, give age information, since

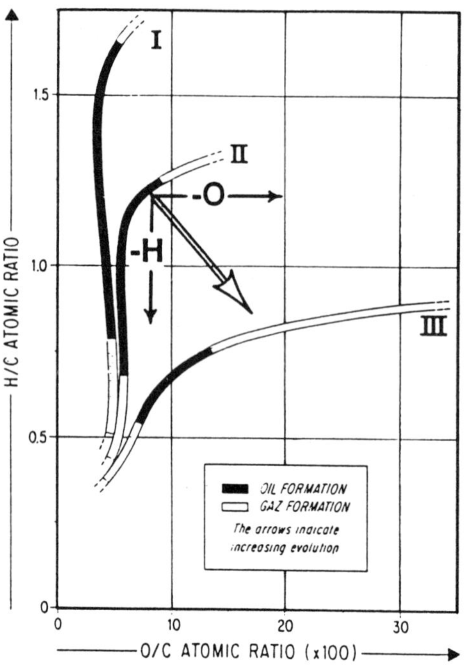

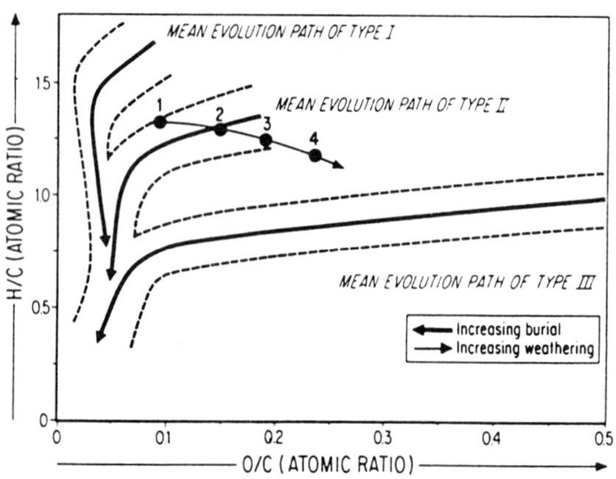

Figure 7.11. Displacement of data points on a van Krevelen diagram due to increased weathering of a Type II kerogen.
(Top) Schematic changes in H and O
(Bottom) Outcrop weathering of organic matter in the Lower Toarcian of the Paris Basin : distance from the surface – **1** : 5.3 m, **2** : 4.8 m, **3** : 1.0 m, **4** : 0.5 m (Durand and Monin, 1980).

oleanane comes from angiosperms that did not evolve until the Late Cretaceous (Moldowan et al, 1994). The source rock for an oil containing oleanane would have to be younger than Late Cretaceous. Similarly, bicadinanes are also associated with certain types of angiosperms that have only been abundant since Late Cretaceous (van Aarssen et al, 1990). Terrestrial oils lack tricyclic terpanes (Peters and Moldowan, 1991), usually have high hopane/sterane ratios (Figure 7.12), and tend to be isotopically light. Dahl et al (1994) found that the percentages of C_{30} and C_{28} 4-desmethyl steranes relative to the total 4-desmethyl steranes correlated with the percentage of amorphous, marine-derived Type II kerogens in the source rock. While there is now some doubt about rearranged steranes being due to acid catalysis (i.e., clastics), the empirical observation that carbonate-related oils are relatively low in rearranged steranes is still valid. Some oils have very characteristic compositions and this is especially true of the middle Ordovician oils which are dominated by the products of *Gloecapsamorpha prisca*. These oils have abundant short chain alkanes (<C-20) with marked odd-even predominance. Pristane and phytane are present in low concentrations, and hopanes are more abundant than steranes (Fowler, 1992). The age of these oils predates the evolution of terrestrial organic matter so that Type III kerogens are not a factor in modeling the maturity of their source rocks.

When several oil accumulations occur in an area it is particularly important to know if they are members of a single family, or whether two (or more) oil types are present. If all the oils are related, then only a single source rock is involved, but if there are multiple oil types there must be more than one source rock. Correlation studies always give the minimum number of source rocks in an area. In areas with several source rocks each

Table 7.2
Some compositional indicators of source.

OLEANANE	: POST-LATE CRETACEOUS
TERPANES FROM CONIFERS	: PERMIAN OR YOUNGER
GAMMACERANE	: HIGH SALINITY ENVIRONMENT
PERHYDROBETACAROTENE	: LACUSTRINE
WAXY CRUDE	: TERRESTRIAL
ISOTOPICALLY LIGHT	: TERRESTRIAL
HIGH DIASTERANES	: CLASTIC ENVIRONMENT
HIGH C-29 STERANES	: TERRESTRIAL

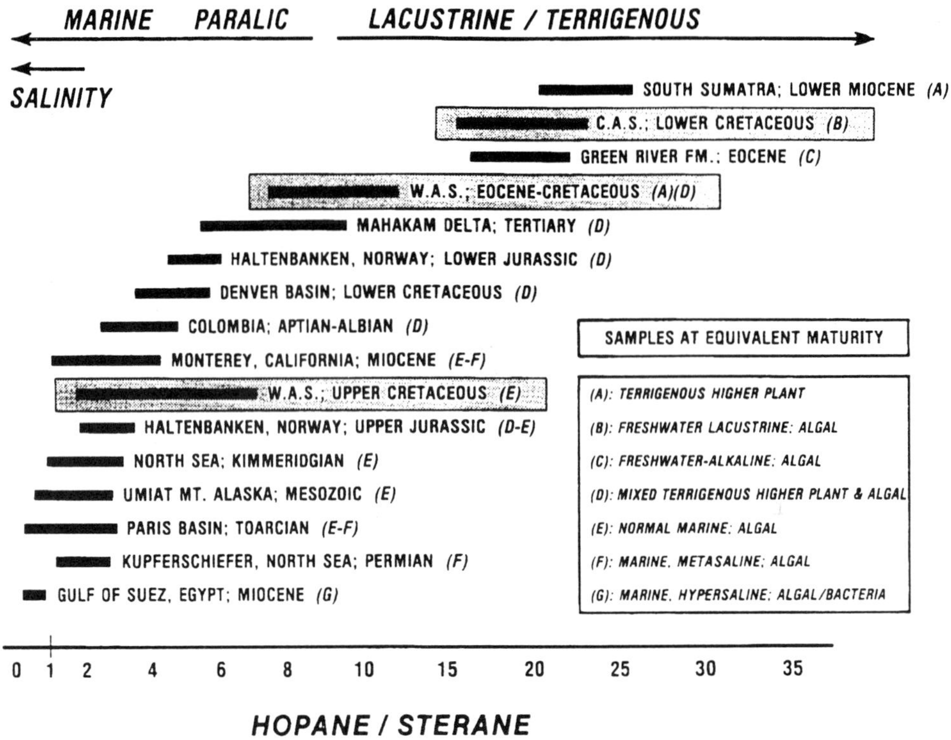

Figure 7.12. Comparative chart showing hopane/sterane ratios determined for oils sourced from marine, paralic, and lacustrine/terrigenous kerogens (Genik, 1993).

one will have to be modeled independently, using the appropriate activation energy, frequency factor, and time-temperature history. If a local high block has two adjacent basins with somewhat different geologies, then it is necessary to establish which one the oil came from, so that the appropriate source rock and burial history can be modeled. For example, in the Oseberg area of the North Sea correlation studies using biomarkers and carbon isotopes showed that there were two oil types which, although probably related, are distinct (Dahl et al, 1987). Combining this information with a deterministic, dynamic basin model suggested that the Oseberg Beta South structure was filled first from a source area in the Huldra/Veslefrikk graben, while the Alpha structure was charged later from the Viking graben.

7.4. TIME INTERVALS

In the rigorous kinetic models, calculations are carried out at various time intervals through the burial history of an area. If the value used for the time increment is very short then the calculation will be repeated many times, and the modeling will use excessive computer time. If, on the other hand, long time intervals are used the reconstruction may have inadequate resolution. The BASINMOD (Platte River Associates) commercial program suggests using 4% of the age of the oldest rock as an appropriate time interval, and this seems to be a reasonable comprise that works well. The BASINMAT program described by Sweeney (1990) used fixed 1 m.y. time increments. In the early stage of model development it can be helpful to use longer time intervals so that the models run faster. The time intervals can be decreased as modeling is refined. For pressure calculations (Section 9.4) much shorter time intervals should be used because pressure can change dramatically over very short time periods.

7.5. CHECKS ON VALIDITY OF MODELS

7.5.1. Introduction

The combination of burial profiles with thermal models is an attempt to reconstruct the geologic history at some location. It is important to check that the predictions from such a reconstruction match everything that is known about the area, and that all of the geologic inputs and the calculated results are self-consistent. If, for example, geochemical oil-rock correlations show that a particular rock is the source of oil, but the thermal modeling shows the rock to be immature, then there is a problem with the model used. Checks on the validity of models are most easily done by comparing predicted present day values with those actually measured. Maturity indicators (TTI, etc) are usually calibrated to equivalent R_0 values, allowing the reflectivities measured in a well to be compared with the calculated values. If the measured and calculated values down the well profile do not agree, then the model is wrong and needs to be modified (Figure 7.13). If the measured and calculated values do agree, then the model may be correct – but this is not guaranteed, since there is always the possibility that the model is incorrect in ways that still happened to give essentially correct R_0 values (a combination of too low a geothermal gradient with burial depths that are too deep, for example). Vitrinite reflectance integrates the total time-temperature exposure and so does not

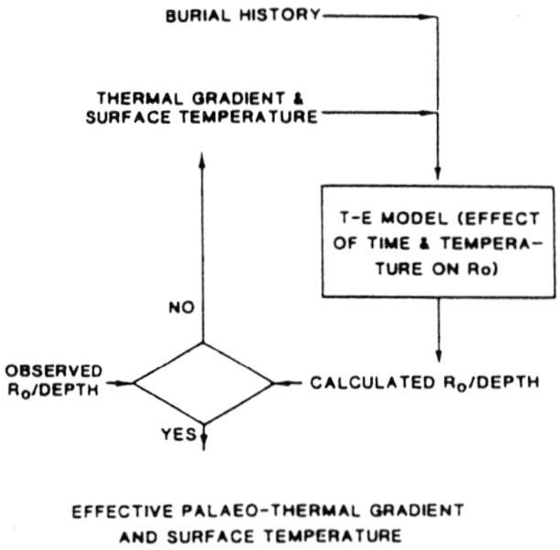

Figure 7.13. Iterative procedure for comparing calculated and measured vitrinite reflectance values (from Yukler and Kokesh, 1984).

give unique solutions. Although vitrinite reflectance is probably the most common way of evaluating models all possible sources of geologic information should be integrated with the model. At the very least there should be reasonable consistency for wells across the study area, and the calculated time of oil generation should post-date the time of trap formation. Glasman et al (1989), for example, dated the time of illite formation in the Brent Field of the North Sea as a way of establishing the timing of hydrocarbon accumulation, and this was used to "provide a check point for geochemical models".

7.5.2. Calculation of Vitrinite Reflectance

The activation energies and frequency factors that control vitrinite maturation (and produce the increase in reflectivity) are not necessarily the same as the E and A values for the organic matter that generates petroleum. It is important to remember that vitrinite is being treated like a thermometer that is removed from the rock as a indicator of cumulative temperature-time effects. Vitrinite is a Type III, wood-derived kerogen and does not generate oil; most oil is derived from Type II kerogens that came largely from marine algae. Because of these differences, the calculated R_o values should be derived using E and A values for vitrinite, while the generation status should be obtained using appropriate E and A values for the organic matter in the potential source rock.

Older methods for calculating present day vitrinite reflectance values for burial histories were either empirical (e.g., derived from TTI), or used only a single value of activation energy (Middleton, 1982; Antia, 1986;

Wood, 1988). They have been replaced by procedures that use a suite of activation energies for parallel first order reactions. The method suggested by Larter (1988) is good up to the end of the oil window (R_o = 1.3) but was extended to R_o of 4.5% by Burnham and Sweeney (1989) using a program called VITRIMAT. This considers a set of 4 independent, parallel, first-order kinetic reactions, each with its own suite of E values. Changes in elemental composition are calculated and compared with well-documented calibration data to give reflectance. Andresen et al (1994) and Throndsen et al (1993) have shown that the model works well even for the rapid heating rates of laboratory pyrolysis. Sweeney and Burnham (1990) later introduced a simplified spreadsheet version of VITRIMAT called EASY%R_o. This still uses parallel first order reactions with appropriate activation energies ranging from 34 to 72 kcal/mole (roughly centered around 55 kcal/mole), and a single A value of 1.0×10^{-13}/sec. They demonstrated the programs usefulness for heating rates ranging from those typical of slow laboratory pyrolysis, through igneous intrusions, to typical burial history rates. As an example the EASY%R_o values were compared to those derived from TTI modeling and Wood's method for pull-apart, foreland, and intracratonic basin models. For all three time-temperature profiles used, EASY%R_o predicts lower values of vitrinite reflectance than either of the other approaches, though there are no systematic difference in %R_o. EASY%R_o appears to have wide applicability, but may slightly overestimate %R_o at low maturity levels (<0.9% R_o) (Morrow and Issler, 1993). Leischner et al (1993) used EASY%R_o to calculate vitrinite reflectance values in the Lower Saxony Basin using a temperature history derived from fluid inclusion homogenization temperatures. Data from inclusions in anhydrite and quartz veins were consistent with the intrusion of magma 90 Ma that was inferred from geologic observations. The calculated R_o values agreed well with the measured values, but when R_o was derived using the TTI method they were overestimated by 0.4%. Suzuki et al (1993) have also proposed a simplified program for calculating R_o called SIMPLE-R_o. For each time step in the burial history it uses a single activation energy, and this changes as a function of R_o to simulate the changing activation energy that occurs in nature as maturity increases.

7.5.3. Comparison of Calculated and Measured R_o Values

Vitrinite reflectance has been most widely used for comparing model predictions with measured values. Sweeney et al (1989) used values from a group of closely spaced wells in Lake Maracaibo, western Venezuela, and found good agreement between their measured values and the calculated

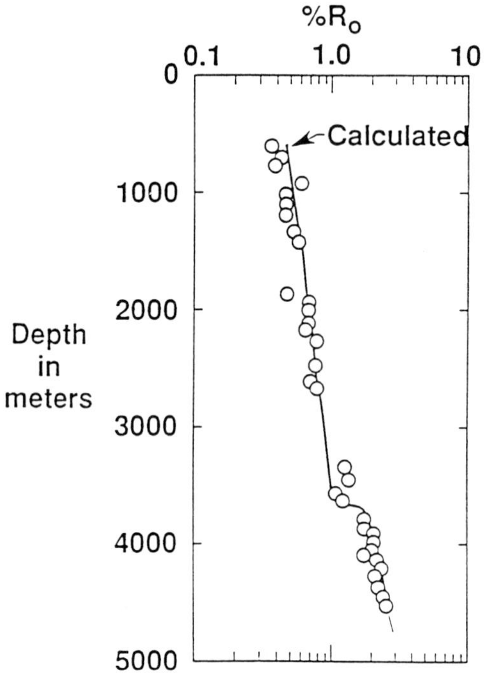

Figure 7.14. Calculated (line) and measured (circles) vitrinite reflectance values for the Cretaceous La Luna carbonate in western Venezuela (Sweeney et al, 1989),

trend. The trend (Figure 7.14) even included a jog to higher values that was coincident with the transition into the enhanced geothermal gradient found below the overpressured Colon shale. For the Scherbenhardt 2 well Espitalie (1984) reported that using the present geothermal gradient of 76.9 °C/km did not produce a good fit between the available data for R_o and the calculated trend, but when a gradient of 63 °C/km was used the fit was very good. This implies that the geothermal gradient had been lower in the past. A similar approach was employed by Yukler and Kokesh (1984) who used geothermal gradients from both corrected and uncorrected BHTs. Although the corrected gradient of 0.0255 °F/ft came closest to fitting the R_o trend, a refinement to 0.0240 °F/ft gave much better agreement (Figure 7.15). Cole et al (1987) found it necessary to adjust the paleogeothermal gradient upward in the Appalachian orogeny (Ohio) to explain their observed vitrinite reflectance data. Lerche (1990) has introduced a *"mean square residual"* (MSR) as a way of quantifying the average degree of

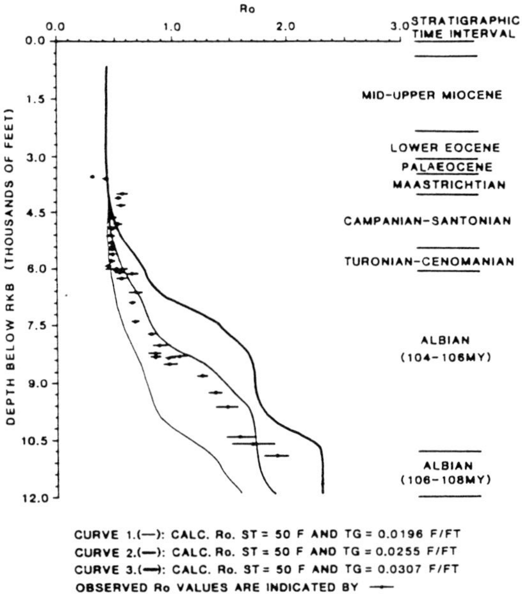

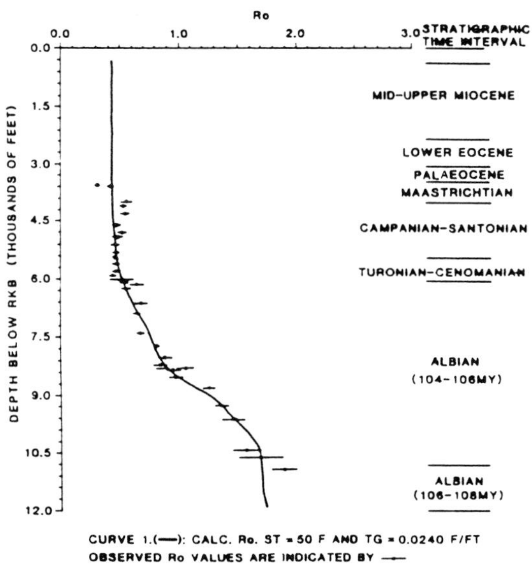

Figure 7.15. (top) Initial attempts to match experimental vitrinite reflectance values with those calculated using three different geothermal gradients, (bottom) refined geothermal gradient value optimizes fit with experimental data (from Yukler and Kokesh, 1984).

mismatch between a series of predicted and measured vitrinite reflectance values. The lower the calculated values of MSR, the better the fit. These studies are typical of the iterative approach that can be used. This is particularly important when key data are missing, such as paleogeothermal gradients or the amount of rock column lost at an unconformity.

In comparing calculated and measured values for vitrinite reflectance it is important to remember that there can be significant errors in the experimentally determined values of R_o. Claypool and Magoon (1985) found $\pm$ 15% variability between laboratories. In the interlaboratory comparison organized by Dembicki (1984) the range was even larger, and on one sample values ranged from 0.62 to 2.29%. He concluded that "extreme caution must be used when comparing source rock data from different laboratories". In general, there is less variation when a single operator makes all the measurements. Uncertainties in the values of R_o measured for the Haltenbanken area, offshore Norway, led Vik and Hermandud (1993) to conclude that they were not sufficiently precise to discriminate between the two thermal models being considered.

Other parameters can be used in addition to R_o. For example, in their study of the Utah-Wyoming thrust belt Deming and Chapman (1989) compared measured heat flow with values predicted by their model. Unfortunately, uncertainties in measuring present day heat flow led to poor "resolution" for this method. Rullkotter and Marzi (1989) applied sterane isomerization and aromatization kinetics to the Michigan basin, U.S.A., and found that the geothermal gradient values suggested by Nunn et al (1985) did not lead to agreement between the calculated and measured sterane ratios. When higher geothermal gradients, and deeper preDevonian burial, were used, as proposed by Cercone (1984), the fit between the experimental and theoretical values was much better (see Section 10.2.3). Some models are set up to hunt for consistency using an iterative procedure where calculated present day values are compared with experimental values. Link and Bustin (1989) used a model of this type that was based on a fixed burial history with various geothermal gradients. Radiometric age and isotope data for diagenetic minerals should also be consistent with interpreted geologic processes. Robinson et al (1993) have integrated this type of data for illite with burial history diagrams and the thermal history in the gas fields of the southern North Sea.

7.5.4. Karst Development

Nearsurface exposure during uplift and erosion can lead to the formation of paleosols, laterites, and "red beds". If a paleosol is encountered in a well (e.g., Hanneman et al, 1994; Wright and Smart,

1994; Dembicki and Machel, 1996), then the burial reconstruction must show that this was at the surface at some appropriate time. Volcanic ash, especially if laterally extensive, also indicates surface exposure. Exposure of carbonates to near-surface conditions may lead to the development of major solution features and the characteristics of karst (Herak and Stringfield, 1972; James and Choquette, 1988). Caves, sinkholes, joint-controlled solution features, karstic collapse breccias, dissolution/vugular porosity, and pendant cements may form within 30-100 m of the surface (Kerans, 1988), and so indicate shallow burial depths at some time in the burial history. Karst features can develop in as little as hundreds of thousands of years. It is important to distinguish features that develop soon after carbonate deposition from those that formed later.

The southern Oklahoma aulacogen was broken up in the Pennsylvanian to form the Wichita, Criner, and Arbuckle uplifts, and the Anadarko, Ardmore and Marietta basins. Burial history plots reconstructed for both the uplifts and the basins by Donovan (1991) show that there may have been as many as three separate episodes of karst formation as the Arbuckle Group carbonates were uplifted. In contrast, none developed in the very deeply buried basinal areas (Figure 7.16).

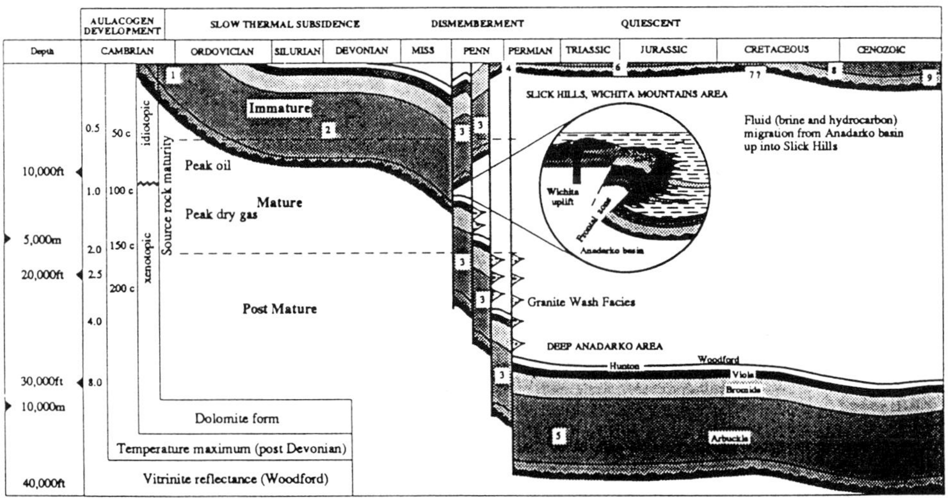

Figure 7.16. Burial history profiles for the Arbuckle Group, Oklahoma, showing an area of late uplift (Slick Hills) and one of late deep burial (Anadarko Basin) (from Donovan, 1991).

7.5.5. Fluid Inclusions Containing Hydrocarbons

If fluid inclusions contain oil then the time of onset of oil generation must precede the time of fluid inclusion development. The presence of oil-filled inclusions has been used most commonly to establish the timing of petroleum migration (Pagel et al, 1986 for the North Sea and Angola; Videtich et al, 1988 for Dubai), but can also provide useful constraints on the time of oil generation. In the vicinity of the Fateh field, Dubai, Videtich et al (1988) reported both oil-bearing and aqueous inclusions in the calcite cements. Temperatures derived from the inclusions were combined with estimates of geothermal gradient to provide the depth range for the formation of the cement window. Comparison of the depth and timing of the cement window with the onset of oil generation calculated from the thermal modeling shows that the modeling is self consistent (Figure 7.17). If oil-bearing inclusions were present in the cements, but the modeling showed that at that depth the organic matter was still immature, some modification of the modeling would be needed.

Fluid inclusions in fracture-filling cements can also be used to infer the time of fracture formation. This is done by combining the temperature obtained from the inclusions in cements with geothermal gradient to establish depth. Comparison with the burial history plot then gives the time. This approach was used by Narr and Burruss (1984) to show that fractures in the Mission Canyon formation at the Little Knife Field, N. Dakota are post-Mesozoic. Fracturing began when the formation was at least as deep as its present depth of burial, and probably continued during uplift and erosional unloading.

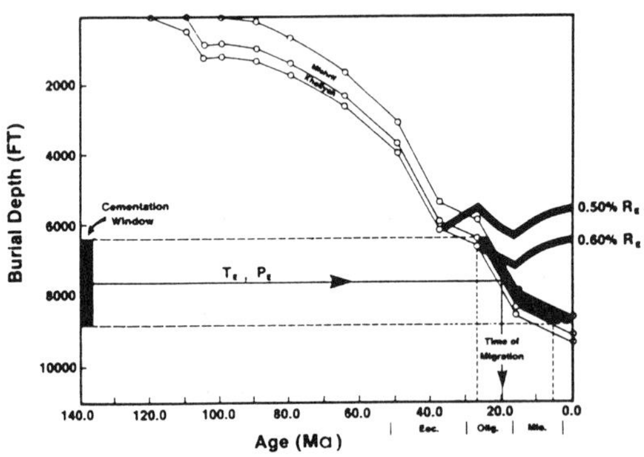

Figure 7.17. Burial plot from Videtich et al (1988) showing onset of maturation for the Khatiyah source rock and the Mishrif reservoir in Dubai. The cementation window derived from fluid inclusion data is superimposed.

In the Brent sandstone of the North Sea, Liewing et al (1987) obtained a radiometric age from K/Ar and Rb/Sr studies of diagenetic illite and then combined this information with the burial history to obtain a depth. Both aqueous and hydrocarbon-bearing fluid inclusions were present in quartz overgrowths and provided a temperature corresponding to the depth. However, the calculated geothermal gradient was much higher than the present day gradient and they were forced to adjust their model for the relative timing of events in this area.

7.5.6. Biodegraded Crude Oils

Bacteria can degrade crude oils in the reservoir (Section 1.8.3). The bacteria that are most effective in causing degradation are aerobic and need free oxygen, and this is brought into the system by meteoric waters that penetrate from the surface. The bacteria cannot tolerate temperatures much above 75 °C, so that the presence of a bacterially degraded crude oil indicates that at some time it must have been cooler than 75 °C. This corresponds to a depth shallower than approximately 2500 m, with the actual depth depending on the geothermal gradient at the time. Obviously the time of petroleum generation must precede the time of biodegradation,

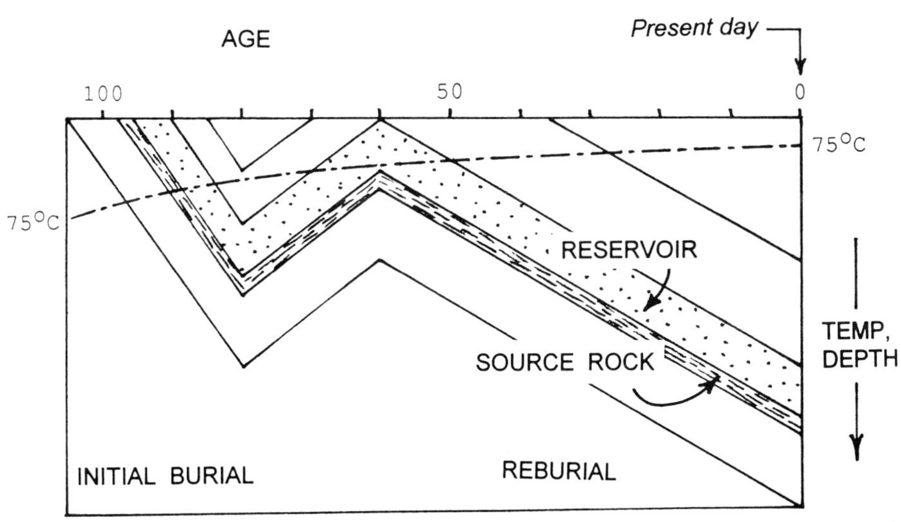

Figure 7.18. Burial history for a source rock-reservoir pair showing that any biodegradation of crude oil must have occurred prior to 50 Ma because temperatures have been above 75 °C since that time. The presence of biodegraded oil in the well would show that oil must have accumulated in the reservoir more than 50 m.y. ago.

and it is this fact which can sometimes help constrain burial histories and thermal models. Figure 7.18 shows the burial history for a system where the reservoir is immediately above the source rock. When drilled today the reservoir temperature was well above 75 °C, but the oil was severely biodegraded. The degradation can only have occurred when the temperature was below 75 °C, and this restricts the time to between 70 and 50 Ma. This in turn implies that petroleum generation must have occurred earlier than 70 Ma, i.e., during the initial burial. If modeling suggests that the source rock did not become mature until, say, 40 Ma, this is clearly not consistent with the presence of biodegraded crude oil in the reservoir. Either the thermal history, or the burial history (possibly the amount of erosion during uplift?), needs to be adjusted so that generation occurs during the initial burial. Alternative possibilities are: (1) the crude oil was generated from the same source rock but in an adjacent area with a different burial history, or (2) the oil was generated from a different, but unrecognized, source rock. Dixon et al (1992) reported biodegraded oil at >2.5 km in the Beaufort/Mackenzie area of Canada and attributed it to biodegradation "during uplift and erosion between the main phases of deltaic deposition".

7.6. RELATIONSHIP TO LOPATIN METHOD

As part of their study of the Newark basin, Walters and Kotra (1990) compared organic maturity levels calculated using data from Sweeney et al (1987), Tissot et al (1987), and the TTI model of Waples (1980). For their particular thermal history the Lopatin Method (TTI) predicted the highest maturity levels and the Sweeney et al data the least maturation. In contrast, Tissot et al (1987) found that Lopatin's method underestimates maturity in young basins (Figure 7.19). This study was based on the deltaic sediments of the Mahakam delta. Wood (1988) also compared the two modeling techniques and found that the Lopatin TTIs tended to underestimate thermal maturity at high heating rates but overestimate it when heating rates were low (significantly less than 1 °C/m.y.). In the Uinta basin, Utah, Sweeney et al (1987) found that the Lopatin method overestimates the maturity level for Type I kerogens. For complex burial histories there are no systematic differences between the two methods. The Lopatin method can underestimate or overestimate the onset of thermal maturity (relative to the kinetic models) depending on the particular E and A values involved, the details of burial history, and the past temperature regimes.

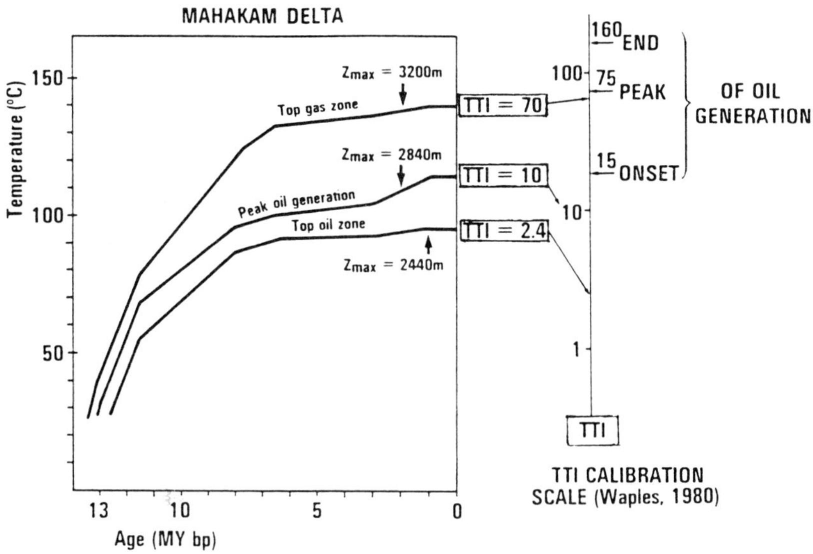

Figure 7.19. A comparison of a kinetic model with the Lopatin method for the Mahakam delta, Kalimantan. This study by Tissot et al (1987) shows that the Lopatin method underestimates maturity in young basins.

CHAPTER 8

USES OF THERMAL MODELS

8.1. INTRODUCTION

An increasing number of "user friendly" thermal modeling programs are becoming available commercially, and are easily accessible to exploration geologists. Thermal modeling is useful at all stages of exploration, and published studies provide examples of how it has been used to support exploration in a wide range of geographic areas. In the early stages of exploration sample control may be very limited (or non-existent), and simple models can be used to answer basic questions, such as the possibility that organic-rich rocks have had sufficient time-temperature exposure to become mature. As more data accumulates, and there is better subsurface control, the models can be refined to provide increasingly detailed information about the timing of generation and the areal extent of source rocks. Ultimately extended 2-D and 3-D models may be used to help constrain the amount of hydrocarbons generated, the development of abnormal pressures, and the direction of migration. These are discussed in Chapter 9. This Chapter will make some general comments on the role of modeling in exploration, and additional specific examples will be given in Chapter 10. Many applications of Lopatin's method have already been published, and a rapidly increasing number of studies using kinetics and advanced modeling techniques are appearing in the literature.

8.2. GENERATION STATUS

8.2.1. Present Day Status

Analyses (such as those discussed in Section 1.3) can establish the current generation status of any potential source rock – but only when samples are available. In the early stages of exploration, the only samples may come from the location of the first wildcat drilled, or from a limited number of wells, or possibly from outcrops. Thermal modeling provides the means to extrapolate away from very limited sample control, either to adjacent undrilled areas or to greater depths.

After oil leaves the source rock its migration is controlled by buoyancy, and so the oil tends to migrate upwards, and reservoirs are commonly located higher in the section than the source rocks which generated them. As a consequence, petroleum exploration frequently leads to the discovery of oil and gas fields before samples of the deeper mature source rocks become available from drilling. In evaluating an exploration play it is important to be able to establish the depth of mature source rocks in the area, so that their locations relative to other traps can be predicted. When well data are available they provide not only good geological control at the well location, but also an opportunity to compare calculated and measured maturity parameters and validate the model. A model is set up for the well location and refined until it is able to reproduce the measured maturity profiles. It can then be used to extrapolate to deeper or laterally equivalent source rocks. The more independent parameters that are used in getting a good match with the well data, the more confidence can be placed in the model. All of the techniques discussed in previous chapters can be used, and heat flow (or paleogeothermal gradients), lithologies and compaction constants, and losses at unconformities can all be adjusted. Although R_0 is the most commonly calculated maturity parameter, Transformation Ratios and biomarker ratios are also used. Fluid inclusion data for overgrowths or fracture-filling cements, apatite fission track information, or clay compositions should be used if available.

Models can be helpful even when there are no samples and little subsurface control. The Etosha basin of Namibia provides an example. This is a large, essentially undrilled basin where seismic lines show big structures, but carrying the seismic reflectors to outcrop suggests that the target reservoirs will be Cambrian in age. The key question in a frontier area such as Namibia, is whether there is still likely to be oil in the reservoir, or whether gas is the sole possibility. Based on geologic setting and basin configuration the Etosha basin appears to be an intracratonic basin, and so geothermal gradient has probably not been extreme in the past. A simple straight line burial plot, an average geothermal gradient, depths estimated from seismic, and a Cambrian age, can be used to build a preliminary model, and this shows that the potential reservoir rocks are still in the oil window. Initial generation of gas is not likely in an area with Ordovician source rocks because they predate the evolution of terrestrial organic matter and there is no Type III kerogen. If the model is refined by making reasonable assumptions for rock lost by erosion at unconformities (which are prominent on the seismic traces), the maturation estimates are not critically changed. In this case, even without subsurface samples, modeling is able to provide a key piece of information that could impact the economics of pursuing exploration in this politically unstable area of West Africa.

In the early stages of exploration off the coast of Norway, subsurface control was very limited and Leadholm et al (1985) developed an innovative technique to relate present day thermal conductivity to interval values obtained from seismic data, or, where available, sonic logs. The thermal data obtained in this way were combined with a thermal modeling program and kinetic data to calculate present day equivalent R_o. Eventually over one hundred wells and pseudo wells were used in the calculation of maturity in the Haltenbanken and Troms areas of the North Sea. This approach is particularly useful in basins with only limited well data, and is significant because it is credited with the discovery of the first oil in the Haltenbanken area – the Tyrihans field. A more recent discussion of the technique has been presented by Ho et al (1994) and this includes a comparison between measured R_o values in selected wells, and values calculated using their proprietary program and using LLNL kinetics (Figure 8.1).

Present day status is, of course, the only data that can be measured directly, and in areas with limited subsurface control it is very important to make sure that the assumptions used to constrain the burial history and the past geothermal gradients lead to calculated present day values that match those actually measured (Section 7.5). If calculated and measured values (say for R_o) are different the model is incorrect and needs to be refined.

8.2.2. Areal Extent of Mature Source Rocks

The most widely used commercial models only deal with 1-D reconstructions that are restricted to a single vertical location. To map the areal distribution of mature source rocks it is necessary to make a large number of individual reconstructions and then combine the data. In the early stages of exploration when there are few wells it is often helpful to set up *"pseudo wells"* or *"dummy wells"* based on seismic data or other information (e.g., Thomsen et al, 1990; Iliffe et al, 1991; Williamson et al, 1994). If possible, seismic lines should run through, or very close to, one or more logged wells, so that 2-way travel times can be related to depth, specific reflectors can be identified, and the depth of potential source rocks traced. Baird (1986), for example, combined data from 156 wells and 136 seismic locations in assembling an adequate data base for modeling in the North Sea.

Ayers et al (1982) used Lopatin cumulative TTI values to establish the distribution of mature Oxfordian-Callovian carbonate source rocks in Saudi Arabia (Figure 8.2). Calculated values were between 10 and 150

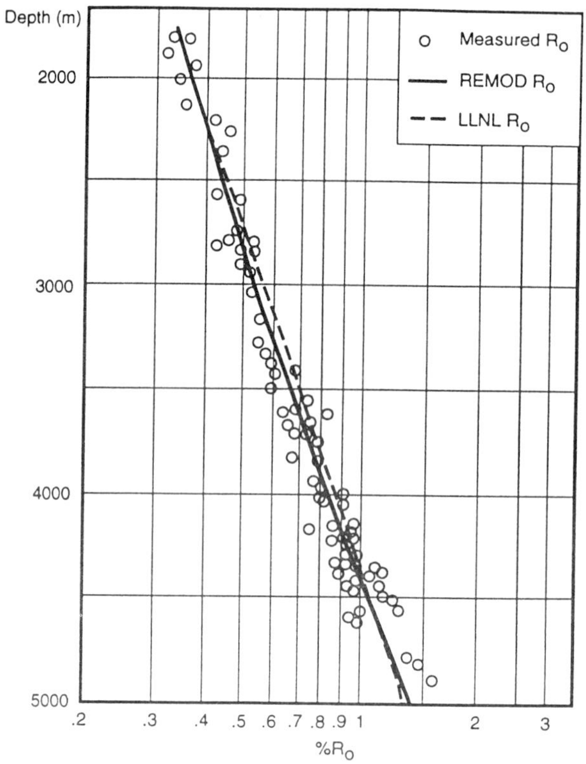

Figure 8.1. R_o versus depth plot based on measured and computed values. REMOD is a Conoco proprietary program and LLNL is the Lawrence Livermore software. (Ho et al 1996).

and, together with supporting data from hydrogen / carbon elemental ratios for the kerogens, defined a mature area of roughly 1.5 x 10^5 km^2. Similar data defined the areal distribution of maturity for Lower Cretaceous organic-rich rocks. It was not possible to use R_o values to establish maturity in this area because there was no terrestrial input into the early Jurassic platform carbonates, and organic matter is dominated by amorphous alginite that follows a Type II track on the van Krevelen diagram. Barker and Dickey (1984) used organic richness and maturity data together with the isopach map from Ayers et al (1982) to calculate the volume of source rocks and hence the volume of oil generated.

The carbonate source rocks and reservoirs discussed by Ayers et al (1982) are almost all in the oil window. In contrast, the Jurassic Smackover Formation carbonates of southern Mississippi, southern

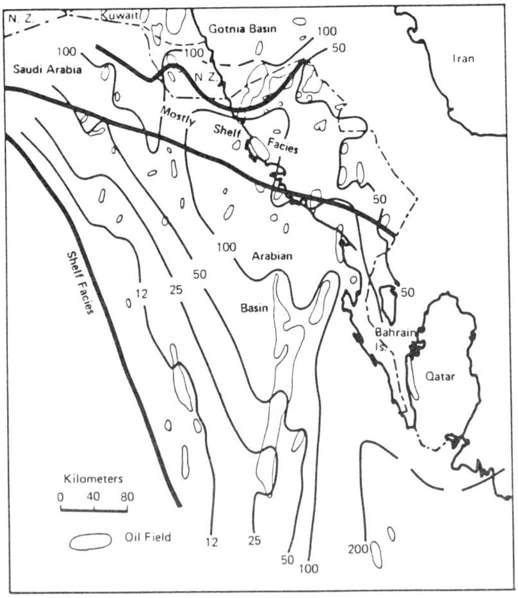

Figure 8.2. Maturation levels calculated using the Lopatin method and contoured as time-temperature index (TTI) for Callovian and Oxfordian source rock facies in Saudi Arabia (Ayers et al, 1982),

Alabama, and the Florida panhandle show a wide range of maturities as they go from outcrop to >6500 m. The lower part of the Smackover Formation is an organic-rich, laminated carbonate that acts as an effective source rock. The upper oolitic parts (together with other formations) are reservoirs. Sassen and Moore (1988) and Nunn and Sassen (1986) combined maturity modeling with organic geochemical studies to define the oil window as well as establish areas of immaturity, condensates, and thermal gas. These areal distributions are shown in Figure 8.3. Crude oil generation from the dominantly Type II kerogens started in the Cretaceous. Later, gas from thermally cracked oil reacted with anhydrites to give a high hydrogen sulfide content, in some areas over 90%.

Carbonate source rocks are also important in generating the large reserves in the Bolivar Coastal Field (BCF) area of western, Venezuela. Initially Talukdar et al (1987) carried out Lopatin-type calculations at 81 well locations to model the maturity distribution in the Cretaceous La Luna limestone. This showed that much of the organic-rich La Luna was currently in the oil generating window. Subsequent modeling by Sweeney et al (1990) and Burnham et al (1992) used the more sophisticated LLNL

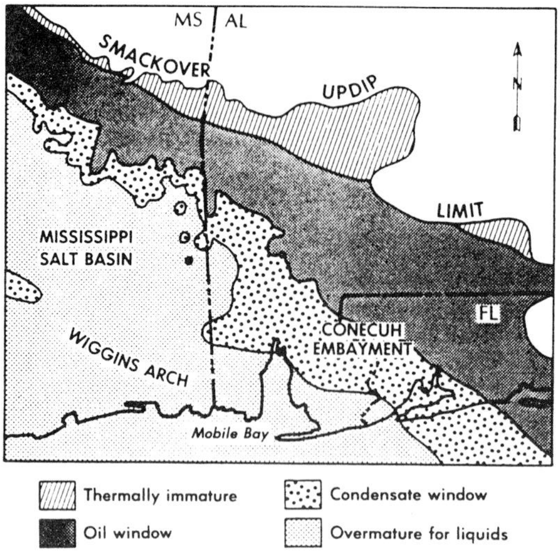

Figure 8.3. Generalized thermal maturity distribution in the eastern Smackover trend along the Gulf coasts of Mississippi, Alabama, and Florida, USA (Sassen and Moore, 1988).

kinetic model and established the distribution of mature LaLuna source rocks in an area of 50,000 km^2 that included the BCF. They developed contour maps showing the percent kerogen converted to oil, both at the present and at 12 Ma and 38 Ma. They also plotted the timing of oil generation (defined as 30% conversion of kerogen to oil), and the beginning of gas generation.

Establishing the areal extent of mature source rocks can be economically important. In the North Sea drilling success in the mature "fairway" has been about 1 in 3, compared with a rate of only 1 in 30 outside this area. Demaison (1984) gave several other examples including the Cooper Basin, Australia, where drilling success ratios were 1 in 2 in the mature Permian source zone, but only 1 in 23 in the immature Permian (Figure 8.4). No producing wells at all have been drilled where the Permian is post-mature. By combining information on the areal extent of mature source rocks with structure it is possible to define local drainage areas in the basin. This puts limits on the potential reserves for individual traps.

In addition to plotting the areal distribution of maturity, it is possible to use values for source rock maturation status calculated at individual locations to give a 2-D cross section. Here also pseudo-wells are useful for

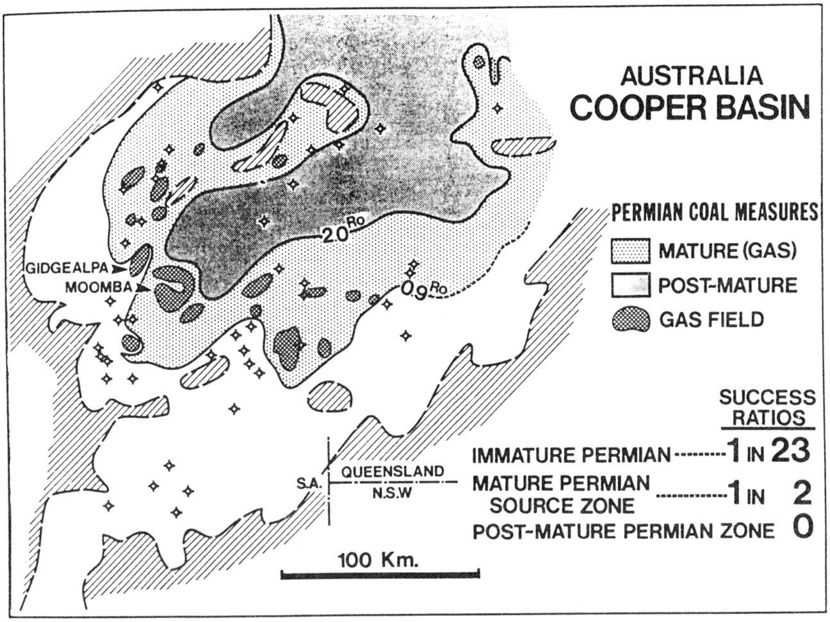

Figure 8.4. The relation between exploration success and source rock maturity in the Cooper Basin, Australia (Demaison, 1984).

developing key points if the seismic control is good enough to establish depths of particular lithologic units.

8.3. TIMING OF GENERATION

8.3.1. Introduction

Analysis of source rock samples provides information on their present maturation status and reflects the integrated past time/temperature exposure. For mature, or post-mature, rocks, present day maturity gives no information about the actual time when they became mature. This is a gap that can often be filled by thermal modeling. In many cases the timing of events is critically important in exploration, and this is especially true for the time of trap formation relative to the time of maturation and migration. Maturity is commonly displayed on burial history diagrams by showing the position of the oil window through time (see Section 6.4). An example was given by Waples (1980) in his original publication on the Lopatin method, and is reproduced here as Figure 8.5. It shows when each rock unit reached maturity, and the time when each one moved out of the liquid window. However, thermal maturity alone is not enough to make a source rock because adequate amounts of organic matter of the right kind are also needed. A well drilled today could penetrate all of the rock units of interest (Figure 8.5), and potentially could provide information on organic type and content. If in this example the only organic rich rock is the one labeled "oil source rock ", then oil will have been generated and migrated in the time interval from 180 to 125 Ma. At 90 Ma there seems to have been some tectonic event that led to deformation and could have produced structural traps. But in this case trap formation postdated the time of generation and any oil formed from the oil source rock would have migrated away before the trap formed. The time of oil generation needs to postdate the time of trap formation. Notice that this does not require the trap to be older than the source rock, just older than the time of generation (Figure 8.6).

It is important to stress that 1-D reconstructions only give information about the modeled location, and it is possible that in some adjacent areas traps formed earlier, or that generation occurred later. It is also important to recognize that the formation of structural traps may be easier to infer from geohistory diagrams, while there may be less evidence for the development of stratigraphic traps.

The time of generation is important because it is the migration pathway at that time that will control where oil goes. Clearly structures located along a migration pathway will be much more prospective (Westcott and Hood, 1994). Mathematical models suggest that the time to migrate even >100 km is negligible, and Thomas and Clouse (1995) concluded from their experiments with scaled physical models that "the rate limiting step in charging a trap is not secondary migration rate, but rather the rate of oil release from the source rock". The development of the Powder River basin, Wyoming and Montana, provides a dramatic example of the way in which migration routes can change radically. Although there are half a dozen source rocks in the basin as it exists today (Momper and Williams, 1984) one type of oil has been correlated to a source rock that is several hundred kilometers to the west and is on the opposite side of a mountain range. The source rock is the Phosphoria shale which started to generate in the Late Jurassic and fed oil through the permeable Permian and Pennsylvanian sandstones into what is now the Powder River basin. The uplift of the Big Horn Mountains and the Casper Arch defined the basin and also cut off the Phosphoria from the basin. This has left one of the major source rocks for oil in the Powder River basin outside of the present day basin.

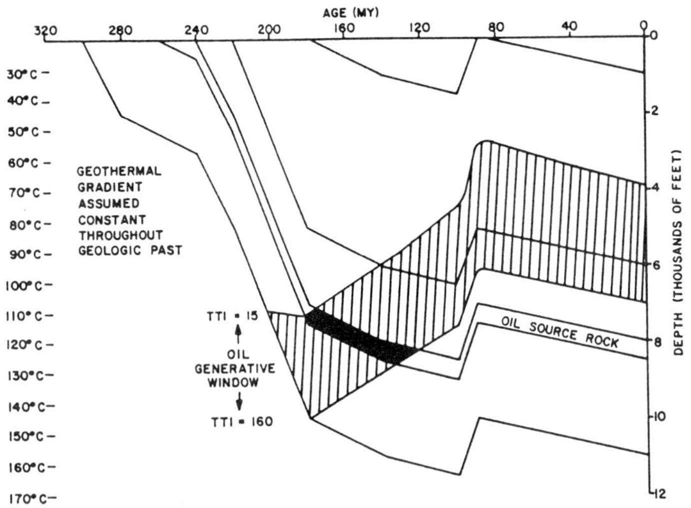

Figure 8.5. Hypothetical burial history showing the oil window through time (shaded) and the time when the source rock was generating (solid). (From Waples, 1980).

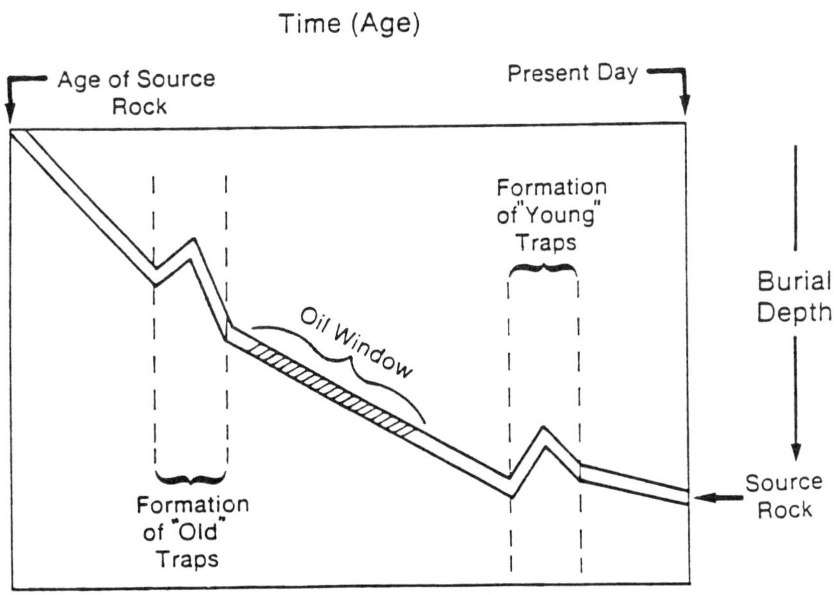

Figure 8.6. Time of trap formation compared with the time of petroleum generation. Effective traps must form prior to petroleum generation (but need not be older than the source rocks).

8.3.2. Timing of Generation and Trap Formation

The importance of the time of trap formation relative to the time of petroleum generation is well illustrated by the distribution of oil and gas in the Amadeus basin of central Australia. In this dominantly Proterozoic-Devonian basin the main production is from Ordovician sands. At Mereenie production is mostly oil with some gas, while Palm Valley has only gas. Geochemical analyses of samples from 23 wells by Jackson et al (1984) showed that potential source rocks in the area include the Ordovician Horn Valley siltstone and part of the Late Proterozoic Bitter Springs Formation. Lopatin-type reconstructions were then used to establish when these rocks generated various petroleum products (see Figure 8.7) and to relate this to the development of traps which occurred around 350 Ma (at the latest). The Bitter Springs Formation was in a post oil-generative stage at both Mereenie and Palm Valley by the time of trap formation, and so only gas could have accumulated in the reservoirs (and only in minor quantities at Palm Valley because of the very high maturity levels in this area). The other source rock, the Horn Valley Siltstone, continued to generate oil after trap formation at Mereenie, but in contrast at the Palm Valley location only gas was being formed. Thus a detailed evaluation of the sequence of events shows that at the Palm Valley site only gas was generated after the trap formed, while at Mereenie oil was generated from one of the source rocks after the reservoir developed. In this case thermal modeling explains how two fields in the same basin, with the same source rocks, could have accumulated very different petroleum products.

Carroll et al (1992) modeled the Junggar basin of northwest China as a foreland basin, and used commercial modeling software to show that the very organic-rich oil shales probably started generating petroleum in the mid-Mesozoic. In the southern Junggar basin potential petroleum traps were formed by Neogene thrusting, which postdated the generation phase. This mismatch in the timing of petroleum generation and trap formation may explain the lack of any significant oil discoveries. The thermal model was constructed to incorporate loss of 5.25 km of Upper Permian-Jurassic section based on local geology and vitrinite reflectance trends; a constant heat flow with a value derived from the contemporary geothermal gradient of 22 °C/km; and reaction kinetics for Type I kerogen.

There have been few commercial petroleum discoveries in the "Late Andean" (Miocene-Pliocene) compressional structures of the Oriente basin, Ecuador. Dashwood and Abbotts (1990) attribute this to their rather recent formation, which post-dates the main phase of primary oil generation and migration. Link and Bustin (1989) found that in northern Canada the generation of oil pre-dated the Laramide orogeny, so that the hydrocarbons were not available to fill the traps formed by the orogenic movements. Another example of the mismatch in timing between oil generation and the availability of reservoirs occurs in the Taranaki Basin of New Zealand. Kamp and Green (1990) explained the 1 Fresne dry hole in the southwest part of the basin by suggesting that the proposed source

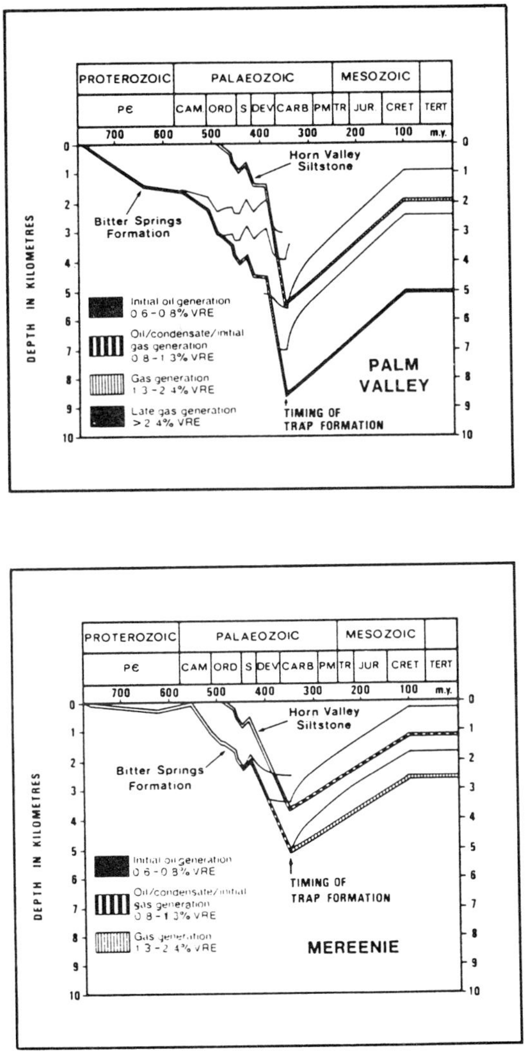

Figure 8.7. Burial and maturation histories for the Palm Valley field (based on Palm Valley 2) (Top), and the Mereenie field (based on East Mereenie 4) (Bottom). In both cases assumed geothermal gradient was 2.9 °C/100 m. (Jackson et al, 1984).

rocks in the area passed through the oil window in the middle Miocene before structural inversion produced potential traps. Their maturity estimates were based on temperature histories obtained from AFTA.

The time of petroleum expulsion may vary considerably, even across a small basin, such as the northern Niigata basin of Japan. Here Nakayama (1987) suggested that in areas of fast burial only minor oil was expelled before oil was cracked to gas. As a consequence reservoirs produce only gas even though gas prone source rocks were not identified. In contrast, the Akita area to the north experienced slower burial and expelled its oil, and it is now an oil-prone area. Major petroleum fields formed when expulsion coincided with the development of structure.

In the Styrian basin heat flow has changed significantly, both areally and through time. The basin is approximately 100 km long and 50 km wide and lies at the western edge of the Pannonian basin where it trends roughly SW-NE. In the southern part of the basin, early-to-mid Miocene volcanism and associated magmatism raised the heat flow to an estimated 300 mW/m^2 within a few kilometers of the volcanic centers, but to only 120 mW/m^2 further away. With diminishing volcanism after the mid Miocene, heat flows decreased to 60 - 85 mW/m^2. Sachsenhofer (1994) modeled thermal maturity using 2-D finite element software (described by Welte and Yalcin (1988) and Wygrala et al (1990)) concluded that the high paleogeothermal gradients near the active volcanic areas generated petroleum. Unfortunately, there were no suitable traps at that time and no commercial accumulations have been found, even though 25 wells have been drilled. Modeling also showed that only small quantities of oil were generated after the Miocene. Calculated temperature distributions are shown in Figure 8.8. This study also provided a good example of the complications due to multiple source rocks with different types of organic matter. In the Styrian basin the top and bottom of the oil window differ by roughly 600 m between rocks with Type III kerogen and the more Type II rich organic matter in the basinal facies (Sachsenhofer, 1994). Activation energy suites for the different kerogen types were determined by laboratory pyrolysis.

8.4. MODELING OIL VERSUS GAS

The economics for producing and marketing oil and gas are quite different. Oil is easier and cheaper to transport than gas, and in frontier areas oil is generally much more desirable. Thermal models can help in

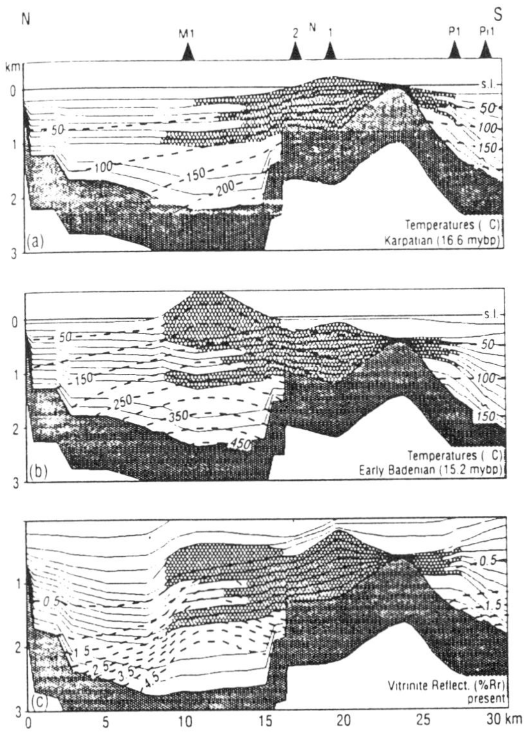

Figure 8.8. Calculated temperature distribution along a north-south cross-section. (a) early Miocene (Karpatian) temperatures; (b) middle Miocene (early Badenian) temperatures; (c) calculated present day vitrinite reflectance values. (From Sachsenhofer, 1994. His Section II).

predicting whether an exploration area is likely to be oil- or gas-prone. The initial control on the nature of the hydrocarbon products is set by the type of organic matter in the source rock. The methods available for establishing organic matter type have been discussed in Section 1.10 and include microscopic examination, elemental analysis (van Krevelen diagram), and pyrolysis. Sometimes it is possible to infer organic matter type from the depositional setting as outlined in Section 7.3.3. Modeling has a role in deciding if, and when, this organic matter is converted to oil or to gas. Rocks containing oil-prone organic matter generate crude oil as they pass through the oil window and then, at higher temperatures, generate gas. Thermal modeling is widely used to define the floor for oil and the onset of gas generation. It can also give the time when gas generation started from gas-prone kerogens. Thermal modeling cannot be used to predict the distribution of biogenic gas.

8.5. PRESERVATION

8.5.1. Oil

Models can be used not only to calculate timing of oil generation in source rocks, but also the thermal cracking of oil to gas in reservoirs. Although in many cases reservoirs are close to source rocks and share very similar thermal histories, in other cases reservoirs may be more than 100 km from the source (e.g., in some continental interior basins), or very much shallower in the section (e.g., high blocks in rifts where source rocks are in the adjacent downthrown grabens). When source rocks and reservoirs have very different histories it becomes necessary to model them separately. In this situation the burial/thermal history of the source rock can be followed until generation/migration occurs, and then, after migration and accumulation, the reservoir is tracked (Figure 8.9). In cases of very long distance migration the burial histories can be quite different.

The oil in the reservoir responds to rising temperature and passing time by thermally evolving, ultimately giving gas and a highly carbonaceous residue. Like the evolution of organic materials in source rocks, this process follows first-order kinetics and can be modeled using appropriate reaction pathways and the relevant E and A suites. This approach is useful for providing economically important information on the depth of the floor for oil. Models using a single E-A set (e.g., Tissot et al, 1987; Quigley et al, 1987; Welte et al, 1988), multiple E values with a single A value (e.g., Mackenzie and Quigley, 1988; Behar et al, 1991), or multiple E-A pairs (e.g., Kuo and Michael, 1994) have all been used. In many cases oil has been treated in the same way as bitumen in a maturing source rock, but it is important to recognize that oil in a reservoir does not have the same overall composition as the organic materials in a source rock. Kuo and Michael (1994) started with oil composition and used the PMOD program developed by Burnham and Braun (1990) and Braun and Burnham (1990, 1992) to model the changes in composition with increasing temperature. PMOD has considerable flexibility and can deal with parallel, serial, competing, and reverse mechanisms. Kuo and Michael used it with seven reactions involving asphaltenes, NSOs, aromatics, saturates, C_2, C_{3-5}, and C_{6-14}, as reactants, and applied it to a Venezuelan and an Indonesian oil (Boscan and Pematang, respectively) to show how

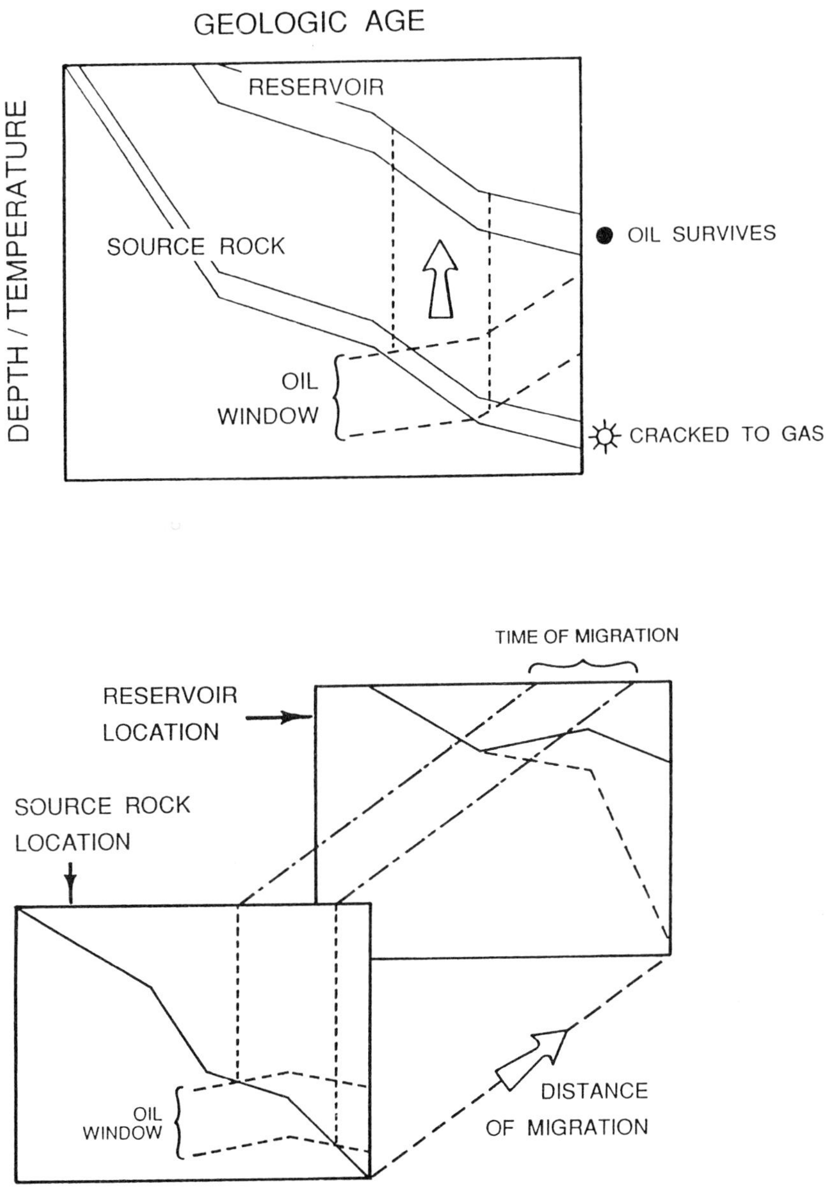

Figure 8.9. Migration from source rock to reservoir at the time the source rock is expelling oil. (Top) vertical migration to a shallower reservoir, and (Bottom) migration over a considerable distance laterally.

composition and API gravity changed with maturity. API gravity was calculated as a mass-weighted harmonic mean of the specific gravities of the individual oil fractions (Braun and Burnham, 1990). Kuo and Michael (1994) found that API gravities tended to converge at higher temperatures regardless of the initial values. Figure 8.10 shows calculated and measured API gravities obtained by applying their model to oils in the southern parts of Alabama and Mississippi. Marquez and Mountjoy (1996) used thermal modeling to establish the onset of thermal cracking of oil so it could be related to the development of overpressuring. They showed that in Upper Devonian reservoirs in the Deep Basin of western Canada microfracturing occurred as a consequence of the high pressures.

Andresen et al (1993) used Well 2/4-14, in the Norwegian part of the Central Graben (just north of Ekofisk) in the North Sea, to evaluate the sensitivity of oil cracking to the E values used. They found that values used for kinetic parameters are critically important, and relatively small changes can lead to considerable changes in the calculated depth of the floor for oil (and end of the oil window) (Table 8.1). They suggested new and improved values for the kinetic parameters controlling the cracking of oil to gas based on laboratory pyrolysis.

The cracking of oil to gas in a reservoir, or the introduction of gas into an oil reservoir from outside, can lead to the precipitation of solid asphaltenes (Section 1.8.2) (Lomando, 1992). These can be important in adversely affecting reservoir permeability and porosity, and if they accumulate at the base of the reservoir as a tar layer they can produce a major permeability barrier. The ability to calculate the timing for gas

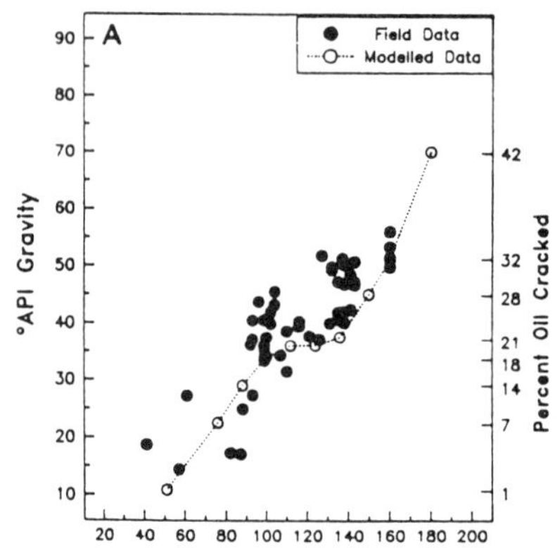

Figure 8.10. API gravities for oils from the Smackover trend of southern Mississippi and Alabama, U.S.A., comparing measured values (solid circles) and predicted values (open circles) (30 m.y. residence time). (From Kuo and Michael, 1994).

Table 8.1
Comparison among different kinetic parameters for cracking oil to gas, and the impact they have on the depth to onset of gas generation for a given thermal history (Andresen et al, 1993). The percent change (Δ) is relative to the Tissot and Welte (1984) data.

DATA SOURCE	E (kcal/mol)	Δ	DEPTH TO ONSET OF GAS GENERATION (M)	Δ
Tissot and Welte	50		4130	
Quigley et al	55	10%	5110	23.7%
KFA study	~67	34%	~6500	~57%

generation (either by cracking oil or from kerogen) can sometimes be used to estimate the timing and regional distribution of reservoir degradation by natural deashphalting. Lomando (1992) provides an example from offshore west Africa where *"reservoir bitumens"* are present only in oil reservoirs that are vertically above areas of gas generation (Figure 8.11). On the eastern side of the studied area the subsalt source rocks are immature, no gas has been generated, and no gas has migrated into the overlying reservoirs. As a consequence, they have no reservoir plugging by precipitated ashphaltenes. In the western area reservoir bitumens are common.

8.5.2. Gas

Deep gas preservation poses a very different problem from those discussed so far because the control on composition is no longer kinetic. Deep gas approaches thermodynamic equilibrium with its surroundings and gas survival depends on free energy considerations for the overall system, including all of the minerals present. Time is no longer a factor (Figure 8.12). Takach et al (1985) have described a thermodynamic computer program that uses minimum free energy criteria to establish equilibria and calculate deep gas composition. In many clastic systems methane is quite stable, but in the presence of sulfur-bearing species it is removed by reaction to form hydrogen sulfide (Figure 8.13) (Barker and Takach, 1992). In limestone and dolomite reservoirs thermal breakdown of the carbonates generates large quantities of carbon dioxide which dilute the methane, so that although methane itself may be thermodynamically stable the final gas composition is uneconomic.

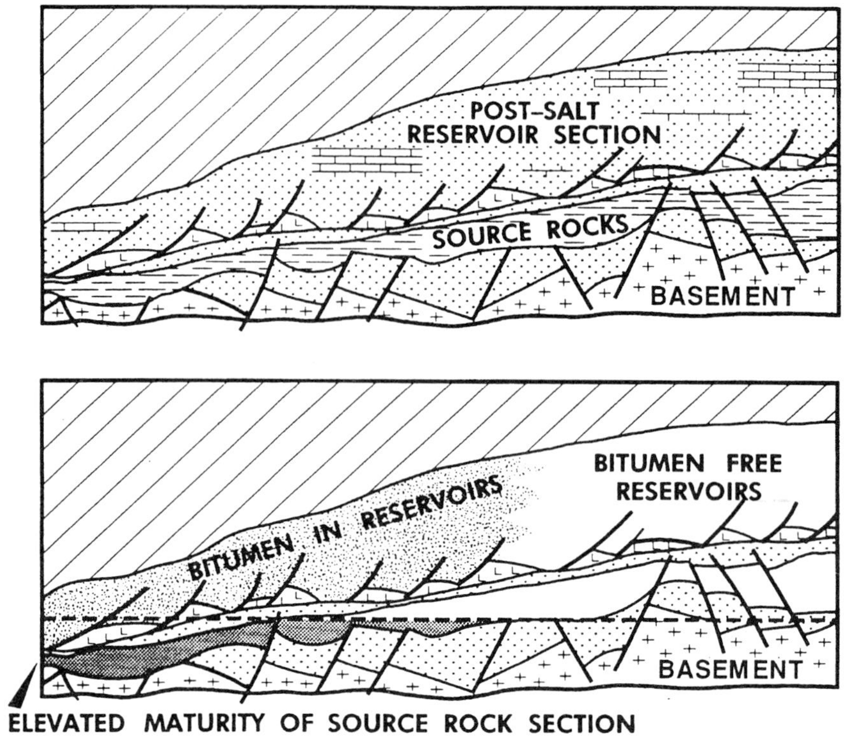

Figure 8.11. Schematic cross section in the Lower Congo basin, west Africa, showing the position of the pre-salt lacustrine source rocks (top) and how reservoir bitumen occurs only in reservoirs above areas where the source rock has been buried below the top of the gas window (dashed line) (Lomando, 1992).

Kinetically-controlled reactions that generate oil and gas are time dependent, while, in contrast, the composition of systems that are in thermodynamic equilibrium are independent of time (Figure 8.12). Thus, the ages of oils, gases, source rocks, and reservoirs do not appear in thermodynamic calculations, and the results are independent of geologic age. At shallow depths, equilibrium will not be reached and thermodynamically unstable materials (such as crude oil) can be found. At present the depth where gas composition is controlled thermodynamically is not known, but the approach to equilibrium is favored by high temperatures and long time intervals, and a thermodynamic approach to calculating deep gas composition is likely to be valid only for depths greater than approximately 8000 m for an average geothermal gradient.

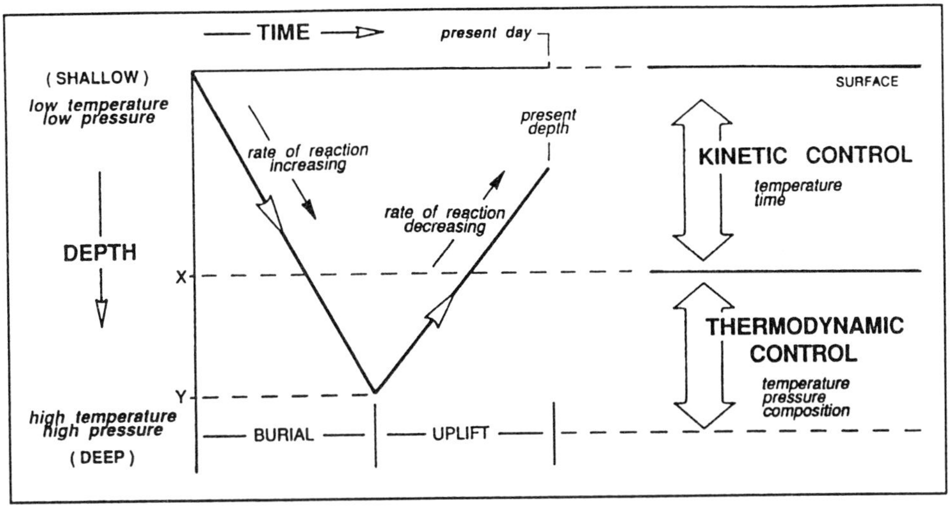

Figure 8.12. Schematic burial history of a gas reservoir (which is now relatively shallow) showing how the change from kinetic to thermodynamic control of gas composition has influenced gas composition at various stages (Barker and Takach, 1992).

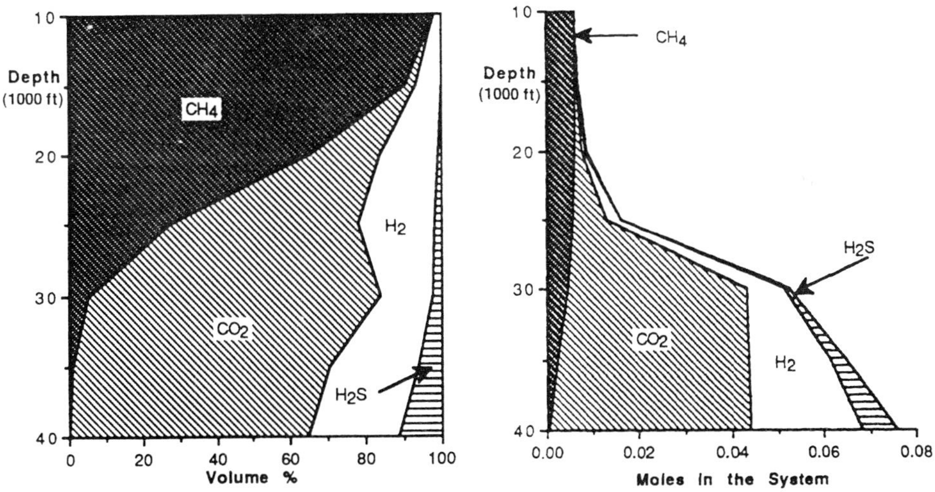

Figure 8.13. Variation of calculated equilibrium gas composition with depth for a sandstone reservoir containing feldspars, iron oxides, calcite, sulfur, and a graphitic residue. Gas composition is given as cumulative volume percent (on a water-free basis) on the left, and as the cumulative number of moles on the right (Barker and Takach, 1992).

Gas that has migrated up from depths greater than about 8000 m, but is currently in shallow reservoirs, will have a gas composition controlled by the mineral assemblage present during deep burial. At depths greater than X in Figure 8.12 the gas composition is controlled thermodynamically, and so depends on temperature, pressure, and rock composition. On uplift the rate at which reequilibration is achieved will steadily diminish, until at some point the kinetics become so unfavorable that the reaction essentially stops, and the gas composition is "frozen in". During uplift from depth Y to depth X (Figure 8.12) gas composition continues to adjust rapidly and remains in thermodynamic equilibrium at each transient set of ambient conditions. Thus, for a given bulk composition, the actual composition at depth X will be the same regardless of the depth represented by Y. This suggests that the gas composition in shallow reservoirs that were once deeply buried is not controlled by the greatest depth of burial, but rather by the shallowest depth at which the gas composition was in thermodynamic equilibrium with the local mineral assemblage (Barker and Takach, 1992).

A minority view holds that gas (and perhaps oil) is generated in the deep crust or upper mantle and later migrates up into conventional reservoirs (Gold, 1993). At the high temperatures found in these deep environments reactions rapidly reach equilibrium and the kinetic methods described throughout this book cannot be used. Thermodynamic methods can be used to calculate the composition of the hydrocarbons being introduced into the crust, but this composition will readjust to declining pressure and temperature as fluids move upwards, and the composition will be controlled by the shallowest depth at which equilibrium is reached with later kinetic modifications as discussed above (Figure 8.12).

8.6. SENSITIVITY TO MODELED CONDITIONS

Errors in the data used as input for thermal modeling will lead to inaccuracies in the final output from the model. It is important to establish whether the uncertainties in the input data are sufficiently large that they change the final conclusions. For example, does the range of possible paleogeothermal gradients lead to source rock maturity values that span the range from mature to immature? It is easy to check the effects of a single parameter by varying the values over the possible range to see how they impact the output. A study of this type was reported by Deming and Chapman (1989) in the Utah-Wyoming thrust belt, and selected data are shown in Table 8.2. Their preferred value for surface temperature, for

Table 8.2
Sensitivity analysis at 70 Ma at the top of the Aspen Formation (Pineview 1 Well). The Table shows sensitivity of the predicted R_o values to changes in model parameters (with constant basal heat flow), and sensitivity of surface heat flow to changes in model parameters (with constant R_o). (From Deming and Chapman, 1989).

PARAMETER	PREFERRED VALUE	RANGE	R_o RANGE (% R_o)	SURFACE HEAT FLOW RANGE (mW/m^2)
Matrix Conductivity				
Adaville Fm	3.50	2.80-4.20	0.57-0.54	36.8-39.5
Hiliard Fm	2.40	1.92-2.88	0.63-0.50	34.2-41.8
Frontier Fm	2.23	2.01-2.45	0.56-0.54	37.6-38.9
Porosity				
Adaville Fm	0.30	0.20-0.40	0.53-0.58	40.0-36.4
Hiliard Fm	0.30	0.20-0.40	0.51-0.61	41.0-35.6
Frontier Fm	0.12	0.07-0.17	0.55-0.56	38.4-38.2
Matrix Thickness				
Adaville Fm	918	618-1218	0.53-0.57	40.3-36.7
Hiliard Fm	1708	1508-1908	0.51-0.59	40.8-36.3
Frontier Fm	674	574-774	0.54-0.57	39.4-37.3
Surface Temperature	15	10-20	0.50-0.61	41.1-35.6
Erosion Rate (mm/yr)	1.0	0.5-2.0	0.56-0.55	37.8-38.3
Time of Thrusting	70	75-65	0.54-0.57	38.5-38.1

example, was 15 °C, but if it was varied from 10 to 20 °C the calculated present day vitrinite reflectance ranged from 0.50 to 0.61% and the calculated present day heat flow fell between 41.1 and 35.6 mW/m^2. Both of these values could be measured and used to constrain the modeling. A similar sensitivity analysis was carried out by Ungerer et al (1986) in the Shoab-Ali area of the Red Sea rift. Using their 2-D kinetic model they varied 9 different parameters independently and calculated the effects on depth and timing of oil generation and migration, as well as on the estimated reserves in the traps. A decrease of 10% in the geothermal gradient delayed the onset of generation by 1.5 - 2.0 m.y. and decreased the amount of oil in the trap by 40%.

Studies of this type permit a sort of "experimental paleo-geochemistry", and the effects of changing the geological model can be evaluated. If there is inadequate geological control to establish the amount of section lost at an unconformity, then the model could be run with a range of values for the amount of section lost to see just how critical this parameter is for the modeled timing of petroleum generation or present day source rock maturity. This approach is particularly useful for studying the role of different kerogen types (e.g., II versus III), or for evaluating poorly-known geothermal gradients. The effects of changing some individual parameters on calculated R_o values are summarized in Figure 8.14.

In practice all of the input parameters will have associated uncertainties and the problem is to establish the combined effects on the model output, and to decide which parameters make significant contributions to the uncertainty (Cao and Lerch, 1990a,b; Nielsen and Dahl, 1991; Waples et al, 1992b). Several studies of this sort have been carried out for the Egersund basin area of the North Sea (Ritter et al, 1987; Ritter, 1988), with the most detailed being reported by Hermanrud et al (1990a). He considered only a single source rock (the Tau Formation) and varied each of 22 key input parameters over an assigned probability distribution. Monte Carlo / Latin hypercube sampling then provided a range of uncertainties for Transformation Ratios and amount of generated

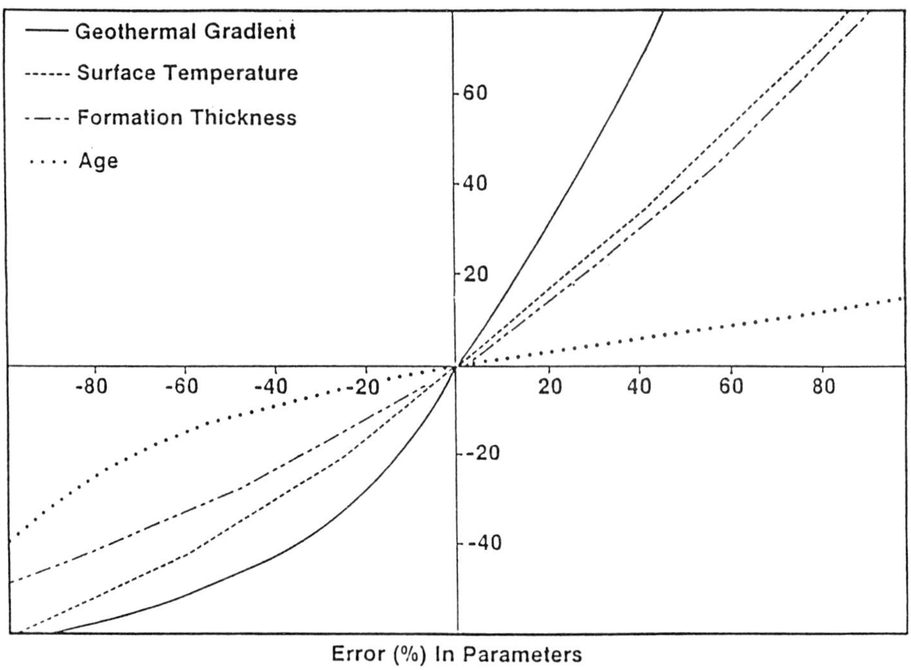

Figure 8.14. Effect of geologic parameters on computed R_o values (T.T.Y. Ho, private communication).

hydrocarbons (Figure 8.15). The Egersund sub-basin is not part of the main extensional zones of the Viking and Central grabens and the source rocks have not been deeply buried. Consequently they are marginally mature or immature in most of the basin, and in these circumstances accurate thermal modeling becomes extremely important. Nielsen (1993) has also used Monte Carlo simulation to consider the interaction of uncertainties in modeling input data for kerogen type and temperature histories, and has applied this approach to the Upper Jurassic Farsund Formation in the North Sea. Although sensitivity analysis is site specific in its details, it is very useful in providing an indication of the accuracy expected in similar areas.

8.7. LIMITATIONS OF THERMAL MODELS

Thermal modeling is proving to be a very powerful tool in petroleum exploration but, like any other tool, it has limitations. Some of these are practical (cannot get good data – or any data!), while others are inherent in the models (global E and A values, single value of A, etc.). All input parameters have some uncertainty associated with them, and this can range from insignificant to major. A key factor in using models is to assemble adequate input data and to use sensitivity analysis to evaluate the impact of poorly constrained parameters.

In addition, there are some specific limitations on modeling, and problems include, among others, the following:

(1) Most thermal models are intended to calculate source rock maturity but the oil may be a very long way from mature areas, especially in continental interior basins where long distance continuity of carrier beds permits long distance migration. In areas where vertical faulting has had a major control on migration (e.g., in rifts) the areal distribution of source rocks and reservoirs may be similar. Extended 2-D and 3-D models that incorporate migration can help deal with this problem. They are discussed in the next chapter.

(2) Thermal modeling can give no information about biogenic gas, and this is a very significant resource that may account for as much as 20% of the world's natural gas (Rice, 1993). For example, the chemical and isotopic compositions of the gas in many of the huge fields in Western Siberian show that they contain biogenic gas. Among these fields is Urengoi, which, with 285.5 Tcf of dry gas, is one of the largest gas fields in the world.

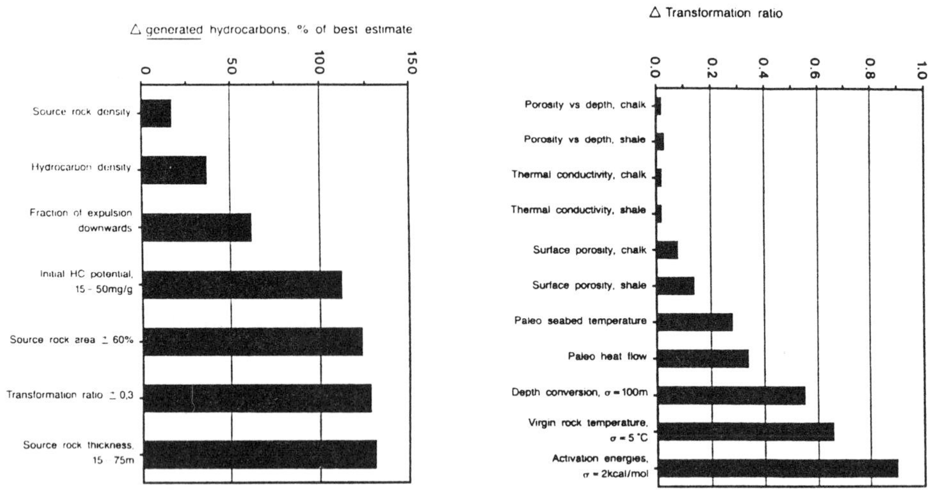

Figure 8.15. The sensitivity of changes (Δ) in generated hydrocarbons (left) and transformation ratio (right) to variations within "extreme" ranges of some input parameters (Hermanrud et al, 1990a).

(3) A careful distinction between maturity and generation status is needed. A source rock may have had sufficient time/temperature exposure to be mature (and this may be indicated by R_o values, and other maturity indicators), but the rate of oil expulsion could be zero. This happens in very lean rocks where there is insufficient organic matter to saturate the pore system and initiate migration. It can also occur on uplift when the rate of generation falls to levels where expulsion is no longer occurring. Rocks with organic maturities in the oil window can be found at the surface today, but they are clearly not expelling oil.

Thermal models should not be used to give "The Answer", but rather as a tool to probe the system (Waples et al, 1992a). This can be done by varying poorly-constrained input parameters over a reasonable range and comparing calculated and measured maturity parameters, such as vitrinite reflectance. These checks and sensitivity analyses have already been discussed. It is very important to remember that modeling is usually being carried out to answer a specific exploration question, rather than to develop a complete reconstruction of the area. Sometimes it is possible to answer the question without having perfect reconstructions of the burial history or the thermal regime. In particular, shallow parts of burial histories are much less critical because temperatures there are lower, and they have only minor impact on maturity increase. A typical concern in exploration is "Did the source rock generate before or after the trap formed?". If the trap formed early there can be considerable uncertainty in the timing of petroleum generation, but as long as it occurred after trap formation the exact time does not matter. If time of generation is very close to the time when the trap developed then it is important to try and refine the modeling to improve accuracy.

CHAPTER 9

EXTENDED MODELS

9.1. INTRODUCTION

The previous chapters have been concerned mainly with the quantitative models developed for establishing source rock maturation status, but petroleum generation is only one of the many factors needed for the accumulation of commercial quantities of oil and gas. In addition to the generation and migration of petroleum, there must be a reservoir rock available, and it must have adequate values of permeability and porosity. The processes that lead to these characteristics, such as winnowing away of fine-grained materials, generally lead to low organic contents. Since reservoirs contain insufficient organic matter to generate the oil and gas now contained in them, it is widely assumed that the petroleum is generated elsewhere in organic rich source rocks and subsequently migrates into the reservoir over distances that may range from a few meters to hundreds of kilometers. This is the so-called *"Source Rock Concept"* (see Section 1.7.1). Clearly, if oil and gas are moving away from the location where they were initially generated, one-dimensional models of the type discussed so far are incapable of adequately describing the overall system. Compaction also involves movement of fluid out of the rock. In Section 3.5 compaction was discussed in some detail, but only as it affects vertical thickness of a rock unit. For compaction to occur water must flow out, and modeling then requires information about the permeability distribution in adjacent and more distant rocks. Again, 1-D models are not adequate for describing the complete system. A variety of 2-D and 3-D models is being developed to cope with the additional complexities. As well as describing petroleum generation status, they deal with water flow, lateral heat transfer, two-phase migration of fluids, and the development of abnormal pressures. This requires considerable additional physical and chemical information as input, and, in particular, it is important to be able to describe the distribution of permeability in some detail. The extended models that are being developed to include these parameters are described in this chapter.

9.2. QUANTITIES GENERATED

9.2.1. Organic Matter Type

Simple thermal models provide estimates of organic maturity through time. While source rock maturity is a requirement for petroleum accumulation, alone it is not sufficient because the formation of commercial quantities of oil and gas in reservoirs also requires adequate organic matter content, significant migration, and correct timing of migration pathways. Migration is discussed below. The convertibility of organic matter to petroleum products varies with organic matter type and is best measured in the laboratory. Pyrolysis techniques, especially RockEval, are widely used to do this, but pyrolysis measurements require immature samples, and these may be difficult to get. Thermal models generally calculate Transformation Ratios (see Section 1.9.4) through time, and this information can be combined with organic content and convertibilities to give the amount of generated petroleum through time.

In practice, the organic matter in a rock may be made up of several different kerogen types that have different generating abilities and different thermal stabilities. This system must be treated as a series of parallel, first order reactions, where the relative amount of each organic matter type is specified. Total petroleum generated at any level of maturation is then obtained by summing the individual reactions. Reactants with lower activation energies are consumed first.

There may be major fluctuations in organic matter type over short depth intervals. For example, Khorasani (1989) reported that inertinite fluctuated from 50% of the organic matter to essentially zero over a depth interval of 6 m in a well drilled in the Haltenbanken area of offshore central Norway. Since inertinite is essentially non-generating, this is equivalent to a hundred percent variation in organic matter content. Such variations are not uncommon. To avoid these problems it is important to sample carefully, and either take a large number of closely-spaced samples, or composite over a suitable depth range. In extrapolating from well control to adjacent undrilled areas, possible changes in organic matter type need to be considered (Barnard et al, 1981; Baird, 1986). Also, in areas were the organic matter is now highly mature it can be very difficult to establish the type of organic matter originally present.

9.2.2. Organic Matter Content

Sample coverage is limited to well locations, but these have been used to contour organic matter distribution and make interpolations (e.g., Dahl et al (1987) in the Oseberg area of the North Sea). Extrapolation away from limited sampling can only be done by relating organic composition and content to the geological setting. Preliminary attempts to do this have been reported by Creaney and Passey (1993) and Katz and Pratt (1993). Pasley et al (1993) studied the relationship between organic content and proximity to sequence stratigraphic surfaces for the Mancos Shale in the San Juan basin of New Mexico, and concluded that organic content is predictable within a sequence stratigraphic framework.

Traditionally organic matter content has been obtained from samples retrieved at intervals down a well. Sample spacing is rarely less than 5 m, and it can be difficult to obtain an appropriate average value from this data, especially when lithology changes rapidly. With scattered sampling there is often a tendency to overemphasize organic-rich beds. On the other hand, very thin, but organic-rich, intervals may be inadequately represented if sampling is restricted to cuttings.

Electric logs provide continuous measurement of vertical sections with a resolution of about 0.25 m, and there have been attempts to relate wireline log response to organic matter content (Ayers et al, 1982; Fertl and Chilingarian, 1988; Herron, 1991; Houzay et al, 1993). Even poorly calibrated logs can give an indication of variability, and therefore some idea of the value of selected sampling locations. Early attempts to use electric logs relied on the empirical relationship between gamma-ray response and organic content for many organic rich shales (Schmoker, 1981), and more detailed studies using gamma-ray spectral logging tools showed that uranium was responsible for the response (Fertl and Rieke, 1980). While these techniques can be calibrated for local areas, there appears to be no universal relationship. Gamma-ray curves have been combined with sonic logs by Dellenbach et al (1983) and Autric and Dumesnil (1985) to improve estimates of TOC. Organic matter is considerably less dense than the mineral matrix and overall rock density decreases as organic content rises. Schmoker (1979) used density log data to estimate TOC, but Schmoker and Hester (1983) pointed out that pyrite is commonly associated with organic matter, and pyrite has a very high density so that a correction may be needed. Resistivity logs have been used by Schmoker and Hester (1989) and others, and this approach relies on recognizing the changing resistivity that accompanies the generation of non-conductive hydrocarbons. It may be more useful for indicating maturity rather than organic matter richness (Meissner, 1978). The most promising methods combine resistivity curves with porosity (Meyer and Nederlof, 1984). Passey et al (1990) developed a procedure for overlaying a properly scaled sonic log (to provide porosity information) with a resistivity curve. These curves are parallel in water-saturated, organic-lean rocks but diverge as organic content and maturity change (Figures 9.1 and 9.2). The

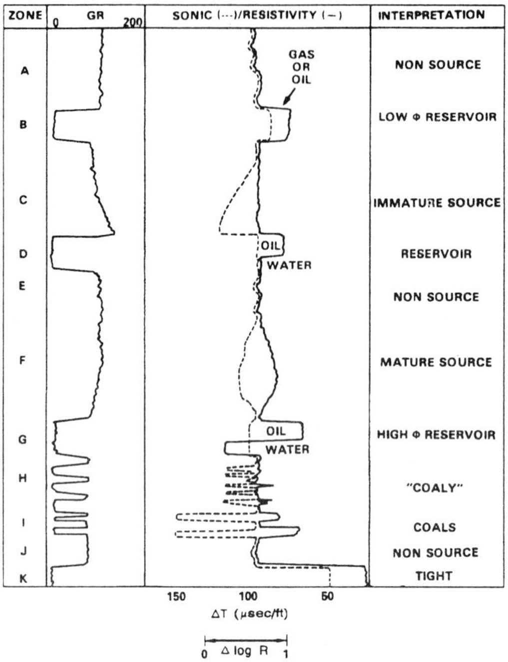

Figure 9.1. Schematic guide for interpreting a wide variety of features observed on ΔlogR overlays (Passey et al, 1990).

changes occur because the porosity curve responds to the presence of kerogen, which gives a low-density, low-velocity response, while the resistivity curve is reflecting the nature of the formation fluid and changes as generated hydrocarbons increase. Separation of the curves was calibrated to TOC and organic maturity. The CARBOLOG(CARBon Organic LOG) method of Carpentier et al (1991) also relies on a combination of sonic and resistivity logs to indicate organic carbon content. Figure 9.3 shows a comparison of log-derived values with those measured at intervals down the well.

Organic matter contents of source rocks are typically given as weight percentages, whereas the quantities of oil and gas generated are calculated as volumes. Kerogen is much less dense that the mineral matter in the rock, and gram for gram occupies a bigger volume. Unfortunately, it is not possible to derive a rigorous conversion factor from weight percent to volume percent because kerogen density varies with organic matter type and with maturity, and the rock density varies with lithology. A conversion factor of 2.5 is generally close.

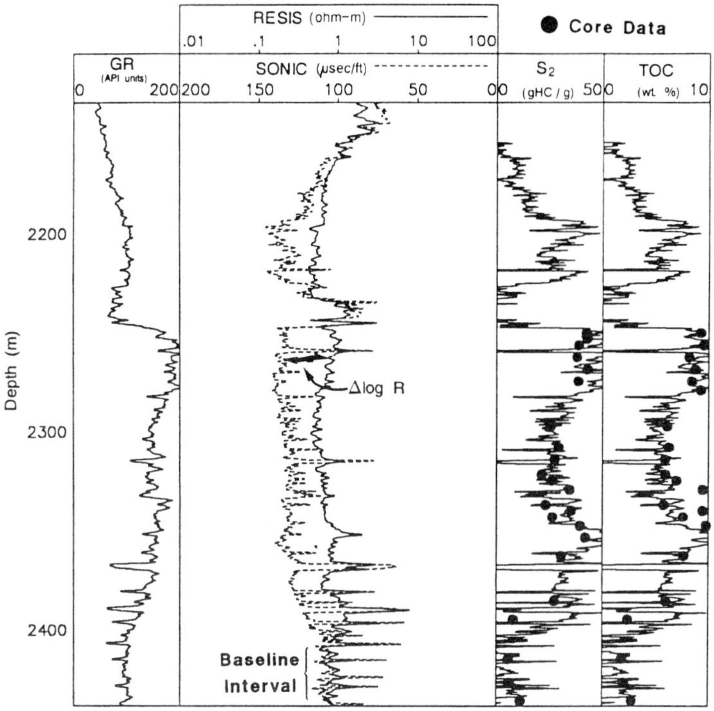

Figure 9.2. Sonic/resistivity overlay showing ΔlogR separation in an organic-rich interval that contains Type II kerogen at a maturity level equivalent to an LOM of 6-7 (from Passey et al, 1990. See their figure for further details).

Organic content is measured on available samples of potential source rocks, and these are not always at low maturity. As rocks thermally mature an increasing amount of kerogen is converted to bitumen, and some may subsequently migrate out of the rock. As a consequence, the residual organic content will decrease and TOC values drop with increasing maturity. When organic content is determined from high maturity rock samples it is necessary to make a correction to get the true quantity that was present when the rock started generating (Cooles et al, 1986). Because the hydrogen content is different for the various types of kerogen, and because migration efficiency varies with organic matter content, it is not possible to calculate a single conversion factor, but an increase by a factor of about 2 for a rock with a R_o above 1.5 % is usually close. A better estimate can sometimes be made by comparing the present Hydrogen Index with the presumed initial value (e.g., Sylta, 1993).

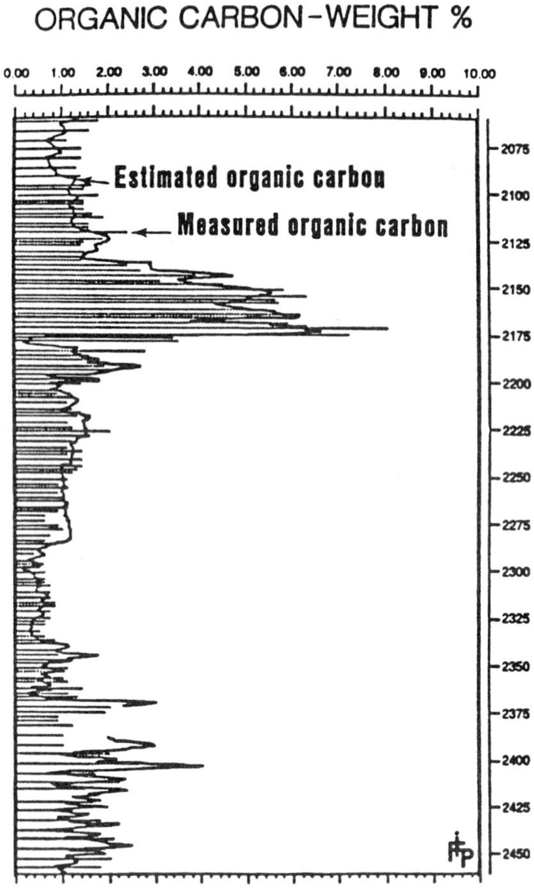

Figure 9.3. Comparison of total weight percent organic carbon determined from the CARBOLOG method with values obtained by analysis of cuttings (Carpentier et al, 1991).

9.2.3. Quantities Generated

In the past there have been many attempts to calculate the quantity of petroleum generated in a basin (Bockmeulen et al, 1983; Barker and Dickey, 1984; Momper and Williams, 1984; Moshier and Waples, 1985; Baird 1986; Comer and Hinch, 1987; Nakayama, 1987, among others). This was usually done by taking the volume of the source rock and its average organic matter content and assuming (or measuring) the amount of bitumen generated. With a further assumption of migration efficiency, it was then possible to estimate the total quantity of expelled oil. An example of this type of calculation is given by Barker and Dickey (1984) for Oxfordian-Callovian source rocks in Saudia Arabia. Ayers et al (1982) had established the thickness and distribution of mature source rock (based in part on Lopatin-type calculations), and this totaled 5000 km^3. They reported organic carbon contents averaging 4 weight percent, or about 10 volume percent, and this leads to a kerogen volume of 500 km^3. Fifteen percent conversion to bitumen gives 75 km^3 of bitumen, and 10% migration gives 7.5 km^3 of migrated oil, or 47.2 x 10^9 bbl. This is probably less than 5% of the oil in Saudi Arabia. No correction was made for kerogen reduction at increased maturities, and in retrospect the values used for generation and migration efficiencies were probably too low, but this calculation shows that it is difficult to account for total reserves with the Oxfordian-Callovian as the only source rock (Jones, 1985).

Williamson et al (1994) calculated the volume of oil generated by maturing source rocks in the Hibernia Drainage area of the Jeanne d'Arc basin, and compared it with the amount of oil in the reservoirs. Based on considerable well and seismic control, the Jeanne d'Arc basin, offshore eastern Canada, has been described by Keen et al (1987) as a classic half graben. It has accumulated and preserved a thick sequence of syn-rift sediments, together with a thick overlying sequence of Cenozoic post-rift sediments (Williamson, 1992). Included in this group is an organic rich (3 - 4% TOC), oil-prone, calcareous shale source rock of Kimmeridgian age, formally defined as the Egret Member of the Rankin Formation. Modeling this source rock showed that the deeper parts started to generate 80 - 100 Ma but the shallower part, including much of the area immediately below the Hibernia Field, did not generate until the last 30 - 40 m.y. Williamson et al (1994) established source rock maturity using BASINMOD (Platte River Associates) and assumed that 20 - 40% of the porosity of the Egret Shale needed to be saturated with oil before oil expulsion could occur. Overall migration efficiency was 20%, and 2% of the petroleum was lost along the migration pathway. Their calculations showed that the volume of mature source rocks in the present Hibernia area is insufficient to account for the volume of oil in the Hibernia structure. An area now located to the north of the Nautilus fault seems to have provided the extra charge. Their modeling results were considerably different when specific Egret activation energies (E) and frequency factor (A) data were used, rather than a North Sea Kimmeridgian source rock default value.

With the more sophisticated modeling tools now available, the quantities of generated hydrocarbons can be accurately calculated for any level of maturity or mix of organic matter types. The progress of generation is commonly reported as the Tranformation Ratio, although it is common to combine estimates of generation with expulsion and report the amount of petroleum produced by a unit volume of the source rock (Larter, 1988). Calculation of total reserves in an area also requires an estimate of mature source rock volume, and precise estimates can be complicated when different parts of the rock are at different stages of maturity.

While it is convenient to talk about "oil" or "gas", in reality they can have a wide range of compositions, and for many subsurface conditions of pressure and temperature they are probably combined into a single phase. Models have been developed to provide more compositional detail, and they treat petroleum either as a range of boiling point fractions (Sweeney et al, 1987) or as carbon number ranges, usually C_1, C_2 - C_5, C_6 - C_{14}, C_{15+} (Ungerer et al, 1988). Light fractions can be generated directly from kerogen or indirectly by cracking oil (Welte et al, 1988), so that in describing the source rock system both primary and secondary reactions must be considered. The kinetic parameters for cracking oil to gas have been derived from laboratory pyrolysis experiments and from experience with oil shale retorting (Burnham and Braun, 1990). Most models assume that any secondary cracking occurs in the source rock pores, and that when oil moves out its composition remains unchanged. An "open" system, where expulsion occurs immediately after generation, is comparable to dry pyrolysis in flowing gas (e.g., RockEval). At the other extreme, in a "closed" system generated materials remain in the source rock and with increasing temperature crack in reactions that lead ultimately to graphite (or "coke") and methane. Clearly neither of these end-member possibilities describes the real system because extractable materials are found in source rocks, but material has also migrated out to provide oil in reservoirs. Models must specify the timing, composition, and percentages of migrated materials, and the treatment of expulsion is a key factor. Expulsion is usually described by the Darcy equation (see below) and this includes a viscosity term for the moving fluid. Viscosity values are best estimated from compositional data. The model developed by Ungerer et al (1988) was tested by Forbes et al (1991), and, with some limitations, it adequately described the petroleum distribution in the Smorbukk Sor area of the North Sea.

9.3. 2-D MODELS

One-dimensional models can be used to calculate the thermal maturation of organic matter because this is essentially independent of the surroundings, and examples have been discussed in the previous chapters. Backstripping can also be carried out successfully for 1-D reconstructions, and paleotemperatures can be reconstructed as long as only conductive heat transfer has to be considered. However, modeling needs to be extended into two dimensions (2-D), or even three dimensions (3-D), when convective heat flow makes a significant contribution to thermal history, or when structural evolution is particularly complex. Although some 1-D models now incorporate a simple version of expulsion which provides a preliminary estimate of the amount and composition of expelled petroleum, 2-D (or 3-D) modeling is essential for describing the fluid flow that characterizes secondary migration of hydrocarbons from source to trap. Commercial 1-D modeling programs have been available for several years and can be run on personal computers. More 2-D and 3-D software is being introduced, but it is more expensive and much more computer intensive. Figure 9.4 shows the general organization of a typical model's input and output files, and the type of information needed for a detailed 2-D study, in this case for the TEMISPACK model (Ungerer et al, 1990). The required input data are a knowledge of the geology of the section; parameters for single-phase fluid flow; heat transfer information; and geochemical data, together with the physical parameters for two-phase migration.

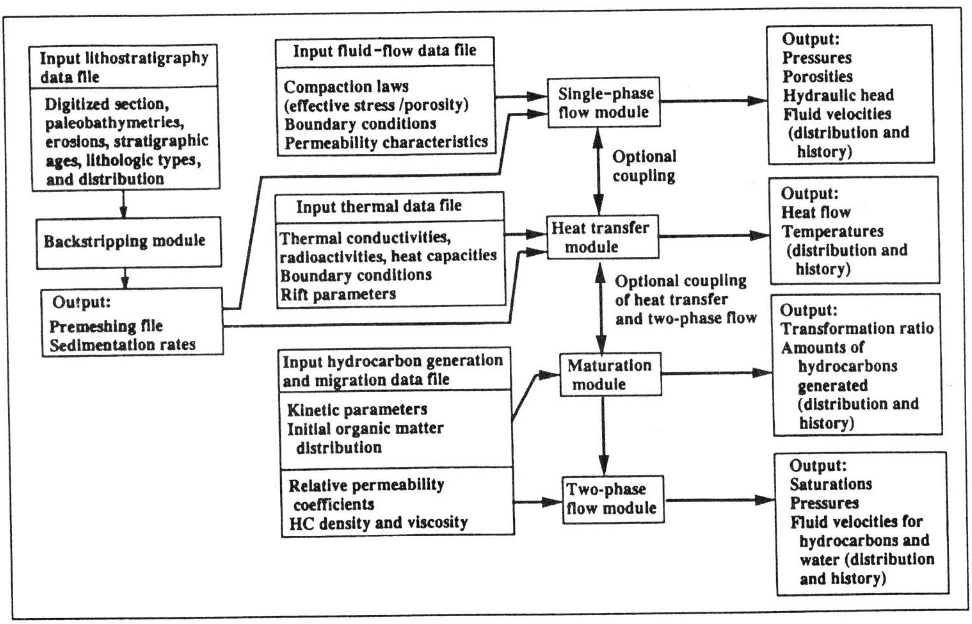

Figure 9.4. General organization of the input and output files for the TEMISPACK model developed by the IFP (Ungerer et al, 1990).

In some cases it may be possible to generate a pseudo 2-D cross-section ("$1^1/_2$-D section") that shows the lateral distribution of maturity by combining the maturity profiles for a group of wells (e.g., Middleton and Falvey, 1983). It is often helpful to do this along a seismic line that is anchored on actual wells. However, when fluids are moving the rate and direction of fluid flow depends on the characteristics of the surrounding rocks, and these have to be specified in any model. The key parameter is generally permeability which varies from $>10^3$ md to $<10^{-6}$ md, usually with considerable anisotropy because horizontal values are almost always greater than vertical ones. Exceptions occur when vertical fracturing or faults are important. Permeability data are difficult to get, even when there is good well control, since they cannot be inferred from electric log response, and measurements on cores or DST determinations are needed. There have been attempts to develop empirical relationships between porosity and permeability, but these commonly give only a very generalized trend. Mudford and Best (1989), for example, suggested:

$$k = A.\phi^B \tag{9.1}$$

for clastics, with $A = 4 \times 10^{-15}$ m^2 and $B = 8$ for shallow shales. Mann and Mackenzie (1990) used the same equation for clastics, but for carbonates used:

$$k = A.10^{B\,\phi} \tag{9.2}$$

while Sylta (1993) has used:

$$\log_{10} k = A\phi + B \tag{9.3}$$

where k is permeability, ϕ is fractional porosity, and A and B are constants. One of the most commonly used relationships between permeability and porosity is the Kozeny-Carman equation. This was developed initially for glass bead packs and works best for clean sands with fairly uniform grain size distributions. For porosities greater than 10% it has the form:

$$k = 0.2\,\phi^3 / \{ S^2 (1 - \phi)^2 \} \tag{9.4}$$

where S is the specific surface area of the rock, which is the surface area to volume ratio obtained by treating grains as spheres and the enclosing volume as a cube. For porosities less than 10% this has been modified (Ungerer, 1990) to:

$$k = 20\,\phi^5 / \{ S^2 (1 - \phi)^2 \} \qquad (9.5)$$

2-D models set up a mesh of connected elements to describe the geology of the area of interest, and the scale and treatment of this mesh is critical to the performance of the model. Petrophysical properties are taken to be uniform within the selected blocks (*"cells"*), and so rock inhomogeneities are described less realistically as the blocks get larger. However, it is important to remember that even for a realistically large number of small blocks, the model is using a grossly oversimplified geological description (see for example, Figure 9.5). Both finite element and finite difference mathematical techniques have been used in modeling. The finite element methods are more computer intensive and this has to be offset by using a smaller number of cells, and they are therefore larger. This leads to reduced geologic resolution. The THEMIS model developed by IFP is a model of this type and an example of a meshed cross section shown in Figure 9.6. The mesh is made up of deformable and mobile angular or quadrangular cells, but these could not be split to account implicitly for fault displacements. Each cell is assigned a specific lithology and appropriate values of porosity, permeability, TOC, etc. Other commercial models have used modified finite difference methods that are capable of handling deforming cells and both the BASINMOD 2 and Lerche models are in this category.

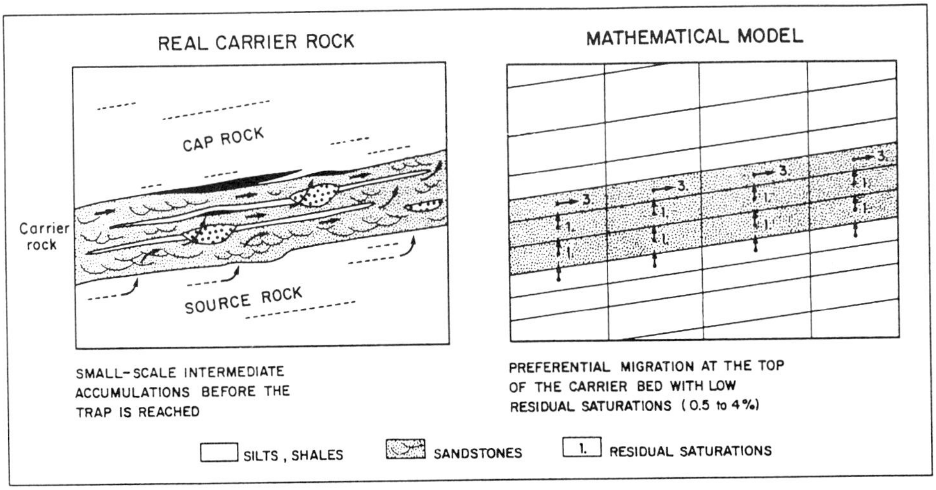

Figure 9.5. Actual geology of a carrier bed and the simplified description needed for modeling. Local heterogeneities cannot be incorporated into the model. (from Ungerer et al, 1990).

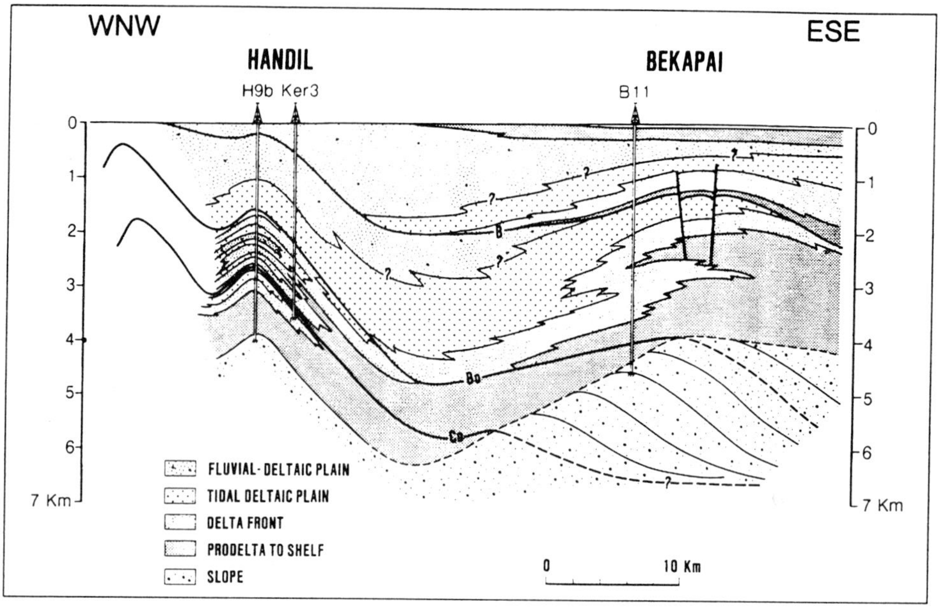

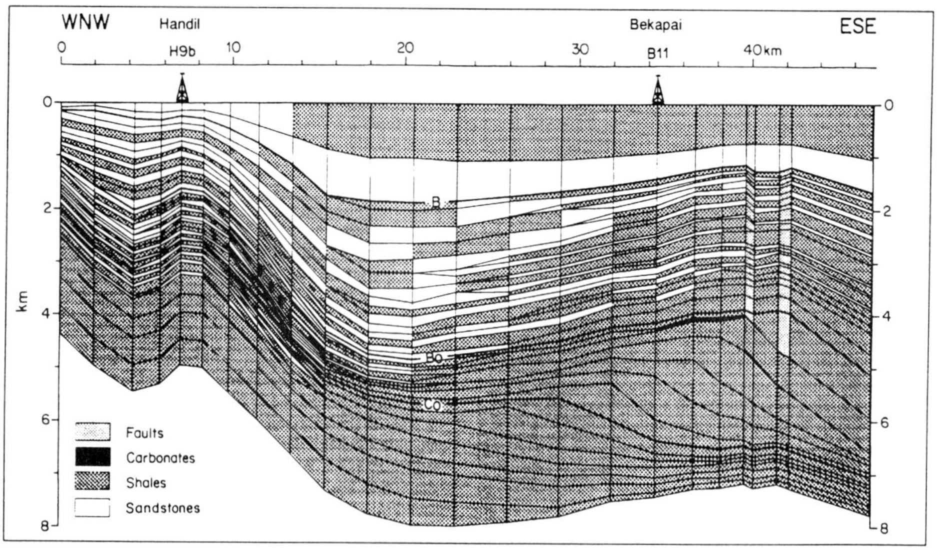

Figure 9.6. Simplified geological cross section of the Mahakam delta, Indonesia, showing the locations of Handil and Bekapai fields (top), and the 2-dimensional mesh used to model the area (bottom) (Ungerer et al, 1990).

True 2-D models can show fluid migration direction and the role of barriers to fluid flow. When fluid flow is not possible, or has a very low rate, then pressures can rise towards the rock load giving abnormal pressures. These are discussed below. Barriers to fluid flow can control the direction of moving water and petroleum, and they also control migration pathways, as well as their termination in traps. Models are set up to be impervious at the bottom layer so fluid flow is restricted to the gridded area.

Nyland (1993) has analyzed the use of models by companies bidding for concessions off Norway (Table 9.1). His table shows that 2-D models were first used in 1985, and this also corresponds to the availability of commercial programs at that time. The Table also shows a move away from using TTI as the calculated maturity parameter. With increasing use of kinetic modeling, calculated R_o and Transformation Ratios are becoming the preferred indicators. Migration modeling is still not widely used. Nyland (1993) was led to comment that "the presence of large structures, which the companies argue they cannot afford not to drill, can result in the prognosis from basin modelling being overridden".

Table 9.1
The use of various types of basin modeling on the Norwegian Shelf (Nyland, 1993). The numbers are based on regional studies presented in a selection of concession rounds.

YEAR/ ROUND	NUMBER OF APPLICANTS	BASIN MOD. STUDIES		INHOUSE/ CONS.	NO SOURCE ROCK EVALUATION	MIGRATION MODELLING	MATURITY PARAMETERS USED			
		1D	2D				TTI/LOM	VR%	TR.	TEMP.
1982/5	20	2	-	2/-	6	-	2	4	-	-
1982/7	19	4	-	4/-	4	1	4	4	1	1
1984/8	19	4	-	4/-	2	2	6	4	-	1
1985/9	19	4	2	4/2	-	2	7	3	-	-
1987/11	13	10	3	9/2	-	4	10	9	1	1
1990/13	21	9	6	9/6	1	3	1	13	7	2

9.4. ABNORMAL PRESSURES

Subsurface pressures are important in driving fluid flow, influencing petroleum expulsion, controlling migration pathways and, if present today, causing drilling problems. They may also influence compaction, and the variable water content (porosity) that results can change thermal conductivity and alter geothermal gradients. Also, in areas where abnormal pressures approach the rock load, induced fracturing can be a significant factor in providing fluid conduits. Normal, or *"hydrostatic"*, subsurface fluid pressure at any depth is equal to the pressure exerted by a column of water reaching to the surface. It varies slightly because water density is affected by salinity, but typically is close to 0.47 psi/ft. Pressures greater than hydrostatic are called *"overpressures"*, *"abnormal pressures"*, or occasionally *"geopressures"*. They have a limiting upper value of 1.0 psi/ft because, in the absence of any tensile strength in the rock, this is equivalent to the pressure exerted by the rock column (= *"overburden"*). If this value is exceeded, the overlying rock column is lifted and a horizontal fracture develops. Vertical fractures can develop at lower pressures corresponding to pressure gradients of 0.70 - 0.90 psi/ft (Du Rouchet, 1981).

Pore fluid pressure (P) is related to sediment load (SL) and effective stress (σ) by the relationship:

$$SL = P + \sigma \tag{9.6}$$

The sediment load is fixed at a value close to 1.0 psi/ft, and if pressure, P, increases the effective stress must decrease. Because effective stress drives compaction, an increase in pore fluid pressure decreases the rate of compaction and leads to abnormally pressured intervals that are commonly undercompacted. Most methods for detecting abnormal pressures rely on recognizing the effects of this undercompaction.

Several processes can cause abnormal pressures (Sahay and Fertl, 1989; Fertl et al, 1994):

(1) *Disequilibrium compaction*: When rock permeability is high fluids are free to move out in response to rising effective stress (SL in Equation 9.6) and P remains normal. If permeability is low fluids cannot move out rapidly enough to maintain equilibrium and the pore fluids take part of the rock load and pore fluid pressure rises. This is frequently the most important overpressuring mechanism.

(2) *Petroleum generation*: The process of petroleum generation converts kerogen to a more mature kerogen plus bitumen and involves a volume increase. If the volume of the system is essentially fixed then

pressure will rise (Law and Dickinson, 1985; Spencer, 1987, 1994; Ungerer et al, 1983). This mechanism of abnormal pressure generation is most likely to occur in rocks with unusually low permeabilities and high organic contents (Burrus et al, 1996).

(3) *Petroleum cracking*: Thermal cracking of oil to gas has the potential for generating high overpressures, and Barker (1990) showed that cracking of 1% of the oil in a closed reservoir will raise the pressure to the rock load. Luo and Vasseur (1996) carried out more detailed modeling of this process and demonstrated that oil-to-gas cracking strongly influences the development of abnormal pressures.

(4) *Thermal expansion of water*: Thermal expansion of water during burial can lead to abnormal pressures in a low permeability system. This is usually called *"aquathermal pressuring"* (Barker, 1972). Kuo et al (1994) found that this became a significant factor in the late development of overpressure in the Delaware basin.

(5) *Miscellaneous mechanisms*: Other mechanisms that may be important in some areas include lateral tectonic compression, clay diagenesis and dewatering, and reverse osmosis.

An empirical 1-D model developed by Mann and Mackenzie (1990) considered vertical fluid flow in a situation where disequilibrium compaction had the major role in producing abnormal pressures. A pseudo 2-D pressure distribution could be obtained by combining calculations for laterally-continuous, high permeability layers that are treated as a set of adjacent pressure cells. Predictions from the model were in good agreement with measured pressures for several areas around the world.

Sediment thickness, compaction constants, and permeability constants were incorporated into a 1-D finite difference model that led Thorne and Watts (1989) to conclude that in the North Sea burial rate was probably the key factor in the development of abnormal pressures, but Mudford and Best (1989) found high gas pressures associated with low sedimentation rates (~17 m/m.y.) on the Scotian shelf of eastern Canada. Mudford (1990) extended 1-D pressure modeling to include two phases. His finite difference model used an unevenly spaced vertical grid, and was applied to the area of the Venture gas field, offshore Nova Scotia (Mudford and Best, 1989; Jansa and Urrea, 1990). Here overpressured reservoirs are not significantly undercompacted, and do not show anomalous sonic and resistivity log responses typical of overpressures in the Gulf of Mexico and many other areas. Overpressuring seems to have developed late and to be due to an increase in sedimentation rate over the last 15 m.y. together with the generation of petroleum. In the absence of detailed geochemical information, generation was assumed to be from Type III kerogen and treated as a single, first-order kinetic reaction using global E and A values of 240 kJ/mol (57.4 kcal/mol) and 10^{16} sec^{-1}, respectively.

Studies in several areas have evaluated the role of petroleum generation in producing abnormal pressures, including those by Bredehoeft et al (1994) in the Uinta basin; Kuo et al (1994) in the Delaware basin; and Spencer (1987) in the Rocky Mountains. Kuo et al (1994) used thermal modeling to establish the time of generation in order to relate this to the observed overpressured intervals.

Most 2-D modeling programs are concerned with fluid flow and calculate pore pressures through time. Ideally all possible overpressuring mechanisms would be evaluated by the model, but in general disequilibrium compaction is most widely considered. The BASIN2 2-D fluid flow model of Bethke (1985) was used by Summa et al (1993) in a study of abnormal pressures in the Gulf of Mexico. They concluded that aquathermal pressuring, clay dewatering, and hydrocarbon generation did not make significant contributions to the generation of overpressures. They also found that the shale permeability values were critically important, and a change by a factor of 10 in permeability increased the depth to the top of the overpressure by 2 km.

9.5. PETROLEUM MIGRATION

9.5.1. Expulsion (Primary Migration)

The first step in the migration of hydrocarbons from source rock to reservoir is the expulsion of part of the generated hydrocarbons out of the thermally mature source. The efficiency of this process of expulsion (or *"primary migration"*) has critically important consequences for the total reserves in an area. Also, the timing of expulsion (relative to the timing of trap development and the orientation of migration pathways) can influence which structures were in a position to accumulate oil and gas. There is also increased interest in trying to predict the composition of the expelled hydrocarbon fluids. The mechanism of expulsion is still not completely understood, and many of the parameters required to describe it in detail are poorly known, but it is now widely accepted that the mechanism is best treated as separate phase migration (Section 1.7.3).

One common approach has been to treat expulsion empirically, and assume that some threshold value for generated hydrocarbons must be exceeded before any are expelled. For a specific organic richness the quantity of hydrocarbons in the source rock pores depends on organic maturity, and it is appropriate to use some measure of generation status as the empirical indicator of migration status. Transformation Ratios or vitrinite reflectance percentages are most commonly employed, and the amount expelled is correlated with the value reached. Alternatively, the quantity of generated bitumen can be compared with available pore space derived from porosity to estimate saturation. Often a fraction of 0.2 is used as the limit, and bitumen in excess of this is assumed to migrate.

More detailed modeling of the process of expulsion has generally assumed that there is pressure driven flow of a separate hydrocarbon phase that can be described by Darcy's Law (Ungerer et al 1984; Doligez et al, 1986; England et al, 1987; Durand, 1988), but Mackenzie et al (1987) concluded that while a model of pressure-driven flow of a petroleum-rich fluid phase could describe their measured distributions of heavier hydrocarbons, some modification was needed to account for diffusion of the gas and gasoline range compounds. The relative permeability for oil is extremely low at low oil saturations, so it is only after oil saturation has built up to a significant percentage of the pore volume that hydrocarbon expulsion becomes efficient (Burnham and Braun, 1990). Values for the minimum critical saturation of oil are not well known. In the absence of constraining data a value of 20% is often used, though Skjervoy and Sylta (1993) varied it from 16% (maximum expelled) to 40% (minimum expelled) to establish the likely range for expulsion. No hydrocarbons will be expelled until the critical saturation is reached, but from then on migration will be close to 100%, unless significant amounts of bitumen are adsorbed by the kerogen (Pepper, 1991). Ungerer (1993) suggested treating saturation thresholds as a user-specified empirical parameter that is calibrated for the local area. Calibration can be based on either the S1 pyrolysis peak or the amount of solvent extract, since both are a measure of the amount of generated material remaining in the rock. As compaction progresses with increasing depth of burial the pore space in the source rock is reduced, but the quantity of hydrocarbons is increasing with rising maturity. For any given level of maturity the ability of the source rock to generate hydrocarbons in the pore space depends on the amount and type of organic matter, and rocks with high TOC values will exceed the threshold values for migration at lower maturity levels and begin to expel earlier. They will also expel a bigger percentage of the generated bitumen (Cooles et al, 1986) (see Figure 1.25).

Most relative permeability curves were developed for use in reservoir rocks and there is considerably uncertainty in extrapolating their use to the very low permeabilities typical of shales. Okui and Waples (1993) proposed a relative permeability curve for source rocks that is more appropriate than those typically used for clastic reservoir rocks, because of the higher surface area/volume ratio for shales compared with sands (the S term in Equations 9.4 and 9.5). However, their simulations showed that most oil expulsion occurred when oil saturation values were high and close to the maximum. They were led to conclude that the detailed shape of the relative permeability curve is probably not of great significance.

The pressure that drives expulsion of a separate hydrocarbon phase can be generated in several different ways. It is most commonly linked to compaction with pressures rising when rocks are loaded faster than the water can escape the low permeability source rocks. Pressure can also increase due to kerogen maturation, or the cracking of oil to gas. If internal pressure approaches the rock load then the rock may respond by fracturing rather than by ductile deformation, and this can be important in providing conduits for petroleum migration (Comer and Hinch, 1987; Talukdar et al, 1987). Most models simulate fracturing by significantly increasing permeability when calculated pore pressure reaches some specified fraction of the lithostatic pressure. This fraction will be 1.0 in the special case of horizontal fractures (which are usually parallel to the bedding planes) (Duppenbecker et al, 1991).

Although petroleum in the pore space of a mature source rock is commonly treated as a single supercritical petroleum phase (Ungerer et al, 1983; Burnham and Braun, 1990), some recent models have considered the possibility that the organic fluids may separate into several phases depending on pressure/temperature/composition conditions (Rudkiewicz and Behar, 1994). These different phases can have quite different mobilities.

In the Paris basin a comparison of calculated amounts of hydrocarbons with those actually present, led to a migration efficiency that increased from 0 to 88% with increasing depth of burial (Espitalie et al, 1988). Migration efficiency was defined as the percentage of the generated hydrocarbons that actually migrate out of the source rock. Cooles et al (1986) also reported that expulsion efficiency could be as high as 90% and Figure 9.7 shows their data relating average bulk expulsion efficiency to average initial petroleum potential. For source rocks with more than 10kg/MT, Mackenzie and Quigley (1988) estimated oil expulsion efficiency to be between 60% and 90%, and Figure 9.8 illustrates their calculated relationship between organic matter type, petroleum generation, and petroleum expulsion efficiency. There are considerable uncertainties about the amount of gas that can be expelled, and Mackenzie and Quigley (1988) concluded that secondary cracking of unexpelled oil leads to efficient expulsion of the gas or gas condensate. Sylta (1993) found that a value of 45% for gas expulsion provided a good match between calculated and observed reserves in the Troll field, North Sea.

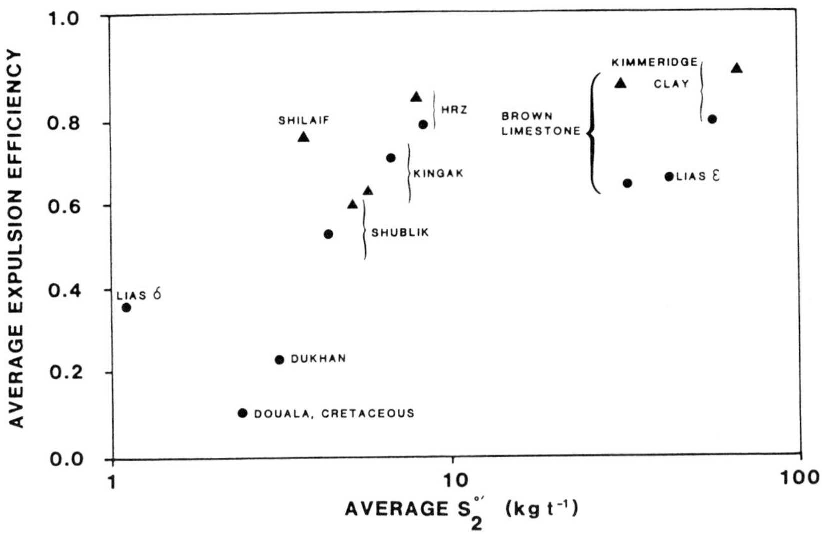

Figure 9.7. Variation of average bulk expulsion efficiency with average initial petroleum potential, obtained either as total solvent extract (circles) or as pyrolysis S1 values (triangles) (Cooles et al, 1986).

9.5.2. Migration through Carrier Beds (Secondary Migration)

Transport of expelled petroleum from source rock to reservoir through carrier beds is driven by buoyancy (Lehner et al, 1987; Sylta, 1987), and is therefore influenced by the oil-water density contrast, slope of the carrier bed, and thickness of the oil column. Oil density is also influenced by thermal maturity and dissolved gas content and, at shallow depths, may be considerably affected by biodegradation. It is usually assumed that migration is instantaneous (i.e., less than 1 m.y.) and occurs normal to the structural contours. (Note that this means the contours at the time the oil is migrating, and also implies a uniform distribution of permeability.) Permeability values for all potential migration pathways are required and these can be difficult to constrain in practice, because permeability cannot be deduced from seismic, or even from well logs. In general, regional scale permeabilities are greater than those on a local scale because large scale fracture networks are not considered in locally determined measurements (Bethke et al, 1991).

Many models treat the carrier beds as homogeneous, but in real rocks the oil almost certainly follows the most porous and permeable routes. In the Paris basin, for example, inferred migration pathways are highly

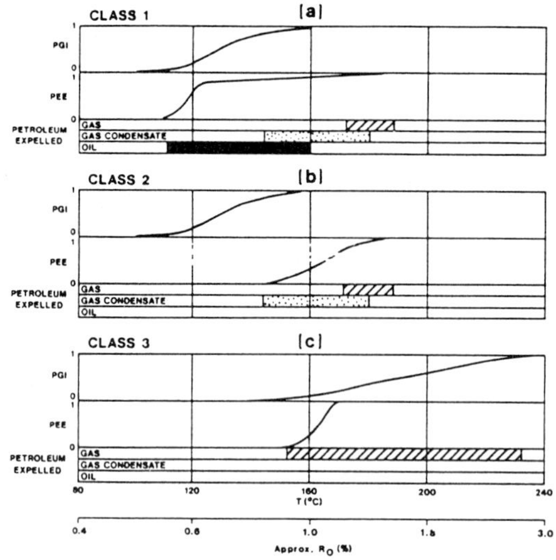

Figure 9.8. Petroleum expulsion efficiency (PEE) and petroleum generation index (PGI) for source rocks heated at 5 °C/m.y. Class 1 organic matter is dominated by labile kerogen present in concentrations greater than 0.01kg/kg of rock (e.g., Kimmeridge Clay formation in the North Sea); Class 2 is a similar organic matter type to that in Class I, but the concentration is below 0.005 kg/kg of rock (examples include source rocks off Nova Scotia, and in the Nile delta); Class 3 organic matter is mainly refractory and expulsion (which is mainly gas) occurs at temperatures above 150 °C (Mahakam delta and Gippsland basin provide examples) (Mackenzie and Quigley, 1988).

irregular, even though the basin as a whole has a rather symmetrical saucer shape (Rhea, 1994). The effect of carrier-bed heterogeneity on secondary oil migration has been investigated by Rhea et al (1994) using two different geostatistical techniques (kriging and conditional simulation). Their study "provides compelling evidence that secondary oil migration is an inherently three-dimensional process".

The role of faults along migration pathways can be difficult to evaluate because they cover the full range from non-sealing faults that provide conduits, to sealing faults that are barriers to migration. Growth faults are especially troublesome because the adjacent lithologies on either side of the fault are changing through time as the fault continues to develop. While many attempts have been made to predict the sealing capacity of faults the impact of faults on basin wide permeability distributions is just starting to be studied. Tectonic fracturing can be modeled for various stages of basin evolution, and Larson et al (1993) have developed a preliminary method that predicts tectonic strain based on the differential subsidence of adjacent nodal blocks in the model. They have shown that in areas of tectonically induced fracturing fluid flow (including hydrocarbon migration) can be radically changed. Abnormal pressures generated by the thermal maturation of organic matter can induce hydrofractures. In their 3-D flow model of the Uinta basin Bredehoeft et al (1994) simulated fracturing by allowing both horizontal and vertical permeabilities to increase by an order of magnitude. In some models major faults, such as growth faults, are treated as a narrow zone of elements with contrasting permeabilities (either high or low for conduits or seals, respectively).

Residual oil can be left distributed along the migration pathway through the carrier beds. Mackenzie and Quigley (1988) proposed a value of 2% for the mean saturation, but Forbes et al (1991) thought their number was poorly constrained and used values of 1% and 3%. Combining this with the estimated pore volume for carrier beds in the vicinity of the Smorbukk Sor field, offshore Norway, Forbes et al (1991) estimated that the amount of petroleum left along the migration pathway was 1.25 to 1.92 times larger than the amount which eventually accumulated in the reservoir. These values may be too high because the total volume of the carrier bed was used in the calculation, and since petroleum moves by buoyancy it may be restricted to just the upper part of the carrier.

As oils move upwards due to buoyancy they move into conditions of reduced pressure and temperature. This can cause phase changes, and an oil may separate into both a liquid and a gas phase. Attempts are being made to use thermodynamic models of fluid equilibria to document these changes (Ungerer, 1993). Migration of oil up into an environment with cooler temperatures can be important in preservation, and the oil is less likely to be cracked to gas on one hand, but more likely to be biodegraded on the other.

A wide range of secondary migration modeling programs are being developed. While most are 2-dimensional, a few are 3-D. Models include those described by Welte and Yukler (1981), Durand et al (1983), Bethke (1989, 1991), Ungerer et al (1985, 1987, 1990), Nakayama (1987), Doligez et al (1986, 1987), Burrus and Audebert (1990), Sylta (1987, 1990), and Lehner et al (1987).

9.6. 3-D MODELS

The ultimate aim of modeling is to integrate the physical and chemical laws that describe the generation, migration, and trapping of petroleum, with the geological and thermal evolution of sedimentary basins in a 3-dimensional, time-dependent simulation. 3-D models are a logical extension of 2-D models, and most of the principles involved are very similar (Yukler and Welte, 1980). However, the demands on computer time are considerably greater since the number of adjacent elements with shared faces or edges in the modeling grid goes from 4 in 2-D models, to 18 in 3-D models (i.e., more than quadruples). Also, a more comprehensive range of input data is needed to make the model geologically realistic. The situation is sometimes simplified by plotting a set of parallel 2-D cross sections to give a "pseudo 3-D", or "2 ½-D" model, but when there is significant migration into or out of the modeled 2-D plane the results may be misleading (Burrus et al, 1996). However, preliminary 2-D modeling can be very helpful in setting up the 3-D model and may save considerable time. Even when true 3-D calculations are made it is easier to display results as cross sections or horizontal depth slices. Sometimes a single unit can be displayed to show (for example) oil saturation (e.g., Yalcin et al, 1988).

Several groups have developed 3-D models including Yalcin et al (1987), Novelli et al (1987), Welte and Yalcin (1987), and Grigo et al (1993). Sylta (1987, 1993) used a 3-D ray tracing technique to model two-phase migration in a single carrier bed. In this case petroleum migration is buoyancy driven and the model is only valid where water flow is not significant. Permeability along the migration pathway can be the most important factor controlling loses of hydrocarbons during secondary migration. Changes in structural orientation significantly modify the spill paths followed by the petroleum, and this is particularly marked in flat-lying areas. Thus the model shows that the drainage area feeding into a particular trap may change through time. This model has been used to study several North Sea areas and three of these are discussed in Section 10.3.4. Figure 9.9 from Lehner et al (1987) shows calculated hydrocarbon column heights at the present, (d), and at various times in the past (a, b, and c). Shifting hydrocarbon supply from the underlying source rocks occurs between 75 Ma and 6 Ma, and the results from the 3-D buoyancy-driven migration model show the distribution of oil as it fills one structure and then spills into a higher, adjacent structure.

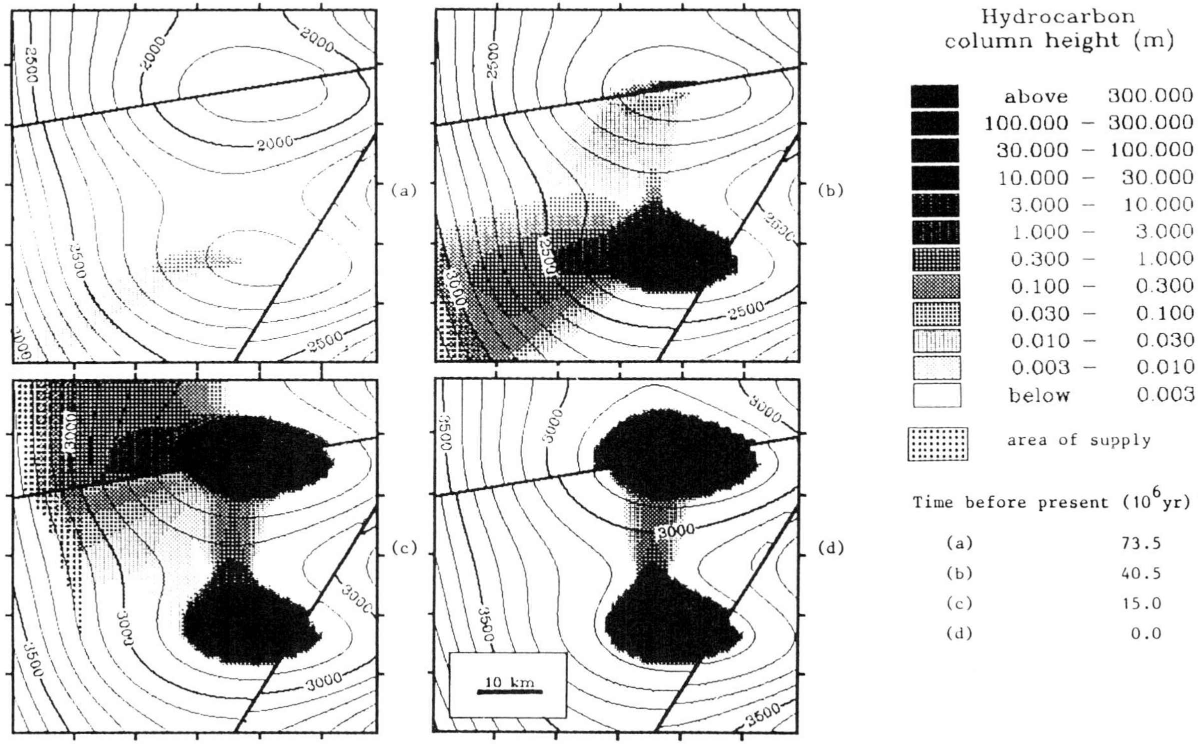

Figure 9.9. Calculated hydrocarbon column heights below a hypothetical cap rock at different times before present (Lehner et al, 1987).

9.7. RESERVOIR QUALITY

Many of the processes that act to control reservoir quality through the destruction or development of porosity are related to temperature, time, and depth of burial. Reservoir quality can be adversely affected as oils mature, and both reservoir permeability and porosity may be reduced by the precipitation of *"reservoir bitumen"* (Rogers et al, 1974). This can make production difficult, or even impossible (Lomando, 1992). While reservoir bitumen can be disseminated throughout the whole reservoir, in some cases it accumulates at the base of the oil as a so-called tar layer. Examples include Oseberg in the North Sea, and Sarir in the Sirte basin of Libya. The precipitation of reservoir bitumen is often called "natural deasphalting". This process may be due either to thermally maturing oil getting steadily lighter, so that the complex, high molecular weight asphaltenes are no longer soluble and precipitate, or it can be induced by the introduction of gas into the reservoir from outside (see Section 1.8.2). The latter is the process usually invoked to explain the tar layer near the base of the Prudhoe Bay field, Alaska. Because the timing of gas generation can be obtained by thermal modeling it may be possible to establish the timing for the introduction of gas and the precipitation of reservoir bitumen, and therefore the timing of the adverse effects on reservoir properties. Lomando (1992) has given an example from west Africa where gas from the sub-salt, overmature source rocks has migrated up into post-salt reservoirs and caused asphalt precipitation. Post-salt sediments that do not lie above areas where the sub-salt source rocks are overmature do not have residual bitumens (see Figure 8.12).

The calculation of rock-fluid interactions as a function of changing depth and temperature is complex (Siever, 1983; Meshri and Ortoleva, 1991) and beyond the scope of this book. However, often it is helpful to overlay generalized information about cementation, clay transformations, or the development of secondary porosity on a burial history diagram (e.g., Schanmugan and Higgins, 1988; Surdam et al, 1989; Burley et al, 1989; Scholle et al, 1991; Martin et al, 1994), so that the time of petroleum generation relative to the time of reservoir development or destruction is clearly shown. In the North Sea, Glasmann et al (1989) provide an example for the Brent Group sandstones, and Cowan (1989) has described diagenesis in the Upper Carboniferous. Thomas (1984) reported an interesting situation in the Perth basin of western Australia where petroleum generation occurs very deep because of the low geothermal gradient, but at these depths sandstone reservoir quality is inadequate for commercial production.

McClimans and Videtich (1989) established a cementation window for the Great Oolite reservoir of southern England based on combined well stratigraphy and pressure-corrected fluid inclusion data. The inferred depths were transferred to burial history diagrams and combined with a thermal history for the Lower Jurassic Lias source rocks. The paleo-thermal history was based on vitrinite reflectance data and an in-house maturity modeling program. Values were 33.9 °C/km before uplift, and 38.6 °C/km afterwards. The calculations showed that in the vicinity of the Godley Bridge and Baxter's Copse wells porosity had been lost prior to the generation of oil and these wells did not produce. In contrast, cementation occurred later in the Storrington well and there was little porosity loss before oil was generated and migrated into the reservoir (Figure 9.10). Notice that, although cementation occurred deeper in the Godley Bridge area, it still predated the time of oil formation. Videtich et al (1988) used a similar approach to constrain the timing of cementation in the Mishrif reservoir, Dubai. Here "the occurrence of oil inclusions in the blocky calcite spar, as well as much of the coarse equant spar, indicated that oil migration was synchronous with that of cementation".

Fluid inclusions were also used in a study of the Alberta Deep basin and showed that calcite precipitated after quartz druse due to cooling on uplift (Tilley et al, 1989). Temperatures derived from oil- and gas-bearing inclusions confirmed that they formed during significant uplift and cooling of the basin. Burley et al (1989) used a similar approach to establish time,

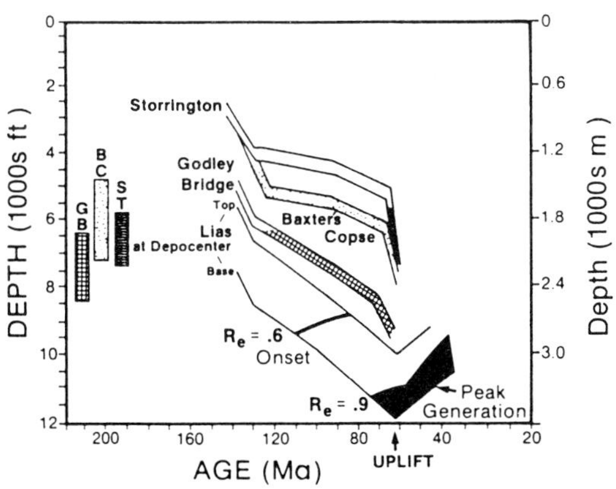

Figure 9.10. Burial and maturation histories for the Great Oolite, southern England, at the locations of the Godley Bridge, Storrington, and Baxter's Copse wells, and for the Lias at the Godley Bridge well. Stipple pattern identifies the "cementation window" calculated from fluid inclusion data. (McClimans and Videtich, 1989).

temperature, and paleoburial depth for diagenesis in the Tartan Reservoir of the U.K. North Sea, and in this case the addition of K-Ar dates for authigenic illites further helped to constrain the timing of events. Fluid inclusions in overgrowths were also used by Saigal et al (1992) to document filling of the Fulmar reservoir in the North Sea.

The introduction of oil into a reservoir can starve crystal surfaces of material and stop cementation, thus preserving the porosity (e.g., Ekofisk, North Sea). Quartz overgrowths may be a less reliable indicator of the timing of oil introduction into a reservoir than K-feldspar albitization (Saigal, 1992) because quartz can continue to dissolve by pressure solution and reprecipitate locally as overgrowths. In contrast, albitization of feldspars requires transport of Na^+ in, and removal of K^+, and is effectively terminated when water is replaced by oil as the reservoir fills.

Information derived form burial histories can be used to make predictions of reservoir quality in undrilled areas. The Kekiktuk sandstones of the North Slope of Alaska are relatively uniform sublitharenites and quartz arenites, and porosity loss is due to silica cementation that is related to burial history. Bloch et al (1990) used outcrop and core data to plot burial profiles for 7 wells, and then established a "synthetic burial history curve" from seismic data for a wildcat location that was 14 miles from the nearest subsurface control. They claimed that for sandstones, porosity predictions from burial history within $\pm$ 2% are possible.

A generalized relationship between sandstone porosity and thermal history (as indicated by TTI or vitrinite reflectance) was suggested by Schmoker and Gautier (1989). It has the form:

$$\phi = A(M)^B \qquad (9.7)$$

where M is the maturity indicator based on time-temperature, and A and B are appropriate constants. Dutton and Diggs (1992) found that it predicted porosities quite well for the Travis Peak area of East Texas when $A = 30$, $B = -0.33$, and M is expressed as TTI.

The ability to document changing porosities with depth can be important in establishing changes that occur in thermal conductivities, because these can lead to significant changes in geothermal gradients.

CHAPTER 10

APPLICATIONS

10.1. INTRODUCTION

Maturity modeling is being used extensively as an exploration tool, and in the last few years has helped solve a wide range of geologic problems. In response to the increased use of geologic models, a variety of commercial modeling software has become available, but many of these programs are concerned with structural modeling, where there is less emphasis on maturation of organic matter. Although they will not be discussed here, these programs can be very useful in developing burial histories for complex geological environments. Burial history plots, either alone or with associated temperature regimes, are now often integrated into many types of geologic studies, and they are not used solely for calculating organic matter maturity levels. They have been useful in documenting the diagenetic reactions that destroy primary porosity, or create secondary porosity. These reactions are strongly influenced by depth and temperature, and burial/thermal histories are a convenient way of evaluating the timing of dissolution of minerals, or the precipitation of authigenic cements. Burial history programs have also been used in studies of the origin and distribution of abnormal subsurface pressures.

In this chapter we will consider the role of thermal modeling in a wide variety of geologic settings, and describe selected examples to provide illustrations of the way the principles outlined in the previous chapters can be applied. For convenience the examples are grouped by basin setting. There is no attempt to apply a rigorous basin classification scheme, or to provide studies from all basin settings. In particular, many basins that are loosely grouped as "intracratonic" may have evolved in a variety of different ways, ranging from aulacogens and forelands, to simple sags. Different basin settings have rather different styles for the generation, migration and accumulation of oil and gas and it is important to take this into account in modeling. Growth faults in tertiary deltas, for example, may present barriers to lateral fluid flow, or they may provide conduits, in which case they favor vertical transmission of fluid. In this basin setting migration direction tends to have a dominantly vertical component. In contrast, small continental interior sag basins are often structurally rather simple and are characterized by long distance continuity of carrier beds.

Here migrating petroleum can move long distances laterally, and migration distances in excess of 100 km are not uncommon. Clearly, it is important to consider basin type in relating the distribution of petroleum reservoirs to the location of mature source rocks. Thermal regimes vary considerably among the different basin types, and the differences between deltas and opening rifts have been mentioned several times. Also, the number of source rocks and their characteristics cover a very wide range, but Figure 10.1 shows that some broad generalizations can be made about the organic matter types in different geologic settings (Klemme and Ulmishek, 1991). While deltas and half sags are dominated by coals and Type III kerogen, platform and linear sag basin styles involve source rocks dominated by kerogen Types I and II. Figure 10.1 also shows that the timing of generation is quite different in the various geologic settings, with basins located on platforms generating the earliest and deltas generating the most recently.

10.2. CONTINENTAL INTERIOR BASINS

10.2.1. Introduction

Basins located in plate interiors contain more than a quarter of the worlds proven and produced reserves (Klemme, 1980). The basins can be quite small and simple ("deceptively simple" – Bally and Snelson, 1980), or large and more complex. Near-circular, flat, saucer-shaped basins less than 300 km in diameter were classed as Type 1 by Klemme (1980). They usually contain a single cycle of sedimentation made up primarily of Paleozoic platform deposits. In general, oil recoveries are low. Klemme's Type 2 interior basins are large, more complex, and range in size from intermontane to subcontinental miogeosynclinal. They contain more than one cycle of sedimentation, usually made up of Paleozoic platform deposits followed by later Paleozoic or Mesozoic orogenic clastics.

Even the large basins are often not very deep, and this coupled with average geothermal gradients results in the source rocks being mature only in the deeper parts. However, the basins are structurally simple with basement-controlled tectonics (apart from deformed margins), so that there is long distance continuity of carrier beds and long distance, lateral migration (Demaison and Huizinga, 1991) (Figure 10.2). Reservoirs may be more than 150 km from mature source rocks. Vertical migration can

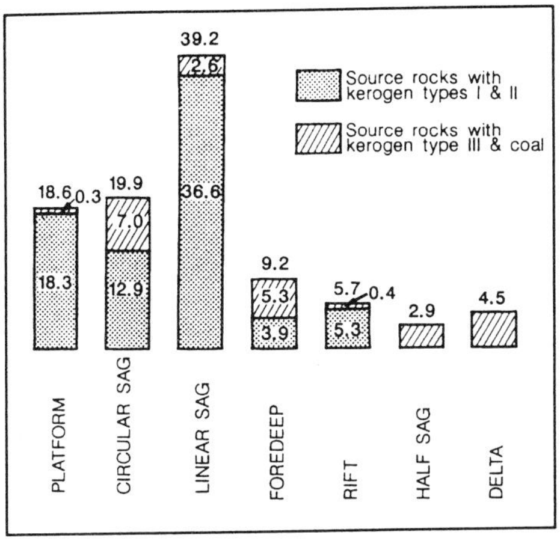

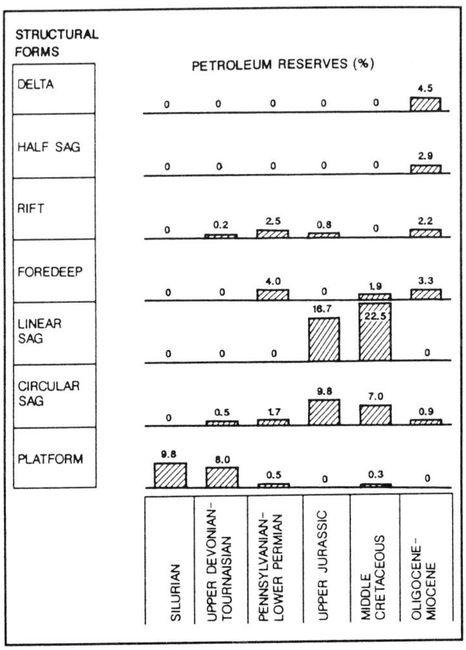

Figure 10.1. Kerogen types and effectiveness of source rocks deposited in various geologic settings. (Top) Kerogen types for source rocks and the percentages each contributes to the original petroleum reserves; (Bottom) Petroleum reserves generated by source rocks in each of six time intervals for each of the basin settings (Klemme and Ulmishek, 1991).

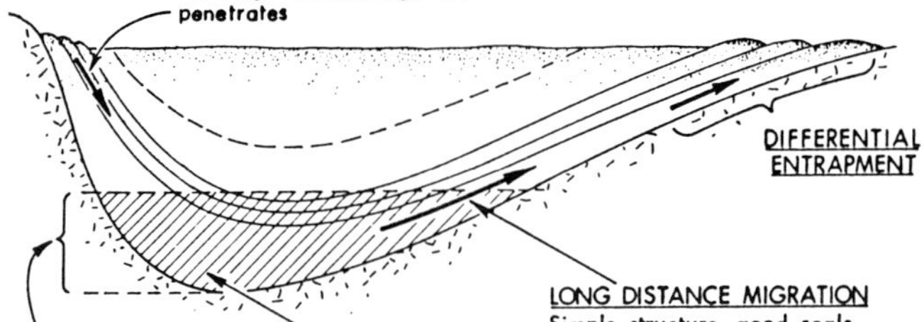

Figure 10.2. Generalized diagram of a simple intracratonic basin showing some of the characteristic features.

also have a role, and has been proposed for the Paris Basin (Espitalie et al, 1987) and the East Texas basin (Wescott and Hood, 1994). In some cases migration is related to fracturing induced by abnormal pressures.

Bordenave (1993) found from a worldwide review of intracratonic basins that the deeper parts contained much better quality source rocks with higher TOC and better hydrocarbon potential. Source rocks can be very rich, and both oil shales and coals occur in continental interior basins. Multiple source rocks are also common, and in the extreme case of the Powder River basin, Wyoming, at least seven different rocks have generated commercial quantities of crude oil (Momper and Williams, 1984). This, of course, complicates thermal modeling because each rock has to be treated separately, and will have a different thermal history, as well as a different mix of organic matter types. Each rock will become mature at a different time for any location in the basin. The combination of several source rocks with long distance migration often produces mixed oils in reservoirs, and this makes oil-source rock correlation more difficult. Further complications occur where recharge on elevated basin margins brings in aerobic bacteria and the oxygen they use. Bacterial degradation of oils is characteristic of the margins of continental interior basins and may make it more difficult to relate an oil to the specific source rock that generated it. In most cases, correlation is possible using compound types that are resistant to biodegradation (see Sections 1.11.2 and 7.3.5).

Penetrating meteoric waters follow the more permeable pathways, and cooling effects can be very localized (Willett and Chapman, 1987) leading to variable geothermal gradients. The cracking of oil to gas is not a major problem in most interior basins because they are too shallow and too cool to reach the necessary temperatures. Much of the gas in these basins is generated from coal (e.g., San Juan basin, U.S.A.; Cooper-Eromanga basin, Australia), or is biogenic (e.g., West Siberia).

10.2.2. Lopatin Modeling

Bond (1984) used the Lopatin method to evaluate Cretaceous source rock maturity on the northern flank of the San Juan basin. This continental interior basin is a roughly circular, asymmetric depression located mainly in New Mexico, but with the northern edge extending into Colorado. In addition to oil the basin contains substantial gas reserves (Rice, 1983). Although Bond (1984) discusses only a single well location in Colorado, he refers to other reconstructions in the general area that are similar. Today the youngest exposed rock is the Eocene San Jose formation, and much of the Cenozoic rock record has been removed by erosion, with a total loss estimated at approximately 4000 - 7000 ft (1220 - 2135 m). A further complication in reconstructing the burial history is establishing the thickness of the Oligocene volcanic cover. The nearest exposure is in the northern Chama basin where there are 4000 - 5000 ft (2135 - 1525 m) of volcanics, but the best estimate for the well location is a loss of close to 2000 ft (610 m). Maximum burial depth could be estimated using bulk density for shales, because shale density steadily increases with depth as compaction decreases porosity and water is squeezed out. Since the process is not reversible, porosity (and hence bulk density) does not revert on uplift, enabling present values to be used in establishing maximum burial depth (see Section 3.6.2). Bond (1984) found that the best fit of San Juan data with normal compaction curves corresponded to 6500 ft (1985 m) of erosional loss (Figure 10.3).

The thermal regime has undergone major changes, and there appears to have been a "heat flash" when source rocks had a brief exposure to high temperatures resulting in rapid maturation. The high temperature was associated with the igneous activity that produced the San Juan Mountains to the northeast of the basin, and the geothermal gradient increased to about 86 °C/km before diminishing. Heat may have been transported away from the volcanic center by laterally-flowing aquifers (Summer and Verosub, 1989). The petroleum generation process itself generated enough hydrocarbons to induce overpressuring and change the thermal conductivity of the rocks, with consequent changes in the geothermal gradient. During generation the shallow gradient was 49 °C/km and the

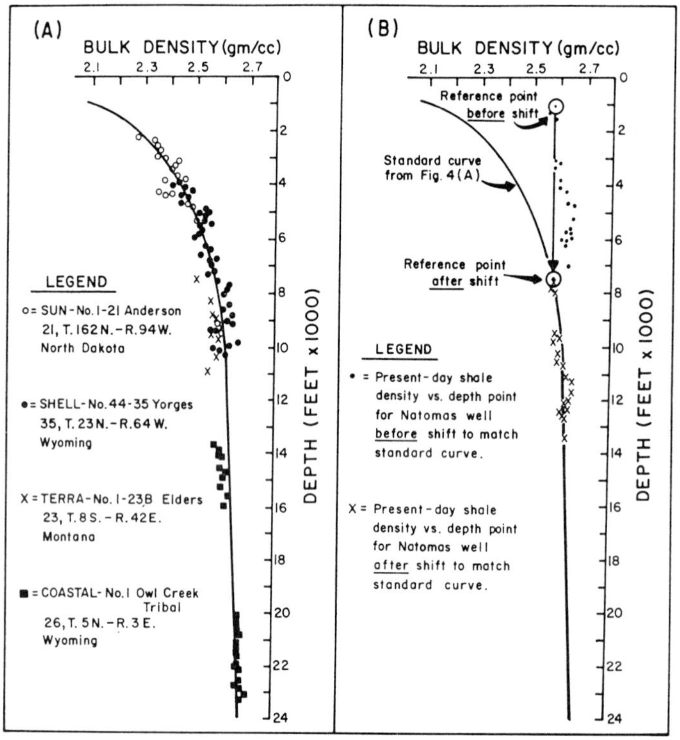

Figure 10.3. (A) Shale bulk density versus depth standard curve for Rocky Mountain Cretaceous shales based on selected wells in the Rocky Mountain region. (B) Present-day shale density versus depth plot for the studied well showing required downward shift of approximately 6,000 ft to achieve best fit with the standard curve (Bond, 1984).

deep gradient 53 °C/km, but the abnormally pressured zone reached 86 °C/km. Burial history and paleotemperatures were combined in the Lopatin procedure and TTIs calculated (Figure 10.4). TTI-R_o calibration data were obtained from Waples (1980), but also compared with data available from the Shell Rumberger No. 5 well (in the Anadarko basin) which gave a generally similar TTI-R_o trend. The modeled and calculated R_o values in the basin were very similar, as shown in Figure 10.4. Based on the thermal modeling, Bond (1984) concluded that the brief exposure to high temperature (the "heat flash") produced rapid maturation and so reconstructing the more normal geothermal gradients was much less important.

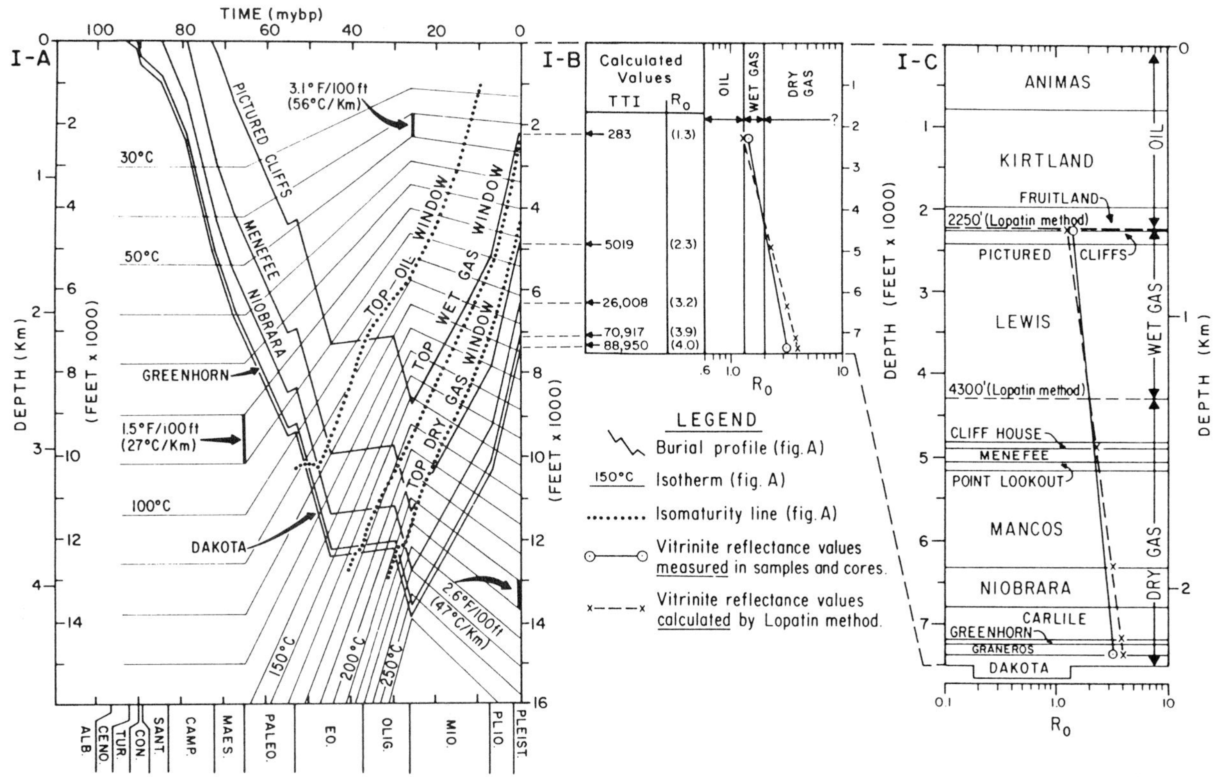

Figure 10.4. (IIA) Burial history and isotherm variation for the Natomas #1-11 Federal, San Juan basin. (IIB and C) TTI values calculated using Lopatin's method; calculated and measured vitrinite reflectance values; and the principle hydrocarbon generation zones related to the stratigraphy. (From Bond 1984).

A similar short-term heating event occurred in the Ordos basin of northcentral China. This Mesozoic interior basin developed over an earlier Paleozoic passive continental margin and accumulated 4 - 5 km of sediments (Zhao, 1990). It is estimated to contain reserves of approximately 15 billion bbl of oil and 140 Tcf of gas. Zhao et al (1996) used a combination of AFTA, vitrinite reflectance, potassium-argon dating, and thermal modeling, to unravel the thermal and tectonic history of the basin, and showed that paleotemperatures higher than present day temperatures were a significant factor in controlling the timing of petroleum generation.

Potassium-argon dating of the <2.0 μm and <0.2 μm clay mineral assemblages from seven core samples showed that the larger size fraction gave ages older than, or close to, the stratigraphic ages, and they decreased with depth. These ages were interpreted as being due to a mixture of authigenic illite with detrital micas and illite. The <0.2 μm fractions all gave very similar ages, and these were independent of depth, suggesting that the diagenetic age of the illites corresponded to a short duration thermal event during the late Middle Jurassic. The magnitude of this event was quantified using vitrinite reflectance data. These were obtained from polished rock chip mounts, and gave unimodal distributions that agreed well with previous values (e.g., Wang and Fu, 1986). Using the Barker and Pawlewicz (1986) calibration to convert R_0 values to maximum temperatures gave 90 - 161 °C for the Triassic in well Q36, 198 - 199 °C for the Lower Permian of well TS1, and 200 - 290 °C for the Permian-Carboniferous coal measures. These values agreed well with corrected fluid inclusion temperatures which were 90 - 173 °C for well Q36, and 124 - 190 °C for well TS1. The R_0 and fluid inclusion temperatures are somewhat higher than those derived from AFTA. The R_0 values show a linear correlation with depth of 0.36%/km through the Triassic corresponding to a paleogeothermal gradient of 57 °C/km (compared with a present day gradient of 22 - 27 °C/km).

Zhao et al (1996) used the thermal and burial reconstructions with the Lopatin (TTI) method to model the maturity of potential source rocks in the Ordos basin. Late uplift and erosion were constrained by R_0 discontinuities (Ren et al, 1994) and AFTA data. For the QC1 well calculated erosional losses were 1900 m and 2000 ± 400 m, respectively. If this uplift occurred in the last 23 m.y. the rate of uplift was 95 m/m.y. Uplift appears to be related to the Himalayan orogeny when the Qinghai-Tibet plateau rose in response to the India-Asia collision. The Q36 well was modeled assuming a 10 - 20 m.y. heat pulse, equivalent to 57 °C/km, due to subsurface magmatic intrusions that were emplaced about 170 - 160 Ma. The thermal modeling showed that although the Upper Triassic Yanchang Formation source rocks in this well were at their maximum depth of burial in the Early Cretaceous, they actually generated oil during the Middle Jurassic thermal event. It appears that oil generation may have occurred before (or possibly during) the formation of structural traps, and Zhao et al (1996) concluded that "the timing of hydrocarbon generation was later than the formation of stratigraphic and lithologic traps that belong to the most important main reservoirs in the basin".

Maturities in the Hedong coal field on the eastern edge of the Ordos basin have been modeled by Dazhen et al (1990). The coals are high volatile to medium volatile bituminous, with locally-occurring higher ranks. The coalification data were used to evaluate four different paleothermal regimes, but the models with slowly decreasing geothermal gradient through time did not fit as well as the results from models with a heat pulse at the end of the Triassic (200-250 Ma). Further local refinements could be made by incorporating heating due to igneous intrusions at ~125 Ma.

In the Denver basin of the central U.S.A., Lee and Bethke (1994) used the Lopatin method with heat flows of 1.2, 1.5, and 1.8 hfu to establish the thermal evolution of the Pennsylvanian source rock. This became mature between 78 and 50 Ma, with a most probable time of 68 Ma, corresponding to the preferred heat flow value of 1.5 hfu. After expulsion the oil moved up by buoyancy into the overlying Lyons Sandstone reservoir where it seems to have been involved in the formation of cements that prevented further migration.

Bethke et al (1991) have modeled migration from the New Albany source rock in the Illinois basin of the U.S.A., where geochemical oil-source rock correlation suggests that migration distance is as much as 150 km (93 miles). The model used was based on previous work by Bethke (1989) and Bethke et al (1988) that considered hydrodynamic water flow in the basin. In the 1991 study the driving force for migration was considered to be a combination of compaction, topographic relief, and buoyancy. The role of compaction was minor and the modeling also showed that abnormal subsurface pressures did not develop. Topographic drive is due to tectonic uplift in the southern part of the basin that drove fluids northwards, and this force was calculated to exceed the buoyancy force when relief was greater than about 500 m. Figure 10.5 shows the calculated ground water regime produced by uplift in the south. Thermal modeling (using Lopatin's method) showed that petroleum generation occurred late in the basin's history.

Lopatin's method has also been used in many other studies of organic maturity in North American basins, including the Illinois basin (Cluff and Byrnes, 1991); Big Horn basin, Wyoming (Hagen and Surdam, 1984); Anadarko basin, Oklahoma (Feinstein, 1981; Schmoker, 1986); Denver basin (Tainter, 1984); Palo Duro basin (Rose, 1986); Williston basin

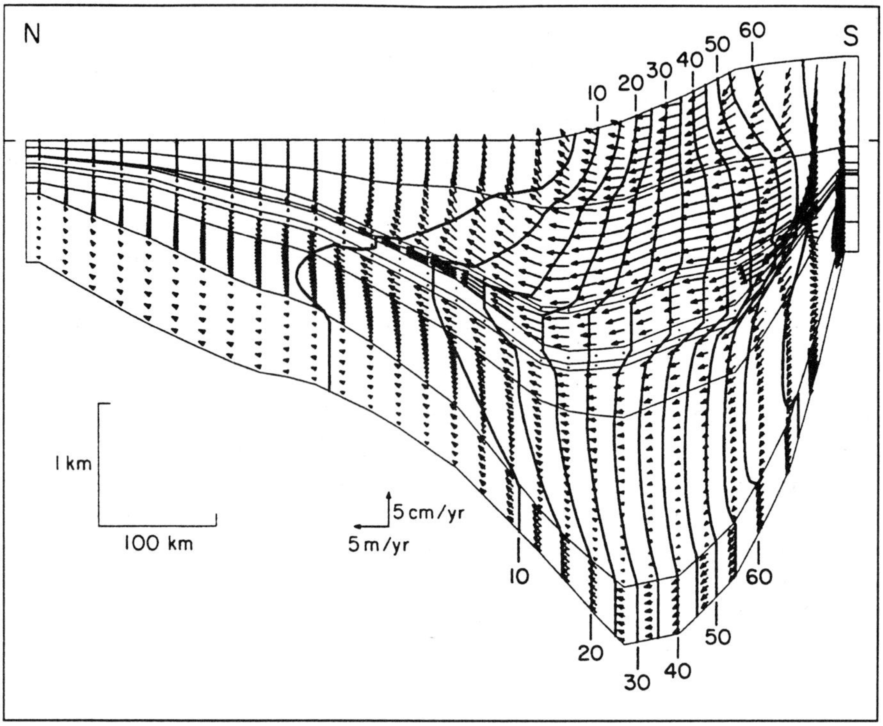

Figure 10.5. Calculated groundwater flow regime resulting from uplift of the Pascola arch in the Mesozoic or latest Paleozoic. Dark lines are equipotentials (MPa, relative to atmospheric pressure at sea level) showing the drive for groundwater flow resulting from topographic relief across the basin (Bethke et al, 1991).

(Webster, 1984); Western Canadian basin (Welte et al, 1984; Moshier and Waples, 1985); and the northern Yukon (Link and Bustin, 1989).

10.2.3. 1-D kinetic Modeling

As modeling procedures have become more sophisticated some of the basins evaluated using Lopatin's method have been reexamined. The San Juan basin, discussed above, is one example. Law (1992) used the BASINMOD program (Platte River Associates) to calculate present day maturities for the Cretaceous and Tertiary sediments. In the absence of detailed kerogen information default values were used. There was reasonable success in matching the R_o profile, even though it has several segments with different gradients.

In the Michigan basin Cercone (1984) made the initial maturity calculations using the Lopatin method, but the results were not completely satisfactory. The Michigan basin is a structurally simple, almost circular, continental interior basin and has accumulated several kilometers of Paleozoic sediments, but little younger material. Although there are several unconformities it is not clear whether they are due to uplift and erosion or to a time of non-deposition. Relatively high levels of maturity in near-surface sediments suggest deeper burial in the past, possibly accompanied by higher geothermal gradients, but there are considerable uncertainties in both burial histories and thermal histories. Cercone (1984) tried to use the best estimates of burial history to infer paleogeothermal gradients. She deduced an uplift of 1000 m from sediment accumulation rates, regional dips, and coal maturities. Figure 10.6 shows a burial history reconstructed for a central location in the basin. Two other locations (on the southern and northern margins) were also modeled, and in all three cases maturity calculations were made using the Lopatin method with either a 45 °C/km geothermal gradient (Model 1) or a 35 °C/km gradient (Model 2). The figure shows that the top of the oil window moves to shallower depths through time for both models. Neither model

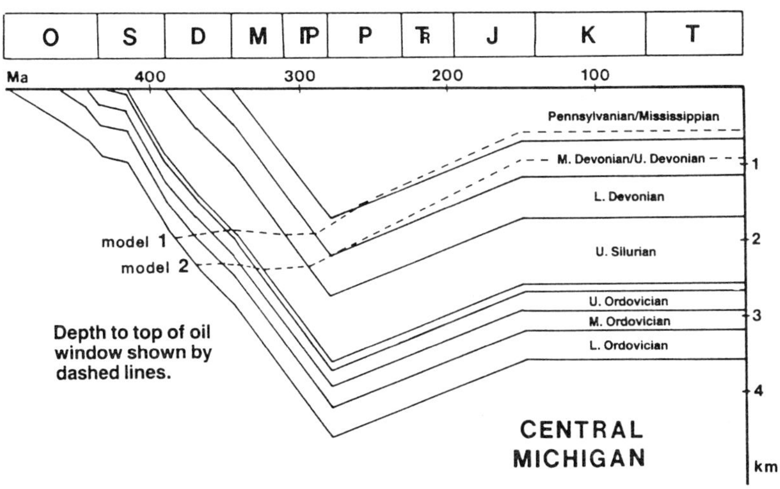

Figure 10.6. Burial history of the central Michigan basin. Model 1 used a 45 °C/km geothermal gradient, and Model 2 used 35 °C/km. All strata below the dashed lines are mature enough to generate hydrocarbons. (From Cercone, 1984).

was able to predict maturities accurately over the whole Paleozoic section, probably because of changing geothermal gradients through time, or because varying thermal conductivities of the different rock types produced non-linear gradients. The 45 °C/km gradient also seems unrealistically high.

Nunn et al (1984) developed a more detailed model of thermal evolution for the basin, and combined this with kinetic modeling to calculate maturities. They found no significant generation in rocks of Silurian age or younger, but this conflicts with geochemical correlations that show Silurian (and possibly Devonian) sourced oils (Rullkotter (1986), and references cited). In attempting to resolve these uncertainties Marzi and Rullkotter (1992) and Rullkotter et al (1992) used sterane isomerization as a maturity indicator. They could not fit their data to the model developed by Nunn et al (1982), and obtained only a partial fit with Cercone (1984). A revised thermal model that had a geothermal gradient increasing to 35 °C/km and then decaying to the present day gradient of 22 °C/km (Figure 10.7) combined with 1800 m of erosional loss (uplift), gave a self-consistent picture where measured and calculated isomerization ratios matched. Other attempts to evaluate thermal evolution of the basin include those by Speece et al, (1985), Vugrinovich (1989), Crowley (1991), and Girard and Barnes (1995), but none of them used the temperature information to calculate maturities. However, the fluid inclusion data from Girard and Barnes (1995) corresponds to geothermal gradients in the range between 34 and 55 °C/km for the late Paleozoic.

The East Texas basin contains one of the largest oil accumulations in North America. In spite of considerable drilling activity and a relatively simple basin architecture many key questions remain. Wescott and Hood (1994) used Amoco's proprietary 1-D kinetic modeling program called SUBSIDE to determine the basin-wide distribution of petroleum by modeling the formation of structural traps, delineating areas of mature source rocks, and establishing migration pathways. The basin evolved from a failed rift, through a restricted environment where evaporites formed, into a marine environment where carbonates developed, and finally became the location for the accumulation of terrigenous clastics. There are 3 major types of oil, with Jurassic, Lower Cretaceous, and Upper Cretaceous source rocks, respectively. The Jurassic is mature throughout the basin and modeling shows that it became mature about 88 Ma, and that vertical migration pathways were necessary to reach many of the reservoirs. The Lower Cretaceous is only mature in the deeper central parts of the basin and began generating about 47 Ma. Most of the organic-rich Upper Cretaceous shales are immature except locally. Wescott and Hood (1994) concluded that lateral migration distances of over 100 km were needed to reach the East Texas giant field. Movement of the Louann salt had a major influence

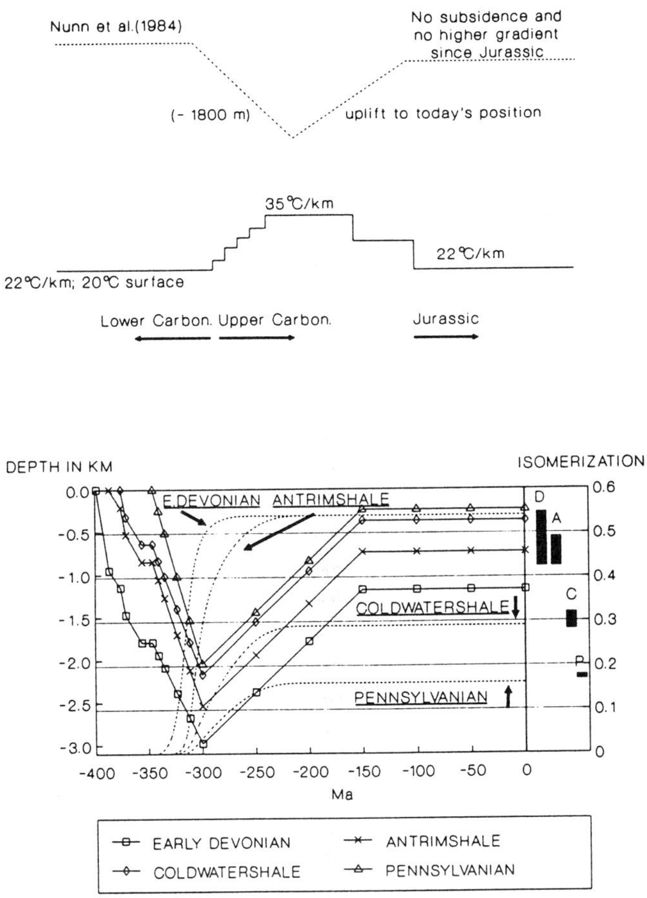

Figure 10.7. Modified subsidence and geothermal histories for the Michigan basin (top) used in calculating evolution of sterane isomerization (bottom). Reconstructions from Marzi and Rullkotter (1992).

on the distribution of petroleum and the deep salt-cored structures are the most prolific hydrocarbon producers in the area.

The Bolivar Coastal Fields of Venezuela are one of the largest petroleum accumulations outside of the Middle East. The intermontane basin is in a foredeep position (Perrodon, 1983), and the main oil-prone source rock is the La Luna formation where limestones and calcareous shales average nearly 4% TOC. Its maturity has been modeled by Blaser and White (1984) and Talukdar et al (1985) using Lopatin's method. The

TTI calculations showed that generation occurred first in the northeastern part of the basin during Middle to Upper Eocene, and in Upper Miocene to Recent in the central, western, and southern parts. The beginning of oil generation appeared to correspond to a TTI of 10 (Talukdar et al, 1985). The time when the La Luna source rock reached maturity is shown in Figure 10.8. Source rocks on the eastern side of Lake Maracaibo are overmature and are now out of the oil window.

Modeling in the Bolivar Coastal Fields was refined by Burnham et al (1992). Their elaborate kinetic model calculated not only the timing of petroleum generation and cracking, but also changes in pore fluid pressures and fluid expulsion. Calculations were made for 80 wells where thermal histories had been refined by matching observed R_o values with those calculated using EASY%R_o (Sweeney and Burnham, 1990). Calculated oil API gravities and sulfur contents agreed well with the measured values. The more advanced PMOD model was used in 22 wells to predict yield as well as oil and gas compositions and pressures. The modeling showed that maturation of organic matter produced a significant fraction of the overpressuring during oil generation. Further developments in their modeling have incorporated various types of laboratory pyrolysis experiments to constrain a model of organic maturation for the Bolivar Coastal Fields (Sweeney et al, 1995).

The Uinta basin of northeastern Utah contains considerable quantities of organic-rich oil shales in the Green River Formation, and retorting has been investigated as a potential commercial process for converting the organic matter to crude oil. In addition to possible economic importance, the Green River oil shales have provided geochemists with a classic example of Type I, hydrogen-rich, amorphous kerogen. Waxy, high pour point, low sulfur crudes in fields such as the giant Altamont-Bluebell have been correlated to the Green River oil shale (Tissot et al, 1978; Anders and Gerrild, 1984), and the timing of maturity has been modeled by Pitman et al (1987) and Sweeney et al (1987). The Green River oil shale is an Eocene lacustrine deposit that ranges from outcrop in the southwest to 3650 m subsurface. A considerable amount of pyrolysis information for the oil shale was obtained during the evaluation of the commercial possibilities of converting oil shale to oil. Much of this work was done at the Lawrence Livermore Laboratory (LLNL), California, and the model they developed for retorting has been modified, extended, and applied to oil generation in nature by Burnham and Braun (1985). Sweeney et al (1987) used the model with a single activation energy of 52.4 kcal/mol (which is reasonable for Type I kerogen, but not appropriate for other kerogen types). The sixteen parallel reactions used provided information on changing oil composition with maturity, as well as predicting the generation and composition of gas.

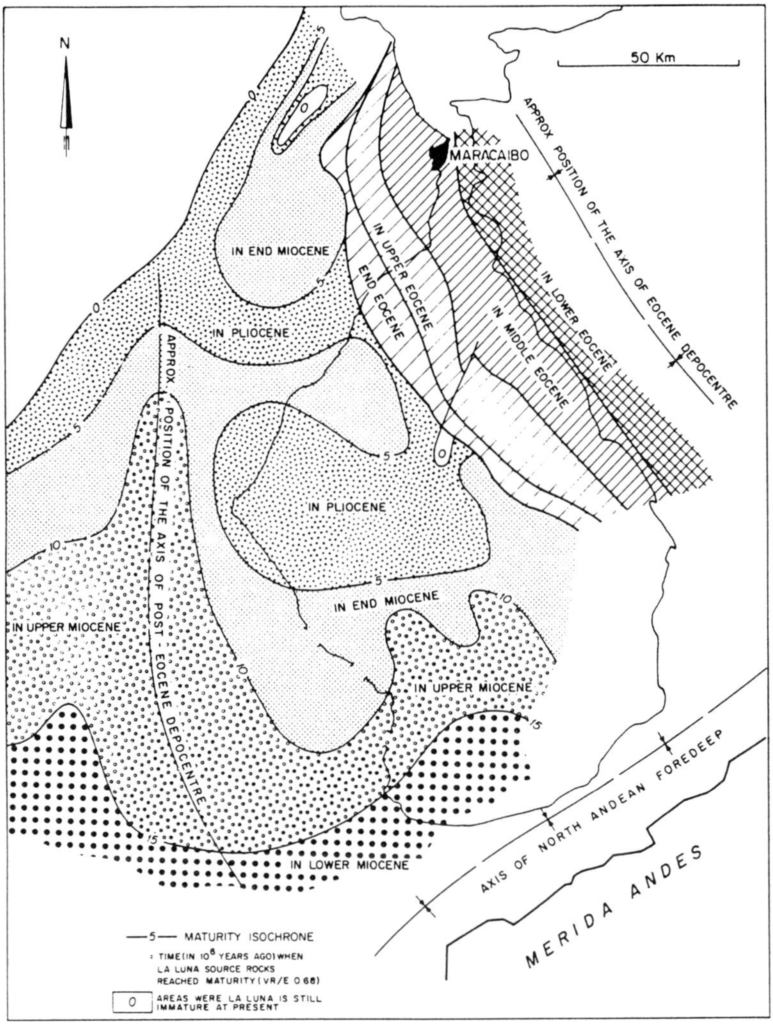

Figure 10.8. General area of Lake Maracaibo, Venezuela, showing the calculated time when the La Luna carbonates reached thermal maturity (Blaser and White, 1984).

The Uinta basin is an asymmetric syncline that trends east-west, and although it is structurally simple it is stratigraphically complex. A major problem in reconstructing the burial history is establishing the amount of uplift and erosion, and previous estimates of rock lost ranged from 510 to

2890 m (Narr and Currie, 1982). Sweeney et al (1987) used trends in sonic log data through the normally-pressured parts of the section because this can be related to the irreversible process of compaction (see Section 3.6.2). Subsequent extrapolation to an assumed zero-depth interval velocity of 200 msec/ft (Magara, 1978) gave estimates of overburden removed in the vicinity of Altamont-Bluebell of 1800 to 2000 m, with an uncertainty of ± 300 m. Tissot et al (1978) had previously used 1780 m, and Pitman et al (1987) used 1006 m. The compaction-corrected burial history was combined with an unchanging geothermal gradient of 25 °C/km (but see Willett and Chapman, 1987) and used with the model to predict both the quantity and composition of generated petroleum. 1-D modeling is, of course, site specific and although it worked well in most of the selected well locations, there were some discrepancies between predicted and measured maturity levels (e.g., Redwash and Pariette Bench areas). These were attributed to migration which could not be evaluated with the 1-D modeling. Calculated boiling point distributions for different locations in the study area showed that the light ends become more abundant in the deeper source rock horizons (Figure 10.9). In the Shell Brotherson well (4004 m) 90% of the oil would distill below 200 °C, while in the 13 Broadhurst (2758 m) it was only 20%, and virtually none would be in the <200 °C fraction for the 13 Broadbent at 1433 m. These calculated trends agree well with measured compositions in the basin.

During marginal marine episodes in the development of interior basins abundant peat may accumulate, and later convert to coal on burial. Garcia-Gonzalez et al (1993) have modeled the thermal maturity of coals at two locations in the Greater Green River Basin, Wyoming. They selected

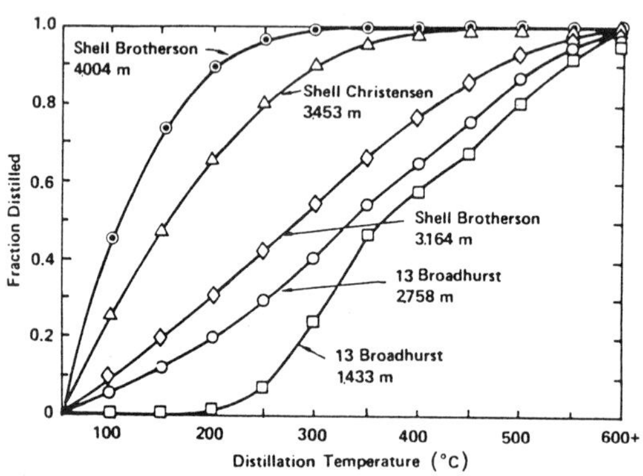

Figure 10.9. Calculated boiling point distributions for oils from different locations in the Uinta basin showing that the lighter fractions become more abundant in the deeper source rock horizons (Sweeney et al, 1987).

one location in the center of the Washakie Basin and one on the Rock Springs Uplift. Modeled coals were part of the Almond Formation. The decompacted burial histories were plotted using R_o values to estimate the amount of uplift, which in the basin center accounted for nearly 6000 ft (1830 m) over the last 40 m.y. Heat flow was calculated with a 1-D steady state model and decreased gradually from the Late Cretaceous to the present. Hydrous pyrolysis experiments provided E and A values for the Almond coals and for the cracking of oil to gas. The maturity modeling showed that in the basin center oil was generated from 50 - 60 Ma, but over the last 40 m.y. most of it has been cracked to gas, so that now only gas can be expected in the basin center. The oil window was defined by calculated Transformation Ratios between 0.1 and 0.4. Oil is expected on the less deeply buried basin flanks.

Several studies in interior basins have employed maturity modeling as a way of getting the time of generation which could then be used to help establish the timing of abnormal pressure development. Luo et al (1994) did this in the Delaware basin of west Texas, and Zhao and Lerche (1993) and Nuccio and Johnson (1984) modeled the Piceance basin of Utah. The Piceance basin formed as a distinct basin when the Laramide orogeny broke up the large Cretaceous seaway into a group of basins. It seems to have experienced a heat pulse associated with Oligocene-Miocene volcanic activity that produced a peak heat flux of 2.7 hfu at 24 Ma. In their study of the basin Zhao and Lerche (1993) did not model hydrocarbon generation because of a lack of kerogen data and because their main emphasis was on abnormal pressure development and thermal evolution. They did calculate R_o for comparison with measured values.

10.2.4. 2-D Kinetic Modeling

The Paris basin, France, has long been an important area for testing and documenting geochemical concepts and has been studied in detail, especially by geochemists at the IFP (Tissot et al, 1974; Espitalie et al, 1987). Espitalie et al (1987) reported some 1-D thermal modeling results, but faced difficulties in the eastern part of the basin (the so-called Saar-Lorraine basin) where erosion during the Hercynian tectonism is poorly constrained, and the Paleozoic geothermal gradients are unknown. The study showed that even in the deepest part of the Paris basin source rocks have not moved out of the oil window. In a subsequent study, the Paris basin was modeled by Espitalie et al (1988) using the TEMISPACK 2-D software (Doligez et al, 1986). This treats flow of water and hydrocarbons independently, and also incorporates regional hydrodynamic flow into the model. A roughly NE/SW cross section with 31 stratigraphic subdivisions was studied and divided into 29 x 31 element grid blocks. The best fit between calculated and observed Transformation Ratios was obtained using an unchanging geothermal gradient of 35 - 40 °C/km, which is slightly higher than the present day gradient of 33 °C/km. Calculated Transformation Ratios of 10 - 90% for the Liassic source rocks are all located in the deeper central part of the basin, with values for the Toarcian shale reaching a maximum of 35%, and the deeper Heltangian/Sinemurian reaching 70%. A total of 2 billion tons of hydrocarbons was expelled. Modeling suggested considerable lateral and vertical migration (through faults) to explain oil accumulations in the stratigraphically shallower Upper Dogger.

TEMISPACK 2-D modeling has also been used by Burrus et al (1996) to model petroleum generation, expulsion, migration, abnormal pressures, and petroleum accumulation in the Williston basin of the northcentral U. S. and southern Canada. They selected a 350 km long section and distinguished 64 geological events and 35 different rock types. With some simplifications this led to the development of the 31 X 38 element-grid shown in Figure 10.10. This figure also gives the locations of major oil accumulations. Present day geothermal gradients show a break reflecting differences in the thermal conductivities between the Mesozoic and

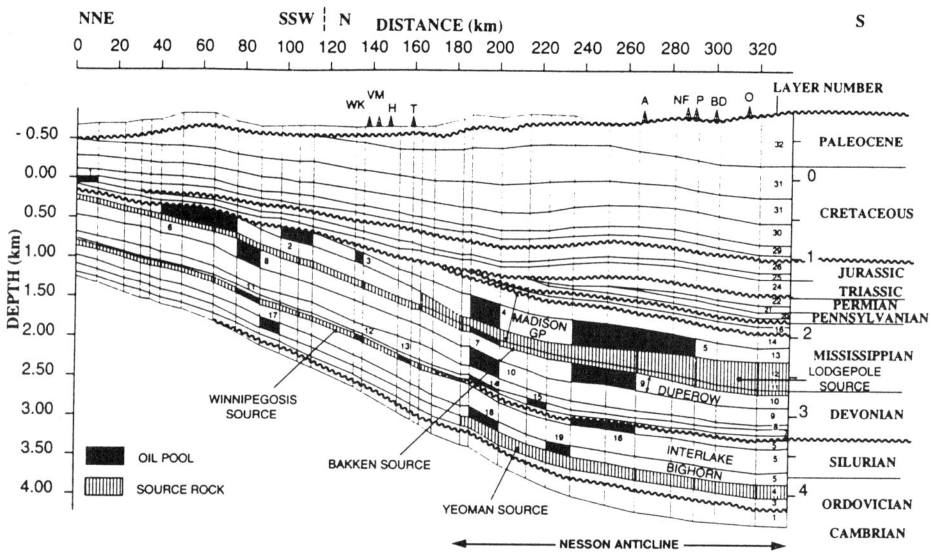

Figure 10.10. Distribution of the 31 x 38 grid of cells used by Burrus et al (1996) in modeling the Williston basin, northcentral U.S.A. Numbers refer to fields included in the study.

Tertiary section, which is mainly shaley, and the Paleozoic section, which is dominated by carbonates. Heat flow is now 55 mW/m^2, except over the Nesson anticline where it is 10 mW/m^2 higher. Most models invoking higher paleo-heat flows do not predict these present day heat flows. Although the Late Cretaceous rift hypothesis that increased heat flow from 55 - 80 mW/m^2 at 75 Ma, with a subsequent linear decrease to 60 mW/m^2, does give good predicted present day heat flows, it fails to provide an accurate prediction of the present maturities for the Bakken shale. Burrus et al (1996) modeled the Williston basin using a constant heat flow of 55 mW/m^2, except for higher values of 65 mW/m^2 over the Nesson anticline.

Previous studies had confirmed the presence of multiple source rocks in the Williston basin, and Burrus et al (1996) considered the Upper Ordovician Yeoman, the Middle Devonian Winnipegosis, the Upper Devonian shale member of the Bakken Formation, and the Mississippian Lodgepole. Maturities of these rock units were modeled using first-order kinetics with E and A values obtained from RockEval pyrolysis. In all cases immature samples were available. The Yeoman gave a narrow distribution of E values dominated by 54 kcal/mol and is a typical Type I kerogen. The other three source rocks contained Type II kerogens. Oil generated from these source rocks was assumed to move through carrier beds by buoyancy, and the necessary permeabilities were obtained from measured values, default values from the literature, or by extrapolation. The very low permeability evaporites divide the basin into four effectively independent flow systems.

Increasing maturity through time, and the onset of petroleum generation for the four source rocks at any location, can be given by the calculated Transformation Ratios (Figure 10.11). The predicted depths for mature source rocks are presented as a cross section in Figure 10.12. and show that, because of the higher heat flow over the Nesson anticline, the oil window was 700 m shallower for Type II kerogen and 1000 m shallower for Type I kerogen. The predicted abnormal pressures for the Bakken Shale were consistent with pressure generated solely by oil generation. Although the development of hydraulic fracturing was not pervasive, and probably not the major mechanism involved in oil migration, about 85% of the generated oil was expelled. Much of this was left along migration pathways in the Madison Group at close to irreducible saturation. Combining migration routes with source rock maturation made it possible to predict the location of oil fields, as well as showing the formation of mixed oil types where two source rocks fed a common conduit. The 2-D model still had limitations, mainly the failure to predict a Lodgepole-sourced oil field at the Madison subcrop. Burrus et al (1996) subsequently modified the model to improve agreement between prediction and

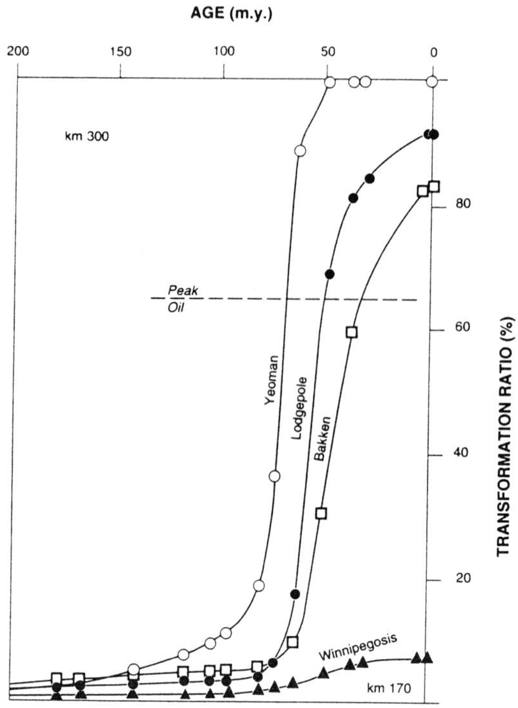

Figure 10.11. Timing of onset of maturity for the Winnipegosis at kilometer 170, and the other three source rocks at kilometer 300. The oil window starts at a Transformation Ratio of 10%. (Burrus et al, 1996).

observation. The first version of the model had overestimated the amount of petroleum generated by the Lodgepole source rock, and refined models were developed in which the Lodgepole was divided into three units of equal thickness. The upper unit was given zero generating potential, and the lower part a much reduced potential. The central unit was given the highest generating potential, and developed a high expulsion efficiency of 90% or more.

This study by Burrus et al (1996) is an interesting example of using thermal modeling in a very mature exploration area. Abundant data provide an opportunity to evaluate the model's ability to explain the present distribution of oils, and illustrates not only the capabilities of the 2-D modeling, but also its limitations. In this case modeling showed the potential for undiscovered Yeoman-sourced oils on the eastern flank of the Williston basin where Ordovician and Silurian rocks are inadequately explored.

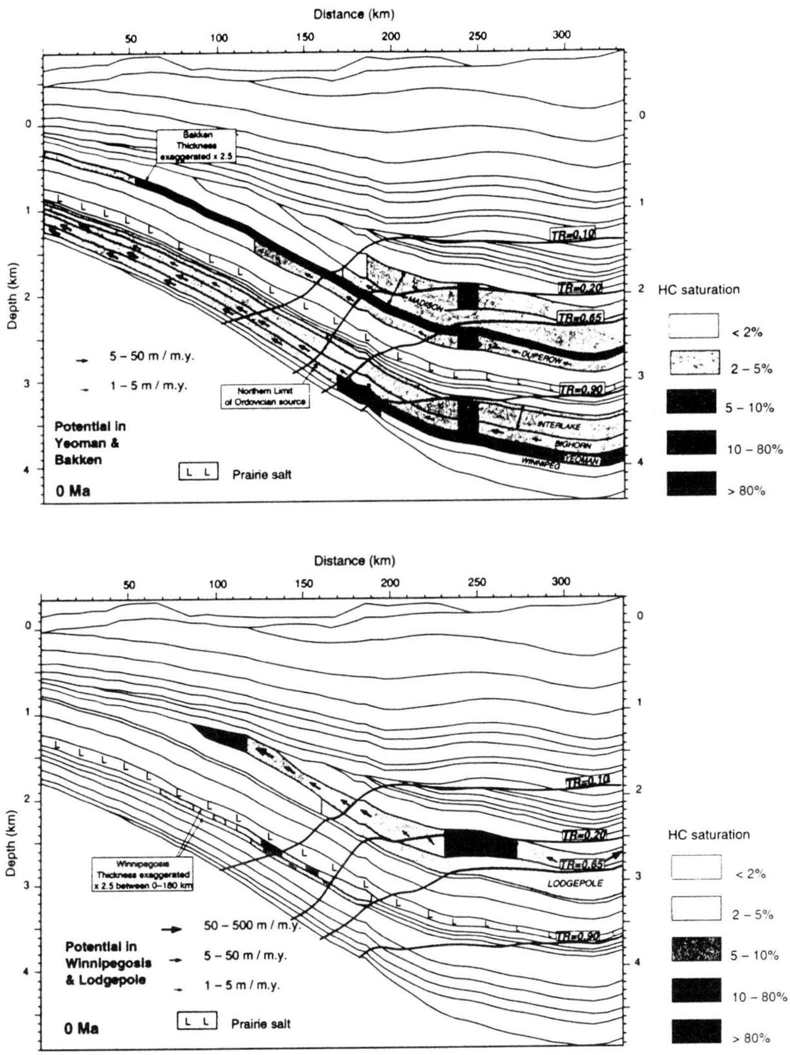

Figure 10.12. Calculated present-day Transformation Ratios (TR) in the Williston basin (Burrus et al, 1996).
Top: system treated as though the Yeoman and Bakken were the only source rocks.
Bottom: system treated as though the Winnipegosis and Lodgepole were the only source rocks.
Saturations between 0.02 and 0.05 represent the drainage system; saturations above 0.1 correspond to accumulation or depleted source rocks. Arrows show active oil migration (Darcy velocities).

10.3. RIFT BASINS

10.3.1. Introduction

Rifts appear as major features of both deep ocean basins and continents. They are commonly formed by crustal extension due to lithospheric stretching. Continental rifting is often the first step in a sequence of events that leads from an initial fault-bounded rift, to a narrow seaway, followed by injection of oceanic crust, and, ultimately the development of a passive margin. Passive margins are discussed as a separate topic later. Based on giant field statistics Klemme (1980) estimated that about 10% of the Worlds petroleum reserves is in rift settings, and that much of this was high gravity, low sulfur crude oil. Table 10.1 shows reserves in some productive rift basins given in barrels of oil equivalent (boe) per square kilometer of basin area. In the richer rifts, such as the North Sea and the Gulf of Suez, volumes exceed 250,000 boe/km^2.

Rifts occur in a wide variety of plate tectonic settings, but details of the initial rifting depends on the processes driving it. It is often not clear whether the rifts develop as a consequence of a mantle hot-spot or convection plume, and in that sense are "active", or whether rifting and magmatic activity are simply the result of regional or local extensional stresses, in which case rifting is "passive". In either case, the top of the aesthenosphere moves to much shallower depths leading to higher heat flows and higher geothermal gradients. Specific values will depend on how much the crust is thinned, and this is conveniently expressed by the β "stretching factor" introduced by McKenzie (1978). It is defined so that the larger the number the greater the lateral extension. In a simple geological situation, when $\beta = 3$ the lithosphere is extended by a factor of 3, and is thinned to $^1/_3$ of its initial thickness. As the hot spot cools both heat flow and geothermal gradient decrease. Changing gradients through time for the Reconcavo basin of Brazil were discussed in Section 5.5 and shown in Figure 5.5. Here the initial gradient was 50 $^{\circ}$C/km, but subsequently cooled to the present day value of 25 $^{\circ}$C/km.

When the thermal effects of the rising aesthenosphere lead to doming at the surface, there are important consequences for sediment accumulation. Doming raises the floor of the rift above sea level, so that the sediments that accumulate are non-marine and are mainly fluvial and lacustrine, but the development of large lakes probably occurs only during the early stages of rift history (Lambiase, 1990). Rift lake environments are very variable, but stratification can lead to anoxic conditions where organic matter is preserved, and organic contents in some present-day

Table 10.1
Reserves in Some Rift-Related Basins (from Grunau, 1990).

AREA	boe/km^2
North Sea rift system	275,234
Gulf of Suez	268,821
Campos basin (Brazil)	112,204
Sirte basin (Libya)	100,252
Gippsland basin (Australia)	93,551
Magdalena basin, upper and middle (Colombia)	70,571
Bohai basin (China)	55,541
Equatorial West Africa miogeocline	43,773
Taranaki basin (New Zealand)	28,549
Phitsanulok basin (Thailand)	7,189
Upper Rhine graben (W. Germany, France)	6,971

lakes exceeds 10% (Demaison and Moore, 1980; Katz, 1995). On burial these organic-rich lake sediments generate petroleum, not only in rift settings (Lambiase, 1995), but also later in passive margins (see below). The organic matter associated with lakes is often Type I kerogen, and when this becomes mature it generates very paraffinic (waxy), high pour point, low sulfur crudes and these are commonly associated with rifts (Table 10.2). Other kerogen types may also occur, often with a systematic distribution in the rift (e.g., Wanli et al, 1985). In some cases doming causes large rivers to drain away from the rift and it is starved of terrigenous sediments. Then, carbonates and evaporites form if climatic conditions are suitable, with the Red Sea and Gulf of Aden providing present-day examples. The horst and graben systems that are common in rifts lead to dramatic short range differences in lithology, because the grabens tend to accumulate larger thicknesses of younger sediments. This makes extrapolation along strike difficult because of the non-synchronous structural development, and its effect on the way river systems disperse their sediments (Nelson et al, 1992). The different thermal conductivities of the various sediment combinations produces considerable local variations in the geothermal gradient. A similar situation arises in rifts where mobile salt produces diapirs. The effects of groundwater flow in continental rifts has been discussed by Person et al (1995).

Table 2
Pour points (°C) and sulfur contents of selected rift-related oils.

RIFT (WELL) REF	°API	POUR POINT	%S	
SUDAN:				
Unity #2	28.2-35.5	21.1-29.4	0.02-0.06	1
Unity #9	26.1-33.4	18.3-29.4	0.01-0.07	1
"Interior"	18-45	45-59	---	2
GABON:				
Onal-1	31.7	36	0.07	3
RECONCAVO:				
Agua Grande	40-41	32	0.04	4
Don Joao	36-38	29	0.05	4
Miranga	38-41	32	<0.1	4
Candeis	29-32	30	0.09	4
Buracica	33-35	28.2-35.	---	4
SIRTE BASIN:				
Sarir	37	13-24	"S-free"	5
Amal	35	18	---	6
CAMBAY BASIN:				
Bombay High	38.5	30-33	0.16	7
SONGLIAO BASIN:				
Daqing	33	25	0.06	8

(1) Hwang, et al (1994); (2) Schell (1988); (3) Kuo (1994); (4) Ghignone and Andrade (1970); (5) Sandford (1976); (6) Roberts (1970); (7) Englen et al (1981); (8) Desheng (1995).

Many of the features of doming, rifting, and development of continental source rocks are well illustrated by the interior Sudan rift basins. These form a major part of the Central African rift system and contain up to 45,000 ft (13,725 m) of non-marine clastics of Jurassic(?)/Cretaceous to Tertiary age. The moderately rich source rocks (average 1.3% TOM) are oil prone and have generated high pour point, low sulfur, paraffinic crudes (Table 10.2). They accumulated in large lakes distal to the primary clastic influx where conditions became suboxic. The present geothermal gradients range from 18 to 33 °C/km and average 26 °C/km. Earliest generation typically begins at about 10,000 ft (3048 m), and vertical migration along fault planes has been an important factor in developing many of the accumulations (Schull, 1988).

A large number of modeling studies have been carried out in rift settings including the North Sea area (Ritter, 1984; Baird, 1986; Ritter et al, 1987; Bailey et al, 1987; and many others); the Gulf of Suez (Ungerer et al, 1986); Bass basin, Australia (Middleton, 1982c; Williamson et al, 1986); Otway basin, Australia (Middleton and Falvey, 1983); Taranaki basin, New Zealand (Pilaar and Wakefield, 1984; Kamp and Green, 1990; Armstrong et al, 1996); Rhine Graben (Espitalie, 1984); Sirte basin, Libya (Gumati and Schamel, 1988; Gumati and Nairn, 1991); Bombay High, India (Pande et al, 1984; Mohan, 1985); Mozambique (Illiffe et al, 1991; Nairn et al, 1991); and central Sumatra (Williams and Eubank, 1995).

10.3.2. Lopatin Modeling

Lopatin modeling has been used to evaluate thermal maturity in a wide range of rift basins around the World including the Bombay High, offshore India (Pande et al, 1984), and onshore U.K. (Kirby et al, 1987; Penn et al, 1987). In the Norwegian North, Sea Baird (1986) mapped the distribution of maturity through time based on maturity calculations at 292 locations, of which 156 were wells and 136 were pseudo well locations selected from seismic sections.

Cohen (1985) used the Lopatin method in the Reconcavo basin. This basin forms the southern part of a failed, mega half-graben that predates continental separation along the Brazilian margin between Salvador and Recife. Present-day geothermal gradients were calculated from measured heat flow and thermal conductivity by Carvalho and Vacquier (1977) (Table 10.3). Heat flows range from 0.87 to 1.25 hfu. Calculated current geothermal gradients are between 15.0 ± 3.1 to 21.4 ± 2.7 °C/km, but past gradients were considerably higher because of the high heat flow typical of the early stages of rifting. Royden et al (1986) considered both an instantaneous necking model and a dike injection model to characterize heat flow variation in an evolving rift setting. For the Reconcavo basin there is very little difference between these two models after the first 20 m.y. of rifting, and Cohen (1985) used the instantaneous necking model to constrain temperature through time (Figure 10.13).

Burial history curves were constructed from well data, but the ages of many of the lithostratigraphic units are poorly known. An apparent decrease in sediment accumulation rate from 140 - 135 Ma is consistent

Table 10.3
Heat flow (Q) and geothermal gradient (GG) history for the Reconcavo basin calculated by Carvalho and Vacquier (1977) and quoted by Cohen (1985).

Time (Ma)	Q (HFU)	GG (K)*
130	4.0	70
125	3.6	60
120	2.9	50
115	2.5	40
110	2.2	40
105	2.0	30
100	1.9	30
95	1.8	30
90	1.7	30
85	1.6	30
80	1.5	30
75	1.5	30
70	1.4	20
65	1.3	20
60	1.3	20
55	1.2	20
50	1.2	20
45	1.2	20
40	1.1	20
35	1.1	20
30	1.1	20
25	1.0	20
20	1.0	20
15	1.0	20
10	1.0	20
5	1.0	20
0	0.9	20

*K = 6×10^{-3} cal cm^{-1} sec^{-1} $°C^{-1}$.

with regional uplift preceding rifting, and the later increase in sediment accumulation rate corresponds with the onset of South American and Africa plate separation. Cohen (1985) combined the burial history information with temperature using the Lopatin method, and cumulative TTIs are shown in Table 10.4 for the top of the Santa Amaro Group. This group includes 300 to 2100 m of lacustrine Candeias Formation. The Candeias became mature after major faulting in the Valanginian, and the faulting juxtaposed the Candeias Formation source rocks with reservoir units in the Sergei Formation. However, later rejuvenation of faults allowed some of this oil to leak up fault planes into younger reservoirs in the Ilhas and Sao Sebastiao Formations. The oil generated by the Candeias Formation is a typical lacustrine product – waxy, high pour point, and low percent sulfur (see Table 10.2).

Table 10.4
Cumulative TTI values calculated by Cohen (1985) for the top of the Brotas Group and the top of the Santa Amaro Group, Reconcavo basin.

Temperature Interval °C	r^n	Δ Time (m.y.)	Interval TTI	Total TTI
Top of Brotas Group				
20-30	2^{-8}	0.715	0.003	0.003
30-40	2^{-7}	0.715	0.006	0.009
40-50	2^{-6}	0.715	0.011	0.020
50-60	2^{-5}	0.715	0.022	0.042
60-70	2^{-4}	0.715	0.045	0.087
70-80	2^{-3}	0.715	0.089	0.176
80-90	2^{-2}	0.715	0.179	0.355
90-100	2^{-1}	0.715	0.358	0.713
100-110	2^{0}	0.715	0.715	1.428
110-120	2^{1}	0.715	1.430	2.858
120-130	2^{2}	0.715	2.860	5.718
130-140	2^{3}	0.715	5.720	11
140-150	2^{4}	0.715	11	23
150-160	2^{5}	0.715	23	46
160-170	2^{6}	0.8	51	97
170-180	2^{7}	0.8	102	199
180-190	2^{8}	0.8	205	404
190-200	2^{9}	0.8	410	813
200-210	2^{10}	0.8	819	1633
210-220	2^{11}	4.0	8192	9825
210-200	2^{10}	3.7	3789	13614
200-190	2^{9}	2.9	1485	15098
190-180	2^{8}	3.1	794	15892
180-170	2^{7}	4.3	550	16443
170-160	2^{6}	4.0	256	16699
160-150	2^{5}	5.5	176	16875
150-140	2^{4}	7.0	112	16987
140-130	2^{3}	11.0	88	17075
130-120	2^{2}	17.5	70	17145
120-110	2^{1}	27.0	54	17199
110-100	2^{0}	31.0	31	17230

Temperature Interval °C	r^n	Δ Time (m.y.)	Interval TTI	Total TTI
Top of Santo Amaro Group				
20-30	2^{-8}	0.57	0.002	0.002
30-40	2^{-7}	0.57	0.004	0.006
40-50	2^{-6}	0.57	0.009	0.015
50-60	2^{-5}	0.57	0.018	0.033
60-70	2^{-4}	0.57	0.036	0.069
70-80	2^{-3}	0.57	0.071	0.140
80-90	2^{-2}	0.57	0.142	0.282
90-100	2^{-1}	13.50	6.75	7.032
90-80	2^{-2}	9.50	2.375	9.407
80-70	2^{-3}	18.0	2.25	11.657
70-60	2^{-4}	50.0	3.125	14.782
60-50	2^{-5}	30.0	0.936	15.718

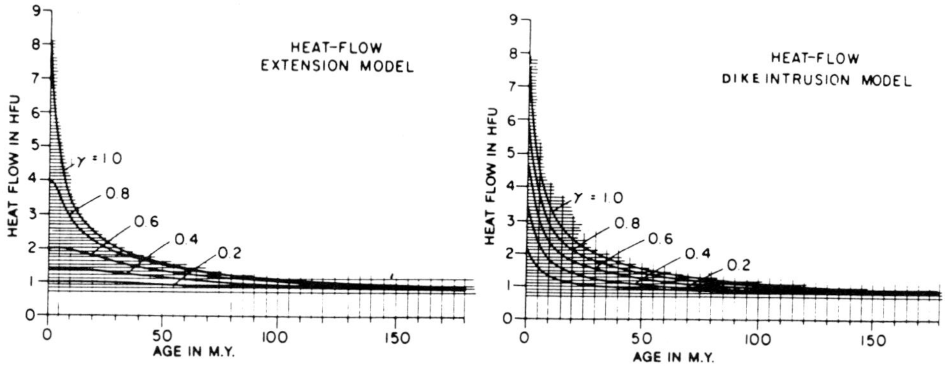

Figure 10.13. Heat flow as a function of time for two models of continental margin evolution: dike injection, and instantaneous extension (from Cohen (1985) after Royden et al (1980)).

The Sirte Basin indents the northern coast of Africa and is wholly within the country of Libya. Approximately East-West lithospheric extension with a stretching factor in the range from 1.1 to 1.75 began in the Late Cretaceous, and basin extension appears to have occurred in several stages (Gumati and Nairn, 1991). Deep troughs were accumulating large quantities of organic rich shales, evaporites, and clastics at the same time that carbonate reefs developed on the high blocks. These contrasting lithologies have different thermal conductivities, and for a given heat flow produce geothermal gradients that are relatively low in the deep, shale-filled troughs, but relatively high on the platforms where carbonates and anhydrite are abundant. Combining uncorrected BHTs from 28 wells gave a present day generalized geothermal gradient of 22 °C/km with a surface temperature of 29 °C. Paleogeothermal gradients were assessed by comparing present-day values with those predicted by Lopatin modeling for various past thermal regimes. Well control in the basin is good and burial histories for 3 selected locations were constructed from formation tops. A reasonably good vitrinite reflectance data base was also available. Geothermal gradients were assumed to be invariant through time, and values between a "fairly hot" 32 °C/km and a "cold" 22 °C/km were used. This modeling showed that in the west a gradient of 32 °C/km gave the best match for the Upper Cretaceous, but only 22 °C/km was needed to match present maturity levels in the Tertiary section, consistent with the expected cooling in an evolving rift. Further refinement suggested that there was significant erosion and loss of section, possibly as much as 3000 m. This is consistent with the limited geological information.

The North American Mid-Continent rift system developed at approximately 1100 Ma and now extends from Lake Superior southwest into Kansas. Because of its great age and limited surface exposure it has not been well explored, and, as a consequence, regional data are limited. At the northern end the syn-rift Nonesuch shale has generated oil, although not in commercial quantities. Newell et al (1993) provide an interesting example from the southern part of the rift where they have used Lopatin modeling to provide a general evaluation of maturity. Their data came mainly from the 1-4 Finn well that was drilled to 3974 ft (1212 m) TD on the western flank of the Nemaha uplift in Marshall County, Kansas. Here the Ordovician Simpson Group lies unconformably on the Precambrian Rice Formation, which is thought to be a potential source rock, and equivalent to the Nonesuch Shale. The Rice Formation is rather lean with only 0.66 wt% TOC, and its present maturity is transitional between oil and wet gas generation. No vitrinite reflectance data are available because the age of the rock predates the time when land-surface organic matter produced the woody, Type III kerogen that includes vitrinite.

Because of the poor understanding of the rift system, and the lack of detailed thermal and burial history information, Newell et al (1993) felt that the use of sophisticated maturity models was not justified, and considered the Lopatin method to be adequate for a preliminary evaluation of the area. Fluid inclusions in calcite veins had homogenization temperatures for oil-filled and aqueous inclusions in the range from 110 to 115 °C. Simple burial was inadequate to account for these temperatures, or for the observed thermal maturity of the organic matter. Either the geothermal gradient was higher in the past, or burial was deeper, implying some erosional loss of section. Given the thermal evolution typical of rifts, and the fact that there is a substantial break between the Precambrian and the overlying Ordovician, it seems likely that both factors were involved. The present day uncorrected BHT and present day surface temperature give a minimum geothermal gradient of 28.4 °C/km, but other data for the area suggest values in the range of 30 - 40 °C/km. Various combinations of burial history and geothermal gradient were used in the Lopatin modeling. They concluded that the Precambrian gray siltstone had sufficient maturity for oil generation and that it became mature near the maximum depth of burial, which occurred in the Precambrian. There was minimal later maturation, and the overlying Middle Ordovician only reached a cumulative TTI of 13 (even when the geothermal gradient was 40 °C/km). The modeling results were consistent with the available analytical data. The authors cautioned that use of TTI methods has been based on experience with Phanerozoic rocks and needs to be extrapolated to Precambrian examples with caution (Newell et al, 1993). Price et al (1996) also used Lopatin modeling in their evaluation of maturity in the North American mid-continent rift system. They combined TTI data with equivalent R_o calculated using the method of Sweeney and Burnham (1990).

In the Danish subbasin burial histories have been derived from seismic sections tied to well control (Sorensen, 1985). In calculating maturities it was assumed that heat flow had increased during tectonically active periods, and 3 different heat flow regimes were combined with the burial history. Figure 10.14 shows the heat flow profiles used, and that the calculated time of onset of generation varied from 40 Ma to 56 Ma. Onset was defined by $R_o = 0.6$, with values being obtained from the Waples TTI-R_o calibration. For the most likely heat flow history (Case II) onset of generation occurred at 50 Ma when the burial depth was 2900 m. Sorensen (1985) concluded that the Lower Jurassic source rocks are oil mature in the studied area of onshore Denmark. Jensen et al (1985) also developed a model for onshore and offshore Denmark where heat flow was higher in the past. Their model varied surface temperature through time based on oxygen isotope data.

10.3.3. 1-D Kinetic Modeling

The Zala basin of western Hungary is part of the Pannonian basin system. It contains more than 5 km of Neogene and Quaternary sediments together with up to 2.5 km of Mesozoic sediments in some parts of the basin. Local faulting formed the Zala subbasin as the main Pannonian basin developed in the Miocene due to lithospheric extension (Royden, 1988). This extension produced high heat flows that are an important factor in the maturation of potential source rocks. At present, heat flows average 80 mW/m^2 and range from 70 - 90 mW/m^2. These values were combined with bottom hole temperatures to calibrate thermal history, with the main heating event assumed to occur following Miocene rifting. The Zala basin has been thermally modeled by Clayton and Koncz (1994) using both the BASINMOD (Platte River Associates) and MATOIL (BEICIP) commercial programs. The kinetic parameters needed were obtained by anhydrous pyrolysis of immature samples, but to get immature samples they had to go to stratigraphically equivalent rocks in the nearby Danube basin and the Transdanubian midmountains. Measured activation energy suites were in the range 49 - 54 kcal/mol or 46 - 49 kcal/mol, and they speculated that the lower range was due to sulfur-rich kerogen. The thermal maturity modeling showed that both Miocene and Triassic source rocks became mature "in the Miocene during rapid subsidence and heating caused by lithospheric extension in the Pannonian basin" (Clayton and

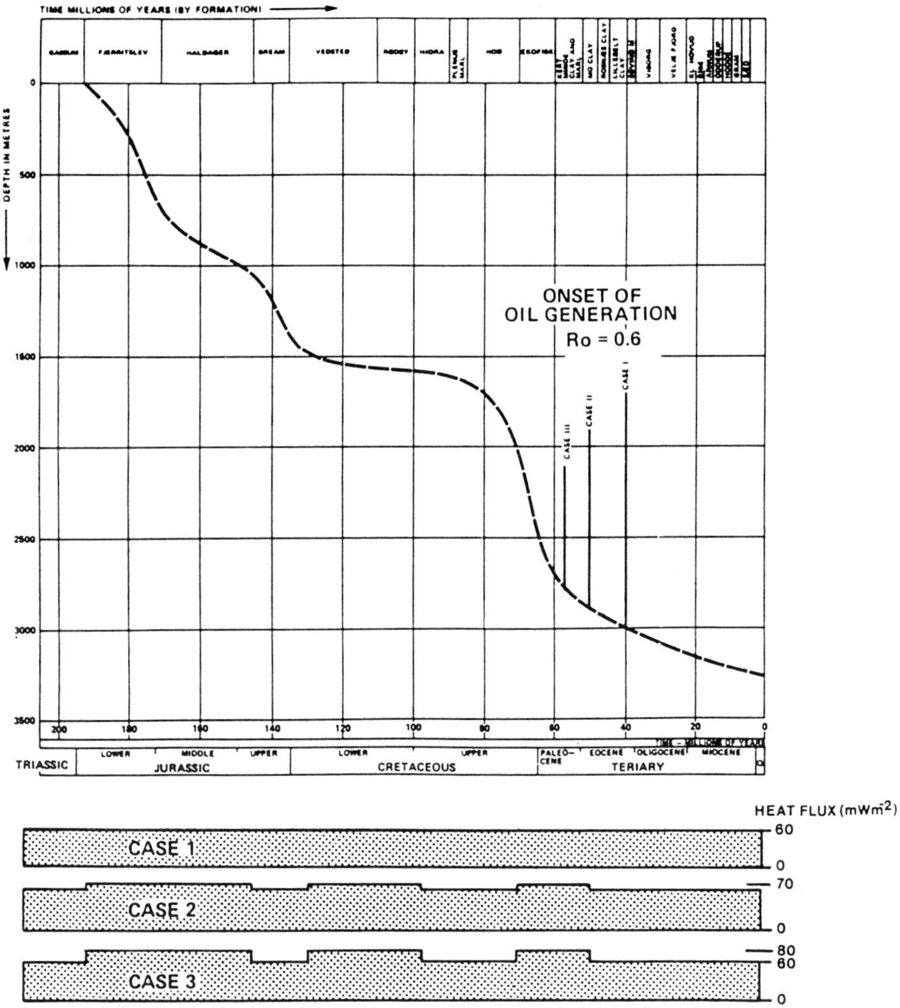

Figure 10.14. Subsidence history showing the depth and time of oil generation in the Danish subbasin for three different histories of heat flow (Sorensen, 1985).

Koncz, 1994). The modeling also showed that below 2500 m source rocks still "have remaining oil or gas potential and are actually generating today". The presence of several mature source rocks was consistent with the geochemical correlation information which showed that the Zala subbasin contained 2 (and possibly 3) oil families.

Rifts and rifted margins in the Norwegian offshore were modeled by Dahl and Augustson (1993) using a 1-D model based on the concepts of Yukler and Welte (1980). Organic maturity (sterane and hopane isomerizations) were calculated using first order kinetics, and R_0 values were obtained from a modified Lopatin technique. Measured and calculated R_0 values showed that the source rocks were mature below 1800 m in the Hammerfest basin, 2600 m in the Viking Graben, and 3100 m in the Halten Terrace. These differences were not thought to be due to different thermal regimes because all three basins have similar geothermal gradients (and similar lithologies, implying similar heat flows). E and A values for the kinetic models were obtained by optimizing the fit of organic maturity to the geological profile for a generalized (pseudo) Viking Graben well. The E and A values obtained could be applied to the Hammerfest basin, but did not work well in the Halten Terrace. This may be due to slight differences in organic matter type, with the Halten Terrace having more of the Type III kerogen which needed a bigger contribution from the higher activation energy reactions to match the measured profile. This explanation is consistent with the measured lower HI values. Differences in hydrocarbon maturation status among the three basins was apparently due to differences in sedimentation rates and erosional losses. In the Viking Graben and the Halten Terrace sedimentation in the Pliocene and Pleistocene was very rapid and produced efficient hydrocarbon generation. In the Hammerfest basin sedimentation was very slow or zero, and the accompanying erosion led to a termination of hydrocarbon generation.

Theis et al (1993) used a 1-D thermal model to establish the floor for oil in the Barents Sea, offshore Norway. This was important in deciding whether a prospect interpreted as a Permian carbonate reef structure was likely to be oil-bearing or gas-bearing. Depth to the potential reservoir was estimated to be 4399 m, and previous work had shown that at this depth source rocks were post-mature for oil. Thermal modeling using a geothermal gradient of 36 ^{o}C/km and an uplift/erosion of 1000 m in the burial history suggested that oil would be cracked to gas at 3700 m. This implies a gas prospect. However, neither the amount of uplift nor the geothermal gradient are known precisely, and so a sensitivity study was performed to evaluate the impact of changes in these parameters. A plot of geothermal gradient (30 - 39 ^{o}C/km) against uplift (0 - 2000 m) was made and contoured in depth to the bottom of the oil window (with contours covering the depth range from 2500 to 5300 m). For a reservoir depth of 4300 m the relevant combination of uplift and geothermal gradient can easily be evaluated. A lower temperature gradient or a smaller amount of uplift would lead to deeper preservation of oil. Mullen (1984) also used 1-D kinetic modeling to evaluate deep gas potential in the Aquitaine basin,

France. Type III kerogen was modeled using E and A values from Tissot and Espitalie (1975).

Most of Indonesia's petroleum reserves are in the back arc rift graben of the Central Sumatra Basin, with the Minas Field (8 billion bbl) and Duri Field (4 billion bbl) accounting for most of them. Organic-rich shales of the Paleogene Pematang Formation provide the major source rocks, and the Pematang Brown Shale Formation has been modeled by Williams and Eubanks (1995). They found that an unchanging geothermal gradient with the present value of 54.3 °C/km led to calculated maturity values which were much higher than those actually observed. The thermal regime seems to be strongly influenced by subduction of the Indian Ocean plate beneath the South East Asian plate. Some late stage local heating by igneous intrusions occurred and the timing was established by dating the intrusives. There is no evidence for a significant role for higher heat flow during the early stages of rifting and models that incorporate high initial values lead to calculated maturities that are too high. Maturity modeling assuming an initial heat flow of 0.9 hfu, with a subsequent increase to 1.1 hfu over the time interval from 24 - 5 Ma, and a final rapid increase to the present value of 1.8 hfu gave a reasonable match between calculated and measured maturity parameters. It showed petroleum generation starting at 10 Ma.

A recent study of the Taranaki basin, New Zealand by Armstrong et al (1996) used the first order reaction kinetics of Sweeney (1990) to quantify petroleum generation processes. Nineteen wells were included in the study which showed that western locations were generally immature, or had very recent generation. While temperatures in the west are now close to maximum, in the southern part of the basin cooling due to uplift has terminated generation. Oil generation and expulsion has occurred in the eastern Taranaki basin, but is geographically variable. Heat associated with Quaternary igneous intrusions may have been a significant factor in enhancing petroleum generation in the northern part of the basin. As expected for an evolving rift, heat flows were initially high. However, Armstrong et al (1996) concluded that this was only a significant factor for rocks buried to >2.5 km prior to 60 Ma.

10.3.4. 2-D Kinetic Modeling

The North Sea system of rifts is a major petroleum province where most of the exploration is fairly recent, and high quality geological, geophysical, and geochemical data are available. Several 1-D modeling studies from the North Sea have been discussed above, and the basin has also been attractive to the various groups developing 2-D models. One of the first groups to use 2-D kinetic models in this area was Ungerer et al

(1985). They applied their model to a study of the central part of the Viking graben in the Frigg area, but it was set up with only 8 x 20 grid elements to cover a cross section of 13.5 km by 6000 m. This required simplification of the geology, for example by ignoring minor faults and assuming that they are not barriers to fluid flow.

Dahl et al (1987) modeled the Oseberg area of the Viking and Huldra/Veslefrikk grabens with a 2-D deterministic, dynamic basin model based on the Yukler and Welte model (1980), but with an improved equation for fluid flow in a system with moving boundary conditions. 27 "events" were recognized since the Early Jurassic (190 Ma) of which 4 were either erosional or non-depositional. Estimated temperatures at the water/sediment interface varied from 9 ^{o}C to 20 ^{o}C, generally decreasing towards the present. Paleowater depths ranged from a maximum of 300 m to zero. Heat flow input for the model was obtained from a combination of R_o, geological history, and present day temperature data combined with 1-D and 2-D modeling. Predictions were tested for consistency with biomarker isomerization, hydrocarbon generation, and clay data. These studies showed that basin heat flow was about 0.05 hfu higher than for the platform areas, and that small increases of about 0.05 hfu occurred during the Eocene.

Combining the various geological and geochemical inputs in the model showed that Draupne Formation source rocks in the Huldra/Veslefrikk Graben started generating in Late Cretaceous (Maastrichtian, ~70Ma), and are now well into dry gas, whereas on the Oseberg structure they are still immature. Secondary migration routes obtained using combined subsurface topography, buoyancy, capillary pressure, and hydrodynamics lead to the conclusion that oils in the Oseberg Alpha structure were most likely sourced from the Viking Graben, while oils in the Oseberg Beta structure came from the Huldra/Veslefrikk Graben to the north. This conclusion was supported by biomarker and isotope data which indicated slightly different oil compositions. Burial history modeling suggested that the Oseberg Alpha structure was very shallow when oils first moved into it, and these oils were bacterially degraded. A later, undegraded oil charge, subsequently filled the structure.

The Oseberg area was also modeled by Doligez et al (1987). They used the THEMIS program developed by IFP (Doligez et al, 1986) and established a 2-D grid extending along a 250 km east-west line that crossed the Oseberg structure and included the Viking Graben and the Horda Platform. Oseberg itself is a rotated fault block developed during the rifting phase. The maturity of the Heather and Kimmeridge Clay shales, and the Brent and Statfjord coals, were calculated using kinetic data for Type II kerogen (based on Paris basin Lower Toarcian shales) and Type III kerogen (based on Lower Cretaceous coals from the North Sea). They did not consider cracking of oil to gas. Temperatures were obtained assuming constant geothermal gradients through time, but with lateral variations that ranged from 30 °C/km in the west to 50 °C/km in the east. Like Dahl et al (above) they concluded that oil was generated in the graben and migrated up onto the Oseberg structure, rather than from vertical migration within the Oseberg block. Pressures calculated by the THEMIS model matched present day values reported by others. High pressures seem to have been a feature of the Oseberg area since the Late Cretaceous and are associated with undercompaction in the graben. Pressures on the Horda platform are normal and this is also consistent with the modeling results. Doligez et al (1987) point out that the match of calculated and measured abnormal pressures requires that the major faults in the graben are impervious to fluid flow.

In another North Sea application of computer modeling the 2-D TEMISPACK software (BEICIP, France) was used to model heat transfer, fluid flow, hydrocarbon generation, and migration in the northern Viking Graben (Ungerer et al, 1990). A 150 km line was studied, and divided into 43 columns and 21 rows with faults treated as thin vertical columns. Thirteen different lithologies were employed, each with its own characteristic physical properties, but they were all made up of various combinations of sand, silt, and shale. Initially heat flow was modeled at 65 mW/m^2 (with half being radiogenic) and at 90 mW/m^2 (with 62 mW/m^2 being subcrustal), and the calculated profiles compared with measured BHTs. The trends (Figure 10.15) clearly show that the 65 mW/m^2 value is more appropriate, and this was used in subsequent modeling. This study provides an interesting example of the effects of different organic matter types when there are multiple source rocks. Marine organic matter dominates the Draupne-Heather-Dunlin rocks, while Brent-Statfjord are mainly continental. Even though the Brent-Statfjord are buried deeper they are less mature (as indicated by Transformation Ratios) than the shallower rocks with marine organic matter (Figure 10.16). Present day Transformation Ratios across the study area are given in Figure 10.17, and this shows the higher maturities in the grabens.

Ungerer et al (1986) have applied a 2-D model to a section across the Red Sea rift that includes the offshore Shoab-Ali oil field. Rifting started in the Miocene and the area now has source rocks and clastic reservoirs in both the pre-rift and post-rift sequences, as well as mid-Miocene evaporites. The evaporites are a prominent feature in the section but have restricted seismic imaging of the deep sediments. A mathematical model of the geodynamic evolution of the Suez rift provided information on the past thermal regime and these data were incorporated into a second model that

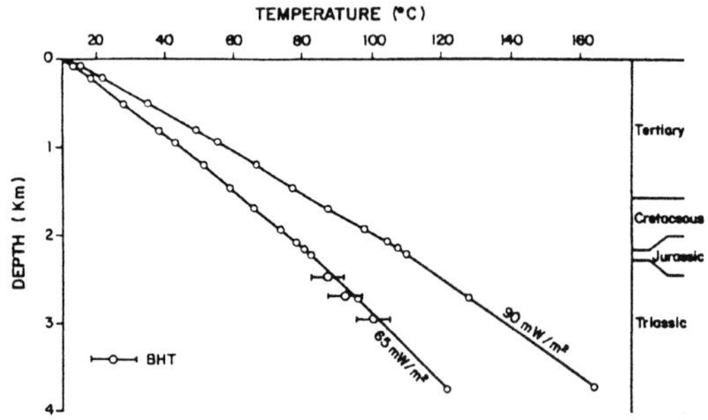

Figure 10.15. Calculated present-day temperature profiles at the top of the Snorre structure for heat flows of 65 and 90 mW/m^2. Measured bottom-hole temperatures (BHT) give a gradient that is consistent with the lower value for heat flow. (Ungerer et al, 1990).

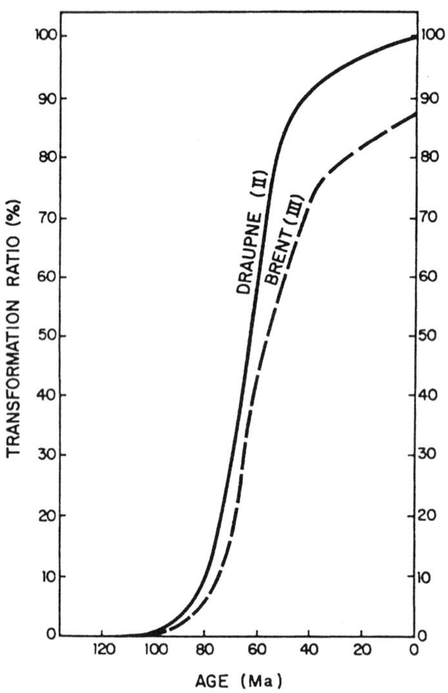

Figure 10.16. Calculated Transformation Ratios through time for two source rocks at the same position in the deep part of the graben. The Draupne, Type II source rock is more mature than the Type III Brent source rock, even though the Brent is buried more deeply. Calculations were made with a heat flow of 65 mW/m^2 (Ungerer et al, 1990).

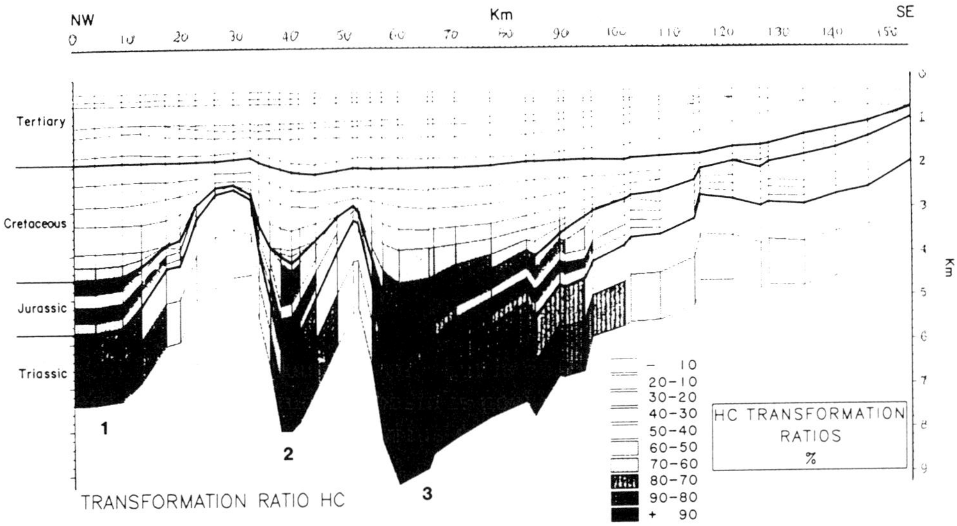

Figure 10.17. Calculated Transformation Ratios for potential source rocks in the Viking Graben showing three mature areas (1, 2, 3). Variations are due to differences in organic matter type and temperature evolution. The oil window corresponds to values of approximately 10 to 60% (Ungerer et al, 1990).

reconstructed petroleum generation and migration. Figure 10.18 shows the geology of the studied cross section and the way it was divided up into 190 elements. Each element has a single lithology and is homogeneous within that element. Generation started at about 6 Ma (as indicated by a Transformation Ratio of 0.1), and significant accumulations have developed from then to the present (Figure 10.19). Calculated oil in place for the Shoab-Ali field was 280 x 10^6 m^3, with an uncertainty of $\pm$ 120 x 10^6 m^3. This is close to the actual reserves which are estimated to be 165 x 10^6 m^3. However, the overestimate by modeling is consistent with the idea that not all of the generated oil is trapped, and some is left along migration pathways. In their sensitivity analysis, Ungerer et al (1986) found that when the geothermal gradient was decreased by 10% it produced a less mature and more viscous oil that did not migrate as efficiently, and that calculated reserves dropped by 40%. Forbes et al (1991) also attempted a similar mass balance in the Smorbukk area of the North Sea.

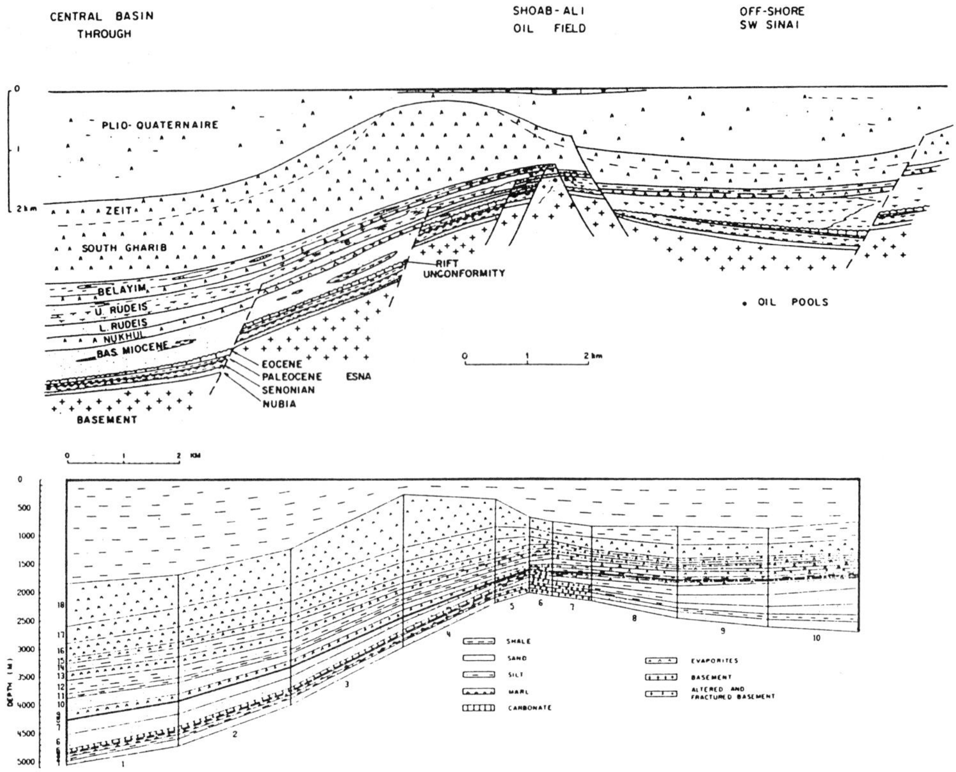

Figure 10.18. Geologic cross section of the Shoab-Ali oil field area, Gulf of Suez (top), and the grid of elements used to model it (bottom). (Ungerer et al, 1986).

10.3.5. 3-D Kinetic Modeling

Several areas of the Norwegian North Sea have been evaluated with a 3-D ray tracing model developed by Sylta (1987, 1993). This was used in trying to predict the reserves for the undrilled traps of the Block 35/9 area on the east flank of the North Viking Graben (Sylta, 1993). Data already available from successful exploration in areas to the north (Agat field), and the south (Troll field) were used to help calibrate the model, and to provide constraints on expulsion efficiencies and oil losses during secondary migration. An extensional model provided β extension factors, and the derived heat flows were used to calculate petroleum generation status. Ten source rock segments were considered, with each one having a fixed HI, but a TOC content and thickness that varied. Boundaries were time-transgressive.

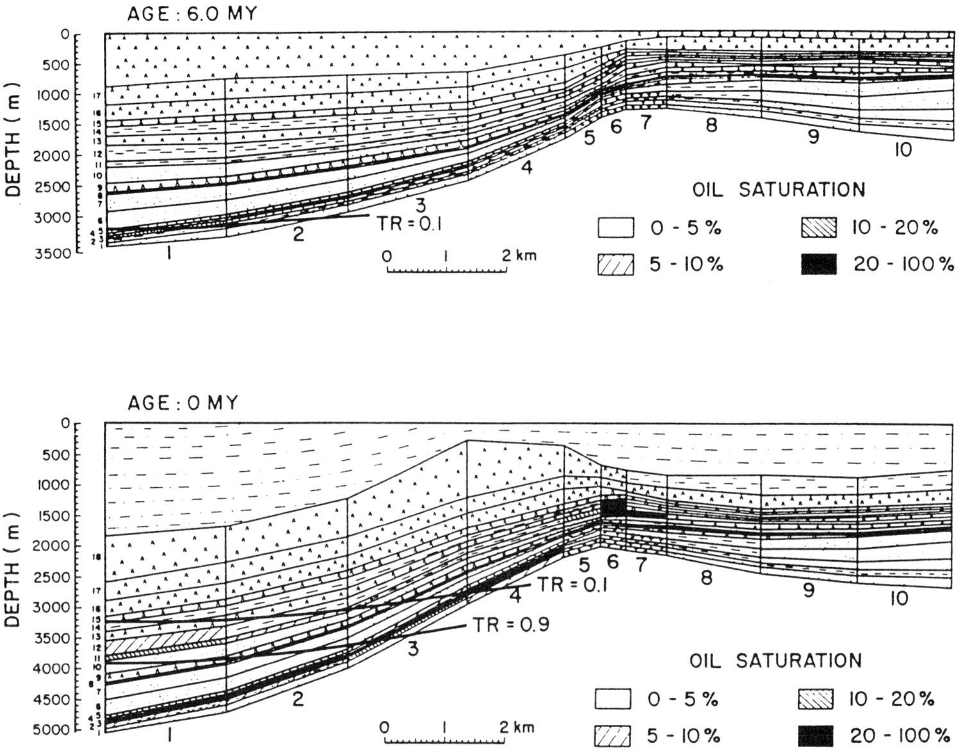

Figure 10.19. Conditions for hydrocarbon generation and migration computed for the Shoab-Ali area at 6.0 Ma and the present day. The TR = 0.1 and TR = 0.9 lines represent the threshold of significant oil generation and the end of the oil generation zone, respectively. Oil saturations are indicated by four ranges, and show the accumulations at the top of the Shoab-Ali structure and on the western side of the horst. (Ungerer et al, 1986).

The permeability values required by the 3-D model are difficult to determine, but can be a key factor in controlling petroleum losses during migration through carrier beds. Values were estimated from the relationship:

$$\log_{10} k = 20\,\phi - b \tag{10.1}$$

where ϕ is fractional porosity, k is permeability in mD, and b is a constant that varies from 2 to 4. Since hydrocarbon migration occurred in a water-wet system along the top of the Brent sequence, where oil saturations are estimated to be 60%, the relative permeability to oil was reduced to 30%. Migration was modeled in a sequence of seven time increments with the basin structure remaining constant during each one. The calculations successfully predicted the commercial accumulation of both oil and gas in the structure on Block 35/9, and this was later confirmed by drilling. Although input parameters were varied somewhat, computer time limitations prevented a complete sensitivity analysis from being carried out.

Schroeder and Sylta (1993) used the SEMI model (Sylta 1987, 1993) to rank undrilled prospects in a 100 km by 250 km area of the North Viking Graben that included Huldra, Veslefrikk, Oseberg, Brage, and Troll fields. As in the 35/9 block study described above, the presence of fields provided the opportunity to calibrate the model in the local area and data from 35 wells were used. Schroeder and Sylta (1993) list the four basic steps in their modeling, and these are as shown in Table 10.5.

Hydrocarbon generation, Step 2, was treated with a 1-D model, and the 35 well locations were modeled and input parameters adjusted until calculated and measured values agreed. Both the Upper Jurassic Draupne "hot" shales and Brent coals were considered to be source rocks. Kinetic data for oil generation, primary gas generation, and secondary gas generation were obtained from Braun et al (1990), MacKenzie and Quigley (1988), and Braun et al (1988), respectively. Heat flow information for generation modeling was obtained from the subsidence history combined with the effects of "anomalous heat flow" associated with crustal thinning during rifting. Predicted temperatures and R_o values were compared with measured values and were considered acceptable for the interval of interest (i.e., below 3000 m).

Many of the parameters required for Step 3 are poorly known (Schroeder and Sylta, 1993) and input parameters were varied over the range of uncertainty to evaluate their effect on the predicted output. Flow paths were obtained by mapping the steepest upward gradient, and hydrocarbons were allowed to fill the nearest trap, and then spill into the next updip structure. Losses along the migration pathway are taken into account using numbers obtained from the calibration with known fields. This computer procedure only considers a single carrier/reservoir unit at any one location through time and so is not strictly a full 3-D model.

The accumulated modeling information was used to rank three prospects in the studied area. One was at a focusing point of hydrocarbon migration, and was thought to have high chance of oil, but the oil/free gas

Table 10.5
Four basic steps in modeling (Schroeder and Sylta, 1993)

(1) COMPILE DATA:	wells:	stratigraphy
		geochemistry
		petrophysics
		temperatures
		reserves
	seismic:	structure
		stratigraphy
		velocities
(2) MODEL HYDROCARBON GENERATION:		subsidence analysis
		thermal history
		maturation history
		source properties and kinetic models
		model oil and gas yield
(3) MODEL SECONDARY MIGRATION:		expulsion efficiency
		migration pathways
		paleostructure
		ray-tracing
		migration efficiency
		trap capacities
		leakage and spillage
(4) EVALUATE PROSPECTS:		generation
		spillage in
		concentration
		timing
		spillage out
		sensitivity

ratio could not be predicted with confidence. A second structure was also oil prone, probably with some gas. The third structure was the least favorable due mainly to immaturity in the drainage area. The structures are currently undrilled and so the predictions remain unverified at present.

The SIMI model (Sylta 1987, 1993) was also used by Skjervoy and Sylta (1993) to investigate a 7500 km^2 area south of the Oseberg/Brage area. Block 30/12 is in the Norwegian part of the North Sea and is located east of the Viking graben on the southwest margin of the Horda platform. The modeling challenges included low relief structures and long distance migration, and the study was designed to establish the drainage area for hydrocarbons migrating into the studied structures. The three phases of the integrated basin modeling study are shown schematically in Figure 10.20. Phase 1 assembles required input data and sets up the burial history and thermal history. These are evaluated by comparing predicted and calculated bottom hole temperatures, and R_o, T_{max}, and HI values. In Phase 2 hydrocarbon generation and oil expulsion are evaluated. Oil is expelled when a critical saturation value is exceeded, and residual oil is cracked to gas. Finally, in Phase 3, secondary migration is modeled to provide information on flow paths and the possibility of filling the prospective structures. The results show that of the 3 potential traps, the smallest two were mutually exclusive in terms of drainage areas. The most optimistic modeling conditions led to oil in the largest structure, some in the medium-sized prospect, and none in the smallest structure. The most pessimistic scenario leads to some gas in the medium-sized prospect but no reserves in either of the other two prospects.

Grigo et al (1993) have used a 3-D model to investigate the Haltenbanken area, offshore Norway. The SEBE-3 finite element model successfully simulates compaction, fluid flow, heat transfer, and petroleum generation, migration, and trapping as the basin evolves. It was used in a 100,000 km^2 area with a 3-D network of cells that have horizontal dimensions of 8 km x 8 km. Based on available well control and seismic information the geology was simplified and described as 14 isochronal events. The potential source rocks in the area are the Middle and Late Jurassic Spekk and Melke Formations, and the Late Triassic to Early Jurassic Are Formation. Hydrous pyrolysis experiments provided E and A values for immature samples of the Spekk and Melke Formations, but because of the difficulty in distinguishing the two units in the modeled area default Type II kerogen values were used. In the case of the Are Formation, no immature sample material was available, and Type III default values were used. A single E value of 57 kcal/mol described generation of gas by cracking.

Well data and preliminary 1-D modeling were used to approximate present day heat flow and showed the need to use a higher present day value to generate the observed geothermal gradient. The 3-D model gives satisfactory values for the computed regional pore pressure, and this was taken as confirming that the permeabilities and compaction trends were essentially correct. The calculations also showed that as the basin evolved temperatures at a given depth varied considerably across the studied area (by as much as 30 °C), and this led to variations in the degree of maturation. The more mature areas are displaced generally northwards, which is consistent with the observed distribution of petroleum (Grigo et al, 1993). Fluid flow was also displaced northwards from the Early Cretaceous to the Early Pliocene, but then changed direction to flow southwards.

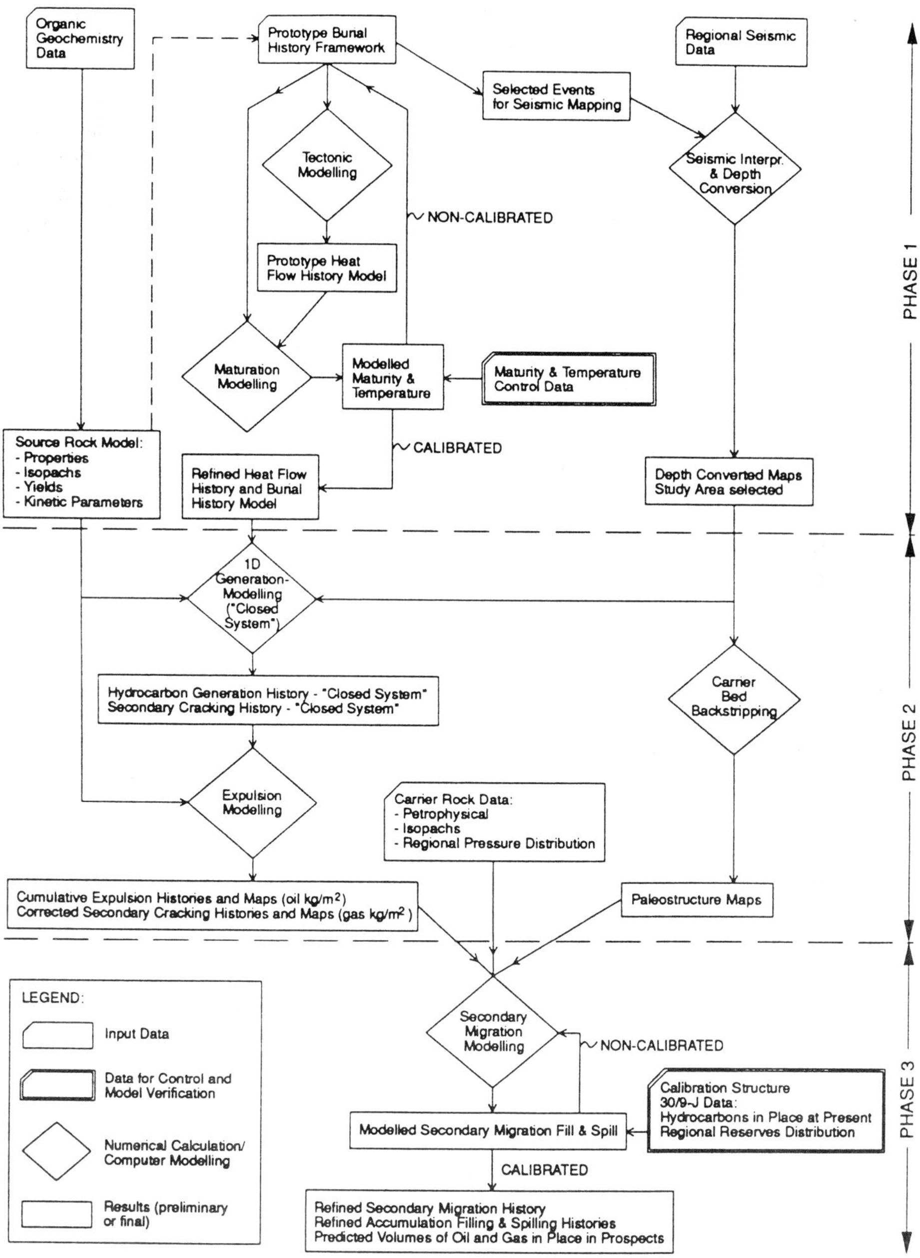

Figure 10.20. Schematic organization of the integrated basin modeling study carried out by Skjervoy and Sylta (1993) in the Oseberg/Braage area of the North Sea.

These directions were consistent with compaction trends in the depocenters.

Most published 3-D modeling has been applied in the North Sea, but Yalcin et al (1988) have used it in the Cambay basin of northwest India. Here all available geological, geochemical, and geophysical data were developed into a conceptual model covering the evolution of the basin over the last 65 m.y. Yalcin et al (1988) regard the conceptual model as "a kind of tailor-made geologic manual to guide the computer simulation of a basin". 1-D modeling at selected locations provided preliminary information on the geologic evolution that helped in designing the 3-D model.

The Cambay basin is a 450 km long, fault-bounded rift that is 55-100 km wide. Rifting started in the late Cretaceous and the basin is now made up of 4 major depressions with separating saddles. Previous geochemical studies had shown that there were several potential source rocks, and they were all in the Eocene. In the Early Eocene a major marine transgression produced the dark gray to black Cambay Shale. This is dominated by Type III kerogen, but has up to 20% Type II in the Ahmedabad depression. A subsequent regressive phase laid down the deltaic sediments of the Kalol Formation in the Middle Eocene. This is up to 400 m thick in the Ahmedabad depression and contains as much as 40% Type II kerogen with the balance being Type III. A later major transgression in the Middle Eocene to Early Oligocene deposited 200 m or more of the Tarapur Shale which contains mainly Type III kerogen. The regressive sequences are more important in the northern part of the study area. Sediments were divided into 15 chronostratigraphic depositional units based on isopach and lithofacies maps, with the Cambay Shale being treated as 5 subunits. In addition, three erosional or non-depositional "events" were recognized. 677 selected grid points provided input for the model (Figure 10.21). Heat flow was used to calculate paleotemperatures, and current values for heat flow range from 1.10 - 1.30 hfu, but locally a recent heat pulse of 1.70 hfu is needed to produce a match with calibration data for pressure, maturity, and temperature. The 3-D model combined burial histories with temperature data to calculate maturities for each source rock, and this information was combined with rock volumes to calculate the total amount of oil and gas generated. Source rocks were treated as closed systems with oil cracking to gas *in situ*. Figure 10.22 shows the calculated hydrocarbon generation potentials (g HC / g TOC) at 5 Ma. As oil is generated and cracked to gas, the gas distribution evolves towards that given in Figure 10.23. As expected, source rocks in the depressions generated earlier and became more mature. The Cambay Shale is the major source rock in the northern part of the rift, generating approximately 10.0 x 10^6 tons/km^2 in

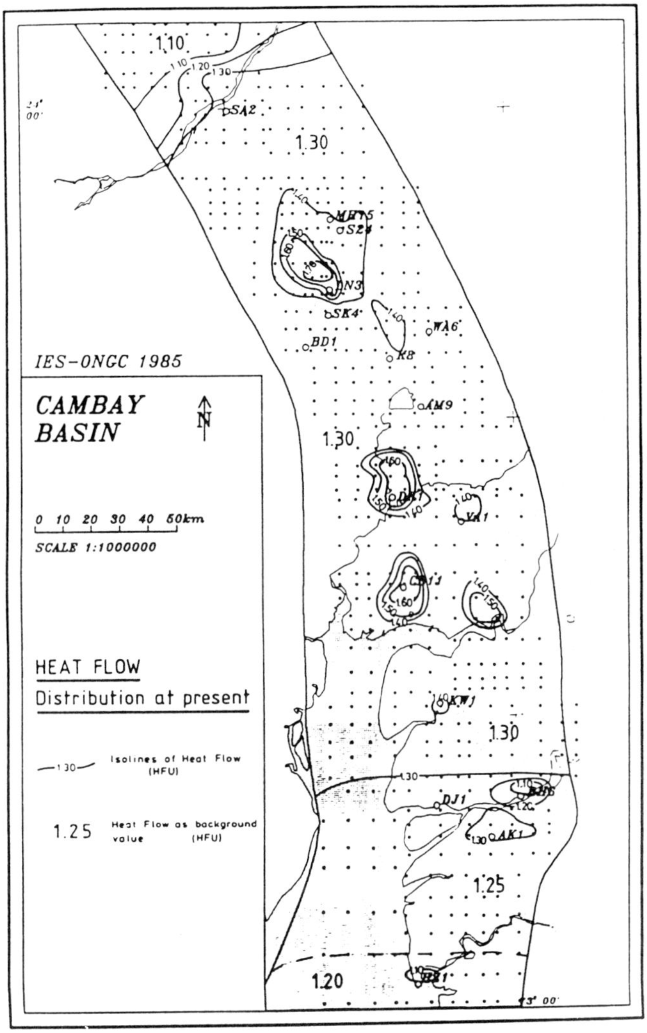

Figure 10.21. Locations of 677 grid points used for modeling in the Cambay basin, northwest India, together with present day distribution of heat flow (Yalcin et al, 1988).

the Almedabad depression, and up to 5.0 x 10^6 tons/km^2 in other areas. The base of the shale (Cambay Shale I) was generating wet gas 43 Ma in the Broach depression, while still immature elsewhere. By 34 Ma it was into dry gas. Oil generation from the Kalol Formation started in the deep

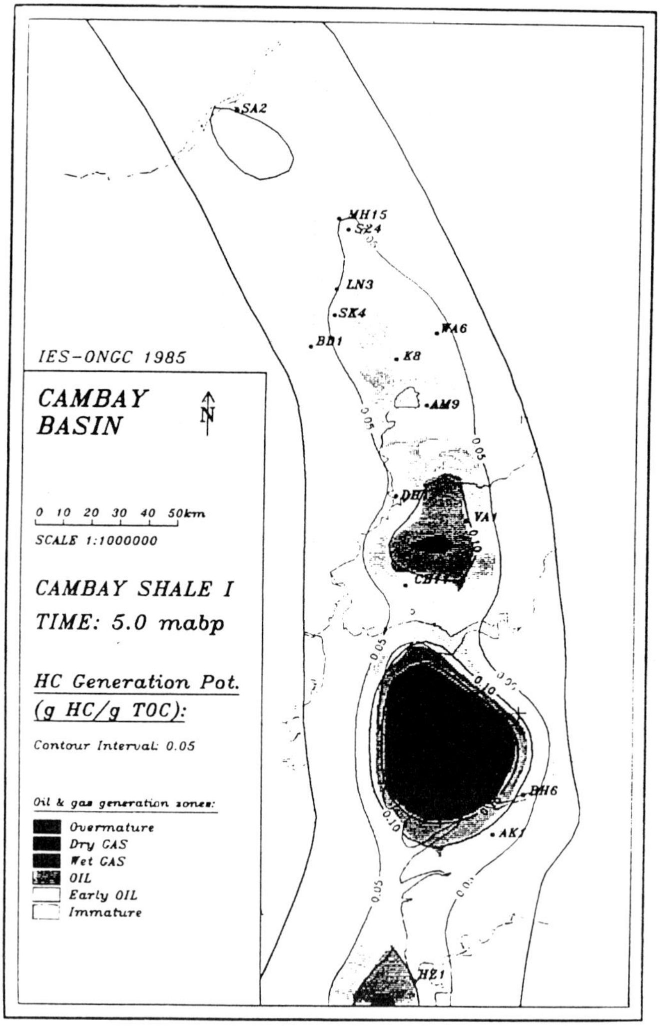

Figure 10.22. Hydrocarbon generation potential and oil-gas generation zones in the Cambay Shale at 5 Ma (Yalcin et al, 1988).

areas and expanded to the present day distribution, with the deepest parts now in the dry gas zone. The Tarapur Shale has only recently reached the oil to wet gas stage in the deepest part of the Broach depression.

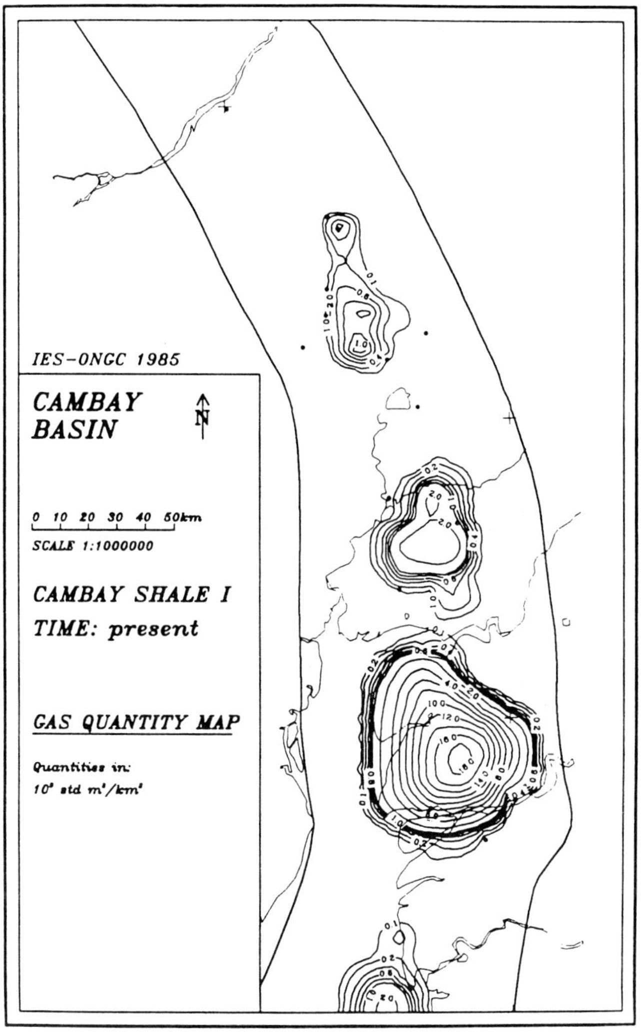

Figure 10.23. Gas quantity map of the Cambay Shale I at the present time (Yalcin et al, 1988).

10.4. PASSIVE MARGINS

10.4.1. Introduction

Rifts where extension continues can evolve into passive margins. Since the passive margins have the rift sediments embedded in them, all the remarks from the previous section are pertinent to the discussion of these basins. As the margin develops by extension, sediments prograde over the rift sequence producing large, linear, asymmetric basins. The high heat flow and high geothermal gradients characteristic of the early stages of rifting have decreased through time giving passive margin temperatures that are typically average to slightly cool. The lacustrine and continental sediments of the rift are overlain by marine sediments containing increased amounts of Type II kerogen. Evaporites that form when the extending rift evolves into a narrow seaway often separate the continental and marine sediments, and salt can be important in controlling migration pathways, and in its effect on thermal regime (see Section 10.7, below). It is also possible to have source rocks both above and below the salt, but these often differ in organic matter type with the post-rifting source rock having more marine affinities (Barker, 1982; Trindade et al, 1992). Mello and Maxwell (1990) estimate that 85% of the oil discovered on the Brazilian passive margin comes from two lacustrine systems of Neocomian to Aptian age that developed during the rift stage of continental breakup. One was freshwater with TOC values up to 6%, and the other was saline with TOC up to 9%. Interestingly, Kuo (1994) also reported 2 subsalt source rocks in Gabon on the matching coast of West Africa.

Thermal modeling studies on passive margins include those carried out for eastern North America by Issler and Beaumont (1987), Issler (1984), Keen (1978), Royden and Keen (1980), Sawyer et al (1982), Siever and Hagar (1981), Streckler and Watts (1978), and Liu et al (1992) (COST G2 well); for the west coast of Africa by Brun and Lucazeau (1988), Jabour and Nakayama (1988), Leith and Rowsell (1979), and Duncan and Roman (1986); for the Gulf of Lion by Chenet (1984), Burrus and Audebert (1990), and Burrus and Bessis (1986). Other modeled margins include Antarctica (Cook and Davey, 1984; Davey, 1985); Alaska (Cao and Lerche, 1989); S. Australia (Hegarty et al, 1988); and Argentina (Fitzgerald et al, 1990).

10.4.2. Lopatin Modeling

The Otway basin on the southern Australian coast developed as a passive margin during the separation of Australia and Antarctica at about 55 Ma. Older Cretaceous and early Paleocene sediments had accumulated in the pre-breakup, continental rift. The thermal evolution of this basin has been summarized by Middleton and Falvey (1983) and combined with an organic maturation model to evaluate source rock potential. They used a slightly modified Lopatin procedure with a reaction rate that doubled every 10.2 °C. Calculations were made both at well locations and for pseudo wells that could be used to develop a maturity cross section. They found that calculated R_o values agreed fairly well with measured values, and that R_o was relatively insensitive to modest variations in paleoheat flow. Much of the basin is in the oil or gas windows, but even though there have been oil shows in exploration wells there is currently no commercial production.

Other modeling studies using the Lopatin method include those by Keen and Lewis (1982) and Thomas and Thomas (1988). Issler (1984) generated TTI -R_o calibration data for eastern Canada that are now widely used.

10.4.3. 1-D Kinetic Modeling

In some locations on passive margins there may be 2 mature source rocks, and these can produce 2 distinct families of oils. In this case each source rock has to be treated individually, not only because they have distinct burial histories, but also because organic matter types may be different. In addition, there is the possibility that mixed oils could form, and these will have compositions that do not simply correlate to either of the source rocks. The Potiguar basin on the northeast Brazilian rift/drift margin provides an example of multiple source rocks. Trindade et al (1992) used both geochemical and geological studies to demonstrate the presence of 2 source rocks, with one being made up of Neocomian and Aptian lacustrine fresh water shales, while the other consists of Aptian marine evaporites, shales, and marls. A combination of burial history and thermal history using MATOIL (BEICIP) commercial modeling software has shown that these 2 rocks become mature at different times. The top of the Neocomian reached the oil window at roughly 56 Ma, whereas the top of the Aptian became mature 28 m.y. later. These conclusions were based on a model that used a present-day geothermal gradient of 30.6 °C/km, and carried out kinetic calculations of maturity using activation energies obtained from laboratory pyrolysis. Values were 53 kcal/mol for the Aptian and 56 kcal/mol for the Neocomian. Oils generated from both source rocks migrated updip and westward to become reservoired in rocks that are thermally immature. Mixing occurred during migration to form oils with intermediate compositions.

Onshore Gabon has 2 subsalt, lacustrine source rocks, and these have been modeled by Kuo (1994) using the 1-D BASINMOD package (Platte River Associates). Generation in both the Kissenda and Melania source rock shales ceased at 50 Ma due to regional tilting, uplift, and erosion. The deeper Kissenda had already reached peak oil generation, but the shallower Melania had only achieved the early oil generation stage before uplift.

The BASINMOD (Platte River Associates) 1-D program was also used by Williamson (1992) to model maturation in the Jeanne D'Arc basin of offshore eastern Canada. This continental margin is a half graben, which accumulated thick Mesozoic syn-rift sediments, and a thinner sequence of Cenozoic post-rift deposits. Subsidence, compaction, and thermal maturity were modeled for 33 exploration wells and 5 pseudo wells to provide a temporal and spatial framework for generation in the Upper Jurassic Egret Member source rock. Figure 10.24 shows the compaction-corrected burial history for the Terra Nova K-08 well based on the porosity-depth data from Figure 10.25. The steady state thermal models used were estimated to lead to uncertainties of the order of 20 Ma for the onset of petroleum generation, with the timing being particularly sensitive to the values selected for thermal conductivity and heat flow. Variations were quantified using two sets of thermal conductivity/heat flow values and the effects due to selected ranges are shown in Figure 10.26 for the Terra Nova K-08 and Nautilus C-92 wells. Further checks on the validity of the modeling used comparisons of calculated and measured R_o values (Figure 10.27). Quantitative estimates of the oil generated by the Egret Shale compared with the oil reserves in Hibernia have been made by Williamson et al (1994), and were discussed in Section 9.2.3.

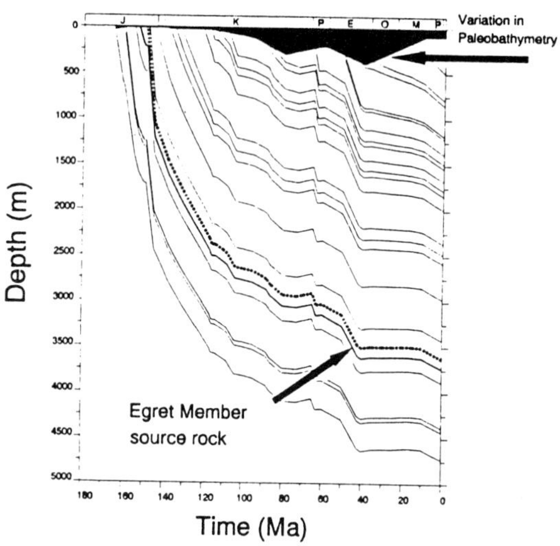

Figure 10.24. Burial history for rock units penetrated by the Terra Nova K-08 well (Williamson, 1992). Thicknesses have been corrected for compaction. Note also the varying water depth (shown solid black).

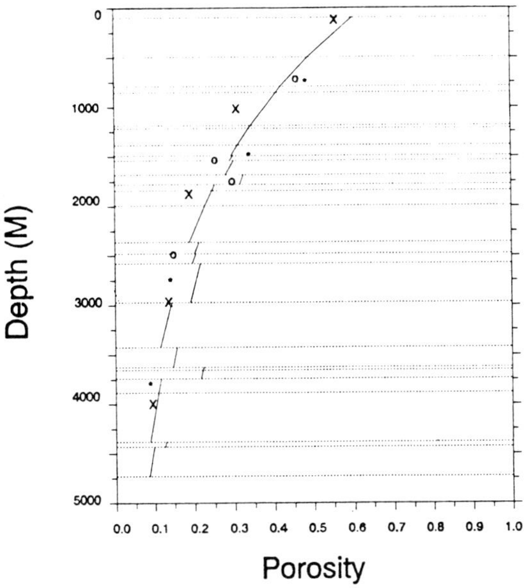

Figure 10.25. Porosity-depth profile used in decompacting sediment thicknesses in the Terra Nova K-08 well (Figure 10.24). The curve was derived from Sclater and Christie (1980). Also plotted are points corresponding to values used in other Grand Banks studies, together with those derived from sonic transit times for the Terra Nova K-08 well. (From Williamson, 1992).

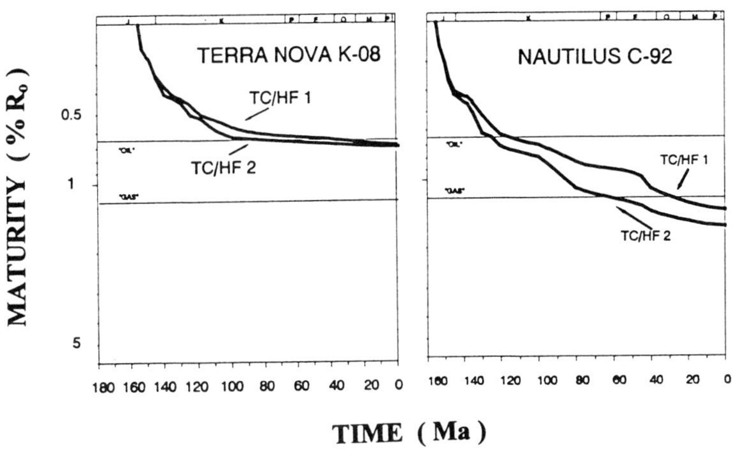

Figure 10.26. Comparison of Egret Member maturity through time at the locations of the Terra Nova K-08 and Nautilus C-92 wells. Data are shown for the 2 sets of thermal conductivities and heat flows used (TC/HF1 and 2).

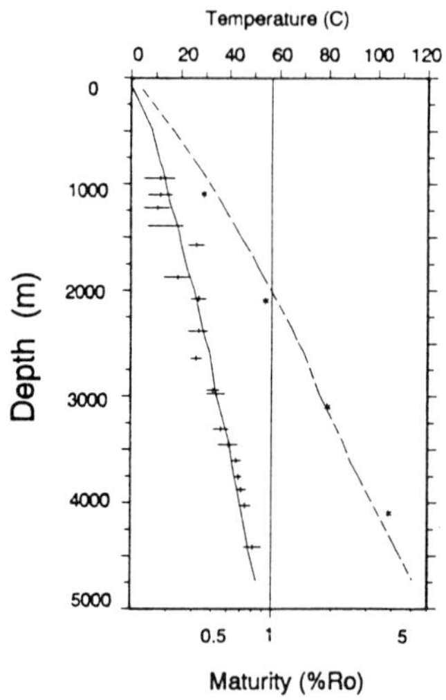

Figure 10.27. Observed present day measured reflectance values (A) and bottom hole temperatures (corrected for fluid circulation) (B) compared with the values calculated from the maturity model conditions corresponding to TC/HF1 for the Terra Nova K-o8 well (Williamson, 1992).

Williamson and Smyth (1992) also report 1-D kinetic modeling on the east coast of Canada. Their study of the Sable basin, offshore Nova Scotia, was concerned mainly with gas generation and the development of abnormal pore pressures.

10.4.4. 2-D Kinetic Modeling

The TEMISPACK 2-D (IFP) model has been used by Forbes et al (1992) in the Venture Field area of offshore Nova Scotia. Here gas is located in overpressured sandstones (Mudford and Best, 1989) and its occurrence had previously been modeled (1-D, but 2 phase) by Mudford (1988, 1990). Much of the emphasis in modeling this part of the passive margin has been directed towards understanding the generation of abnormal pressures. Forbes et al (1992) did calculate gas generation and the distribution of gas saturation, using temperature data from corrected BHTs and DSTs. No immature samples occur in the area, and they were unable to generate kinetic information for the Type III kerogen and so used default values. They concluded that gas generation did not have a major role in producing abnormal pressures, and that, in spite of the low sedimentation rates, disequilibrium compaction was probably the major overpressuring mechanism.

10.5. DELTAS

10.5.1. Introduction

Sandstones deposited in delta systems are frequently important petroleum reservoirs (Whateley and Pickering, 1989), but in many cases the source rocks that generated the petroleum are not part of the delta. Even though individual deltaic sands can be hundreds of feet thick, the total sand may represent only a minor part of the basin fill, and the sediment pile would not be appropriately modeled as a delta. For a substantial pile of delta sediments to accumulate and fill a basin (like the large Tertiary delta systems) the river must remain in a relatively stable position and continue to supply sediment for a long time. The course of the Niger River, for example, is controlled by the Benue rift, and the sediment load is being deposited to form the Niger Delta at its seaward end. This rift is an aulacogen that formed in the late Cretaceous, and the delta has been developing since that time.

Here discussion will be restricted to the large Tertiary delta systems that are "vertically integrated" in the sense that source rocks, reservoirs, and migration pathways develop as aspects of the same overall depositional process. Such deltas commonly form at the transition from continental to oceanic crust (Audley-Charles et al, 1977) and basement tectonics can have a significant role in the structural evolution of the prograding delta. Productive delta provinces contain 4.5% of the oil generated by the Worlds major source rock stratigraphic intervals (Klemme and Ulmishek, 1991) (Figure 10.1), and virtually all of the generation and accumulation occurred post-Oligocene.

The large Tertiary delta systems are very gas prone and production from some, like the Nile and Bangladesh, is dominated by gas, with little oil production. When oil is present it is usually a high-paraffin, low sulfur, waxy crude, commonly with a high pour point (although bacterially degraded equivalents have lowered pour points). Examples include the Niger, Mahakam, Mackenzie and Baram deltas, and the Gippsland basin. The Mississippi delta is the only major oil-producing Tertiary delta system that is not dominated by waxy crudes.

10.5.2. Sediments and Thermal Regimes

The major Tertiary deltas are large accumulations of clastic sediments with few, if any, carbonates. The shape of the delta that develops, and the distribution of sands and shales in it, depend on the interaction of tide and wave energy with the sediments brought by the river. In general grain size decreases seawards from the coarse-grained fluvial, beach, and near-shore sands to fine-grained, prodelta muds deposited in deep waters that are frequently anoxic. The most productive parts of deltas are where sands and shales interfinger to provide reservoir/seal combinations (Ejedawe, 1981).

Accumulation of sediment is rapid and may exceed 500 m/m.y. (Ejedawe et al, 1984; Issler and Snowdon, 1990). Slumping in this unstable sediment pile can lead to penecontemporaneous growth faults with associated rollover anticlines. The anticlines formed in this way deform whole packages of sands and shales and lead to multiple, superimposed reservoirs. This situation is common in the Mississippi, Mackenzie, and Niger deltas, but growth faulting has not occurred in the Mahakam delta. Sediment loading and subsequent readjustments can produce shale-cored anticlines where the local contrasts in thermal properties lead to lateral heat transmission and *"heat refraction"* (Issler and Snowdon, 1990).

The high rate of sediment accumulation has some important consequences for modeling:

(1) In the delta's active depocenter the cold sediments accumulate so quickly that they do not warm up fast enough to reach thermal equilibrium. Dixon et al (1992) found this to be the case in the Mackenzie delta where the top 4 - 5 km of the sediments had been deposited in the last 5 m.y. In this situation, heat flow and sediment thermal conductivity cannot be used to calculate geothermal gradients. However, if river position switches and the delta is abandoned, or if the depocenter is overrun to build a new depocenter further seaward, then the original low temperatures start to rise through time and onset of petroleum generation becomes steadily shallower through time. Figure 10.28 shows schematically the changing rates of sediment accumulation in a prograding delta. In the 20 m.y. between 50 Ma and 30 Ma about 10,540 m of sediments were deposited (XY), but as the delta continues to prograde only 1460 m of sediments (YZ) accumulated at the well location in the next 20 m.y. In this example sediment accumulation rates have dropped from 527 m/m.y. to 73 m/m.y. and the old sediments (50 - 30 Ma) will continue to warm up to the regional average as the delta progrades seawards.

(2) Loading of muds squeezes out water and causes compaction. When the loading is very rapid, water may not be able to flow out fast enough to maintain equilibrium because of the low permeabilities of mud rocks. In the process of disequilibrium compaction the pore fluids support part of the rock load and pressures become abnormal (see Section 9.4). These rocks are often undercompacted and the high water content depresses the thermal conductivity and leads to locally high geothermal gradients. In addition high pore pressures that approach the rock load may induce fracture development.

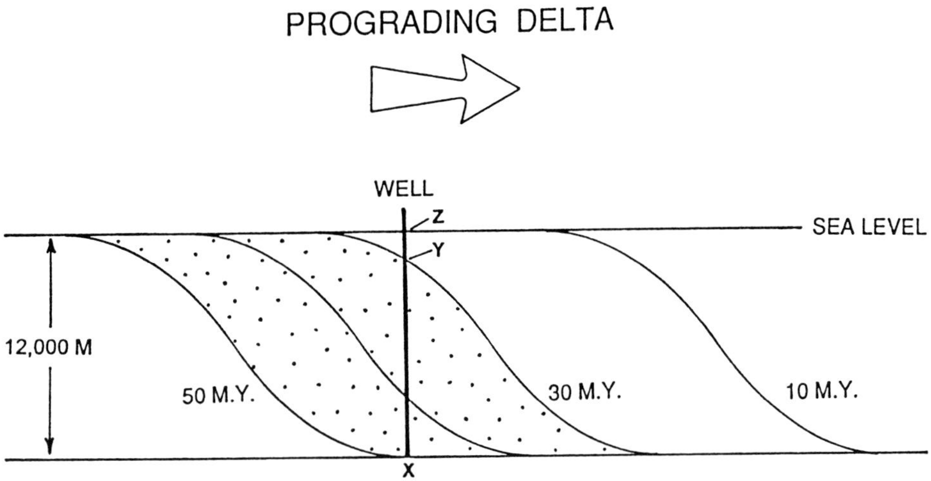

Figure 10.28. Schematic cross section of a prograding delta. At the well location sediment accumulation is initially very rapid (XY in 20 m.y.) but decreases (YZ in 20 m.y.) as the delta builds seaward.

(3) The fast accumulation of clastics usually acts to dilute organic materials leading to low TOC values. Organic contents are rarely much above 1 percent in the main progradational part of the delta (e.g., Bustin, 1988; Dixon et al, 1992) (but can be much higher in interdistributary bays). When organic contents drop below some critical minimum they are unable to generate enough oil to cause migration out (Section 9.5.1), and it is only later when oil is cracked to gas that the volume increases sufficiently to expel gas (Cooles et al, 1986). This implies that a lean source rock with Type II kerogen will only be an effective gas source rock, and this may be part of the reason why deltas are gas prone environments.

10.5.3. Organic Materials

Deltas incorporate organic matter from both the land surface and aquatic environments. Terrestrial material, including vegetation that grows in the delta plain, often forms a major contribution and includes not only the woody, structural material, but also the waxy surface coatings (*"cuticle"*), resins and spores. While the lignin-rich wood is gas-prone, Type III material, the other terrestrial materials are hydrogen-rich, oil-prone organics that are likely to generate highly paraffinic, waxy crudes. Terrestrial aquatic organic matter is derived from fresh water algae and bacteria that grow in lakes and swamps.

Nutrients brought by the river stimulate the growth of marine algae and phytoplankton off the delta front and in the pro-delta areas, and these contribute to the organic content of the sediments. The distribution of terrestrial and aquatic organic matter in deltas is not uniform. Terrestrial material has its highest concentration nearshore, at the point of input, and is steadily diluted by aquatic material with increasing distance from shore. This trend is shown by the increasingly heavier carbon isotope values offshore (Kennicutt et al, 1987), by the increasing organic matter density with distance offshore (Nwachukwu and Barker, 1985), and by changing biomarker ratios, such as the C_{27}, C_{28}, and C_{29} steranes (McCaffrey et al, 1994). Figure 10.29 schematically illustrates the trend in organic matter types for 3 time slices in a prograding delta. As the delta builds out into its receiving basin, a vertical profile from terrestrial (T) to aquatic (A) with increasing depth develops. An example from the Mississippi is shown in Figure 10.30 from Dow and Pearson (1975) with the "structured, coaly, gas type" (i.e., T) organic matter shallow, and the "non-structured, sapropelic, oil type" (i.e., A) deep. This profile also shows the low TOC values typical of many delta systems. In selecting appropriate E and A values for kinetic models it is important to recognize these vertical and lateral changes in organic matter type.

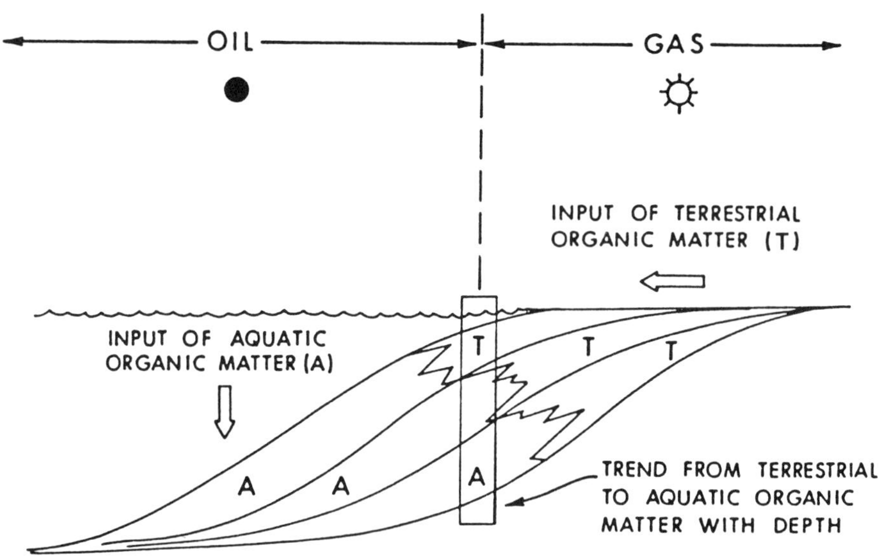

Figure 10.29. Three "time slices" through a prograding delta showing how the distribution of terrestrial (T) and aquatic (A) organic matter varies (Barker, 1979).

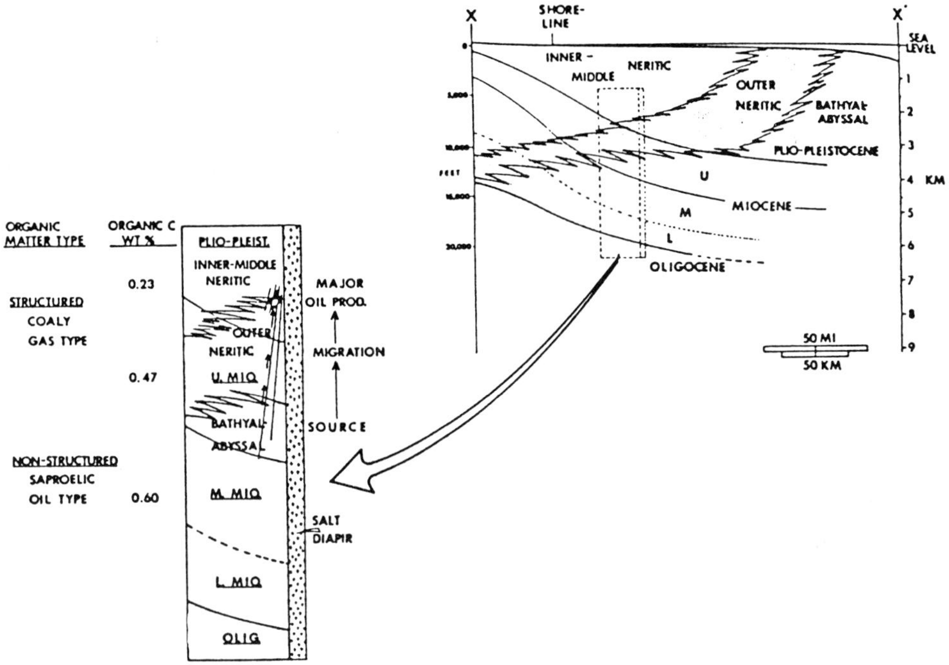

Figure 10.30. Profile of organic matter types in the Mississippi delta (Dow and Pearson, 1975).

Deltaic clastic sediments are derived by weathering, transporting, and redepositing pre-existing rocks. If these rocks contain organics some of them may be recycled and incorporated into the new sediments. This material has already been buried once and at least partially matured, and then during transport it has been further weathered and oxidized, so that little, if any, petroleum generating ability remains. Sediments may also incorporate fossil charcoal derived from forest fires. Neither of these materials contribute to petroleum generation yet both will be included in a TOC determination. It is important to recognize this non-generating, recycled contribution from visual examination or pyrolysis. Almost all clastic rocks contain some recycled organic material, and in exceptional cases it can exceed 80% of the total.

As the mud-rich delta piles compact there is slow subsidence near-shore and interdistributary bays and marshes form. Peat (and eventually coals) develop in these environments and may accumulate or be reworked into the prograding delta. van Krevelen diagrams are commonly used to establish organic matter type, and coaly material will plot close to the Type III track because it is dominated by the gas-prone, lignin type of material.

The van Krevelen diagrams show only the weighted average of all organic matter types in the kerogen and terrestrial input may include several percent of Type I kerogens such as cuticle, exinite, etc. This may move the data points slightly above the idealized Type III trend line, but it may still be interpreted as solely gas-prone material and activation energies for Type III kerogen used in modeling. However, a rock with a few percent of hydrogen-rich Type I kerogen can be an oil source rock regardless of whether the rest of the rock is shale or coal, and it is important to recognize this contribution and model its changing maturity with appropriate E and A values.

10.5.4. Petroleum Generation and Accumulation

Petroleum generation requires both temperature and time, but in the Tertiary deltas there is not much of either. They are young and the available time for maturation is often less than 50 m.y. They are cool because of the rapid accumulation of cold sediments. Both these factors lead to deep onset of generation. This trend is further enhanced because much of the oil is generated from Type I kerogen which needs a higher maturity level for the onset of generation than the Type II kerogens. Also, deltas are generally areas of steady accumulation with minor periods of hiatus or uplift and erosion.

Although mature source rocks may be deep, much of the production from deltas is relatively shallow. This implies extensive vertical migration. Growth faults (or other faults) are usually invoked as the routes for migration (Demaison and Huizinga, 1991), but they can be difficult to model because they appear to be conduits at some times, and barriers to fluid flow at others (Weber, 1987). A single fault can show both characteristics at different locations, and this may be related to the angle of the fault, the local sand/shale ratio, local pressure regimes, and the timing of fault movement. If rollover anticlines formed adjacent to growth faults are full to the spill point with oil and gas they may spill off the bottom leading to differential entrapment with the oil moving updip, which is generally towards the shore line (Barker, 1979). Natural distillation may also occur through the fault system leading to shallow condensates and very light oils (Ross and Ames, 1988).

Many of the waxy crudes common in the Tertiary delta systems have high pour points because of abundant long chain alkanes. These compounds are the most susceptible to bacterial degradation and their removal produces a dramatic decrease in pour point, often by more than 50 °C. In the Niger delta many of the crudes at depths corresponding to temperatures less than about 75 °C are biodegraded and have low pour points, but there is an abrupt transition with depth to high pour point crudes (Evamy et al, 1978). In such situations it is important to recognize that there is a single oil family with different degrees of biodegradation (and not two oil families). A single oil family shows, of course, that there is only one source rock. If biodegradation, or other degradation processes such as water washing, were not correctly identified, then the presence of two apparently different oil types may be incorrectly interpreted as implying the presence of at least two source rocks. These issues also become important in using oil composition to establish organic matter type in a source rock.

10.5.5. Lopatin Modeling

The simple Lopatin model has been applied to several delta systems including the Nile. The Nile delta is predominantly gas-prone with minor oil (Sestini, 1989), and the Abu Madi 2 and El Qara 2 locations have been modeled by Abu El-Ella (1990). He assumed an unchanging geothermal gradient through time, and used values that ranged from 36.6 to 26.8 °C/km for individual formations but recognized that "no single geothermal gradient can describe the entire section". Deep burial has occurred in the last 8 m.y. and over this short time period unchanging local surface temperatures that ranged from 15.5 °C offshore to 26.7 °C onshore were justified. Present day subsurface temperatures came from log headings, and empirical corrections of +15 °C at 50 °C; +30 °C at 100 °C; and +35 °C at 150 °C were applied. At El Qara 2 the cumulative TTI is only 4.64 at a TD of 3460 m, but Abu Madi 2 has just reached the oil window at 4154 m with a calculated cumulative TTI of 15.17 (Table 10.5). This is consistent with the measured R_o values of 0.50 - 0.58%.

The Niger delta is a major petroleum province and its petroleum geology has been summarized by Weber and Daukoru (1975), Evamy et al (1978), Ekweozor and Okoye (1980), and Ejedawe (1981). The deep marine shales of the Akata Formation are overlain by the paralic Agbada Formation and the continental Benin Formation, with total sediment thickness varying from 9000 - 12,000 m. Most oil and gas is reservoired in the alternating sand and shale sequence of the Agbada, but the deeper Akata appears to be the main source rock. Current geothermal gradients are lowest in the active depocenter (Figure 10.31). Growth faults provide the avenues for migration. Ejadawe et al (1984) and Ejadawe and Coker (1984) have used the Lopatin method to evaluate the depth and timing of petroleum generation. They assumed a simple progradational model for delta development, with each depocenter accumulating sediment at an average rate of 500 m/m.y. until overrun as the next depocenter developed seaward. Burial rates were based on values obtained from 6 fields in the delta. The burial histories were combined with average geothermal

Table 10.5
Maturation history data for six formations occurring in two Nile delta wells (Abu El-Ella, 1990). GG – geothermal gradient.

WELL AND FORMATION	MAXIMUM DEPTH (M)	TTI	LOM	R_o (%)	GG (°C/100m)	T_{MAX}
EL QARA 2						
Mit Ghamr	1069	n/a	4.9	0.20	4.10	72
El Wastani	1331	n/a	5.4	0.22	3.75	77
Kafr El Sheikh	3095	0.85	7.8	0.37	3.16	125
Abu Madi	3413	1.64	8.1	0.56	3.25	134
Qawasim						
Sidi Salem	3460	4.64	8.2	0.58	3.10	136
ABU MADI 2						
Mit Ghamr	958	n/a	4.6	n/a	3.05	56
El Wastani	1217	n/a	5.0	0.20	3.15	65
Kafr El Sheikh	3171	0.74	7.8	0.34	3.12	126
Abu Madi	3306	1.87	8.0	0.46	3.15	131
Qawasim	3602	4.08	9.1	0.55	3.08	138
Sidi Salem	4154	15.17	9.7	0.66	3.02	153

gradients ranging from 22 to 51 °C/km and a surface temperature of 27 °C to produce a series of nomographs. The oil window was bracketed by cumulative TTI values of 15 and 160, and onset of generation corresponds to a temperature of 140 - 146 °C, a time of 7 - 11 m.y., and a depth of 3000 - 5200 m. Through time, the onset of generation moves to lower temperatures and shallower depths, and the oil window rises up through the sedimentary sequence (Figure 10.32).

Tang and Lerche (1993) have modeled the Mackenzie delta and reconstructed burial histories for 7 wells in and near the delta. They used breaks in R_o profiles to obtain erosional losses, which in some cases were as much as 2000 m. Temperature history was obtained by inversion of vitrinite reflectance data and combined with burial history to establish thermal maturity levels of potential source rocks in the Jurassic and Tertiary section. Waples (1980) calibration data with a cumulative TTI of 15 were used to indicate onset of oil generation. The kinetic models were also used to simulate the rate of hydrocarbon generation.

The Lopatin method should be used with caution in delta systems. The commonly-used TTI calibration data are appropriate only for Type II

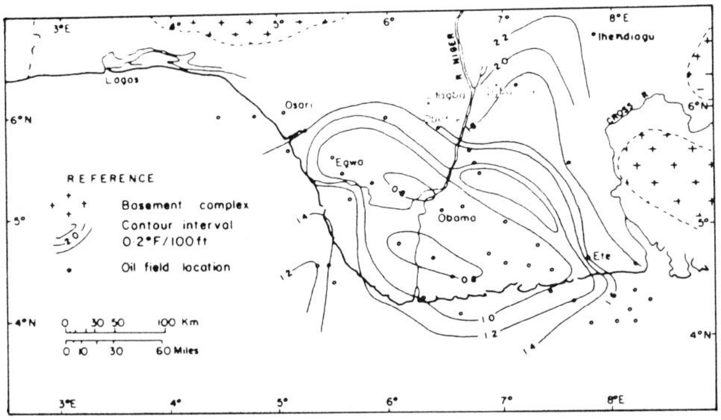

Figure 10.31. Corrected geothermal gradient map for the Niger delta (Nwachukwu, 1976)

kerogens buried in areas with average sedimentation rates and average geothermal gradients. In contrast, deltas are gas-prone (Type III kerogen), with common waxy crudes that are generated from Type I kerogen. Also, during active sediment accumulation deltas are characterized by rapid burial and low geothermal gradients.

10.5.6. 1-D Kinetic Modeling

Modeling based on the kinetic equations is better able to deal with the range of organic materials incorporated into deltas, and is a more reliable method for coping with major changes in sedimentation rate, as well as the deep formation of gas. It does, of course, require additional information, and E and A values must be specified. In their study of the Beaufort-Mackenzie basin, Issler and Snowdon (1990) used dry pyrolysis to derive activation energies and frequency factors for Type III kerogens in about 100 immature samples. Nominal heating rates of 5, 25, and 50 ^{o}C/min were used, and the curves fitted with suites of E and A values, with E values being constrained to 10 kJ/mole increments. There was a rather small range of values for all the Cretaceous to Oligocene sediments with the samples showing a major contribution at 230 kJ/mole (55.0 kcal/mole).

Oil was discovered in the Beaufort-Mackenzie basin in 1970, although the on-shore area is dominantly gas with major reserves at Taglu, Parsons Lake, and Niglintgak (Lane and Jackson, 1980; Dixon et al, 1985). The basin developed as a consequence of rifting when the Mackenzie River built a substantial delta into the basin. Temperature and burial histories for thermal modeling were derived from this rifting/delta progradation model, where initially high geothermal gradients decayed to a minimum

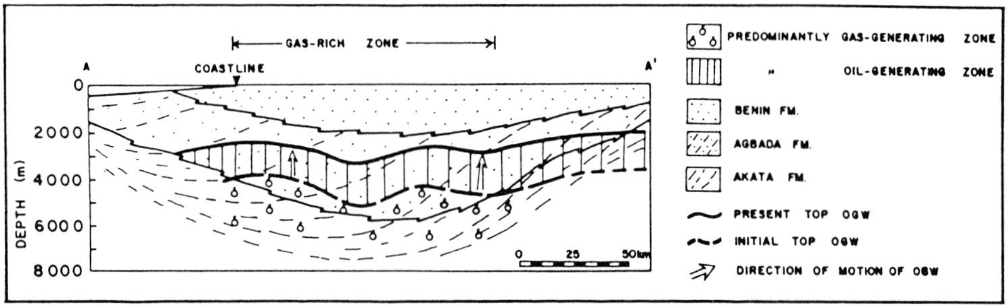

Figure 10.32. Cross section of the Niger delta showing how the top of the oil window moves into shallower sediments through time (Ejedawe et al, 1984).

before subsequent warming. The thermal regime was complicated by (1) the development of Pleistocene permafrost, (2) thermal blanketing due to rapid sediment accumulation that caused non-equilibrium transient stages of heating, and (3) lateral heat refraction from shale diapirs that produced temperature differences of as much as 40 °C between the modeled wells. In spite of extrapolating laboratory pyrolysis data 10 orders of magnitude to geological conditions, Issler and Snowdon claimed good success with their thermal modeling. It explained the lack of mature source rocks, even when temperatures approached 140 °C, because geologically rapid heating rates associated with enhanced Plio-Pleistocene sedimentation have shifted the oil window to greater depths. They projected that the depth of the oil window will be increased in areas of thick Plio-Pleistocene sediments. This situation is very similar to that for other deltas, including the Niger discussed above, where rapid burial, short time periods, and low geothermal gradients combine to produce deep onset of petroleum generation.

In a later study (Dixon et al, 1992) the effects of using MATOIL (BEICIP) default E and A values for Type III kerogen were compared with results obtained using the laboratory pyrolysis data. Figure 10.33 shows that the calculated time for the beginning of the oil window is considerably earlier for the measured values than for those derived by using standard default parameters.

10.5.7. 2-D Kinetic Modeling

In deltas, 1-D models cannot adequately describe the development of abnormal pressures, petroleum migration and entrapment, and heat refraction in the vicinity of shale-cored anticlines. 2-D models are

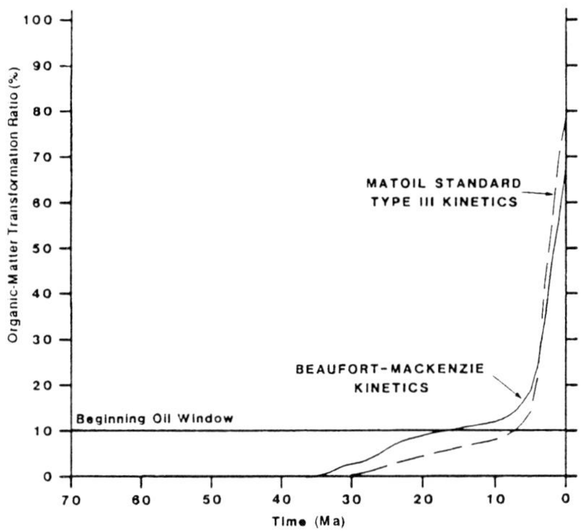

Figure 10.33. Calculated hydrocarbon generation as a function of time for both Beaufort-Mackenzie kinetic parameters and standard default parameters used by MATOIL (BEICIP) for Type III organic matter. Early generation for Beaufort-Mackenzie parameters results from a higher than normal initial potential for low activation energy reactions. (From Dixon et al, 1992).

necessary to describe the movement of water out of compacting shales and the migration of petroleum from source rock to reservoir. The THEMIS 2-D model developed by the French Petroleum Institute (IFP) (Doligez et al,1987) has been extended (and called TEMISPACK) and used in the Mahakam delta by Ungerer et al (1990). The Mahakam Delta is an active delta system today, and appears to have begun in the middle Eocene with rapid sediment accumulation and the development of thick shales until the Oligocene. After a long transgressive phase, alternating transgressions and regressions continued up to the present. The basin is typical of a young delta system although it does lack major growth faults. The basement has never been drilled or imaged by seismic. The geology along a 47 km WNW-ESE cross section that crosses the Handil and Bekapai fields has been shown in Figure 9.6. The first step in modeling the area is to represent the geology by a two dimensional grid that is a compromise between the complex geology and the constraints of the THEMIS model. This is a major challenge, but eventually 1196 grid elements were selected in 23 columns along 52 rows as shown in Figure 9.6. Combinations of the basic sandstone, shale, siltstone and carbonate lithologies were used to represent the fluvial-deltaic, delta plain, delta front and prodelta facies.

Sandstone-shale intercalations of less than 30 m could not be incorporated into the model.

Paleotemperature reconstructions depended on heat flows, but while a value of 80 mW/m^2 at 25 km depth provided a consistent match between calculated and observed data for the Handil field, it did not fit Bekapai which needed a value of 60 mW/m^2. Ungerer et al (1990) concluded that a purely conductive thermal regime could not account for the data in the whole section, and there must be either convection or lateral changes in basal heat flow.

Abnormal pressures are common in the Mahakam delta and the model reproduced present values quite well. Transition zones from normal to abnormal pressures were successfully calculated, and the observed changes in depth of the transition zone on the flanks of the Handil structure were reproduced. Migration of fluids was modeled using the two-phase migration module in the THEMIS program together with viscosities appropriate for gas. This predicted oil accumulations at depths similar to those actually observed at both Handil and Bekapai (Figure 10.34), and even the stratigraphic accumulations on the flanks of Handil were qualitatively reproduced. The efficiency of migration is strongly dependent on the distribution of sandstones as well as the volume of generated hydrocarbons and the properties of the hydrocarbons. Most petroleum was generated in the syncline, and accumulation in the Handil structure occurred at about 7 Ma, and at about 3 Ma near Bekapai. The modeling suggested that accumulations southeast of Handil were in both stratigraphic and structural traps, and that the spill point had been reached for some of the reservoirs implying that there may have been contributions to the fields west of Handil. Modeling was less successful when viscosities and densities for oil were used (instead of gas), and the model predicted no accumulation at Bekapai and the fit at Handil was poor. This probably reflects the difficulty of accounting for the role of C_{1-4} hydrocarbons in promoting the migration of the C_{5+} oil-like hydrocarbons.

Other 2-D modeling studies in deltaic environments include those by Maubeuge and Lerche (1993, 1994) (with a strong emphasis on the development of abnormal pressures).

10.5.8. Complicating Factors

In developing models to evaluate delta systems several potential complicating factors need to be considered. These include:

(1) *Temperatures near faults* : Growth faults, or other faults, in deltas can act as conduits for hotter fluids moving up from depth. Many of the petroleum accumulations in deltas are in rollover anticlines close to

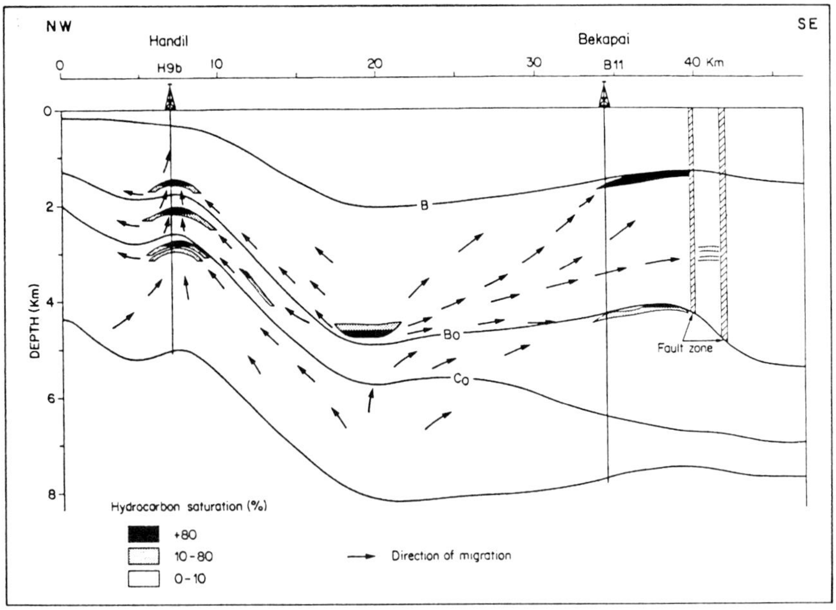

Figure 10.34. Calculated gas saturations and migration directions on a present cross section of the Mahakam delta (Ungerer et al, 1990).

growth faults, so that the temperatures measured in exploration wells drilled to test these structures may be too high (Cooper, 1977; Bodnar and Sharp, 1988) (see Figure 4.18). Other factors that complicate the determination of paleotemperatures have been discussed in Chapter 5.

(2) *Transient overpressures* : The deepest part of every major delta system in the world appears to be abnormally pressured. Since rapid loading and disequilibrium compaction play a major role in generating these high pressures the sediments are often undercompacted. This reduces thermal conductivity and increases geothermal gradient. If the high pressures subsequently bleed off and return to normal there may be little present day evidence of the past high temperature gradients.

(3) *Coals nearshore* : Coals commonly form from peats that accumulate in the interdistributary bays of delta systems (Haszeldine, 1985). The coals can generate considerable quantities of gas, and in some cases may also generate oils that are typically very paraffinic with high pour points and low sulfur contents (Scott and Fleet, 1994). The gas generating part of coal should be modeled using values for Type III

kerogens, while the oil prone component is more appropriately modeled using Type I kerogen kinetics.

(4) *Biogenic Gas* : Isotopically light, dry gas produced by methanogenic bacteria is common in the shallower parts of deltas and can accumulate in commercial quantities (Mattavelli et al, 1983; Rice, 1993). The gas is not formed from organic matter in response to rising temperature and passing time, and so cannot be modeled by the kinetic equations commonly used.

10.6. OVERTHRUST BELTS AND FORELAND BASINS

10.6.1. Introduction

Large scale compressive forces can produce overthrust terrains where sheets of the order of 4 to 6 km thick are thrust 100 km or more laterally. Rates of movement range from 5 - 6 cm/yr, which is comparable to the rates observed in plate tectonics. The overriding sheet is uplifted and cooled, while the underlying sheet is buried more deeply and becomes hotter, just how much hotter depending on the rate of uplift versus the rate of erosion. Because of the uplift and cooling, maturation slows or stops in the overthrusted sheet and it retains a record of prethrusting events. In contrast the buried slab contains the entire record of thermal evolution. The major problems in modeling overthrust belts are (1) reconstructing the thermal regime through time, and (2) constraining the balance between the rate of uplift and rate of erosion in the overthrust sheet.

Most attempts to reconstruct thermal regimes have followed the approach used by Oxburgh and Turcotte (1974) and Angevine and Turcotte (1983) where the instantaneous emplacement of the overthrust sheet produces a sawtooth temperature profile that decays through time to restore a smooth trend (Figure 10.35). The overthrust sheet is cooled, especially in the lower part, by conduction of heat into the underlying slab. For the example illustrated in Figure 10.35, temperature in the overthrust slab decreased immediately after thrusting, and at a depth of 3.75 km was only 70 °C after 0.1 m.y., but had risen to 75 °C by 1 m.y., to 104 °C after 10.0 m.y., and eventually reequilibrated to its prethrusting value of 115 °C after a long elapsed time.

The evolving thermal regime is strongly influenced by the balance between the rate of uplift and the rate of erosion (Warner and Royse, 1987). In extreme cases, the rocks elevated by overthrusting have been eroded completely and no sedimentary record remains. While this limits the techniques available for estimating erosional loss it may still be possible to

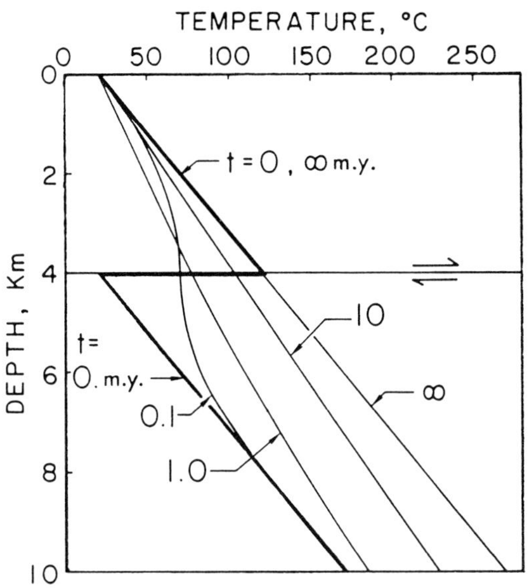

Figure 10.35. Temperature profile in a thrust plate and the underlying sediments. The heavy line is the temperatures after instantaneous emplacement, and the profiles show temperatures at 0.1,1.0, and 10 m.y., and at infinite time. (Based on a figure from Angevine and Turcotte, 1983).

infer thermal regime using AFTA, because these techniques can be applied to crystalline rocks. Omar et al (1994) provide an example from the shear zone of the Beartooth overthrust in Montana and Wyoming. Here temperatures deduced from AFTA were higher during the Cenozoic than those in the upper and lower sheets, this was attributed to the flow of higher temperature groundwater along the thrust. Fluid inclusions have also been used to constrain paleo-temperatures in overthrusts (Yonkee et al, 1989; Montanez, 1994).

The large contact area between the moving sheets suggests that frictional heating may be important, but the evidence does not seem to support this and any heating effects are very localized (Bustin, 1983). It is possible that friction effects are severely reduced by the development of abnormal pressures caused by the rapid loading. This has been invoked to explain movement of thrust sheets, especially when they are very long or relatively thin. While overpressures do not seem to be essential for overthrusting they will obviously facilitate the process (Wygrala et al, 1990).

Large scale compression and overthrusting that leads to mountain building may influence fluid flow in at least two ways. The increased hydraulic head can drive water flow into the associated foreland basin (Garven and Freeze, 1984), while the additional load can enhance compaction and expel fluids (Oliver, 1986) (Figure 10.36). Nunn and Deming (1991), Deming and Nunn (1991), and Deming et al (1990) have used 2-D numerical models to evaluate the thermal impact of these two processes. They concluded that sediment compaction induced by overthrusting can only produce Darcy velocities on the order of cm/year, which is too low to cause major changes in thermal regime. Two-dimensional finite-element modeling by Ge and Garven (1989) also lead to flow velocities in the range of cm/year or less, and they showed that it produced only short term thermal anomalies. Continuing foreland basin thrusting may develop episodic fluid flow. Deming and Nunn (1991) found that topographically-driven recharge can generate flow velocities on the

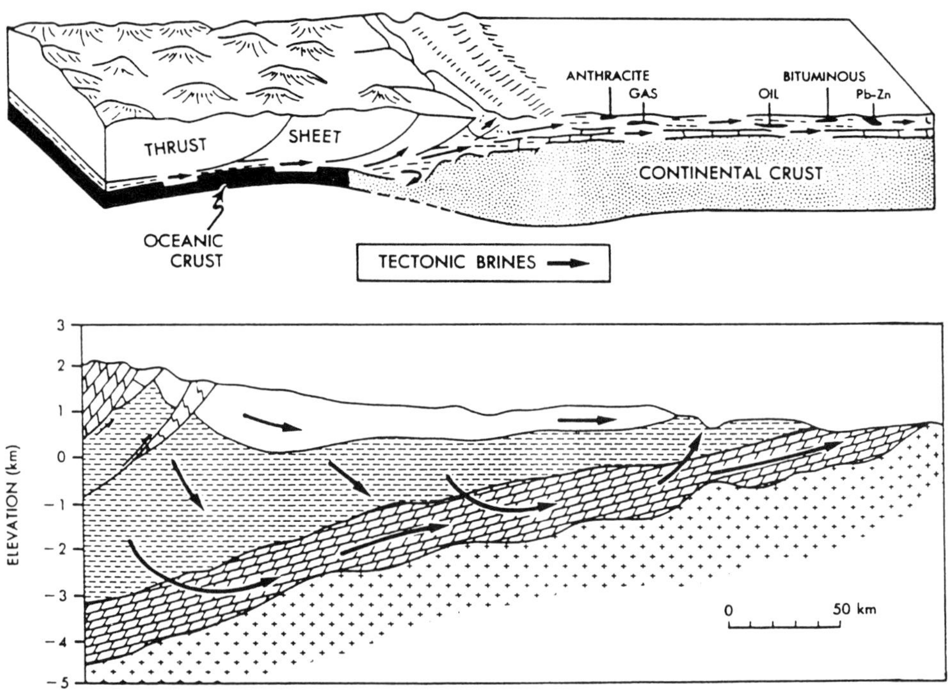

Figure 10.36. Schematic illustration of hypotheses proposed to explain fluid migration in overthrusts. *Top*: the "squeegee" model of Oliver (1986), and *Bottom*: topographically driven recharge (Garven and Freeze, 1984).

order of m/year, but that this is still probably too low to cause major temperature changes unless water flow is focused into confined channels. The Middle Devonian Presqu'ile seems to provide just such a focusing barrier that has channeled fluids in the western Canadian basin, and Qing and Mountjoy (1992, 1994) have found fluid inclusion evidence for higher temperatures. In many cases water flow tends to follow near-horizontal permeable beds where the isotherms often parallel the bedding, but faulting may provide an avenue for the upward movement of heated waters. Figure 10.37 presents one set of calculated isotherms for topographically-driven recharge in an overthrust/foreland basin system and clearly shows the cooling expected under the overthrust. Many of the geological and geochemical studies of thrust belt/foreland basin systems have suggested an important role for water flow in modifying temperature regimes. Burtner and Nigrini (1994) used AFTA, R_o, and T_{max} pyrolysis data to reconstruct paleotemperatures regimes for the U.S. western overthrust along the Idaho-Wyoming border, and concluded that gravity-driven flow raised temperatures and led to earlier maturity of the Phosphoria Formation source rocks. Flow patterns could be changed rapidly by later thrusting, and areas of above normal surface heat flow could become areas where it was below normal. For one location they presented evidence for a decrease in geothermal gradient from 60 to 20 ^{o}C/km in 10 m.y. Deming et al (1992) report geothermal gradients of 22 ^{o}C/km in the foothills of the overthrusted Brooks Range of Alaska increasing to a high of 53 ^{o}C/km out on the coastal plain to the north. This temperature distribution was consistent with a topographically-driven ground water flow system. Willet and Chapman (1987) found cooling on recharge in the Uinta Basin, and downflow of water in the Llanos sub-Andean foreland basin in Colombia has also led to cooling, but with very variable geothermal gradients (Bachu et al, 1995). Weir and Furlong (1987) have also discussed the effect of penetrating waters on thermal regimes.

10.6.2. Lopatin Modeling

The so-called Western Overthrust Belt of the U.S. has been an important exploration play that runs through Wyoming, Utah, and Idaho. The locations of the major thrusts and some of the large petroleum accumulations are shown in Figure 10.38. Edman and Surdam (1984) used Lopatin's method to evaluate the timing of maturation for the Phosphoria Formation at four well locations that penetrated the Meade, Darby, and Absaroka thrust sheets. Their burial reconstructions and temperature changes through time assume a geothermal gradient that was considerably

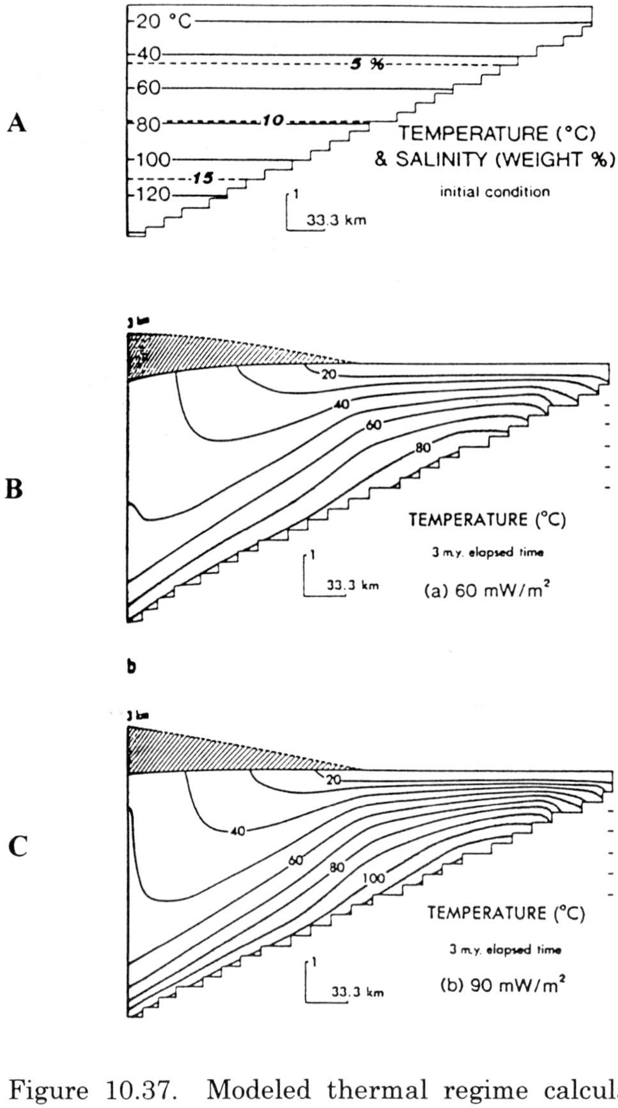

Figure 10.37. Modeled thermal regime calculated as a consequence of loading when the initial temperature (and salinity) conditions were those shown in A. Temperature is shown after 3 m.y. for surface heat flows of 60.5 mW/m^2 (B) and 90.5 mW/m^2 (C). (From Deming and Nunn, 1991).

depressed at the time of thrusting, but recovered to present day values that are close to the pre-thrusting values. Calculated TTIs showed that in the Absaroka thrust sheet location, the Phosphoria source rock had passed through the oil window before major thrusting occurred. Warner and Royse also modeled wells that penetrated the Absaroka thrust sheet and Figures 10.39 and 10.40 show their reconstructions of the burial histories at Ryckman Creek field and Whitney Canyon field. Well locations are shown in Figure 10.41. The calculated and observed vitrinite reflectance values agreed quite well (Table 10.6).

Mussman et al (1988) modeled the overthrusted area of southwest Virginia where Alleghenian thrusting during arc-continent collision led to horizontal displacements of tens of kilometers. Their simple model applied Lopatin's method to a non-decompacted burial history and used TTI values of 15 and 160 to define the oil window. Geothermal gradient was assumed to have remained constant at the present value of 30 °C/km. They were interested in the timing of the oil window because it allowed them to infer the timing of release of water from the conversion of smectite to illite.

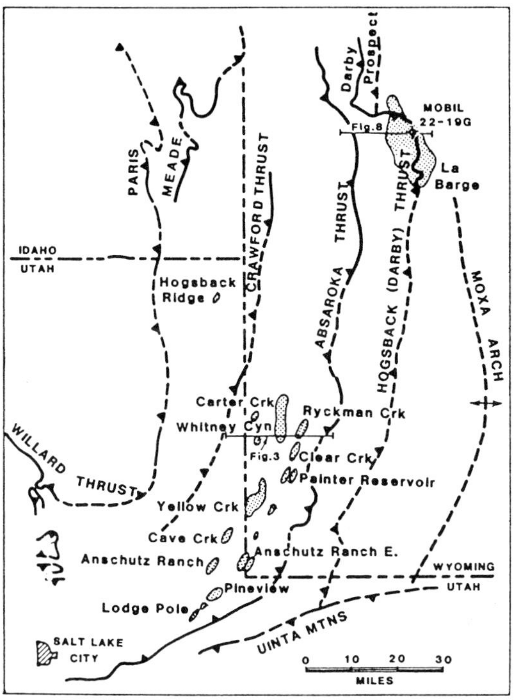

Figure 10.38. Part of the U.S. "western overthrust" in southwest Wyoming and northern Utah showing surface traces of major thrust faults and the positions of some major oil and gas fields (Warner and Royse, 1987).

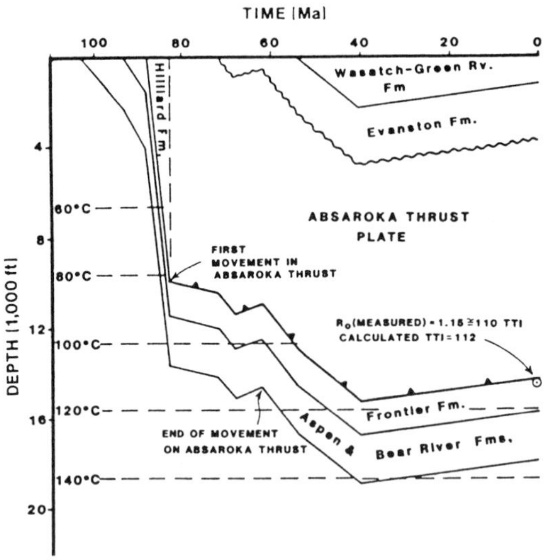

Figure 10.39. Lopatin-type diagram for the subthrust Cretaceous source rock sequence at Ryckman Creek field. The positions of the isotherms were calculated from a 22 °C/km geothermal gradient with a surface temperature of 15 °C (Warner and Royse, 1987).

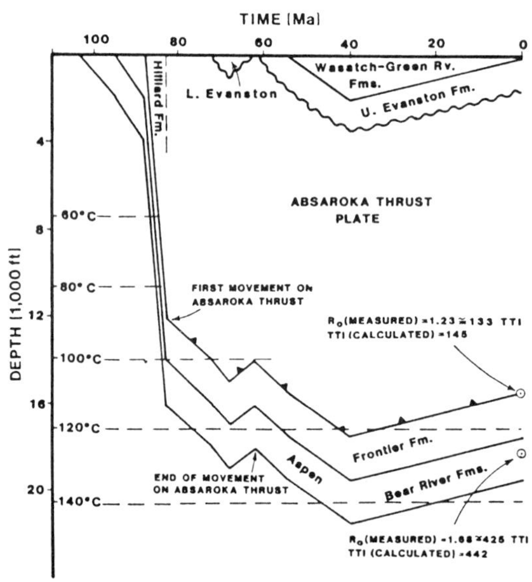

Figure 10.40. Lopatin-type diagram for the subthrust Cretaceous source rock sequence at Whitney Canyon field. The positions of the isotherms were calculated from a 22 °C/km geothermal gradient with a surface temperature of 15 °C (Warner and Royse, 1987).

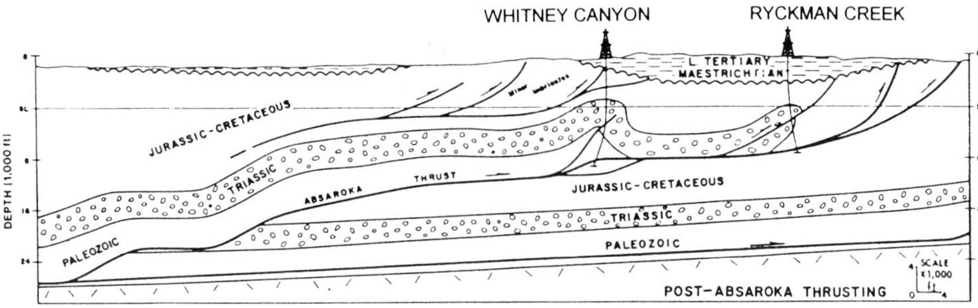

Figure 10.41. Cross section showing the Absaroka Thrust and the relative positions of the Ryckman Creek and Whitney Canyon fields (Warner and Royse, 1987).

Rocks up through Middle Ordovician (approximately 520 Ma) entered the oil window before Carboniferous overthrusting buried them an additional 2.5 km in about 20 m.y. The most likely source rocks are the organic-rich shales of the Cambrian Conasauga Group. Fluid inclusions used to establish diagenetic fluid temperatures in the area (Montanez, 1994) show a maximum temperature in the range 115 - 165 °C (not pressure corrected), with a subsequent decrease through time. Supporting temperature data were also obtained using oxygen isotope values for the dolomite cements, but is was recognized that this assumes that there were no significant changes in the oxygen isotope values of the fluids responsible for dolomitization. The increase in temperatures through time followed by a decrease was interpreted as due to emplacement of thrust sheets, followed by uplift and erosion (Montanez, 1994).

Part of the northern edge of Tethys that formed the old European continental margin, was overthrusted in the late Miocene and produced the western French Alps. Butler (1991) modeled a single line of cross section through the Annecy district of this classic example of a foreland fold and thrust belt. He calculated thermal maturity using the Lopatin method with an oil window defined by cumulative TTI values of 15 and 150. Burial histories were constructed in detail and these fairly sophisticated models included decompaction. In contrast, thermal models were simple and used an unchanging prethrusting geothermal gradient of 28 °C/km and a post-thrust gradient of 30 °C/km, which has continued to the present. Maturity calculations showed that source rocks in the Borres and Semnoz thrust sheets had generated petroleum before thrusting, but that the lower Subalpine sheet was in the oil window at the time of thrusting. This implies that trapping configurations developed as a consequence of

Table 10.6
Comparison of measured and calculated thermal maturity for vitrinite samples taken beneath the Absaroka thrust (Warner and Royse, 1987).

LOCATION	RYCKMAN CREEK	WHITNEY CANYON	WHITNEY CANYON
DEPTH, ft	14,000	15,300	18,200
m	4,300	4,660	5,550
SAMPLE DATA			
measured R_o	1.15	1.23	1.68
TTI equivalent	110	133	430
LOPATIN DIAGRAM (constant temperature gradient)			
TTI, calculated	112	145	412
R_o, equivalent	1.16	1.27	1.69
LOPATIN DIAGRAM (transient temperature gradient)			
TTI, calculated	610	1980	5245
R_o, equivalent	1.85	2.33	2.88

thrusting will not be prospective for Berriasian-sourced oil. Recognizing the limits of the Lopatin method, Butler (1991) also considered peak temperature exposure as suggested by Quigley and Mackenzie (1988), but this did not change the conclusions significantly. Figure 10.42 shows a burial reconstruction (post 60 Ma) for the Borres massif. Temperature values are superimposed on the Berriasian shale, which is the potential source rock. Lopatin modeling shows that it had entered and left the oil window before thrusting occurred at 11 Ma.

Lopatin's method was used by Mani et al (1988) to evaluate maturity of potential source rocks in the Assam basin, so that this information could be integrated into the existing framework of geology and geophysics. The basin is in a foreland setting associated with the foldbelts of northeast India and appears to have cooled off through time with current values being the lowest. Present day temperature gradients were constrained using corrected bottom hole temperatures and combined with geothermal gradients at two times in the past to represent the changing thermal histories of selected areas. Subsidence curves for different locations in the basin were based on both well control and seismic data, and absolute ages

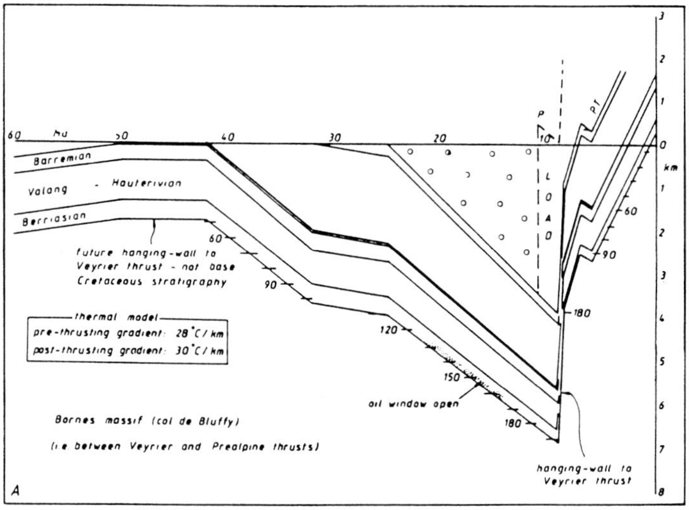

Figure 10.42. Time-temperature plot for the Bornes massif in the west Alpine thrust belt. Oil window is defined for the preserved base of the Berriasian only. P : prealpine emplacement; PT : prealpine thrust. (From Butler 1991).

for rock units were obtained by relating them to regional unconformities which were then related to the dated global cycles of sea level change. Maps prepared using the TTI data showed the areal distribution of maturity in various stratigraphic sequences, including the organic-rich Kopili Formation. This has good to very good source potential and appears to have generated about 3.5 - 5.0 Ma in the Safrai depression, and is at peak oil generation in the subthrust area below the Schuppen belt.

Other overthrust maturation studies using Lopatin's method include those in western Canada by England and Bustin (1986a,b) and in the Junggar basin by Graham et al (1990).

10.6.3. 1-D Kinetic Modeling

Overthrusting involves significant mass transport with burial rates of the order of 50,000 m/m.y. -- which is more than 100 times faster than normal sediment accumulation rates. In addition there are rapid temperature changes, especially at the bottom of the overriding slab. The Lopatin method is not well suited for dealing with these rapid burial and heating rates, and kinetic models for calculating organic maturities are more appropriate. One of the first studies to use first order kinetics to evaluate organic matter maturity, and the generation and cracking of oil, was by Angevine and Turcotte (1983) in the Absaroka thrust belt of western Wyoming. Maturity calculations were made assuming Type III kerogen and using E and A values from Tissot and Espitalie (1975). They found that burial of the footwall sequence by the overriding sheet raised temperatures and promoted oil generation and its subsequent cracking. The details depend on thrust plate thickness. For a plate thickness of 8 km, oil was destroyed only 5 m.y. after emplacement, while it survived 10 m.y. under a 6 km thick plate, and had a long residence time for a 4 km thick plate.

The Lower Miocene Carapita Formation of the fold thrust belts of the Maturin Subbasin, eastern Venezuela, was modeled by Tocco et al (1994) using the MATOIL (BEICIP) software. They showed that the organic-rich Carapita Formation entered the oil window 6.4 Ma when the depth was approximately 15,900 ft (4850 m). In making the calculations, the top of the oil window was defined by a (Hydrocarbon)/(Organic Matter) ratio of 10%, and geothermal gradient was set at 1.3 °F/100 ft. Calculated onset of maturation occurred after thrusting, and was associated with tectonic loading that predated the early Miocene. The modeled sequence of events was consistent with RockEval data, as was the correlation of the continental organic matter in the Carapita Formation with a produced oil that had strong terrestrial affinities.

Cusiana is the largest field in Columbia. It formed in the foreland/foothills setting of the Llanos basin, with the frontal fold belt and thrust belt developing in the last 5 m.y. (Cooper et al, 1995). The source rock is thought to be the Gacheta Formation (which is time equivalent to the LaLuna in western Venezuela). The upper Eocene Mirador Formation that forms most of the reservoir was rapidly buried as 12,000 ft (3660 m) of the Pliocene Guayabo Formation accumulated. These rocks never achieved thermal equilibrium during their rapid burial. Erosion of the Guayabo Formation started immediately after thrusting and is still continuing, and its pre-erosion thickness was estimated from the maximum thickness currently observed in the present day foredeep of the Llanos foreland. Cazier et al (1995) calculated source rock maturity with the model described by Pepper and Corvi (1995) using a typical heat flow of 35-45 mW/m^2 for the foreland stage. Temperatures in the Mirador were constrained by corrected fluid inclusion homogenization temperatures in quartz overgrowths. These ranged from 100 to 150 °C. The maturity modeling showed that oil was generated in the Miocene and that the "oil charge to the Cusiana structure began no earlier than 5 Ma, but probably occurred during the last 2 m.y." (Cazier et al, 1995).

The Junggar (Zhungaer) basin of northwest China is probably a foreland basin downwarped by continental collision, but there have been alternative suggestions involving rifting or tectonic extension during or after the Late Permian. Although the Paleozoic history is not well known it is assumed that Permian black calcareous shales, such as the Fencheng and Uerho formations along the northern border (King et al, 1994), are the effective source rocks. Preliminary modeling for the location of the Aican 1 well on the northwest margin of the basin by Graham et al (1990) used a decompacted burial history with Lopatin's method and suggested that the Jurassic and younger rocks had never become mature (as defined by a cumulative TTI of 15). The base of the Upper Permian began to generate at roughly 150 Ma. This first attempt at maturity modeling used a constant geothermal gradient of 22 °C/km because this gave a reasonable fit to the R_o data and bottom hole temperatures. Present day gradients are in the general range of 20 - 23 °C/km. Graham et al (1990) felt that more sophisticated modeling was not justified given the poorly constrained nature of the area.

King et al (1994) were able to improve estimates of paleoheat flow and refine the thermal modeling based on analyses of additional sample materials from the Aican 1 well, and samples from the Caican 4 well in the eastern and shallower part of the basin. A variety of techniques was used to develop a self-consistent history of heat flow, including vitrinite reflectance, biomarkers, and illite/smectite analyses. Temperatures through time were estimated from heat flow and thermal conductivities calculated from rock matrix default values combined with the conductivity of water-filled porosity. These were then used to calculate R_o using EASY%R_o (Sweeney and Burnham, 1990), and to calculate sterane and hopane isomerization ratios using the kinetic parameters suggested by Mackenzie and McKenzie (1983) (but recognizing the complications pointed out by Abbott et al, 1990). From a comparison of calculated and observed values, King et al (1994) developed a consistent picture of changing thermal gradients which showed heat flow decreasing through time from 54 to 22 mW/m^2 at Aican 1, and from 82.5 to 27 mW/m^2 at Caican 4. These lead to a steadily decreasing geothermal gradient through time, eventually reaching the low values observed today. The temperature history was used with the BASINMOD program (Platte River Associates) to estimate source rock maturity through time. Unfortunately, the E and A values needed for kinetic modeling could not be determined for Upper Permian samples retrieved from the Aican 1 well because of their high maturity levels, but immature samples were available from the Caican 4 well and were used with RockEval pyrolysis to give the kinetic parameters. Although the main emphasis of the paper was on establishing paleoheat flow, the modeling does show that the Upper Permian is mature at the Aican 1 location on the western flank of the Junggar basin (Figure 10.43). This is in the general area of the giant Karamay oilfield.

On the southern margin of the Junggar basin Carroll et al (1992) modeled a generalized burial history using the MATOIL (BEICIP) program. A major problem in reconstructing burial of the Upper Permian potential source rocks in this area is the erosional loss of overburden, since the Triassic is now exposed at the surface. R_o values were extrapolated to a surface value of 0.27% (the approximate value of immature peats) and showed that 5.25 km had been removed. By combining this information with a constant geothermal gradient of 22 °C/km and kinetic parameters appropriate for a Type I lacustrine kerogen, maturity could be modeled using MATOIL. This showed that the formation of potential traps by thrusting in the Neogene occurred after the main phase of hydrocarbon generation. This is consistent with the lack of significant oil discovery in this part of the Junggar basin.

Thermal modeling studies that used Lopatin's method to estimate organic maturities in the Western Overthrust Belt of the U.S. have been discussed above. These have been extended to include 1-D kinetic modeling by Furlong and Edman (1989). Source rock maturity was

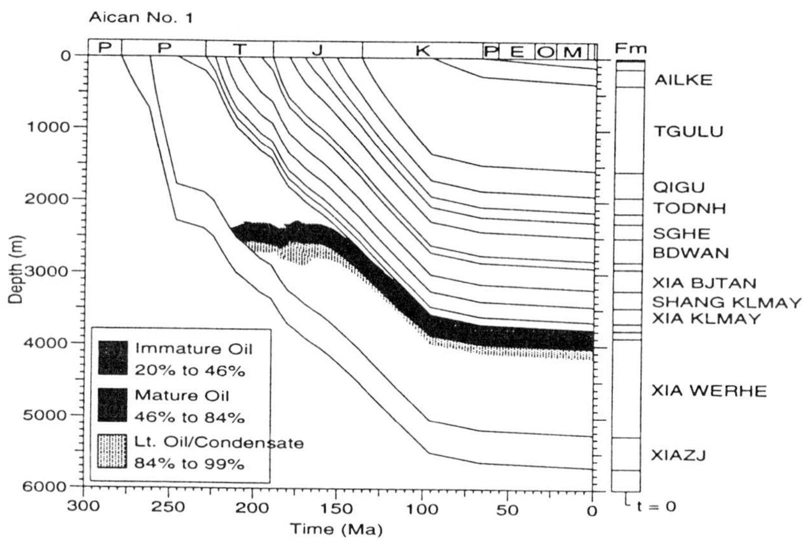

Figure 10.43. Position of the oil window through time calculated for the Upper Uerho formation in the Aican No. 1 well using the kinetic parameters obtained for a sample from a depth of 2146 m in the Caican No. 4 well. The geothermal gradient has changed through time and decreased towards the present. (King et al, 1994).

modeled at a Whitney Canyon site and for the Tip Top (Mobil 22-19G) site (see Figure 10.38 for location) using kinetic parameters from Ungerer and Pelet (1987) for Type II kerogen, and an assumed constant basal heat flow of 50 - 70 mW/m^2. Cumulative TTI values were also calculated and compared with the empirical calibration data of Waples (1989) to provide R_o values. These are shown in Figure 10.44 together with the measured values at Whitney Canyon. A clear "sawtooth" distribution of values is apparent. Furlong and Edman (1989) recognized that 1-D models cannot incorporate horizontal heat flow or fluid flow, but they considered the available well data to be inadequate to constrain a 2-D model.

Deming and Chapman (1989) have provided a useful description of the way data are assembled in modeling an overthrust area and have evaluated the sensitivity of calculated values to uncertainties in input parameters (see Table 8.2).

10.6.4. 2-D Kinetic Modeling

Wygrala et al (1990) report one of the few attempts to use a 2-D model in describing overthrusting. This is based on IES simulation software, such as that described by Yukler and Kokesh (1984), Yalcin et al (1988), and Yalcin and Welte (1988). It treats a 5 km thick sediment pile as ten 500 m thick layers, and then moves part of the slab up a 45° slope in

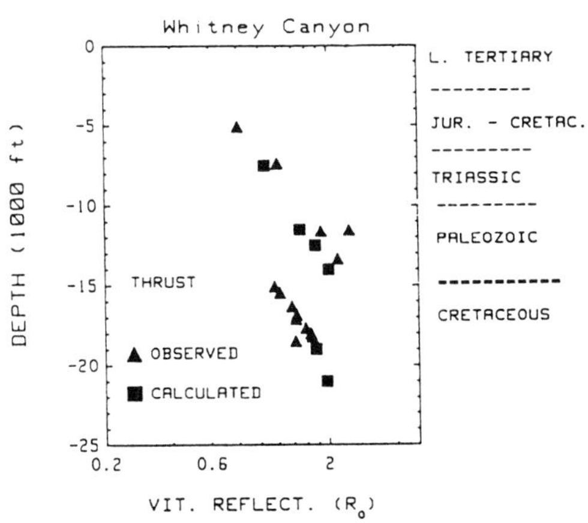

Figure 10.44. Vitrinite reflectance profile for the Whitney Canyon wellsite. Triangles are observed values (Warner and Royse, 1987) and the squares are calculated values (Furlong and Edman, 1989). Calculated values are omitted for the depth range 15,000 to 20,000 ft because they overlie the observed values.

a series of 10 increments to simulate overthrusting. Mass balance is conserved, either by complete isostatic readjustment, by partial readjustment, or with no readjustment. However, all three versions lead to very similar thermal regimes. The model was not applied to a specific geographic area, but used to make some general observations on temperatures and petroleum generation during overthrusting. The role of thermal properties, thrust sheet velocities, heat flow, erosion during and after thrusting, and the location of model grid points were all evaluated. Figure 10.45 shows the temperature distribution that has developed 100,000 yr after rapid overthrusting. The way this temperature readjusts towards equilibrium is given in Figure 10.46 and shows that reequilibration is essentially complete in 10 m.y. for both slow and rapid sheet emplacement. Sawtooth-shaped temperature profiles that are a consequence of the instantaneous emplacement used in the Angevine and Turcotte (1983) model do not appear in the modeling by Wygrala et al (1990). With more geologically reasonable emplacement models the temperature starts to reequilibrate during the thrusting, but for high thrusting rates (4 - 5 cm/yr or more) temperature inversions can still occur (see Figures 10.45 and 10.46). If petroleum generation has started before thrusting it may be possible to generate two maxima in the trend of oil generating potential with depth, as shown in Figure 10.47. The shallower peak in the overthrusted slab is in a sequence that has been cooled on uplift, and as a consequence the rate of oil generation is low and may not be significant.

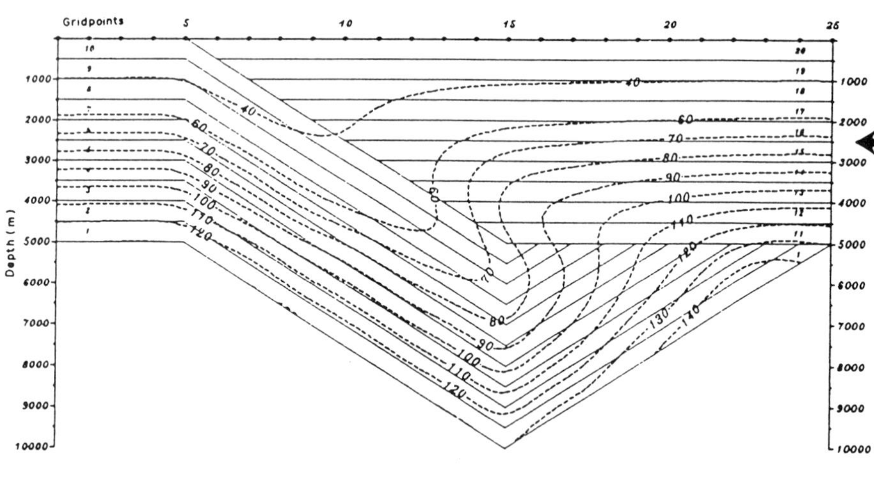

Figure 10.45. Temperature distribution 100,000 years after rapid overthrusting calculated using ten steps in the model described by Wygrala et al (1990).

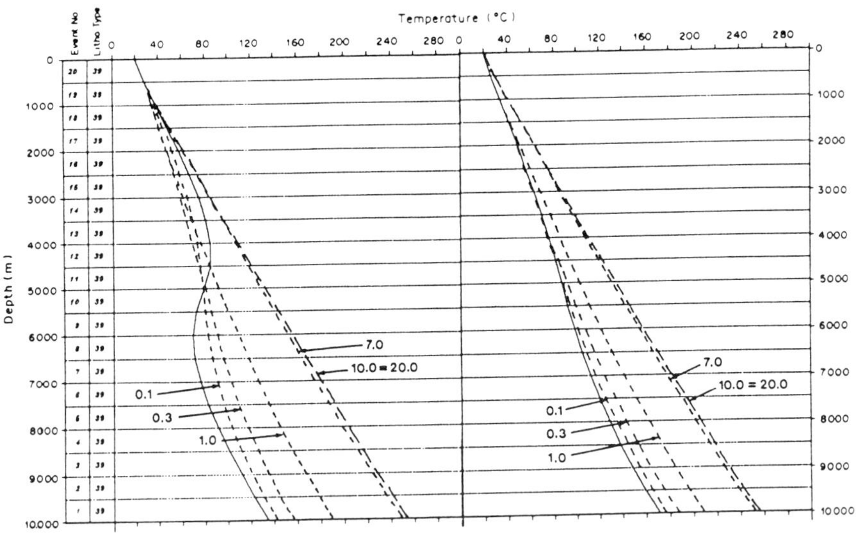

Figure 10.46. Vertical temperature profiles for various times after overthrusting. The solid line shows the initial profile immediately after thrusting, and the numbers identify profiles millions of years later. Profiles for rapid overthrusting are given on the left, and for slow overthrusting on the right (Wygrala et al, 1990).

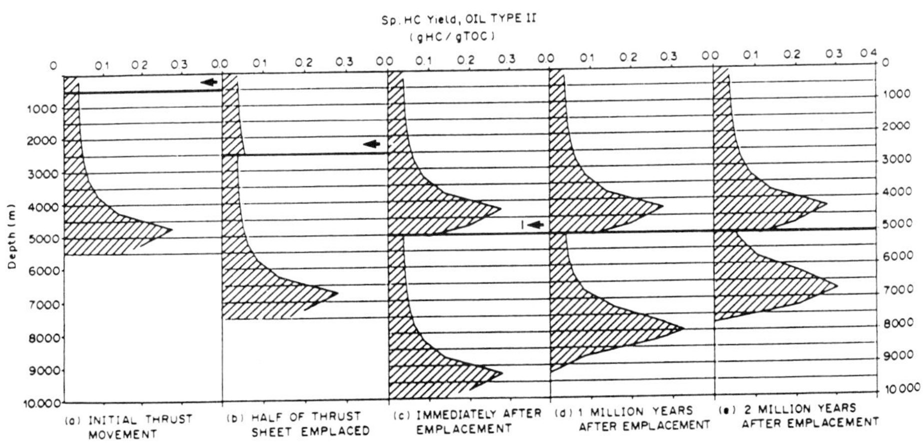

Figure 10.47. Oil generation potential versus depth for various times during and after overthrusting (Wygrala et al, 1990).

10.7. EVAPORITES

10.7.1. Introduction: Role in Petroleum Geology

Evaporites can accumulate in any location with a moisture deficit, and they commonly develop in restricted environments, such as those that form in lakes, lagoons, restricted seaways, opening rifts, isolated arms of the sea, and continental interior basins. Although evaporites contribute to the sediments in many basins, there have been few comprehensive basin-scale studies. The Zechstein basin of central and northwestern Europe, the Miocene Mediterranean basin, and the Upper Silurian of the Michigan basin are well known, and some other basin studies have been described by Peryt (1987). Increasing interest in the subsalt play in the Gulf of Mexico has focused recent attention on evaporites in this area. Evaporites occur as part of the sedimentary fill in a wide variety of basin types and need to be integrated with other sedimentary units in modeling these basins. However, evaporites have some unique properties, and because their presence in a basin impacts almost all aspects of modeling petroleum generation and accumulation (Lerche and Petersen, 1995), it is worth discussing them as a separate topic. Evaporitic rocks can play important roles in sourcing petroleum, controlling migration pathways, creating traps, and sealing reservoirs. They can also affect sedimentation and sediment accumulation (Lerche and Petersen, 1995).

Potential source rocks form because the environments where evaporites develop are extremely productive biologically, exceeding even the richest oceanic settings. Salt itself accumulates very rapidly, with accumulation rates often exceeding 1000 m/m.y., and this results in extreme inorganic dilution giving low organic contents (Hofmann et al, 1993). Associated carbonates and shales accumulate much more slowly and develop high values for TOC (Evans and Kirkland, 1988; Huc and Sinninghe Damste, 1993; Benali, 1995). The ability of salt to flow and reheal small fractures makes it an excellent sealing material, and it is effective not only in reservoirs, but also on a regional scale where it can be important in controlling migration pathways by forming laterally extensive permeability barriers. Although other evaporitic minerals, such as anhydrite, do not form such large accumulations as halite, they are also effective seals and barriers to fluid flow. Finally, the mobility of salt leads to deformation of adjacent sediments and the possibility of developing trapping configurations.

10.7.2. Burial History Reconstruction

The properties of evaporites most relevant to thermal modeling are those that influence the thermal regime and the burial history. Because halite accumulates rapidly and is very mobile in the subsurface it has a significant influence on surrounding sediments. Salt accumulation rates exceeding 1000 m/m.y., and salt thicknesses of more than 1 km are well known in many areas. The rapidly thickening salt causes fast burial of the underlying sediments.

Salt is unique in the way its thickness can change dramatically, either decreasing or increasing over relatively short time periods (see Figure 3.28). It is very mobile in the subsurface, and with a low density of only 2.2 g/cm^3 tends to "float" upwards, progressing from low amplitude swells to diapirs. Subsequent post-diapiric movement can further deform the surrounding sediments (Halbouty, 1979). Movement can occur very rapidly and diapiric growth rates of 150 - 530 m/m.y. (lasting 10 - 30 m.y.) have been reported for Mississippi and Louisiana by Seni and Jackson (1983) (Table 10.8), but lower rates of 10 - 40 m/m.y. for pillow growth in the North Sea were obtained by Hughes and Davidson (1993). Differences in overburden pressure from one area to another can also cause lateral movement of salt. In the Gulf of Mexico this has produced a gradual basinward movement, and in areas of salt withdrawal additional sediments have accumulated. The process of salt intrusion can deform overlying beds and originally uniform layers above a diapir may be internally deformed leading to significant variations in layer thickness. Later differential deposition over the growing structure produces considerable variations in bed thickness (Brewer and Groshong, 1993). Flanks of piercement salt domes or diapirs commonly have adjacent strata tipped up against them and most petroleum production associated with salt domes is from the flanks rather than the cap rocks.

The salt that goes into the rising diapir is withdrawn from the "mother layer" and a rim syncline forms. This accumulates additional thickness of sediments as the system develops. An additional characteristic that sets evaporites apart from other sediments is their high solubilities in water. If salt is removed by solution it can trigger the collapse of overlying beds. Also, the conversion of gypsum to anhydrite with the loss of water leads to a thinning of evaporite beds (see Section 3.5.2) that may affect overlying sediments.

A major application of maturity modeling is in relating the time of generation to the time of trap formation. At the time hydrocarbons were being generated in the East Texas basin all the salt domes were in a diapiric or post-diapiric stage (Wescott and Hood, 1994). The well-developed rim synclines around them precluded the migration of large

Table 10.8
Comparison of published methods and rates of salt-dome growth (Seni and Jackson (1983); see their publication for other references).

Author	Growth Rate $\dot{G}$(m/Ma)	Location	Method	Comments
Ewing and Ewing, 1962	1,000 (net)	Sigsbee Knolls, Gulf of Mexico	Uplift of dated marker bed	Abyssal-plain seamounts project above flat floor. Pleistocene sediments have been uplifted 150 to 300 m
Trusheim, 1960 Sanneman, 1968 Jaritz, 1980	300 (average) 100 to 500–diapir stage ~20–later stages	Zechstein salt, Germany	Unknown	Average. Consistent long-term rate, same for pillows and domes.
Netherland, Sewell, and Associates, 1976	6 (low average) 44 (high average)	Various domes, East Texas basin	Uplift/time	General decrease in growth rates for individual domes through time. Maximum calculated rate of growth is 54 m/Ma for Mt. Sylvan dome in Early Cretaceous. Minimum calculated rate of growth is 3 m/Ma for Mt. Sylvan and Bullard domes in post-Eocene. Maximum estimated growth rate is 153 m/Ma in Late Jurassic.
Kupfer, 1976	5 to 10	Vacherie dome, N. Louisiana salt basin	Volume of salt/time	General decrease in dome growth rate through time. Maximum growth in Early Cretaceous (0.4 km^3/Ma), sharp decline in Late Cretaceous to 0.04 km^3/Ma.
Kumar, 1977	1 to 42 (net)	Various domes, N. Louisiana salt basin	Stratigraphic thinning and uplift/time	General decrease in dome growth through time. Maximum growth in Early Cretaceous is 42 m/Ma, maximum growth in Late Cretaceous is 19 m/Ma and minimum growth rate in Cenozoic is 1 m/Ma.
This study	13 to 104 (net)	Various domes, East Texas basin	Net rate of pillow growth—rate of sediment thinning	
	26 to 222 (net)	Various domes East Texas basin	Net rate of dome growth—maximum rate of sedimentation in withdrawal basin	
	10 to 530 (gross)	Various domes, East Texas basin	Gross rate of diapir growth—volume of salt divided by cross-sectional area of salt stock	

quantities of oil into the traps adjacent to the diapirs, and these diapiric structures are not expected to yield substantial reserves. Wescott and Hood (1994) related other salt features, including salt-cored structures and turtle-structure anticlines, to the timing of source rock maturation and to the development of migration pathways. The deep salt-cored structures are the most prolific locations in contrast to the diapiric structures which contain the lowest reserves.

10.7.3. Thermal Reconstruction

The high thermal conductivity of salt that makes it a good conductor leads to low geothermal gradients through salt beds (see Section 4.1.4). Salt diapirs that have punched through overlying sediments provide a conduit for the conduction of heat upwards, and can produce shallow temperatures that are locally higher than normal (Keen, 1983; Petersen and Lerche, 1995). This withdrawal of heat depresses the deeper sediment temperatures in the vicinity of the diapir (O'Brien and Lerche, 1984; Petersen and Lerche, 1995) with obvious consequences for the generation and preservation of oil (Mello et al, 1995). O'Brien and Lerch (1987) modeled the thermal regime around a salt dome and used the information to calculate organic matter maturity. A comparison of the local and regional maturity profiles showed that near the diapir maturities were higher at shallow depths, but lower at greater depths (Figures 10.48 and 10.49). This effect can only be expected when the height of the diapir is less than about 5 km because the thermal conductivity of salt decreases with increasing temperature (Lerche and Torudbakken, 1996).

Isotherms across a diapir in the Gulf of Mexico are shown in Figure 10.50 (taken from Cassidy and Ranganathan, 1992), and in this case temperatures near the salt increase as much as 10 °C. Jensenius (1987) has reported apparent geothermal gradients up to 52 °C/km over a diapir in Skjold field in the Danish North Sea. At Sable Island, off the east coast of Canada, the locally elevated temperatures associated with a salt diapir have been sufficient to mature the oil-prone source rocks and produce a commercial oil field (Rashid and McAlary, 1977; Rashid, 1978), though Keen (1983) suggested that the upward migration of hot fluids immediately adjacent to the diapir may have played a major role. In contrast, at most other locations along the east coasts of Canada and the U.S. only the deeper sediments that contain predominantly terrestrially-derived organic materials are mature, and this is generally a gas-prone passive margin (Barker, 1982).

Diapirism and salt withdrawal (or collapse features) produce rapid lateral variations in salt thickness, and as a consequence geothermal gradients vary over short distances. In modeling these areas it is important to consider carefully the exact locations of the temperature data being used. It is also important to review the changing local gradients through time as the diapirs and other structures developed.

Sediments deposited in the restricted environments formed by opening rifts often include evaporites. In the Provencal basin rift of the Gulf of Lions more than 2 km of salt and evaporites accumulated during the post-rift Messinian between 7 and 5 Ma (Burrus and Audebert, 1990). This rift, which forms part of the southern end of the West European rift system, and is located in the northwestern Mediterranean, is still in a

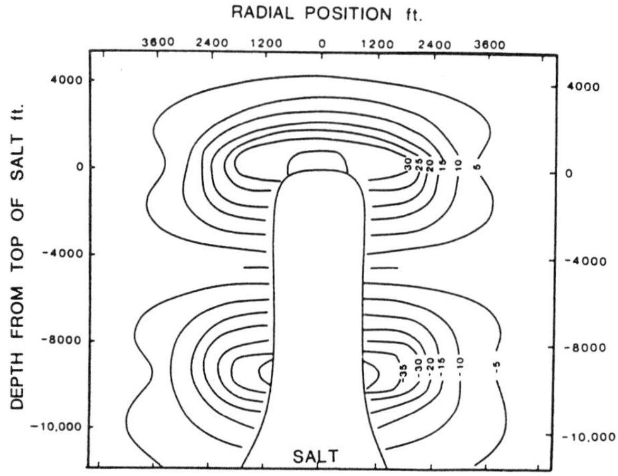

Figure 10.48. A contour plot of the predicted subsurface temperature anomalies (°C above or below the regional geothermal gradient) for a salt dome of height 9000 m and radius 600 m buried under 1600 m of sediment (O'Brien and Lerch, 1987).

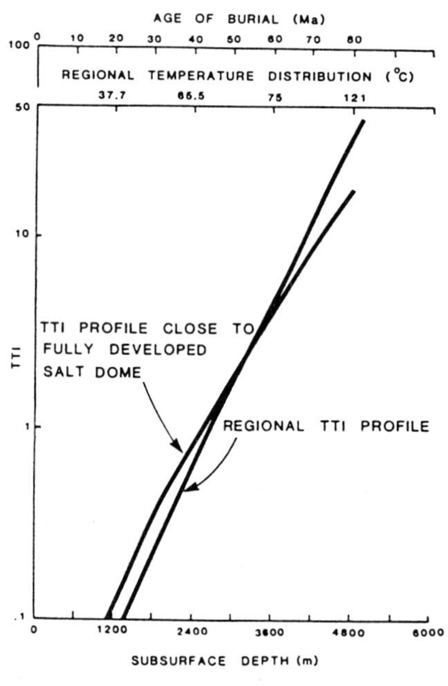

Figure 10.49. Comparison of the regional TTI profile and the profile close to a rising salt dome. Deposition rate was 60 m/m.y., and regional geothermal gradient was 23 °C/km. (O'Brien and Lerch, 1987)

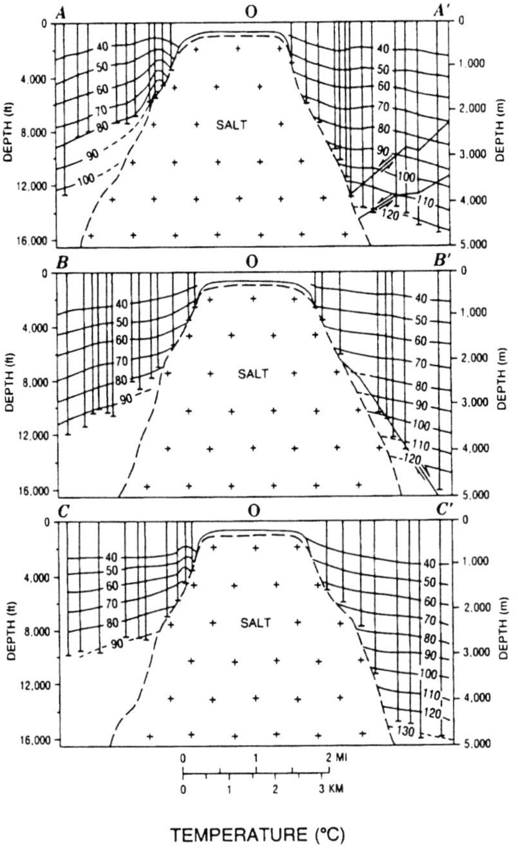

Figure 10.50. Cross sections showing temperatures (°C) across a salt dome (Cassidy and Ranganathan, 1992)

transient thermal regime. Basin evolution, thermal regime, pressure regime, and source rock maturity were modeled by Burrus and Audebert (1990) using the TEMISPACK software (described by Doligez, 1986, and Ungerer et al, 1990). The 2-D finite difference model was applied to a 275 km cross section of the crust and mantle. The thick salt layer has modified temperatures, and shales below the salt are 20 - 25 °C cooler than they would be in the absence of salt (see Chapter 4). The salt also has an indirect effect on temperature because it acts as a permeability barrier that prevents normal shale compaction, and this leads to undercompacted shales with more water-filled porosity and a somewhat increased temperature. Figures 10.51 and 10.52 show that this disequilibrium

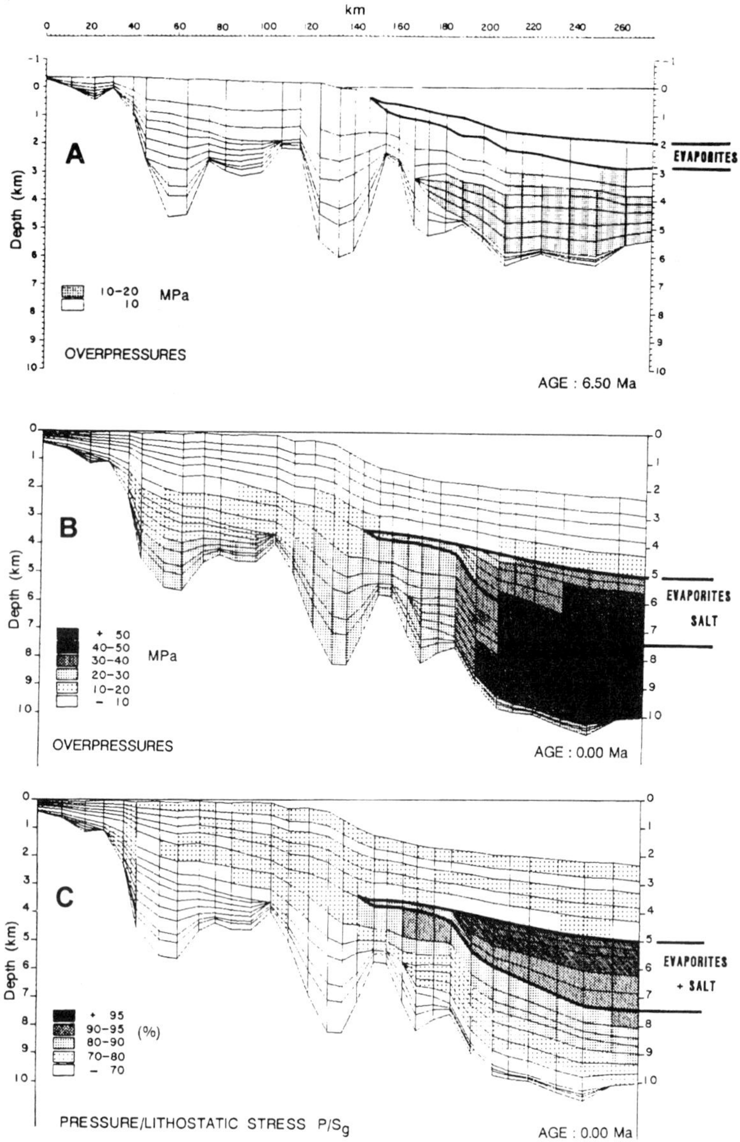

Figure 10.51. Computed pattern of geopressures due to compaction disequilibrium. (a) Overpressures are restricted to the shales below the depositing evaporites, 6.5 Ma; (b) overpressures at present below the evaporites and salt, and at depth in the shales; (c) pressure/lithostatic stress computed (P/S_g) ratio, which, when above 75%, indicates the area is affected by hydraulic fracturing. (From Burrus and Audebert, 1990).

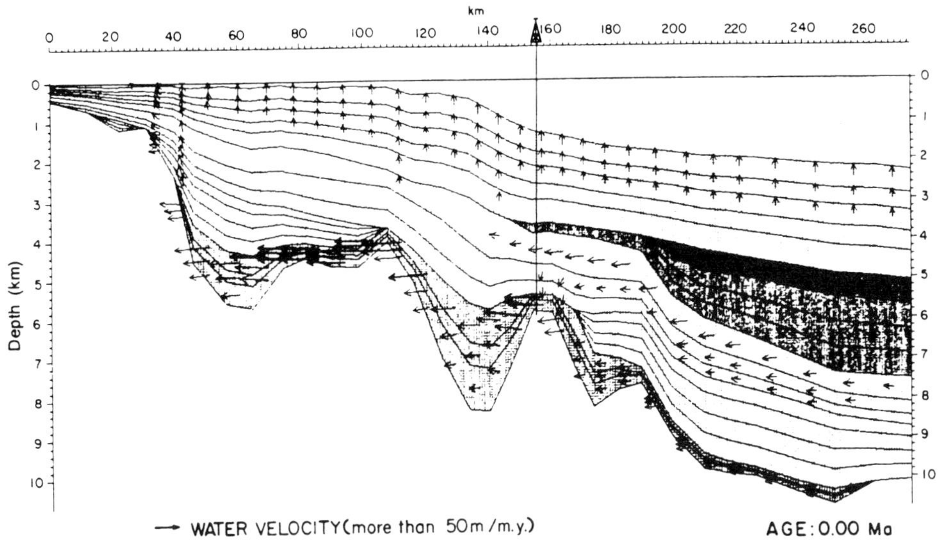

Figure 10.52. Computed pattern of geopressures due to compaction disequilibrium. (a) Overpressures are restricted to the shales below the depositing evaporites, 6.5 Ma; (b) overpressures at present below the evaporites and salt, and at depth in the shales; (c) pressure/lithostatic stress computed (P/S_g) ratio, which, when above 75%, indicates the area is affected by hydraulic fracturing. (From Burrus and Audebert, 1990).

compaction in the shales has generated abnormal pressures under the salt and caused lateral fluid flow. Where pressures have approached the rock load hydraulic fracturing appears to be occurring today. In the shallower sediments above the salt, or in areas without salt, fluid flow is vertical. Burrus and Audebert (1990) combined temperature data with kinetic parameters to model maturity in a hypothetical source rock. They used a single A value with eight E values centered around 54 kcal/mol. The maturity calculation showed that maturity is underestimated if linear geothermal gradients are used or if lithologic character is ignored.

APPENDIX 1

ABBREVIATIONS

α	orientation of a side group or atom on a molecule below the viewing plane
β	orientation of a side group or atom on a molecule above the viewing plane
β	crustal stretching factor
A	Arrhenius constant, frequency factor, or pre-exponential factor
AFTA	apatite fission track analysis
A-I	aromatization-isomerization (biomarker plots)
bbl	barrel
BHP	bottom hole pressure
BHT	bottom hole temperature
BHFT	bottom hole flowing temperature
bp	before present
BSR	bottom simulating reflector
btu	British Thermal Unit
CAI	conodont alteration index
CCD	carbonate compensation depth
DST	drill stem test
E	activation energy
esr	electron spin resonance
Ga	billion years ago
GOR	gas-oil ratio
GWC	gas-water contact
HFU	heat flow units
HI	Hydrogen Index (pyrolysis)
IFP	French Petroleum Institute (Paris)
I-I	isomerization-isomerization (biomarker plots)
I/S	illite/smectite
k	rate constant for reaction
KB	Kelly bushing
LLNL	Lawrence Livermore National Laboratory (California)
LOM	level of organic metamorphism
Ma	million years ago

md	millidarcy
MSL	Mean Sea Level
NSOs	nitrogen-sulfur-oxygen compounds
OI	Oxygen Index (pyrolysis)
OWC	oil water contact
PDB	Pee Dee belemnite (carbon isotope standard)
R	(rectus) right-handed form of a biomarker
R	gas content
RFT	repeat formation tester
R_o	vitrinite reflectance (in oil)
ROP	rate of penetration
S	(sinister) left handed form of a biomarker
S	solidity (complement of porosity)
S1	pyrolysis peak 1
S2	pyrolysis peak 2
SCI	spore coloration index
SIM	single ion monitoring (mass spectrometry)
SL	sea level
SMOW	Standard Mean Ocean Water (oxygen isotope standard)
sp	spontaneous potential (electric log response)
TAI	thermal alternation index
TD	total depth
T_h	homogenization temperature for fluid inclusion
T_m	melting temperature for fluid inclusion contents
T_{max}	temperature corresponding to top of pyrolysis Peak 2 (S2)
TOC	total organic carbon
TOM	total organic matter
TR	transformation ratio
TTI	time-temperature index
TVD	true vertical depth
TVT	true vertical thickness
VR	vitrinite reflectance

APPENDIX 2

GLOSSARY

Abnormal Pressure Any subsurface fluid pressure that is above or below hydrostatic pressure.
Activation Energy (E) The energy that must be absorbed by a molecule, or molecular complex, to break the bonds and form new products.
Aerobic Bacteria Bacteria that require free oxygen.
Aliphatic non-aromatic hydrocarbon
Alkane See *Paraffin*
Alkene See *Olefin*
Amino Acid Organic compound which contains both an acid group and an amino group ($-NH_2$). They are the building blocks (monomers) of proteins.
Aromatic Hydrocarbons containing one or more benzene rings. Polycyclic aromatic hydrocarbons (PAH) contain several rings with two or more carbon atoms shared between rings.
Arrhenius Constant See *Frequency Factor*
Asphalt Black to dark brown solid or semisolid bitumens that gradually liquefy when heated, composed principally of the elements carbon and hydrogen but containing appreciable quantities of nitrogen, sulfur, and oxygen and largely soluble in carbon disulfide.
Asphaltenes Asphaltic constituents of crude oil that are soluble in benzene or carbon disulfide but insoluble in petroleum ether or n-pentane.
Asymmetric Carbon A carbon atom located at a position in a molecule where it is bonded to four different organic groups.
Bacterial Gas See *Biogenic gas*
Base of Crude Oil The "base" of a crude oil is descriptive of the chemical nature of its main constituents. A paraffin-base oil contains predominantly paraffinic hydrocarbons. An intermediate- or mixed-base crude contains roughly equivalent mixtures of paraffins and naphthenes (cycloparaffins). A naphthenic-base crude contains predominantly naphthene hydrocarbons. An asphalt-base crude is one containing a relatively high proportion of nonhydrocarbon constituents such as nitrogen, sulfur, and oxygen compounds. (The term aromatic base is not used because there are no oils known to contain predominantly aromatic hydrocarbons)
Biodegradation (Bacterial Degradation) A process by which organic matter or oil is altered by microbial activity.

Biogenic Gas Methane formed by anaerobic bacteria (methanogens). It is isotopically very light and contains essentially no higher hydrocarbons.
Biomarkers (Biological Markers) Organic compounds whose carbon structure, or skeleton, is formed by living organisms and is sufficiently stable to be recognized in crude oil or the organic matter of ancient sediments.
Bitumen The liquid and/or solid organic material in sediments which is soluble in organic solvent; the "extractables".
Bubnoff Unit A rate of sediment accumulation of 1 m / million years.
Carbohydrates Biopolymers made up from repeated sugar monomers. They have the general formula $(C.H_2O)_n$ and include starch and cellulose.
Carotenoids Plant pigments composed of mono- and dicyclic tetraterpenes. (The yellow, brown, and red colors of leaves in the fall are caused by carotenoids.)
Catagenesis The process by which organic material in sediments is thermally altered by increasing temperature. Catagenesis covers the temperature range between diagenesis and rock metamorphism, approximately 50-200°C (122-392°F).
Chemical Fossils See *Biological Markers*
Chiral Center See *Asymmetric carbon.*
Chromatography A group of analytical methods for separating individual chemical compounds or groups of compounds from complex mixtures. It relies on a differential rate of movement over some suitable substrate.
Clathrate See *Gas Hydrate*
Coal Maceral Microscopically recognizable constituents of coal that can be differentiated by their morphology. Macerals are analogous to the minerals of inorganic rocks but differ in having less uniform chemical composition and physical properties. The carbon content of macerals increases with increasing temperature. The major maceral groups are vitrinite, exinite and inertinite (see listings in Glossary).
Coaly Kerogen lack solid organic matter derived from subaerial oxidation or metamorphism of kerogen. Usually consists of inertinite or vitrinite.
Condensate A hydrocarbon mixture that is gaseous in the ground but condenses into fluid when produced. Its gravity ranges from 60 °API upward. The term is often used loosely for any high API gravity oil.
Connate Water Fossil reservoir water that has not been in contract with the atmosphere since deposition. It is high in chloride and calcium and frequently contains more than 100,000 ppm total dissolved solids.
Cracking The breaking-up of complex chemical compounds into simpler ones, usually as a result of heating (thermal cracking).

Cuticle The waxy outer layer of terrestrial plant cells.

Cuttings Gas Hydrocarbons in the range C_1 to C_4 recovered from cuttings.

Cycloalkane See *Naphthene*

Cycloparaffin See *Naphthene*

Exinite See *Liptinite*

Expulsion See *Primary migration*

Fission Track A trail of damage produced in a mineral by recoiling heavy nuclei formed when uranium-238 decays by fission. Etching produces a track visible under a microscope.

Free Radical A chemical compound which processes an unpaired electron.

Frequency Factor The constant ("A") in the kinetic rate equation. (Also called the "Arrhenius Constant" or sometimes simply the "Pre-exponential Factor").

Gas Hydrates Crystalline compounds in which the ice lattice of water expands to form cages than can physically trap gas molecules. Methane hydrates occur offshore in many areas.

Gasoline Range Hydrocarbons Hydrocarbons in the range of C_4 to C_7.

Herbaceous Kerogen Solid organic matter consisting of such plant materials as spore, pollen, cuticle, and other membranous substances; also called liptinite or exinite.

Heteroatom An atom other than carbon or hydrogen; in organic geochemistry the common heteroatoms are nitrogen, oxygen, and sulfur.

Heterocyclic Compounds Ring compounds in which one (or more) of the atoms in the ring is nitrogen, sulfur, or oxygen.

Homologue A member of a homologous series, which is a group of compounds that have similar structures but differ by the addition or removal of $-CH_2-$ units.

Hopane A five-ring naphthene having four 6-membered rings and one 5-membered ring that are fused together. They are one of the most commonly analyzed biomarkers.

Humic Organic Matter The decomposition and polymerization products of the lignin, tannins, and cellulose of plant cell and wall material plus carbonized organic matter deposited in swamps and soils under aerobic conditions with partial restriction of oxygen.

Hydrocarbon A compound composed only of the elements hydrogen and carbon. These may be arranged in straight chains, branched chains, or rings.

Hydrostatic Gradient The pressure increase with depth in a liquid column that is connected to the surface. Gradients are generally close to 0.47 psi/ft but vary slightly with salinity.

Inertinite The coal maceral group that shows little or no reaction during the coking process. Inertinite includes fusinite (fossil charcoal), sclerotinite derived from fungal remains, and other high-carbon materials with a low-hydrogen content and a high reflectance.

Inspissation Thickening of an oil deposit as a result of the evaporation of gases and light fractions.

Isomers Molecules that have the same number and kinds of atoms but with different arrangements. Structural isomers differ in the way atoms are linked together. Stereoisomers differ in the spatial arrangements of groups; Optical isomers are stereoisomers with nonidentical mirror images (comparable to right- and left-hands).

Isoprenoids A general term for hydrocarbons that are built up from repeated isoprene monomers. This important group includes chains (e.g., pristane, phytane) and rings (e.g., steranes, hopanes).

Isotopes Atoms of an element that have different weights (due to different numbers of neutrons but constant number of protons). Isotopes show essentially the same chemical reactions.

Kerogen The naturally occurring, disseminated organic matter in rocks that is insoluble in nonoxidizing acids, bases, and organic solvents.

Lignin The complex polymeric substance, composed of aromatic phenols and alkyl groups, which gives structural support to woody tissue.

Lipids A broad term that includes all oil-soluble, water-insoluble substances such as fats, waxes, fatty acids, steroids, pigments, and terpenoids.

Liptinite (Exinite) A coal maceral group that is rich in hydrogen. Liptinite macerals include sporinite, cutinite, resinite, and alginite, which are derived from spores and pollen, cuticles, resins, and algae respectively.

Maceral Organic constituents of coal (analogous to minerals in rocks) consisting of three major groups; vitrinite, exinite (liptinite) and inertinite. (see Coal Maceral)

Mercaptans (Thiols) Compounds containing the -SH group. They are sulfur analogs of the alcohols in that the oxygen in the alcohol (-OH) group is replaced by sulfur.

Meteoric Water Fresh surface water entering a subsurface sedimentary section through permeable outcrops. It is high in sodium, bicarbonate and sulfate and usually contains less than 10,000 ppm total dissolved solids.

Micelle A configuration of dissolved surfactants in which the polar and nonpolar ends of the molecules are separated. This separation results in the enhanced solubility for these molecules in water.

Naphthene (also called Cycloalkane, Cycloparaffin) A group of compounds containing one or more saturated rings.

Naphtheno-Aromatic Hydrocarbon A polycyclic hydrocarbon that contains both aromatic and naphthenic rings.
Natural Gas A mixture of varying proportions of gaseous hydrocarbons such as methane, ethane, propane, isobutane, and occasionally containing small amounts of liquid hydrocarbon such as pentanes and hexanes, and nonhydrocarbon gases such as carbon dioxide, hydrogen sulfide, nitrogen, hydrogen, and helium.
Oil Floor Bottom of the oil window.
Oil Shale A compact rock of sedimentary origin with an ash content of more than 33 percent and containing organic matter that yields oil when destructive distilled but not appreciably when extracted with petroleum solvents.
Oil Window (Liquid Window) Depth interval in which liquid hydrocarbons are generated in source rocks or preserved in reservoir rocks.
Olefin (Alkene) A straight or branched-chain hydrocarbon that has one (or more) carbon-carbon double bond(s).
Optical Rotation The property of certain organic compounds to rotate plane polarized light.
Paraffin (Alkane) A saturated hydrocarbon with the molecular formula C_nH_{2n+2}. It includes normal straight-chain paraffins, branched paraffins, and naphthenes.
Petroleum A naturally occurring and usually complex mixture of predominantly hydrocarbon substances in the gas and/or liquid state which constitutes commercial crude oils and natural gas.
Phytane A saturated isoprenoid chain containing 20 carbon atoms, with a branch on every fifth carbon atom.
Porphyrins Complex compounds originating in living materials (mainly from chlorophyll) and having a basic structure consisting of four interconnected rings, each ring containing four carbon atoms and one nitrogen atom.
Pour Point The temperature at which crude oil will not flow when a tube containing it is first heated in a bath to dissolve all wax and then cooled slowly. At cooling intervals of 3^oC (5^oF), the tube is held horizontal until there is no flow for 5 seconds.
Pre-Exponential Factor See *Frequency Factor*
Primary Migration (Expulsion) The movement of oil and gas out of the fine-grained source rocks into the permeable reservoir rocks.
Pristane A saturated isoprenoid chain containing 19 carbon atoms with a branch on every fifth carbon atom.
Proteins High-molecular-weight polymers built up from amino acid monomers. They account for more than 50 percent of the dry weight of animals.

Pyrobitumen Black to dark brown, hard bitumens that decompose before they melt.
Pyrolysis Thermal decomposition of chemical compounds by heating in an inert atmosphere.
Resin Petroleum resins are the fraction of residuum that is insoluble in liquid propane but soluble in normal pentane.
Resinite A coal maceral in the liptinite group. Resinites are fossil tree resins derived from balsam, mastic, latex, and some plant gums, waxes, oils and fats.
Sapropelic Organic Matter The decomposition and polymerization products of high-lipid organic materials, such as spores and planktonic algae deposited in subaquatic muds (marine or lacustrine) under predominately anaerobic conditions.
Saturates (Saturated Hydrocarbon) A general term including normal and branched alkanes and naphthenes (paraffins and cycloalkanes). Saturates contain no multiple C–C bonds and have the maximum number of hydrogen atoms for the structure.
Secondary Migration he movement of fluids within the permeable rocks that eventually leads to the segregation of oil and gas into accumulations in certain parts of these rocks.
Sour Gas A gas containing noticeable quantities of volatile sulfur compounds such as hydrogen sulfide and mercaptans.
Source Rock A fine-grained sediment that in its natural setting has generated and released enough hydrocarbons to form a commercial accumulation of oil and gas.
Soxhlet Extraction A technique for exhaustive extraction of bitumens from rocks by repeated extractions with freshly distilled organic solvent.
Stable Isotope An isotope that does not spontaneously decay into another isotope, i.e., is not radioactive.
Steranes A four-ring naphthene having three 6-carbon rings and one 5-carbon ring fused together with several side chains attached to the rings. Steranes are one of the most commonly analyzed biomarkers.
Terpenoid An isoprenoid polymer usually containing 10, 15, 20, 30 or 40 carbon atoms (mono, sesqui-, di-, tri-, and tetra-, respectively). Terpenoids include hydrocarbons, alcohols, and acids. Among the common terpanes (hydrocarbons) are the hopanes.
Tertiary Migration The remobilization of an oil or gas accumulation into another reservoir.
Thermal Alteration Index A quantitative measure of the changing color of organic matter (from pale yellow, through light browns, to dark browns, and finally to black).

Transformation Ratio A measure of the progress of petroleum generation or pyrolysis given as the fraction (or percent) of conversion to products.
Triterpanes Class of isoprenoid hydrocarbons which are naphthenic (i.e., have saturated rings); hopanes are triterpanes that are commonly used in oil correlations
Vitrinite A component (maceral) of coal and/or kerogen derived from the cellulosic and lignitic tissue of terrestrial plants.
Vitrinite Reflectance A technique in which the reflectivity of vitrinite is measured; values are expressed in units of percent R_o and indicate source rock maturity or coal rank.
Waxes Long-chain compounds formed by the bonding of a long-chain fatty acid and a long-chain alcohol; also refers to long-chain normal paraffins in oils (e.g., waxy oils).
Wet Gas Natural gas consisting of methane and significant quantities of higher hydrocarbons generally with natural gas liquid vapors amounting to 4 liters or more per 100 m^3 (0.3 or more gallons per 1000 ft^3).

APPENDIX 3

GEOLOGIC TIME SCALE

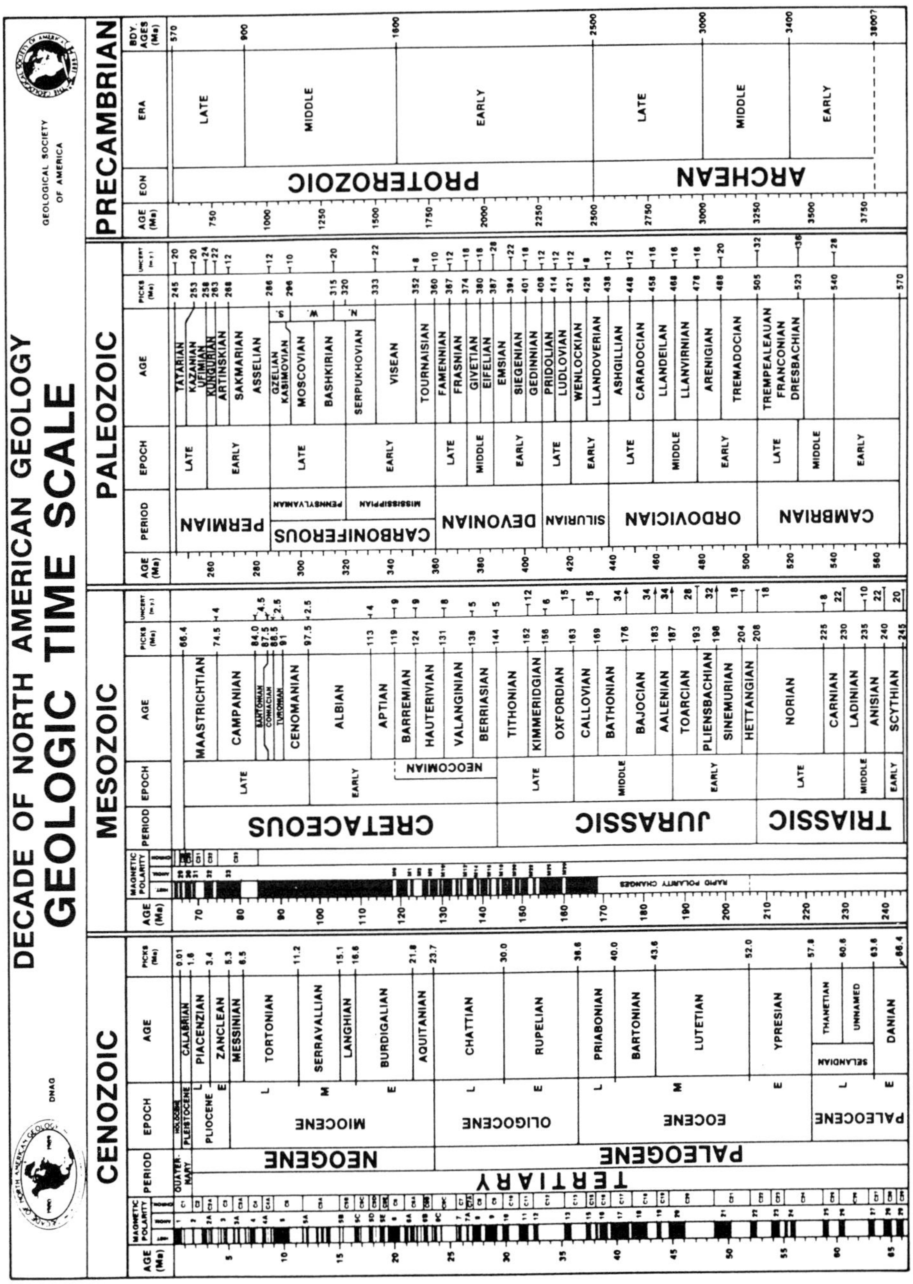

The Decade of North American Geology
1983 Geologic Time Scale

Compiled by
Allison R. Palmer
Centennial Science Program Coordinator
Geological Society of America, P.O. Box 9140, Boulder, Colorado 80301

Preparation of the 27 synthesis volumes of *The Geology of North America* for the Decade of North American Geology (DNAG) is now in progress. In order to encourage uniformity among DNAG authors in the citation of numerical ages for chronostratigraphic units of the geologic time scale, an ad hoc Time Scale Advisory Committee was established by the DNAG Steering Committee in 1982. This advisory committee, consisting of Z. E. Peterman (Chairman) and J. E. Harrison, U.S. Geological Survey; R. L. Armstrong, University of British Columbia; and W. A. Berggren, Woods Hole Oceanographic Institution, was asked to evaluate numerical dating schemes that were either recently published or in press and to provide recommendations for the best numbers to use in preparation of a DNAG time scale. The chart on the opposite side of this page was developed from the recommendations of the Time Scale Advisory Committee.

Geochronometric ages (Ma, Ga) assigned to chronostratigraphic boundaries are subject to several uncertainties in addition to those introduced by the numerical dating methods themselves; boundary stratotypes for many units are not yet chosen, so disagreement exists about exact biostratigraphic placement and correlation of a boundary; and many materials that can be numerically dated are not known in good context with biostratigraphic data, so extrapolation to a chronostratigraphic boundary is commonly required. Furthermore, with respect to the late Mesozoic and the Cenozoic, differing numerical age calibrations of the magnetic polarity-reversal scale based on differing choices of scattered isotopically dated tie points, differing interpretations of the positions of biostratigraphic boundaries with respect to the polarity-reversal scale, and uncertainties in the meaning of isotopic ages derived from glauconies lead to disagreement about ages assigned to some chronostratigraphic boundaries.

With these caveats, the numerical ages given in this chart represent interpretations acceptable to the DNAG Time Scale Advisory Committee. The uncertainty bars for Paleozoic and Mesozoic ages are from data in Harland and others (1982). Uncertainty bars for the Cenozoic are not available.

Sources for the numerical ages and for the chronostratigraphic nomenclature are given below.

CENOZOIC

Berggren, W. A., Kent, D. V., and Van Couvering, J. A., 1984, Neogene geochronology and chronostratigraphy; *in* Geochronology and the geologic record: Geological Society of London (in press).

Berggren, W. A., Kent, D. V., and Flynn, J. J., 1984, Paleogene geochronology and chronostratigraphy, *in* Geochronology and the geologic record: Geological Society of London (in press).

MESOZOIC

Base of Campanian to end of Cretaceous

Berggren, W. A., Kent, D. V., and Flynn, J. J., 1984, Appendix, *in* Geochronology and the geologic record: Geological Society of London (in press).

Base of Aptian to base of Santonian

Harland, W. B., Cox, A. V., Llewellyn, P. G., Picton, C.A.G., Smith, A. G., and Walters, R., 1982, A geological time scale: Cambridge, Cambridge University Press, 128 p.

Base of Hettangian to base of Barremian (dating and chronostratigraphic correlation of the "M" series)

Kent, D. V., and Gradstein, F. M., 1984, A Jurassic to Recent chronology, *in* Tucholke, B. E., and Vogt, P. R., eds., The Western Atlantic region, Volume M of The geology of North America: Boulder, Colorado, Geological Society of America (in press).

Note: Rhaetian has been eliminated from the Late Triassic chronostratigraphic scale following Tozer, E. T., 1979, Latest Triassic ammonoid faunas and biochronology, western Canada: Geological Survey of Canada Paper 79-1B, p. 127–135.

Base of Ladinian to base of Norian

Armstrong, R. L., 1982, Late Triassic–Early Jurassic time scale calibration in British Columbia, Canada, *in* Odin, G. S., ed., Numerical dating in stratigraphy: New York, John Wiley & Sons, p. 509–513.

Base of Scythian to base of Anisian

Webb, J. A., 1982, Triassic radiometric dates from eastern Australia: *in* Odin, G. S., ed., Numerical dating in stratigraphy: New York, John Wiley & Sons, p. 515–521.

PALEOZOIC

All numerical ages except those for the upper and lower boundaries of the Paleozoic are derived from Harland and others (see above, 1982, p. 52–55). Late Carboniferous numbers are for continentally based ages (N = "Namurian"; W = Westphalian; S = Stephanian). The marine-based ages are from Harland and others (1982, Fig. 5.6). The earlier estimate for the base of the Cambrian at 570 Ma is retained.

PRECAMBRIAN

Harrison, J. E., and Peterman, Z. E., 1982, North American Commission on Stratigraphic Nomenclature, Report 9, Adoption of geochronometric units for divisions of Precambrian time: American Association of Petroleum Geologists Bulletin, v. 66, p. 801–802.

REFERENCES

AAPG, 1974. Geothermal gradient maps of North America (set of 39 maps).

Abbott, G. D., Wang, G. Y., Eglington, T. I., Home, A. K., and Petch, G. S., 1990. The kinetics of sterane biological marker release and degradation processes during the hydrous pyrolysis of vitrinite kerogen: Geochimica et Cosmochimica Acta, v. 54, p. 2451-2461.

Abu El-Ella, R., 1990. Maturation history of Neogene-Quaternary sediments, Nile Delta Basin, Egypt: AAPG Bulletin, v. 74, p. 77-84.

Aizawa, J., 1990. Paleotemperatures from fluid inclusions and coal rank of carbonaceous material of the Tertiary formations in northwest Kyusku, Japan: Journal of Mineralogy, Petrology, and Economic Geology, v. 85, p. 145-154.

Akramkhodzhayev, A. M. and Kirshin, A. V., 1983. Temperature-time relation in the alteration of the parent organic matter of oil and gas: p. 158-162 in "Petroleum Geochemistry, Genesis and Migration - International Geology Review 1968-1981" (Edited by H. D. Hedberg and J. F. Mason). AGI Reprint Series: VI.

Al Shdidi, S., Thomas, G., and Delfaud, J., 1995. Sedimentology, diagenesis, and oil habitat of Lower Cretaceous Qamchuqa Group, northern Iraq: AAPG Bulletin, v. 79, p. 763-779.

Albrecht, P., Vandenbroucke, M., and Mandengue, M., 1976. Geochemical studies on the organic matter from the Douala Basin (Cameroon) - I. Evolution of the extractable organic matter and the formation of petroleum: Geochimica et Cosmochimica Acta, v. 40, p. 791-799.

Alexander, R., Larcher, A. V., Kagi, R. I., and Price, P. L., 1992. An oil-source correlation study using age-specific plant-derived aromatic biomarkers: p. 201-221 in "Biological Markers in Sediments and Petroleum" (Edited by J. M. Moldowan, P. Albrecht, and R. P. Philp). Prentice Hall, Engelwood Cliffs, New Jersey.

Allen, P. A. and Allen, J. R., 1990. "Basin Analysis: Principles and Practice" Blackwell Scientific Publications, Oxford. 416 pp.

Anders, M. H., Krueger, W. W., and Sadler P. M., 1987. A new look at sedimentation rates and the completeness of the stratigraphic record: Journal of Geology, v. 95, p. 1-14.

Andersen, B., Throndsen, T., Barth, T., and Bolstad, J., 1994. Thermal generation of carbon dioxide and organic acids from different source rocks: Organic Geochemistry, v. 21, p. 1229-1242.

Anderson, N. L. and Knapp, R., 1993. An overview of some of the large scale mechanisms of salt dissolution in western Canada: Geophysics, v. 58, p. 1375-1387.

Anderson, R. N., He, W., Hobart, M. A., Wilkinson, C. R., and Nelson, H. R., 1991. Active fluid flow in the Eugene Island area, offshore Louisiana: Geophysics: The Leading Edge of Exploration, v. 10, no. 4, p. 12-17.

Andresen, P., Mills, N., Schenk, H. J., and Horsfield, D., 1993. The importance of kinetic parameters in modelling generation by cracking of oil to gas - a case study in 1-D from well 2/4-14: p. 563-571 in "Basin Modeling: Advances and Applications" (Edited by A. G. Dore et al) Norwegian Petroleum Society Special Publication 3, Elsevier, Amsterdam.

Andrews-Speed, C.P., Oxburgh, E. R., and Cooper, B. A., 1984. Temperatures and depth-dependent heat flow in western North Sea: AAPG Bulletin, v. 68, p. 1764-1781.

Angevine, C. L. and Turcotte, D. L., 1983. Oil generation in overthrust belts: AAPG Bulletin, v. 67, p. 235-241.

Angevine, C. L., Heller, P. L., and Paola, C., 1990. "Quantitative Sedimentary Basin Modeling" AAPG Continuing Education Course Note Series No. 32. 133pp.

Anjos, S. M. C., 1986. Absence of clay diagenesis in Cretaceous Tertiary marine shales, Campos basin, Brazil: Clays and Clay Minerals, v. 34, p. 424-434.

Ankara, K., 1996. Two-dimensional basin analysis for petroleum exploration: Ph.D. Dissertation, University of South Carolina, Columbia, S.C. 214pp.

Anon, 1987. Big strikes elude Soviet exploration in eastern Siberia: Oil and Gas Journal, v. 85, no. 9, p. 15-17.

Antia, D. D., 1986. Kinetic method for modeling vitrinite reflectance: Geology, v. 14, p. 606-608.

Aoyagi, K. and Asakawa, T., 1984. Paleotemperature analysis by authigenic minerals and its application to petroleum exploration: AAPG Bulletin, v. 68, p. 903-913.

Armagnac, C., Bucci, J., Kendall, C. G. St. C., and Lerche, I., 1989. Estimating the thickness of sediment removed at an unconformity using vitrinite reflectance data: p. 217-238 in "Thermal History of Sedimentary Basins" (Edited by N. D. Naeser and T. H. McCulloh). Springer-Verlag, New York.

Armagnac, C., Kendall, C.G. St.C., Kuo, C., Lerche, I., and Pantano, J., 1988. Determination of paleoheat flux from vitrinite reflectance data, and from sterane and hopane isomer data: Journal of Geochemical Exploration, v. 30, p. 1-18.

Armstrong, P. A., Chapman, D. S., Funnell, R. H., Allis, R.G., and Kamp, P. J. J., 1996. Thermal modeling and hydrocarbon generation in an active-margin basin: Taranaki basin, New Zealand: AAPG Bulletin, v. 80, p. 1216-1241.

Arne, D. C., 1991. Evidence from apatite fission-track analysis for regional Cretaceous cooling in the Ouachita Mountain fold belt and Arkoma Basin of Arkansas: AAPG Bulletin, v.76, p. 392-402.

Arne, D. C., 1991. Regional thermal history of the Pine Point Area, Northwest Territories, Canada, from apatite fission-track analysis: Economic Geology, v. 86, p. 428-435.

Athy,L. F., 1930. Density, porosity, and compaction of sedimentary rocks: AAPG Bulletin, v. 14, p. 1-24.

Audley-Charles, M. G., Curray, J. R., and Evans, G., 1977. Location of major deltas: Geology, v. 5, p. 341-344.

Autric, A. and Dumesnil, P., 1985. Resistivity, radioactivity and sonic transit time logs to evaluate the organic content of low permeability rocks: The Log Analyst, v. 26, p. 36-45.

Avbovbo, A. A., 1978. Geothermal gradients in the Southern Nigeria basin: Bulletin of Canadian Petroleum Geology, v. 26, p. 258-274.

Ayers, M. G., Bilal, M., Jones, R. W., Slentz, L. W., Tartir, M., and Wilson, A. O., 1982. Hydrocarbon habitat in main producing areas, Saudi Arabia: AAPG Bulletin, v. 66, p. 1-9.

Bachu, S. and Cao, S., 1992. Present and past geothermal regimes and source rock maturation, Peace River Arch area, Canada: AAPG Bulletin, v. 76, p. 1533-1549.

Bachu, S., Ramon, J. C., Villegas, M. E., and Underschultz, J. R., 1995. Geothermal regime and thermal history of the Llanes basin, Colombia: AAPG Bulletin, v. 79, p. 116-129.

Bailey, N. J. L., Walko, P., and Sauer, M. J., 1987. Geochemistry and source rock potential of the west of Shetlands: p. 711-721 in "Petroleum Geology of North West Europe" (Edited by J. Brooks and K. Glennie). Graham and Trotman, London.

Baird, R. A., 1986. Maturation and source rock evaluation of Kimmeridge Clay, Norwegian North Sea: AAPG Bulletin, v. 70, p. 1-11.

Baker, J. C. and deCaritat, P., 1992. Postdepositional history of the Permian sequence in the Denison trough, eastern Australia: AAPG Bulletin, v. 76, p. 1224-1249.

Baldwin, B. and Butler, C. O., 1985. Compaction curves: AAPG Bulletin, v. 69, p. 622-626.

Bally, A. W. and Snelson, S., 1980. Realms of subsidence: p. 9-94 in "Facts and Principles of World Petroleum Occurrences" (Edited by A. D. Miall). Canadian Society of Petroleum Geologists Memoir 6, Calgary.

Bandy, O. L. and Arnal, R. E., 1969. Middle Tertiary basin development, San Joaquin Valley, California: Geological Society of America Bulletin, v. 80, p. 783-820.

Bandy, O.L., 1960. The geologic significance of coiling ratios in the foraminifer *Globigerina pachyderm* (Ehrenberg): Journal of Paleontology: v. 34 (4), p. 671-681.

Barker, C., 1972. Aquathermal pressuring - role of temperature in development of abnormal pressure zones: AAPG Bulletin, v. 56, p. 2068-2071.

Barker, C., 1974. Pyrolysis techniques for source rock evaluation: AAPG Bulletin, v. 58, p. 2349-2361.

Barker, C., 1979. "Organic Geochemistry in Petroleum Exploration" AAPG Continuing Education Course Note Series, No. 10. 159pp.

Barker, C., 1979. Generation and accumulation of oil and gas in deltas: p. 83-96 in "Pennsylvanian Sandstones of the Mid-Continent" (Edited by N. J. Hyne). Tulsa Geological Society Special Publication No. 1.

Barker, C., 1980. Primary migration: The importance of water-mineral-organic matter interactions in the source rock: AAPG Studies in Geology No. 10, p. 19-31.

Barker, C., 1982. Oil and gas on passive continental margins: AAPG Memoir 34, p. 549-565.

Barker, C., 1985. Origin, composition and properties of petroleum: p.11-45 in "Enhanced Oil Recovery, 1" (Edited by E.C. Donaldson, G.V. Chilingarian, and T.F. Yen), Elsevier, Amsterdam.

Barker, C., 1987. Development of abnormal and subnormal pressures in reservoirs containing bacterially generated gas: AAPG Bulletin, v. 71, p. 1404-1413.

Barker, C., 1990. Calculated volume and pressure changes during the thermal cracking of oil to gas in reservoirs: AAPG Bulletin, v. 74, p. 1254-1261.

Barker, C., 1996. Nomenclature in the petroleum industry: Kirk-Othmer Encyclopedia of Chemical Technology, Fourth Edition, Volume 18, p. 342-346.

Barker, C. and Dickey, P. A., 1984. Hydrocarbon habitat in main producing area, Saudi Arabia: Discussion: AAPG Bulletin, v. 68, p. 108-109.

Barker, C. and Takach, N. E., 1992. Prediction of natural gas composition in ultradeep sandstone reservoirs: AAPG Bulletin, v. 76, p. 1859-1873.

Barker, C. and Wang, L., 1988. Applications of pyrolysis in petroleum geochemistry: a bibliography: Journal of Analytical and Applied Pyrolysis, v. 13, p. 9-61.

Barker, C. E., 1983. Influence of time on metamorphism of sedimentary organic matter in liquid-dominated geothermal systems, western North America: Geology, v. 11, p. 384-392.

Barker, C. E., 1988. Geothermics of petroleum systems: implications of the stabilization of kerogen thermal maturation after a geologically brief heating duration at peak temperature: p. 26-29 in "Petroleum Systems of the United States" (Edited by L. B. Magoon), U.S. Geological Survey Bulletin 1870.

Barker, C. E., 1989. Temperature and time in the thermal maturation of sedimentary organic matter: p. 75-98 in, "Thermal History of Sedimentary Basins - Methods and Case Histories," (Edited by N. D. Naeser and T. H. McCulloh), Springer-Verlag, New York.

Barker, C. E., 1991. Implications for organic maturation studies of evidence for a geologically rapid increase and stabilization of vitrinite reflectance at peak temperature: Cerro Prieto geothermal system, Mexico: AAPG Bulletin, v. 75, p. 1852-1863. (see also AAPG Bulletin, v 77, p. 665-578).

Barker, C. E., 1996. A comparison of vitrinite reflectance measurements made on whole-rock and dispersed organic matter concentrate mounts: Organic Geochemistry, v. 24, p. 251-256.

Barker, C. E. and Goldstein, R. H., 1990. Fluid-inclusion technique for determining maximum temperature in calcite and its comparison to the vitrinite reflectance geothermometer: Geology, v. 18, p. 1003-1006.

Barker, C. E. and Halley, R. B., 1986. Fluid inclusion, stable isotope, and vitrinite reflectance evidence for the thermal history of the Bone Spring Limestone, southern Guadalupe Mountains, Texas: Society of Economic Paleontologists and Mineralogists Special Publication No. 38, p. 189-203.

Barker, C. E., and Bone, Y., 1995. The mineral response to contact metamorphism by the Devonian Buchan Caves Limestone, Buchan Rift, Victoria, Australia: Organic Geochemistry, v. 22, p. 151-164.

Barker, C. E. and M. J. Pawlewicz, 1986. The correlation of vitrinite reflectance with maximum temperature in humic organic matter: p. 79-93 in "Paleogeothermics", Lecture Notes in Earth Sciences, v. 5, (Edited by G. Buntebarth and L. Stegena), Springer-Verlag.

Barker, C. E. and W. A. Elders, 1981. Vitrinite reflectance geothermometry and apparent heating duration in the Cerro Prieto Geothermal Field: Geothermics, v. 10, p. 207-223.

Barnard, P. C., Collins, A. G., and Cooper, B. S., 1981. Identification and distribution of kerogen facies in a source rock horizon - examples from the North Sea Basin: p. 271-282 in "Organic Maturation Studies and Fossil Fuel Exploration" (Edited by J. Brooks), Academic Press, New York.

Barnes, D. A., Girard, J. P., and Aronson, J. L., 1992. K-Ar dating of illite diagenesis in the Middle Ordovician St. Peters Sandstone, central Michigan Basin, USA: Implications for thermal history: p. 35-48 in "Origin, Diagenesis, and Petrophysics of Clay Minerals in Sandstones" (Edited by D. W. Houseknecht and E. D. Pittman), Society of Economic Paleontologists and Mineralogists (Society for Sedimentary Geology) Special Publication No. 47.

Barth, T. and Nielsen, S. B., 1993. Estimating kinetic parameters for generation of petroleum and single compounds from hydrous pyrolysis of source rocks: Energy Fuels, v. 7, p. 100-110.

Bass, R., 1989. "Oil Notes" Houghton Mifflin/Seymour Lawrence, Boston. 712 pp.

Be, A. W. H. and Hamlin, W. H., 1967. Ecology of Recent planktonic foraminifera. Part 3 - Distribution in North Atlantic during the summer of 1962: Micropaleontology, v. 13 (1), p. 87-106.

Beaumont, C. and Tankard, A. J. (Editors), 1987. "Sedimentary Basins and Basin-Forming Mechanisms", Canadian Society of Petroleum Geologists, Memoir 12 and Atlantic Geoscience Society, Special Publication 5. 527 pp.

Beaumont, C., Boutilier, R., Mackenzie, A. S., and Rullkotter, J., 1985. Isomerization and aromatization of hydrocarbons and the paleothermometry and burial history of Alberta foreland basin: AAPG Bulletin, v. 69, p. 546-566.

Beck, A. E. and Shen, P. Y., 1989. On a more rigorous approach to geothermic problems: Tectonophysics, v. 164, p. 83-92.

Behar, F., Kressmann, S., and Rudkiewicz, J. L., 1991. Thermal evaluation of crude oils in sedimentary basins: experimental simulation in a confined system and kinetic modeling: Review Institute Francaise du Petrol, v. 46, p. 151-181.

Belford, D. J., 1989. Planktonic foraminifera and age of sediments, west Tasmanian margin, South Tasman Rise and Lord Howe Rise: BMR Journal of Australian Geology and Geophysics, v. 11, p. 37-62.

Benali, S., Schreiber, B. C., Helman, M. L., and Philp, R. P., 1995. Characterization of organic matter from a restricted/evaporative sedimentary environment: Late Miocene of Lorca basin, southeastern Spain: AAPG Bulletin, v. 79, p. 816-830.

Bennett, S. S. and Hanor, J. S., 1987. Dynamics of subsurface salt dissolution at the Welsh Dome, Louisiana Gulf Coast: p. 653-677 in "Dynamical Geology of Salt and Related Structures" (Edited by I. Lerche and J.J. O'Brien), Academic Press.

Benson, R. H., 1975. The origin of the psychrosphere as recorded in changes of deep-sea ostracode assemblages: Lethaia, v. 8, p. 69-83.

Berggren, W. A., Kent, D. V., and Flynn, J. J., 1985. Paleogene geochronology and chronostratigraphy: p. 141-195 in "The Chronology of the Geologic Record" (Edited by N. J. Snelling), Geological Society of London, Memoir 10.

Bessis, F., 1986. Some remarks on the study of subsidence of sedimentary basins. Application to the Gulf of Lions margin (Western Mediterranean): Marine and Petroleum Geology, v. 3, p. 37-63.

Bethke, C. M. and Marshak, S., 1990. Brine migrations across North America - the plate tectonics of groundwater: Annual Review of Earth and Planetary Sciences, v. 18, p. 287-315.

Bethke, C. M., 1985. A numerical model of compaction driven groundwater flow and heat transfer and its application to the paleohydrology of intracratonic sedimentary basins: Journal of Geophysical Research, v. 90, p. 6817-6828.

Bethke, C. M., 1989. Modeling subsurface flow in sedimentary basins: Geologische Rundschau, v. 78, p. 129-154.

Bethke, C. M., Harrison, W. J., Upson, C., and Altaner, S. P., 1988. Supercomputer analysis of sedimentary basins: Science, v. 239, p. 261-267.

Bethke, C. M., Reed, J. D., and Oltz, D. F., 1991. Long-range petroleum migration in the Illinois basin: AAPG Bulletin, v. 75, p. 925-945.

Bishop, R. S., 1979. Calculated compaction states of thick abnormally pressured shales: AAPG Bulletin, v. 63, p. 918-933.

Blackwell, D. D. and Steele, J. L., 1989. Thermal conductivity of sedimentary rocks: measurement and significance: p. 13-36 in "Thermal History of Sedimentary Basins" (Edited by N. D. Naeser and T. H. McCulloh). Springer-Verlag, New York.

Blackwell, D. D., and Steele, J. L., (Compilers), 1992. Geothermal Map of North America: Geological Society of America, Decade of North American Geology. Continental-scale map 006.

Blaser, R. and White, C., 1984. Source-rock and carbonization study, Maracaibo basin, Venezuela: AAPG Memoir 35, p. 229-252.

Bloch, S., McGowen, J. H., Duncan, J. R., and Brizzolara, D. W., 1990. Porosity prediction, prior to drilling, in sandstones of the Kekiktuk Formation (Mississippian), North Slope of Alaska: AAPG Bulletin, v. 74, p. 1371-1385.

Bockmeulen, H., Barker, C., and Dickey, P. A., 1983. Geology and geochemistry of crude oils, Bolivar Coastal Fields, Venezuela: AAPG Bulletin, v. 67, p. 242-270.

Bodenhausen, J. W. A. and Ott, W. F., 1981. Habitat of the Rijswijk oil province, onshore, The Netherlands: p. 301-309 in "Petroleum Geology of the Continental Shelf of North-West Europe" (Edited by L. V. Illing and G. D. Hobson), Heyden Press, London.

Bodnar, R. J., 1994. Philosophy of fluid inclusion analysis: p. 1-6 in "Fluid Inclusions in Minerals: Methods and Applications" (Edited by B. DeVito and M. L. Frezzotti). Short Course Notes of the Working Group "Inclusions in Minerals", International Mineralogical Association.

Bodnar, R. J. and Bethke, P. M., 1984. Systematics of stretching of fluid inclusions; fluorite and sphalerite at 1 atmosphere confining pressure: Economic Geology, v. 79, p. 141-16.

Bodnar, R. J., Binns, P. R., and Hall, D. L., 1989. Synthetic fluid inclusions IV: quantitative evaluation of the decrepitation behaviour of fluid inclusions in quartz at one atmosphere confining pressure: Journal of Metamorphic Geology, v. 7, p. 229-242.

Bodnar, D. P. and Sharp, J. M., Jr., 1988. Temperature variations in South Texas subsurface: AAPG Bulletin, v. 72, p. 21-32.

Boltovskoy, E. and Wright, R., 1976. "Recent Foraminifera" W. Junk, The Hague. 515pp.

Bond, W. A., 1984. Application of Lopatin's method to determine burial history, evolution of the geothermal gradient, and timing of hydrocarbon generation in Cretaceous source rocks in the San Juan Basin: p. 433-447 in "Hydrocarbon Source Rocks of the Greater Rocky Mountain Region" (Edited by J. Woodward, F. F. Meissner, and J. L. Clayton). Rocky Mountain Association of Geologists, Denver.

Bordenave, M. L., (Editor), 1993. "Applied Petroleum Geochemistry", Editions Technip, Paris.

Borer, J. M. and Harris, P. M., 1991. Lithofacies and cyclicity of the Yates Formation, Permian Basin: Implications for reservoir heterogeneity: AAPG Bulletin, v. 75, p. 726-779.

Bosscher, H. and Schlager, W., 1993. Accumulation rates of carbonate platforms: Journal of Geology, v. 101, p. 345-355.

Bostick, N. H., 1973. Time as a factor in thermal metamorphism of phytoclasts: Compte Rendu, v. 2, p. 183-193.

Bostick, N. H., 1979. Microscopic measurement of the level of catagenesis of solid organic matter in sedimentary rocks to aid exploration for petroleum and to determine former burial temperatures - a review: p. 17-43 in "Aspects of Diagenesis" (Edited by P. A. Scholle and P. R. Schluger), Society of Economic Paleontologists and Mineralogists Special Publication No. 26.

Bostick, N. H., and Pawlewicz, M. J., 1984. Paleotemperatures based on vitrinite reflectance of shales and limestones in igneous dike aureoles in the Upper Cretaceous Pierre Shale, Walsenburg, Colorado: p. 387-392 in, "Hydrocarbon Source Rocks of the Greater Rocky Mountain Region" (Edited by J. Woodward, F. F. Meissner, and J. L. Clayton) Rocky Mountain Association of Geologists, Denver.

Bostick, N. H., Cashman, S. M., McCulloch, T. H., and Waddell, C. T., 1978. Gradients in vitrinite reflectance and present temperatures in the Los Angeles and Ventura Basins, California: p. 65-96 in "Low Temperature Metamorphism of Kerogen and Clay Minerals" Society of Economic Paleontologists and Mineralogists (Pacific Section). Symposium in Geochemistry.

Braun, R. L. and Burnham, A. K., 1990. Mathematical model of oil generation, degradation, and expulsion: Energy and Fuels, v. 4, p. 132-146.

Braun, R. L. and Burnham, A. K., 1992. PMOD: a flexible model of oil and gas generation, cracking and expulsion: Organic Geochemistry, v. 19, p. 161-172.

Braun, R. L. and Burnham, A. K., 1987. Analysis of chemical reaction kinetics using a distribution of activation energies and simpler models: Energy and Fuels, v. 1, p. 153-161.

Braun, R. L., Burnham, A. K., and Sweeney, J. J., 1988. Petroleum Geochemistry Project, Annual Report, October 1987 to September 1988. LLNL Internal Report, UCID-21-303-88-4.

Braun, R. L., Burnham, A. K., Reynolds, J. G., and Clarkson, J. E., 1990. Pyrolysis kinetics for lacustrine and marine source rocks by programmed micropyrolysis: Energy and Fuels, v. 5, p. 192-204.

Bredehoeft, J. D., Wesley, J. B., and Fouch, T. D., 1994. Simulations of the origin of fluid pressure, fracture generation, and the movement of fluids in the Uinta basin, Utah: AAPG Bulletin, v. 78, p. 1729-1747.

Brewer, R. C. and Groshong, R. H., Jr., 1993. Restoration of cross sections above intrusive salt domes: AAPG Bulletin, v. 77, p. 1769-1780.

Brigaud, F., Chapman, D. S., and Le Douaran, S., 1990. Estimating thermal conductivity in sedimentary basins using lithologic data and geophysical well logs: AAPG Bulletin, v. 74, p. 1459-1477.

Briggs, N. D., Naeser, C. W., and McCulloh, T. H., 1981. Thermal history of sedimentary basins by fission track dating: Nuclear Tracks, v. 5, p. 235-237.

Brooks, J. M., Cox, H. B., Bryant, W. R., Kennicutt, M. C., II, Mann, R. G., and McDonald, T. J., 1986. Association of gas hydrates and oil seepage in the Gulf of Mexico: Organic Geochemistry, v. 10 p. 221-234.

Brooks, J. M., Kennicutt, M. C., II, Fay, R. R., McDonald, T. J., and Sassen, R., 1984. Thermogenic gas hydrates in the Gulf of Mexico: Science, v. 225, p 409-411.

Brown, R. W., 1991. Backstacking apatite fission-track "stratigraphy": A method for resolving the erosional and isostatic rebound components of tectonic uplift histories: Geology, v. 19, p. 74-77.

Brun, M. V. L. and Lucazeau, F., 1988. Subsidence, extension and thermal history of the West African margin in Senegal: Earth and Planetary Science Letters, v. 90, p. 204-220.

Brune, J. N., Henyey, T. L., and Roy, R. F., 1969. Heat flow, stress, and rate of slip along the San Andreas fault, California: Journal of Geophysical Research, v. 74, p. 3821-3827.

Buiskool Toxopeus, J. M. A., 1983. Selection criteria for the use of vitrinite reflectance as a maturity tool: p. 295-307 in "Petroleum Geochemistry and Exploration of Europe" (Edited by J. Brooks), Blackwell Scientific Publications, Oxford.

Bulat, J. and Stoker, S. J., 1987. Uplift determination from interval velocity studies, UK southern North Sea: p. 293-305 in "Petroleum Geology of North West Europe" (Edited by J. Brooks and K. W. Glennie), Graham and Trotman, London.

Buntebarth, G. and Stegena, L. (Editors), 1986. "Paleogeothermics: Evaluation of Geothermal Conditions in the Geological Past." Lecture Notes in Earth Sciences, v. 5, Springer-Verlag. 234 pp.

Buntebarth, G. and Stegena, L., 1986. Methods in paleogeothermics: p. 5-39 in "Paleogeothermics" (Edited by G. Buntebarth and L. Stegena), Springer-Verlag, Berlin.

Burley, S. D., Mullis, J., and Matter, A., 1989. Timing diagenesis in the Tartan Reservoir (UK North Sea): constraints from combined cathodoluminescence microscopy and fluid inclusion studies: Marine and Petroleum Geology, v.6, p. 98-120.

Burnham, A. K. and Braun, R. L., 1990. Development of a detailed model of petroleum formation, destruction, and expulsion from lacustrine and marine source rocks: Organic Geochemistry, v. 16, p. 27-39.

Burnham, A. K. and Sweeney, J. J., 1989. A chemical kinetic model of vitrinite reflectance and maturation: Geochimica et Cosmochimica Acta, v. 53, p. 2649-2657.

Burnham, A. K. and Sweeney, J. J., 1991. Modeling the maturation and migration of petroleum: p. 55-63 in "Source and Migration Processes and Evaluation Techniques" (Edited by R. K. Merrill), AAPG Treatise of Petroleum Geology/Handbook of Petroleum Geology.

Burnham, A. K. and Sweeney, J. J., 1991. Reply to comments by S. B. Nielsen and T. Barth on "A chemical kinetic model of vitrinite maturation and reflectance": Geochimica et Cosmochimica Acta, v. 55, p. 643-644.

Burnham, A. K., 1994. Comment on "The effects of the mineral matrix on the determination of kinetic parameters using modified Rock-Eval pyrolysis" by H. Dembicki, Jr., and the resulting comment by R. Pelet: Organic Geochemistry, v. 21, p. 985-986.

Burnham, A. K., Braum, R. L., Sweeney, J. J., Reynolds, J. G., Vallejos, C., and Talukdar, S., 1992. Kinetic modeling of petroleum formation in the Maracaibo Basin: Final Report: U.S. Department of Energy, Report DOE/BC/92001051 (DE92001051).

Burnham, A. K., Braun, R. L., and Samoun, A., 1987. Further comparison of methods for measuring kerogen pyrolysis rates and fitting kinetic parameters: Pergamon Press, New York. also, Organic Geochemistry, v. 13, p. 839-845.

Burnham, A. K., Braun, R. L., Gregg, H. R., and Samoun, A. M., 1987. Comparison of methods for measuring kerogen pyrolysis rates and fitting kinetics parameters: Energy and Fuels, v. 1, p. 452-458.

Burns, T. R., 1982. Structure and petroleum potential of the continental margin between Cross Sound and Icy Bay, Northern Gulf of Alaska: U.S. Geological Survey Open-File Report No. 82-929.

Burrus, J. (Editor), 1987. "Thermal Modelling in Sedimentary Basins" Gulf Publishing Co., Houston. 624 pp.

Burrus, J. and Audebert, F., 1990. Thermal and compaction processes in a young rifted basin containing evaporites: Gulf of Lions, France: AAPG Bulletin, v. 74, p. 1420-1440.

Burrus, J. and Bessis, F., 1986. Thermal modeling in the Provencal Basin (NW-Mediterranean): p. 393-413 in "Thermal Modeling in Sedimentary Basins" (Edited by J. Burrus), Editions Technip, Paris.

Burrus, J., Osadetz, K., Wolf, S., Doligez, B., Visser, K., and Dearborn, D., 1996. A two-dimensional regional basin model of Williston basin hydrocarbon systems: AAPG Bulletin, v. 80, p. 265-291.

Burruss, R. C. and Goldstein, R. H., 1980. Time and temperature of hydrocarbon migration: Fluid inclusion evidence from the Fayetteville Formation, NW Arkansas: Geological Society of America, Abstracts with Programs, v.12, p.396.

Burruss, R. C. and Hollister, L. S., 1979. Evidence from fluid inclusions for a paleo-geothermal gradient at the geothermal test well sites, Los Alamos, New Mexico: Journal Volcanology and Geothermal Research, v. 5, p. 163-177.

Burruss, R. C., 1987. Diagenetic paleotemperatures from aqueous fluid inclusions: re-equilibration of inclusions in carbonate cements by burial heating: Mineralogical Magazine, v. 51, p. 477-481.

Burruss, R. C., 1989. Paleotemperatures from fluid inclusions: advances in theory and technique: p. 119-131 in "Thermal History of Sedimentary Basins" (Edited by N. D. Naeser and T. H. McCulloh), Springer-Verlag, New York.

Burruss, R. C., Cercone, K. R., and Harris, P. M., 1983. Fluid inclusion petrography and tectonic-burial history of the Al Ali No. 2 well: Evidence for the timing of diagenesis and oil migration, northern Oman Foredeep: Geology, v. 11, p. 567-570.

Burruss, R. C., Cercone, K. R., and Harris, P. M., 1985. Timing of hydrocarbon migration: Evidence from fluid inclusions in calcite cements, tectonics and burial history: p. 277-389 in "Carbonate Cements" (Edited by N. M. Schneidermann and P. M. Harris), Society of Economic Paleontologists and Mineralogists, Special Publication No. 36.

Burst, J. F., 1969. Diagenesis of Gulf Coast clayey sediments and its possible relation to petroleum migration: AAPG Bulletin, v. 53, p. 73-93.

Burtner, R. L. and Nigrini, A., 1994. Thermochronology of the Idaho-Wyoming thrust belt during the Sevier orogeny: A new, calibrated, multiprocess thermal model: AAPG Bulletin, v. 78, p. 1586-1612.

Burtner, R. L., Nigrini, A., and Donelick, R. A., 1994. Thermochronology of Lower Cretaceous source rocks in the Idaho-Wyoming thrust belt: AAPG Bulletin, v. 78, p. 1613-1636.

Bustin, R. M., 1983. Heating during thrust faulting in the Rocky Mountains: Friction or fiction? Tectonophysics, v. 95, p. 309-328.

Bustin, R. M., 1988. Sedimentology and characteristics of dispersed organic matter in Tertiary Niger Delta: Origin of source rocks in a deltaic environment: AAPG Bulletin, v. 72, p. 277-298.

Butler, R. W. H., 1991. Hydrocarbon maturation, migration and tectonic loading in the Western Alpine foreland thrust belt: p. 227-244 in "Petroleum Migration" (Edited by W. A. England and A. J. Fleet). Geological Society (London) Special Publication No. 59.

Calvert, S. E., 1983. Sedimentary geochemistry of silicon: p. 143-186 in "Silicon Geochemistry and Biogeochemistry" (Edited by S. R. Aston), Academic Press, London.

Cao, S. and Lerche, I., 1989. Geohistory, thermal history and hydrocarbon generation history of Navarin Basin COST No. 1 Well, Bering Sea, Alaska: Journal of Petroleum Geology, v. 12, p. 325-352.

Cardott, B. J., and Lambert, M. W., 1985. Thermal maturation by vitrinite reflectance of Woodford Shale, Anadarko Basin, Oklahoma: AAPG Bulletin, v. 67. p. 1982-1998.

Carlson, P. R., Golan-Bac, M., Karl, H. A., and Kvenvolden, K. A., 1985. Seismic and geochemical evidence for shallow gas in sediment on Navarin continental margin, Bering Sea: AAPG Bulletin, v. 69, p. 422-436.

Carpentier, B., Bessereau, G., and Huc, A. Y., 1991. Well logs and source rocks: Assessing organic carbon content by CARBOLOG technique: The Log Analyst, v. 32, p. 279-297.

Carroll, A. R., Brassell, S. C., and Graham, S.A., 1992. Upper Permian lacustrine oil shales, southern Junggar basin, northwest China: AAPG Bulletin, v. 76, p. 1874-1902.

Carstens, H. and Finstad, K., 1981. Geothermal gradients of the northern North Sea basin, 59-62°N: p. 152-161 in "Petroleum Geology of the Continental Shelf of Northwestern Europe" (Edited by L. V. Illing and G. D. Hobson), Heyden Press, London.

Carter, A., Bristow, C., and Hurford, A. J., 1995. Thermal histories of Permain and Triassic basins in Britain derived from fission track analysis: p. 41-56 in "Permian and Triassic Rifting in Northwest Europe", Geological Society (London) Special Publication No. 91.

Carvalho, H. S. and Vacquier, V., 1977. Method for determining terrestrial heat flow in oil fields: Geophysics, v. 42, p. 584-593.

Cassidy, D. P. and Ranganathan, V., 1992. Groundwater upwelling, near Bay St. Elaine salt dome in southeastern Louisiana, as inferred from fluid property variations: AAPG Bulletin, v. 76, p. 1560-1568.

Castano, J. R. and Sparks, D. M., 1974. Interpretation of vitrinite reflectance measurements in sedimentary rocks and determination of burial history using vitrinite reflectance and authigenic minerals: Geological Society of America Special Paper 153, p. 31-52.

Castano, J. R., 1985. Uses of vitrinite reflectance in determining thermal history in sedimentary basins (Abstr.): AAPG Bulletin, v. 69, p. 243.

Castelli, A., Chiaramonte, M. A., Beltrame, P. L., Carniti, P., DelBianco, A., and Stroppa, F., 1990. Thermal degradation of kerogen by hydrous pyrolysis. A kinetic study: Organic Geochemistry, v. 16, p. 75-82.

Cathles, L. M. and Nunns, A. G., 1991. A temperature probe survey on the Louisiana shelf: Effects of bottom-water temperature variations: AAPG Bulletin, v. 75, p. 180-186.

Cazier, E. C., Haywood, A. B., Espinosa, G., Velandia, J., Mugniot, J-F., and Leel, W. G., Jr., 1995. Petroleum geology of the Cusiana Field, Llanos basin foothills, Colombia: AAPG Bulletin, v. 79, p. 1444-1463.

Cercone, K. R., 1984. Thermal history of Michigan basin: AAPG Bulletin, v. 68, p. 130-136.

Cercone, K. R. and Lohman, K. C., 1987. Late burial diagenesis of Niagaran (Middle Silurian) pinnacle reefs in Michigan Basin: AAPG Bulletin, v. 71, p. 156-166.

Cercone, K. R., Deming, D., and Pollack, H. N., 1996. Insulating effect of coals and black shales in the Appalachian Basin, Western Pennsylvania: Organic Geochemistry, v. 24, p. 243-249.

Cerveny, P. R., Naeser, N. D., Zeitler, P. K., Naeser, C. S., and Johnson, N. M., 1988. History of uplift and relief of the Himalaya during the past 18 million years: Evidence from fission track ages of detrital zircons from sandstones of the Siwalik Group: p. 43-61 in "New Perspectives in Basin Analysis" (Edited by K. L. Kleinspehn and C. Paola), Springer-Verlag, New York.

Chapman, D. S., Keho, T., Bauer, M., and Picard, M. D., 1984. Heat flow in the Uinta basin determined from bottom hole temperature (BHT) data: Geophysics, v. 49, p. 453-466.

Chapman, R. E., 1983. "Petroleum Geology", Elsevier, Amsterdam. 415pp.

Chenet, P. Y., Bessis, F., Ungerer, P., Nogaret, E., and Perrin, J. F., 1983. How geological mathematical models can reduce exploration risks: Proceedings of the Eleventh World Petroleum Congress, v. 2, London.

Chenet, P. Y., 1984. Thermal transfer in sedimentary basins. Paleotemperature reconstruction and maturation studies in the Gulf of Lion margin: p. 257-269 in "Thermal Phenomena in Sedimentary Basins" (Edited by B. Durand). Editions Technip, Paris.

Cheung, P. K., 1978. The geothermal gradient in sedimentary rocks in Oklahoma: Unpublished M.S. Thesis, Oklahoma State University, Stillwater. 55 pp.

Cifelli, R., 1971. On the temperature relationships of planktonic foraminifera: Journal of Foraminferal Research, v. 1(4), p. 170-177.

Clark, S. P., 1966. Handbook of Physical Constants; Geological Society of America Memoir No. 97. 587 pp.

Clarkson, G. and Reiter, M., 1987. The thermal regime of the San Juan Basin since Late Cretaceous times and its relationship to San Juan Mountains thermal sources: Journal of Volcanology and Geothermal Research, v. 31, p. 217-237.

Claypool, G.E. and Magoon, L.B., 1988. Oil and gas source rocks in the National Petroleum Reserve in Alaska: p. 451-481 in "Geology and Exploration of the National Petroleum Reserve in Alaska, 1974 to 1982" (Edited by G. Gryc), U.S. Geological Survey Professional Paper 1399.

Clayton, J. L. and King, J. D., 1987. Effects of weathering on biological marker and aromatic hydrocarbon composition of organic matter in Phosphoria shale outcrop: Geochimica et Cosmochimica Acta, v. 51, p. 2153-2157.

Clayton, J. L. and Koncz, I., 1994. Petroleum geochemistry of the Zala basin, Hungary: AAPG Bulletin, v. 78, p. 1-22.

Clayton, J. L. and Ryder, R. T., 1984. Organic geochemistry of black shales and oils in the Minnelusa Formation (Permian and Pennsylvanian), Powder River Basin, Wyoming: p. 231-253, in "Hydrocarbon Source Rocks of the Greater Rocky Mountain Region" (Edited by J. Woodward, F. F. Meissner, and J. L. Clayton), Rocky Mountain Association of Geologists, Denver.

Clayton, J. L. and Swetland, P. J., 1978. Subaerial weathering of sedimentary organic matter: Geochimica et Cosmochimica Acta, v. 42, p. 305-312.

Clayton, J. L., and Swetland, P. J., 1980. Petroleum generation and migration in Denver Basin: AAPG Bulletin, v. 64, p. 1613-1633.

Cluff, R. M. and Byrnes, A. P., 1991. Lopatin analysis of maturation and petroleum generation in the Illinois basin (Abstr.): AAPG Bulletin, v. 75, p. 1393.

Cohen, C. R., 1985. Role of fault rejuvenation in hydrocarbon accumulation and structural evolution of Reconcavo Basin, northeastern Brazil: AAPG Bulletin, v. 69, p. 65-76.

Cole, G. A., Drozd, R. J., Sedivy, R. A., and Halpern, H. I., 1987. Organic geochemistry and oil-source correlations, Paleozoic of Ohio: AAPG Bulletin, v. 71, p. 788-809.

Coleman, M. L., Curtis, D. C., and Irwin, H., 1979. Burial rate a key to source and reservoir potential: World Oil, March 1979, p. 83-92

Comer, J. B. and Hinch, H. H., 1987. Recognizing and quantifying expulsion of oil from the Woodford Formation and age-equivalent rocks in Oklahoma and Kansas: AAPG Bulletin, v. 71, p. 844-858.

Connan, J., 1974. Time-temperature relation in oil genesis: AAPG Bulletin, v. 58, p. 2516-2521.

Connan, J., 1984. Biodegradation of crude oils in reservoirs: p.300-330 in “Advances in Petroleum Geochemistry, Volume 1” (Edited by J. Brooks and D. H. Welte), Academic Press, London.

Connolly, C. A., 1989. Thermal history and diagenesis of the Wilrich Member shale, Spirit River Formation, northwest Alberta: Bulletin of Canadian Petroleum Geology, v. 37, p. 182-197.

Conrad, K. T. and C. E. Griffith, 1983. Thermal maturation model of late Proterozoic-Paleozoic Amadeus basin, Central Australia (Abstr): AAPG Bulletin, v. 67, p. 441

Cook, R. A. and Davey, F. J., 1984. The hydrocarbon exploration of the basins of the Ross Sea, Antarctica, from modelling of the geophysical data: Journal of Petroleum Geology, v. 7, p. 213-226.

Cooles, G. P., Mackenzie, A. S., and Quigley, T. M., 1986. Calculation of petroleum masses generated and expelled from source rocks: Organic Geochemistry, v. 10, p. 234-245.

Cooper, B. S., 1977. Estimation of the maximum temperatures attained in sedimentary rocks: p. 127-146 in "Developments in Petroleum Geology, Vol. 1" (Edited by G. D. Hobson), Applied Science Publishers, London.

Cooper, B. S., Coleman, S. H., Barnard, P. C., and Butterworth, J. S., 1975. Paleotemperatures in the northern North Sea: p. 487-492 in "Petroleum on the Continental Shelf of North West Europe Vol. 1", (Edited by A. Woodland), Applied Science, Baking.

Cooper, M. A., Addison, F. T., Alvarez, R., Coral, M., Graham, R. H., Hayward, A. B., Howe, S., Martinez, J., Naar, J., Penas, R., Pulham, A. J., and Taborda, A., 1995. Basin development and tectonic history of the Llanos basin, Eastern Cordillera, and Middle Magdalena Valley, Columbia: AAPG Bulletin, v. 79, p. 1421-1443.

Cope, M. J., 1986. An interpretation of vitrinite reflectance data from the southern North Sea Basin: p. 85-98 in "Habitat of Paleozoic Gas in N. W. Europe" (Edited by J. Brooks, J. C. Goff, and B. van Horn), Geological Society Special Publication No. 23, London.

Cornford, C., Morrow, J. A., Turrington, A., Miles, J. A., and Brooks, J., 1983. Some geological controls on oil composition in the U.K. North Sea: p. 175-194 in "Petroleum Geochemistry and Exploration of Europe" (Edited by J. Brooks), Blackwell Scientific, Oxford.

Correia, M., 1971. Diagenesis of sporopollenin: p. 569-620 in "Sporopollenin" (Edited by Brooks, Grant, Muir, Van Gijzel and Shaw), Academic Press, London.

Cowan, G., 1989. Diagenesis of Upper Carboniferous sandstones: southern North Sea Basin: p. 57-73 in "Deltas: Sites and Traps for Fossil Fuels" (Edited by M. K. G. Whately and K. T. Pickering), Geological Society Special Publication No. 41. Blackwell Scientific Publications, Oxford.

Creaney, S. and Passey, Q. R., 1993. Recurring patterns of total organic carbon and source rock quality within a sequence stratigraphic framework: AAPG Bulletin, v. 77, p. 386-401.

Cross, T. A. (Editor), 1990. "Quantitative Dynamic Stratigraphy" Prentice Hall, West Nyack, New York. 625 pp.

Crowley, K. D., 1991. Thermal history of Michigan basin and southern Canadian Shield from apatite fission track analysis: Journal of Geophysical Research, v. 96, no. B1, p. 697-711.

Crowley, T. J. and North, G. R., 1991. "Paleoclimatology", Oxford University Press, Oxford. 338 pp.

Curiale, J. A., Larter, S. R., Sweeney, R. E., and Bromley, B. W., 1989. Molecular thermal maturity indicators in oil and gas source rocks: p. 53-72 in "Thermal History of Sedimentary Basins" (Edited by N. D. Naeser and T. H. McCulloh), Springer-Verlag, New York.

Curiale, J. A., 1993. Oil to source rock correlation: Concepts and case studies: p.473-490 in "Organic Geochemistry" (Edited by M. H. Engel and S. A. Macko), Plenum Press, New York.

Currie, J. B. and Nwachukwu, 1974. Evidence of incipient fracture porosity in reservoir rocks at depth: Bulletin of Canadian Petroleum Geology , v. 22, p. 42-58.

D'Heur, M., 1991. West Ekofisk field - Norway Central Graben, North Sea: AAPG Bulletin, v. 75, p. 946-968.

Dahl, B. and Augustson, J. H., 1993. The influence of Tertiary and Quaternary sedimentation and erosion on hydrocarbon generation in Norwegian offshore basins: p. 419-431 in "Basin Modeling: Advances and Applications" (Edited by A. G. Dore et al), Norwegian Petroleum Society Special Publication 3, Elsevier, Amsterdam.

Dahl, B. and Yukler, A., 1991. The role of petroleum geochemistry in basin modeling of the Oseberg area, North Sea: p. 65-85 in "Source and Migration Processes and Evaluation Techniques" (Edited by R. K. Merrill), AAPG Treatise of Petroleum Geology/Handbook of Petroleum Geology.

Dahl, B., Nysaether, E., Speers, G. C., and Yukler, A., 1987. Oseberg area-integrated basin modelling: p. 1029-1038 in "Petroleum Geology of North West Europe" (Edited by J. Brooks and K. Glennie), Graham and Trotman, London.

Dahl, J. E. P., Moldowan, J. M., Terrman, S. C., McCaffrey, M. A., Sundararaman, P., and Stelting, C. E., 1994. Source rock quality determination from oil biomarkers I: A new geochemical technique: AAPG Bulletin, v. 78, p. 1507-1526.

Dallenbach, J., Espitalie, J., and Lebreton, F., 1983. Source rock logging: Transactions of the Eight European Society of Petroleum Well Log Analysts Symposium, paper D.

Dashwood, M. F. and Abbotts, I. L., 1990. Aspects of the petroleum geology of the Oriente Basin, Ecuador: p. 89-117 in "Classic Petroleum Provinces" (Edited by J. Brooks), Geological Society (London).

Davey, F. J., 1985. The Antarctic margin and its possible hydrocarbon potential: Tectonophysics, v. 114, p. 443-370.

Dazhen, T., Fermont, W. J. J., Qi, Y., and Zhigui, P., 1990. Maturity evolution and hydrocarbon generation in the Hedong coal-field, central North China: International Symposium on Organic Petrology, Mededelingen Rijks Geologische Dienst, v. 45, p. 122-137.

DeBremaeker, J. C., 1983. Temperature, subsidence and hydrocarbon maturation in extensional basins: a finite element model: AAPG Bulletin, v. 67, p. 1410-1414.

deCaritat, P., 1989. Note on the maximum upward migration of pore water in response to sediment compaction: Sedimentary Geology, v. 65, p. 371-377.

Degens, E. T., 1965. "Geochemistry of Sediments - a Brief Survey", Prentice-Hall. 342 pp.

Delvaux, D., Martin, H., Leplat, P., and Paulet, J., 1990. Geochemical characterization of sedimentary organic matter by means of pyrolysis kinetic parameters: Organic Geochemistry, v. 16, p. 175-187.

Demaison, G. and Huizinga, B. J., 1991. Genetic classification of petroleum systems: AAPG Bulletin, v. 75, p. 1626-1643.

Demaison, G. J. and Moore, G. T., 1980. Anoxic environments and oil source bed genesis: AAPG Bulletin, v. 64, p. 1179-1209.

Demaison, G., 1984. The generative basin concept: AAPG Memoir 35, p. 1-14.

Dembicki, E. A. and Machel, H. G., 1996. Recognition and delineation of paleokarst zones by the use of wireline logs in the bitumen-saturated Upper Devonian Grosmont Formation of northeastern Alberta, Canada: AAPG Bulletin, v. 80, p. 695-712.

Dembicki, H., Jr., 1984. An interlaboratory comparison of source rock data: Geochimica et Cosmochimica Acta, v. 48, p. 2641-2649.

Dembicki, H., Jr., 1992. The effects of the mineral matrix on the determination of kinetic parameters using modified Rock Eval pyrolysis: Organic Geochemistry, v. 18, p. 531-539.

Deming, D. and Chapman, D. S., 1989. Thermal histories and hydrocarbon generation: example from Utah-Wyoming thrust belt: AAPG Bulletin, v. 73, p. 1455-1471.

Deming, D. and Nunn, J. A., 1991. Numerical simulations of brine migration by topographically driven recharge: Journal of Geophysical Research, v. 96, no. 82, p. 2485-2499.

Deming, D., Nunn, J. A., and Evans, D. G., 1990. Thermal effects of compaction-driven groundwater flow from overthrust belts: Journal of Geophysical Research, v. 95, no. B5 p. 6669-6683.

Deming, D., Sass, J. H., Lachenbruch, A. H., and de Rito, R. F., 1992. Heat flow and subsurface temperature as evidence for basin scale ground-water flow, North Slope of Alaska: Geological Society of American Bulletin, v. 104, p. 528-542.

Demongodin, L., Vasseur, G., and Brigaud, F., 1993. Anisotropy of thermal conductivity in clayey formations: p. 209-217 in "Basin Modeling: Advances and Applications" (Edited by A. G. Dore et al) Norwegian Petroleum Society Special Publication 3, Elsevier, Amsterdam.

Deroo, G., Powell, T. G., Tissot, B., and McCrossan, R. G., 1977. The origin and migration of petroleum in the western Canadian sedimentary basin, Alberta. A geochemical and thermal maturation study: Geological Survey of Canada, Bulletin No. 262, 136 pp.

Dickin, A. P., 1995. "Radiogenic Isotope Geology", Cambridge University Press. 452 pp.

Dickinson, G., 1953. Geological aspects of abnormal reservoir pressures in Gulf Coast Louisiana: AAPG Bulletin, v. 37, p. 410-432.

Ditchfield, P. and Marshall, J. D., 1989. Isotopic variation in rhythmically bedded chalks: paleotemperature variation in the Upper Cretaceous: Geology, v. 17, p. 842-845.

Dixon, J., Dietrich, J. R., McNeil, D. H., McIntyre, D. J., Snowden, L. R., and Brooks, P. W., 1985. "Geology, biostratigraphy and organic geochemistry of Jurassic to Pleistocene strata, Beaufort-Mackenzie area, Northwest Canada", Canadian Society of Petroleum Geologists Course Notes. 150 pp.

Dixon, J., Dietrich, J., Snowdon, L.R., Morrell, G., and McNeil, D. H., 1992. Geology and petroleum potential of upper Cretaceous and Tertiary strata, Beaufort-Mackenzie area, northwest Canada: AAPG Bulletin, v. 76, p. 927-947.

Doligez, B., Bessis, F., Burrus, J., Ungerer, P., and Chenet, P. Y., 1986. Integrated numerical simulation of the sedimentation heat transfer, hydrocarbon formation and fluid migration in a sedimentary basin: The Themis model: p. 173-195 in "Thermal Modeling in Sedimentary Basins" (Edited by J. Burrus), Editions Technip, Paris.

Doligez, B., Ungerer, P., Chenet, P.Y., Burrus, J., Bessis, F., and Bessereau, G., 1987. Numerical modelling of sedimentation, heat transfer, hydrocarbon formation and fluid migration in the Viking Graben, North Sea: p. 1039-1048 in "Petroleum Geology of North West Europe" (Edited by J. Brooks and K. Glennie), Graham and Trotman, London.

Donovan, R. N., 1991. The Arbuckle Group - An aide de memoire: p. 199-208 in "Arbuckle Group Core Workshop and Field Trip" (Edited by K. S. Johnson), Oklahoma Geological Survey Special Publication 91-3.

Dore, A. G., et al (Editors), 1993. "Basin Modelling: Advances and Applications" Norwegian Petroleum Society Special Publication 3, Elsevier, Amsterdam. 675 pp.

Dow, W. G., 1977. Kerogen studies and geological interpretations: Journal of Geochemical Exploration, v. 7, p. 79-99.

Dow, W. G and O'Connor, D. I., 1982. Kerogen maturity and type by reflected light microscopy applied to petroleum exploration: p. 133-157 in Society of Economic Paleontologists and Mineralogists Short Course No. 7, "How to Assess Maturation and Paleotemperatures."

Dow, W. G. and Pearson, D. B., 1977. Organic matter in Gulf Coast sediments: p. 85-94 in Preprints, Offshore Technology Conference, Houston, Texas.

Dowdle, W. L. and Cobb, W. M., 1975. Static formation temperature from well logs - an empirical method: Journal of Petroleum Technology, v. 27, p. 1326-1330.

Downey, M. W., 1984. Evaluating seals for hydrocarbon accumulations: AAPG Bulletin, v. 68, p. 1752-1763.

Drewry, G. E., Ramsey, A. T. S., and Smith, A. G., 1974. Climatically controlled sediments, the geomagnetic field, and the trade wind belts in Phanerozoic time: Journal of Geology, v. 81, p. 531-553.

Dumitru, T. A., Duddy, I. R., and Green, P. F., 1994. Mesozoic-Cenozoic burial, uplift, and erosion history of the west-central Colorado Plateau: Geology, v. 22, p. 499-502.

Duncan, W. and Roman, C., 1986. New plays in the offshore Aaiun basin: Oil and Gas Journal, June 23rd, p. 58-63.

Dunoyer de Segonzac, G., 1970. The transformation of clay minerals during diagenesis and low-grade metamorphism: a review: Sedimentology, v. 15, p. 281-346.

Dunton, M. L. and Hunt, J. M., 1962. Distribution of low molecular weight hydrocarbons in recent and ancient sediments: AAPG Bulletin, v. 46, p. 1-6.

Durand, B. (Editor), 1984. "Thermal Phenomena in Sedimentary Basins", Gulf Publishing Co., Houston. 326 pp.

Durand, B. and Monin, J. C., 1980. Elemental analysis of kerogens (C, H, O, N, S, Fe): Chapter 4 (p. 113-142) in "Kerogen" (Edited by B. Durand), Editions Technip, Paris.

Durand, B. and Nicaise, G., 1980. Procedures for kerogen isolation: p. 35-53 in "Kerogen" (Edited by B. Durand), Editions Technip, Paris.

Durand, B., 1988. Understanding of hydrocarbon migration in sedimentary basins (present state of knowledge): Organic Geochemistry, v. 13, p. 445-460.

Durand, B., Ungerer, P., Chiarelli, A., and Oudin, J. L., 1993. Modelisation de la migration de l'huile; application a deux exemples de bassins sedimentaires: p. 3-15 in Proceedings of the Eleventh World Petroleum Congress, PDI (1). Wiley and Sons, New York.

DuRouchet, J., 1981. Stress fields, a key to oil migration: AAPG Bulletin, v. 65, p. 74-85.

Dutton, S. P. and Diggs, T. N., 1992. Evolution of porosity and permeability in the Lower Cretaceous Travis Peak Formation, East Texas: AAPG Bulletin, v. 76, p. 252-269.

Dykstra, J., 1987. Compaction correction for burial history curves: application to Lopatin's method for source rock maturation determination: Geobyte, v. 2, no. 4, p. 16-23.

Eadington, P. J., Hamilton, P. J., and Green, P., 1989. Hydrocarbon fluid history in relation to diagenesis in the Hutton Sandstone, south-west Queensland: p. 601-618 in "The Cooper and Eromanga Basins, Australia" (Edited by B. J. O'Neil), Proceedings Petroleum Exploration Society of Australia, Society of Petroleum Engr., Australian Society of Exploration Geophysis. (SA Branches), Adelaide.

Edman, J. D. and Furlong, K. P., 1987. Thrust faulting and hydrocarbon generation: Reply: AAPG Bulletin, v. 71, p. 890-896.

Edman, J. D. and Pitman, J. K., 1986. Lopatin modeling as tool for regional basin analysis (Abstr.): AAPG Bulletin, v. 70, p. 585-586.

Edman, J. D. and Surdam, R. C., 1984. Influence of overthrusting on maturation of hydrocarbons in Phosphoria Formation, Wyoming-Idaho-Utah Overthrust Belt: AAPG Bulletin, v. 68, p. 1803-1817.

Edman, J. D. and Surdam, R. C., 1986. Organic-inorganic interactions as a mechanism for porosity enhancement in the Upper Cretaceous Ericson sandstone, Green River basin, Wyoming: Society of Economic Paleontologists and Mineralogists, Special Publication No. 38, p. 85-109.

Ejedawe, J. E., 1981. Patterns of incidence of oil reserves in Niger Delta Basin: AAPG Bulletin, v. 65, p. 1574-1585.

Ejedawe, J. E. and Coker, S. J. L., 1984. Dynamic interpretation of organic - matter maturation and evolution of oil-generative window: AAPG Bulletin, v. 68, p. 1024-1028.

Ejedawe, J. E., and Coker, S. J. L., 1986. Dynamic interpretation of organic-matter maturation and evolution of oil-generative window: Reply: AAPG Bulletin v. 70, p. 1011-1014.

Ejedawe, J. E., Coker, S. J. L., Lambert-Aikhionbare, D. O., Alofe, K. B., and Adoh, F. O., 1984. Evolution of oil-generative window and oil and gas occurrence in Tertiary Niger Delta Basin: AAPG Bulletin, v. 68, p. 1744-1751.

Ekweozor, C. M., Okoye, N. V., 1980. Petroleum source-bed evaluation of Tertiary Niger Delta: AAPG Bulletin, v. 64, p. 1251-1259.

Elliott, R. E., 1985. Quantification of peat to coal compaction stages, based on phenomena in the East Pennine Coalfield, England: Proceedings of the Yorkshire Geological Society, v. 45, p. 163-172.

Elliott, W. C., Aronson, J. L., Matisoff, G., and Gautier, D. L., 1991. Kinetics of the smectite to illite transformation in the Denver Basin: clay mineral, K-Ar data and mathematical model results: AAPG Bulletin, v. 75, p. 436-462.

Elliot, W.C. and Matisoff, G., 1996. Evaluation of kinetic models for the smectite to illite transformation: Clays and Clay Minerals, v. 44, p. 77-87.

Ellyard, D. J., 1984. Oil migration in the Northern Canning Basin – a regional review: Geological Society of Australia Inc., Perth; p. 359-375.

Elphick, R. Y., 1989. Lopatin modelling for hydrocarbon maturity, Part 1: Geobyte, v. 4, no. 2, p. 49-54.

Elphick, R. Y. and Reed, R. J., 1989. Lopatin modeling for hydrocarbon maturity, Part 2 - DOS and Mac programs: Geobyte, v. 4, no.3, p. 28-36.

Emery, D. and Robinson, A., 1993. "Inorganic Chemistry: Applications to Petroleum Geology" Blackwell Scientific Publications,Oxford. 254 pp.

Engel, M. H. and Macko, S. A. (Editors), 1993. "Organic Geochemisty Principles and Applications", Plenum Press, New York. 861 pp.

England, T. D. J. and Bustin, R. M., 1986. Effect of thrust faulting on organic maturation in the southeastern Canadian cordillera: Organic Geochemistry, v. 10, p. 609-616.

England, T. D. J. and Bustin, R. M., 1986a. Thermal maturation of the western Canadian sedimentary basin south of the Red Deer River. (I) Alberta Plains: Bulletin of Canadian Petroleum Geology, v. 34, p. 71-90.

England, W. A., Mackenzie, A. S., Mann, D. M., and Quiqley, T. M., 1987. The movement and entrapment of petroleum fluids in the subsurface: Journal of the Geological Society of London, v. 144, p. 327-347.

England, W. A., Mann, A. L., and Mann, D. M., 1991. Migration from source to trap: p. 23-46 in "Source and Migration Processes and Evaluation Techniques" (Edited by R. K. Merrill), AAPG Treatise of Petroleum Geology/Handbook of Petroleum Geology.

Epstein, A. G., Epstein, J. B., and Harris, L. D., 1977. Conodont color alteration - an index to organic metamorphism: U.S. Geological Survey Professional Paper 995, 27 pp.

Erdogan, L. T. and Akgul, A., 1981. An oil migration and re-entrapment model for the Mardin Group reservoirs of Southeast Anatolia: Journal of Petroleum Geology, v. 4, p. 57-75.

Erez, J., 1978. Vital effect on stable-isotope composition seen in Foraminifera and coral skeletons: Nature, v. 273, p. 199-202.

Ericson, D. B., 1959. Coiling direction of *Globoratalia pachyderma* as a climatic index: Science, v. 130, p. 219-220.

Eslinger, E. V., Savin, S. M., and Yeh, H-W., 1979. Oxygen isotope geothermometry of diagenetically altered shales: p. 113-124 in "Aspects of Diagenesis" (Edited by P. A. Scholle and P. R. Schluger), Society of Economic Paleontologists and Mineralogists, Special Publication No. 26.

Espitalie, J., 1984. Tentative reconstruction of geothermal paleogradients in some wells of the Rhine Graben: p. 147-159 in "Thermal Phenomena in Sedimentary Basins" (Edited by B. Durand), Gulf Publication Co., Houston.

Espitalie, J., Laporte, J. L., Madec, M., Marquis, F., Leplat, P., Paulet, J., and Boutefeu, A., 1977. Methode rapide de characterisation des roches meres, de leur potential petrolier et de leur degre d'evolution: Review of the French Petroleum Institute, v. 32, p. 23-42.

Espitalie, J., Marquis, F., and Barsony, I., 1984. Geochemical logging: p. 274-304 in "Analytical Pyrolysis - Techniques and Applications" (Edited by K. J. Vorhees), Butterworth, Boston.

Espitalie, J., Marquis, F., and Drouet, S., 1993. Critical study of kinetic modelling parameters: p. 233-242 in "Basin Modeling: Advances and Applications" (Edited by A. G. Dore et al), Norwegian Petroleum Society Special Publication 3, Elsevier, Amsterdam.

Espitalie, J., Marquis, F., and Sage, L., 1987. Organic geochemistry of the Paris Basin: p. 71-86 in "Petroleum Geology of North West Europe" (Edited by J. Brooks and K. Glennie), Graham and Trotman, London.

Espitalie, J., Maxwell, J. R., Chenet, Y., and Marquis, F., 1988. Aspects of hydrocarbon migration in the Mesozoic in the Paris Basin as deduced from an organic geochemical survey: Organic Geochemistry, v. 13, p. 467-481.

Espitalie, J., Senga, M. K., and Trichet, J., 1984. Role of the mineral matrix during kerogen pyrolysis: Organic Geochemistry, v. 6, p. 365-384.

Essene, E. J. and Peacor, D. R., 1995. Clay mineral thermometry - A critical perspective: Clays and Clay Minerals, v. 43, p. 540-553.

Evamy, B. D., Haremboure, J., Karerling, P., Knaap, W. A., Malloy, F. A., and Rowlands, P. H., 1978. Hydrocarbon habitat of Tertiary Niger delta: AAPG Bulletin, v. 62, p. 1-39.

Evans, R. and Kirkland, D. W., 1988. Evaporitic environments as a source of petroleum: p. 256-299 in "Evaporites and Hydrocarbons" (Edited by B. C. Schreiber), Columbia University Press, New York.

Evans, T. R., 1977. Thermal properties of North Sea rocks: Log Analyst, v. 18, no. 2, p. 3-12.

Falvey, D. A. and Deighton, I., 1982. Recent advances in burial and thermal geohistory analysis: APEA Journal, v. 22, p. 65-81.

Falvey, D. A. and Middleton, M. F., 1982. Passive continental margins: evidence for a pre-breakup deep crustal metamorphic subsidence mechanism: Colloquium on Geology of Continental Margins (C3), Oceanologica Acta, v. 4(Suppl.), p. 103-114.

Faure, G., 1986. "Principles of Isotope Geology" (Second Edition), John Wiley, New York. 589 pp.

Feary, D. A., Davies, P. J., Pigram, C. J., and Symonds, P. A., 1991. Climatic evolution and control on carbonate deposition in northeast Australia: Palaeogeography, Palaeoclimatology, Palaeoecology (Global and Planetary Change Section), v. 89, p. 341-361.

Feazel, C. T. and Aram, R. B., 1990. Interpretation of discontinuous vitrinite reflectance profiles: Discussion: AAPG Bulletin, v. 74, p. 91-93.

Feinstein, S., 1981. Subsidence and thermal history of southern Oklahoma aulacogen: implications for petroleum exploration: AAPG Bulletin, v. 65, p. 2521-2533.

Feinstein, S., Issler, D.R., Snowdon, L. R., and Williams, G.K., 1996. Characterization of major unconformities by paleothermometric and paleobarometric methods: Application to the Mackenzie Plain, Northwest Territories, Canada: Bulletin of Canadian Petroleum Geology, v. 44, p. 55-71.

Fertl, W. H., 1985. Logs help approximate reservoir temperatures: Oil and Gas Journal, April 29th, p. 89-92.

Fertl, W. H., Chapman, R. E., and Hotz, R. F. (Editors), 1994. "Studies in Abnormal Pressures", Elsevier, Amsterdam.

Fertl, W. H. and Chilingarian, G. V., 1988. Total organic carbon content determined from well logs: Society of Petroleum Engineers Formation Evaluation, v. 3, p. 407-419.

Fertl, W. H. and Wichmann, P. A., 1977. How to determine static BHT from well log data: World Oil, v. 184, no. 1, p. 105-106.

Fertl, W. J. and Rieke, H. H., 1980. Gamma-ray spectral evaluation techniques identify fractured shale reservoirs and source-rock characteristics: Journal of Petroleum Technology, v. 31, p. 2053-2062.

Field, M. E. and Kvenvolden, K. A., 1985. Gas hydrates on the northern California continental margin: Geology, v. 13, p. 517-520. (see also "Comment and Reply" by D. A. Warnke, Geology, v. 14, p. 536-538).

Finkel, E. A. and Wilkinson, B. H., 1990. Stylolitization as source of cement in Mississippian Salem Limestone, West-Central Indiana: AAPG Bulletin, v. 74, p. 174-186.

Fischer, A. G., 1969. Geological time-distance rates: the Bubnoff unit: Geological Society of America Bulletin, v. 80, p. 549-551.

Fitzgerald, M. G., Mitchum, R. M., Jr., Uliana, M. A., and Biddle, K. T., 1990. Evolution of the San Jorge Basin, Argentina: AAPG Bulletin, v. 74, p. 879-920.

Fitzgerald, P. G. and Stump, E., 1991. Early Cretaceous uplift in the Ellsworth Mountains of west Antarctica: Science, v. 254, p. 92-94.

Fitzgerald, P. G., 1994. Thermochronologic constraints on post-Paleozoic tectonic evolution of the central Transantarctic Mountains, Antarctica: Tectonics, v. 13, p. 818-836.

Fjeldskaar, W., Prestholm, E., Guargena, C., and Stephenson, M., 1993. Mineralogical and diagentic control on the thermal conductivity of the sedimentary sequences in the Bjornoya Basin, Barents Sea: p. 445-453 in "Basin Modeling: Advances and Applications" (Edited by A. G. Dore et al), Norwegian Petroleum Society Special Publication 3, Elsevier, Amsterdam.

Forbes, P. L., Ungerer, P. M., Kuhfuss, A. B., Riis, F., and Eggen, S., 1991. Compositional modeling of petroleum generation and expulsion: Trial application to a local mass balance in the Smorbukk Sor Field, Haltenbanken area, Norway: AAPG Bulletin, v. 75, p. 873-893.

Forbes, P. L., Ungerer, P., and Mudford, B. S., 1992. A two-dimensional model of overpressure development and gas accumulation in Venture field, eastern Canada: AAPG Bulletin, v. 76, p. 318-338.

Frakes, L. A., 1979. "Climates Throughout Geologic Time", Elsevier, Amsterdam. 310 pp.

Frakes, L. A., Francis, J.E., and Syktus, J. I., 1992. "Climate Modes of the Phanerozoic": Cambridge University Press. 274 pp.

Frakes, L. A., Probst, J. L., and Ludwig, W., 1994. Latitudinal distribution of paleotemperature on land and sea from early Cretaceous to middle Miocene: C. R. Acad. Sci. Paris, v. 318, p. 1209-1218.

Frandsen, N., Vejboek, O. V., Moller, J. J., and Michelsen, O., 1987. A dynamic geological model of the Danish Central Trough during the Jurassic-Early Cretaceous: p. 453-468 in "Petroleum Geology of North West Europe" (Edited by J. Brooks and K. Glennie), Graham and Trotman, London.

Fred, R. L. and Peacor, D. R., 1989. Variability in temperature of the smectite/illite reaction on Gulf Coast sediments: Clay Minerals, v. 24, p. 171-180.

Freeman, E. S. and Carroll, B., 1958. The application of thermoanalytical techniques to reaction kinetics. The thermogravimetric evaluation of the kinetics of the decomposition of calcium oxalate monohydrate: Journal of Physical Chemistry, v. 62, p. 394-397.

Friedman, I. and O'Neil, J. R., 1977. Compilation of stable isotope fractionation factors of geochemical interest: U. S. Geological Survey Professional Paper 440K, 12 pp.

Friend, P. F., Johnson, N. M., and McRae, L. E., 1989. Time-level plots and accumulation patterns of sediment sequences: Geological Magazine, v. 126, p. 491-198.

Furlong, K. and Edman, J. D., 1984. Graphical approach to determination of hydrocarbon maturation in overthrust terrains: AAPG Bulletin, v. 68, p. 1818-1824.

Furlong, K. P. and Edman, J. D., 1989. Hydrocarbon maturation in thrust belts: Thermal considerations: p. 137-144 in "Origin and Evolution of Sedimentary Basins and Their Energy and Mineral Resources" (Edited by R. A. Price), Geophysical Monograph 48, IUGC Volume 3 American Geophysical Union and the International Union of Geodesy and Geophysics.

Garcia-Gonzalez, M., MacGowan, D. B., and Surdam, R. C., 1993. Coal as a source rock of petroleum and gas - a comparison between natural and artificial maturation of the Almond Formation coals, greater Green River Basin in Wyoming: p. 405-437 in "The Future of Energy Gases" (Edited by D. G. Howell), U.S. Geological Survey Professional Paper 1570.

Garven, G. and Freeze, R. A., 1984. Theoretical analysis of the role of groundwater flow in the genesis of stratabound ore deposits: American Journal of Science, v. 284, p. 1085-1174.

Ge, S. and Garven, G., 1989. Tectonically induced transient groundwater flow in foreland basin: p. 145-157 in "Origin and Evolution of Sedimentary Basins and Their Energy and Mineral Resources" (Edited by R. A. Price), Geophysical Monograph 48, IUGC Volume 3 American Geophysical Union and the International Union of Geosesy and Geophysics.

Genik, G. J., 1993. Petroleum geology of Cretaceous-Tertiary rift basins in Niger, Chad, and Central African Republic: AAPG Bulletin, v 77, p. 1405-1434.

Geoff, J. C., 1983. Hydrocarbon generation and migration from Jurassic source rocks in the E. Sheltand Basin and Viking Graben of the North Sea: Journal of the Geological Society (London), v. 140, p. 445-474.

Gernant, R. E. and Kessling, R. V., 1966. Foraminiferal paleoecology and paleoenvironmental reconstruction of the Oligocene Middle Frio in Chambers County, Texas: Transactions of the Gulf Coast Association of Geological Societies, v. 16, p. 131-158.

Geyh, M. A. and Schleicher, H., 1990. "Absolute Age Determination: Physical and Chemical Dating Methods and Their Application", Springer-Verlag, New York. 514 pp.

Ghignone, J. I. and Andrade, G., 1970. General geology and major oil fields of Reconcavo basin, Brazil: p. 337-358 in “Geology of Giant Petroleum Fields” (Edited by M.T. Halbouty) AAPG Memoir No. 14.

Gildner, R. F., 1990. STRIPPER: an interactive backstripping, decompaction and geohistory problem: p. 164-180 in "Quantitative Dynamic Stratigraphy" (Edited by T. A. Cross), Prentice Hall, Englewood Cliffs, New Jersey.

Gill, D., 1979. Differential entrapment of oil and gas in Niagaran pinnacle-reef belt of northern Michigan: AAPG Bulletin, v. 63, p. 608-620.

Girard, J. P. and Barnes, D. A., 1995. Illitization and paleothermal regimes in the Middle Ordovician St. Peter sandstone, central Michigan basin: K-Ar, oxygen isotope, and fluid inclusion data: AAPG Bulletin, v. 79, p. 49-69.

Girard, J. P. and Onstott, T. C., 1991. Application of $^{40}Ar/^{39}Ar$ laser-probe and step-heating techniques to the dating of diagenetic K-feldspar overgrowths: Geochimica et Cosmochimica Acta, v. 55, p. 3777-3793.

Girard, J. P., Aronson, J. L., and Savin, S. M., 1988. Separation, K/Ar dating and $^{18}O/^{16}O$ ratio measurements of diagenetic K-feldspar overgrowths: An example from the Lower Cretaceous arkoses of the Angola margin: Geochimica et Cosmochimica Acta, v. 52, p. 2207-2214.

Glasmann, J. R., Clark, R. A., Larter, S., Briedis, N. A., and Lundegard, P. D., 1989a. Diagenesis and hydrocarbon accumulation, Brent sandstone (Jurassic), Bergen High area, North Sea: AAPG Bulletin, v. 73, p. 1341-1360.

Glasmann, J. R., Larter, S., Briedis, N. A., and Lundegard, P. D., 1989b. Shale diagenesis in the Bergen High area, North Sea: Clays and Clay Minerals, v. 37, p. 97-112.

Gleadow, A. J. W., Duddy, I. R., and Lovering, J. F., 1983. Fission track analysis: a new tool for the evaluation of thermal histories and hydrocarbon potential: Australian Petroleum Exploration Association (APEA) Journal, v. 23, p. 93-102.

Gleadow, A. J. W., Duddy, I. R., Green, P. F., and Hegarty, K. A., 1986. Fission track lengths in the apatite annealing zone and the interpretation of mixed ages: Earth and Planetary Science Letters, v. 78, p. 245-254.

Gleadow, A. J. W., Duddy, I. R., Green, P. F., and Lovering, J. F., 1986. Confined fission track lengths in apatite: a diagnostic tool for thermal history analysis: Contributions to Mineralogy and Petrology, v. 94, p. 405-415.

Gleadow, A. J. W., McKelvey, B. C., and Ferguson, K. U., 1984. Uplift history of the Transantarctic Mountains in the Dry Valleys area, southern Victoria Land, Antarctica, from apatite fission track ages: New Zealand Journal of Geology and Geophysics, v. 27, p.457-464.

Glennie, K. W. and Boegner, P. L. E., 1981. Sole Pit inversion tectonics: p. 110-120 in "Petroleum Geology of the Continental Shelf of North-West Europe" (Edited by L. V. Illing and G. D. Hobson), Institute of Petroleum, London.

Goff, J. C., 1983. Hydrocarbons generation and migration from Jurassic source rocks in East Shetland basin and Viking graben of the northern North Sea: Journal of the Geological Society of London, v. 140, p. 445-474.

Gold, T., 1993. The origin of methane in the crust of the earth: p. 57-80 in "The Future of Energy Gases" (Edited by D. G. Howell), U.S. Geological Survey Professional Paper 1570.

Goldstein, R. H., 1987. Reequilibration of fluid inclusions in low-temperature calcium-carbonate cement: Geology, v. 14, p. 792-795.

Goldstein, R. H. and Reynolds, T. J., 1994. "Systematics of Fluid Inclusions in Diagenetic Minerals", SEPM Short Course No. 31. 199 pp.

Golubic, S. O., Perkins, R. D., and Lucas, K. J., 1975. Boring microorganisms in carbonate substrates: p. 229-259 in "The Study of Trace Fossils" (Edited by R. W. Frey), Springer-Verlag, New York.

Goodarzi, F., Gentzis, T., Harrison, C., and Thorsteinsson, R., 1992. The significance of graptolite reflectance in regional thermal maturity studies, Queen Elizabeth Islands, Arctic Canada: Organic Geochemistry, v. 18, p. 347-357.

Gordon, W. A., 1975. Distribution by latitude of Phanerozoic evaporite deposits: Journal of Geology, v. 83, p. 671-684.

Goy, G., 1979. Les "schistes cartons" (Toarcien Inferieur) du bassin de Paris: Unpublished doctoral thesis, Pierre and Marie Curie University.

Grace, J. D. and Hart, G. F., 1986. Giant gas fields of northern West Siberia: AAPG Bulletin, v. 70, p. 830-852.

Graham, S. A., Brassell, S., Carroll, A. R., Xiao, X., Demaison, G., McKnight, C. L., Liang, Y., Chu, J., and Hendrix, M. S., 1992. Characteristics of selected petroleum source rocks, Xianjiang Uygur Autonomous Region, northwest China: AAPG Bulletin, v. 74, p. 493-512.

Green, K. E., 1960. Ecology of some Arctic foraminifera: Micropaleontology, v. 6(1), p. 57-78.

Green, P. F., Duddy, I. R., Gleadow, A. J. W., Tingate, P. R., and Laslett, G. M., 1986. Thermal annealing of fission tracks in apatite 1. A qualitative description: Chemical Geology (Isotope Geoscience Section), v. 59, p. 239-253.

Green, P. F., Duddy, I. R., Gleadow, J. W., and Lovering, J. F., 1989. Apatite fission-track analysis as a paleotemperature indicator for hydrocarbon exploration: p. 181-195 in "Thermal History of Sedimentary Basins" (Edited by N. D. Naeser and T. H. McCulloh), Springer-Verlag, New York.

Green, P. F., Duddy, I. R., Laslett, G. M., Hegarty, K. A., Gleadow, A. J. W., and Lovering, J. F., 1989. Thermal annealing of fission tracks in apatite: 4-quantitative modeling techniques and extension to geological timescales: Chemical Geology (Isotope Geosciences Section), v. 79, p. 155-182.

Gretener, P. E. and Curtis, C. D., 1982. Role of temperature and time on organic metamorphism: AAPG Bulletin, v. 66, p. 1124-1149.

Gretener, P. E., 1981. "Geothermics: Using Temperature in Hydrocarbon Exploration", AAPG Education Course Note Series No. 17, 156 pp.

Grigo, D., Marangna, B., Arienti, M. T., Fiorani, M, Parisi, A., Marrone, M., Squazzero, P., and Uberg, A. S., 1993. Issues in 3-D sedimentary basin modelling and application to Haltenbanken, offshore Norway: p. 455-468 in "Basin Modeling: Advances and Applications" (Edited by A. G. Dore et al), Norwegian Petroleum Society Special Publication 3, Elsevier, Amsterdam.

Grist, A. M., Zentilli, M., and Ryan, R. J., 1995. The thermal evolution and timing of hydrocarbon generation in the Maritimes basin of eastern Canada: Evidence from apatite fission track data: Bulletin of Canadian Petroleum Geology, v. 43, p. 145-155.

Grunau, H. R., 1990. Rift systems can point way to hydrocarbon richness: World Oil, November 1990, p. 96-98.

Guidish, T. M., Kendall, C.G.St C., Lerche, I. Toth, D. J., and Yarzab, R. F., 1985. Basin evalution using burial history calculations: an overview: AAPG Bulletin, v. 69, p. 92-105.

Gumati, Y. D. and Nairn, A.E.M., 1991. Tectonic subsidence of the Sirte Basin, Libya: Journal of Petroleum Geology, v.14, p. 93-102.

Gumati, Y. D. and Schamel, S., 1988. Thermal maturation history of the Sirte Basin, Libya: Journal of Petroleum Geology, v. 11, p. 205-218.

Gussow, W. C., 1954. Differential entrapment of oil and gas; a fundamental principle: AAPG Bulletin, v. 38, p. 816-853.

Gussow, W. C., 1968, Migration of reservoir fluids: Journal of Petroleum Technology, v. 20, p. 353-363.

Guthrie, J. M., Houseknecht, D. W., and Johns, W. D., 1986. Relationships among vitrinite reflectance, illite crystallinity, and organic geochemistry in Carboniferous strata, Ouachita Mountains, Oklahoma and Arkansas: AAPG Bulletin, v. 70, p. 26-33.

Habicht, J. K. A., 1979. “Paleoclimate, Paleomagnetism, and Continental Drift”, AAPG Studies in Geology No. 9. 31 pp.

Hacquebard, P. A. and Donaldson, J. R., 1970. Coal metamorphism and hydrocarbon potential in the Upper Paleozoic of the Atlantic Provinces, Canada: Canadian Journal of Earth Sciences, v. 7, p. 1139-1163.

Hacquebard, P. A., 1977. Rank of coal as an index of organic metamorphism for oil and gas in Alberta: p. 11-22 in "The Origin and Migration of Petroleum in the Western Canadian Sedimentary Basin, Alberta", Geological Society of Canada Bulletin No. 262.

Hagen, E. S. and Surdam, R. C., 1984. Maturation history and thermal evolution of Cretaceous source rocks of the Bighorn basin, Wyoming and Montana: p. 321-338 in "Hydrocarbon Source Rocks of the Greater Rocky Mountain Region" (Edited by J. Woodward, F. F. Meissner and J. L. Clayton), Rocky Mountain Association of Geologists, Denver.

Hagen, E. S. and Surdam, R. C., 1989. Thermal evolution of Laramide-style basins: Constraints from the northern Big Horn basin, Wyoming and Montana: p. 277-295 in "Thermal history of Sedimentary Basins" (Edited by N. D. Naeser and T. H. McCulloh), Springer-Verlag, New York.

Hailwood, E. A., 1989. "Magnetostratigraphy", Geological Society (London) Special Report No. 19. 84 pp.

Halbouty, M. T., 1979. "Salt Domes, Gulf Region, United States and Mexico", Gulf Publishing Company, Houston.

Hambrey, H. A. and Harland, W. B. (Editors), 1981. "Earth's Pre-Pleistocene Glacial Record", Cambridge University Press, Cambridge.

Hamilton, P. J., Fallick, A. E., Macintyre, R. M., and Elliott, S., 1987. Isotopic tracing of the provenance and diagenesis of Lower Brent Group sands, North Sea: p. 939-949 in "Petroleum Geology of North West Europe" (Edited by J. Brooks and K. Glennie), Graham and Trotman, London.

Handyman, D. L., Wisdom, C. J., and Halvers, J. W., 1994. Calcic paleosols: their use in subsurface stratigraphy: AAPG Bulletin, v. 78, p. 1360-1371.

Hanor, J. S., 1980. Dissolved methane in sedimentary brines: potential effect on the PVT properties of fluid inclusions: Economic Geology, v. 75, p. 603-617.

Haq, B., Hardenbol, J., and Vail, P., 1987. Chronology of fluctuating sea levels since the Triassic (250 million years ago to present): Science, v. 235, p. 1156-1167.

Hardie, L. A., 1991. On the significance of evaporites: Annual Review of Earth and Planetary Sciences, v. 19, p. 131-168.

Harland, W. B., Cox, A. V., Llewellyn, P. G., Pickton, C. A. G., Smith, A. G., and Walters, R., 1982. " A Geologic Time Scale", Cambridge University Press, New York, 128 pp.

Harland, W. B., et al., 1990. "A Geologic Time Scale 1989" Cambridge University Press, 274 pp.

Harrison, T. M. and Be, K., 1983. $^{40}Ar/^{39}Ar$ age spectrum analysis of detrital microclines from the southern San Joaquin Basin, California: an approach to determining the thermal evolution of sedimentary basins: Earth and Planetary Science Letters, v. 64, p. 244-256.

Harrison, T. M. and Burke, K., 1988. $^{40}Ar/^{39}Ar$ thermochronology of sedimentary basins using detrital feldspar: examples from the San Joaquin Valley, California, Rio Grande Rift, New Mexico, and North Sea: p. 141-155 in "Thermal History of Sedimentary Basins" (Edited by N. D. Naeser and T. H. McCulloh), Springer-Verlag, New York.

Harrison, W. E., Luza, K. V., Prater, M. L., and Cheung, P. K., 1983. "Geothermal Resource Assessment in Oklahoma", Oklahoma Geological Survey Special Publication 83-1. 42 pp.

Haszeldine, R. S., 1989. Coal reviewed: depositional controls, modern analogues and ancient climates: p. 289-308 in "Deltas: Sites and Traps for Fossil Fuels" (Edited by M. K. G. Whateley and K. T. Pickering), Geological Society Special Publication No. 41. Blackwell Scientific Publications, Oxford.

Haszeldine, R. S., Samson, I. M., and Cornford, C., 1984. Dating diagenesis in a petroleum basin, a new fluid inclusion method: Nature, v. 307, p. 354-357.

Hayba, D. O. and Bethke, C. M., 1995. Timing and velocity of petroleum migration in the Los Angeles basin: Journal of Geology, v. 103, p. 33-49.

Heald, M. T., 1955. Stylolites in sandstones: Journal of Geology, v. 63, p. 101-114.

Heasler, H. P. and Surdam, R. C., 1983. A thermally subsiding basin model for the maturation of hydrocarbons in the Pismo Basin, California: p. 69-74 in "Petroleum Generation and Occurrence in the Miocene Monterey Formation, California" (Edited by C. M. Isaacs and R. E. Garrison), Society of Economic Paleontologists and Mineralogists (Pacific Section).

Heasler, H. P. and Surdam, R. C., 1985. Thermal evolution of coastal California with application to hydrocarbon maturation: AAPG Bulletin, v. 69, p. 1386-1400.

Heasler, H. P. and Kharitonova, N. A., 1996. Analysis of sonic well logs applied to erosion estimates in the Bighorn basin, Wyoming: AAPG Bulletin, v. 80, p. 630-646.

Hedberg, H. D., 1980. Methane generation and petroleum migration: AAPG Studies in Geology No. 10, p. 179-206.

Hegarty, K. A., Weissel, J. K., and Mutter, J. C., 1988. Subsidence history of Australia's southern margin: constraints on basin models: AAPG Bulletin, v. 72, p. 615-633.

Hein, J. R., Scholl, D. W., Barron, J. A., Jones, M. G., and Miller, J., 1978. Diagenesis of Late Cenozoic diatomaceous deposits and formation of the bottom simulating reflector in the southern Bering Sea: Sedimentology, v. 25, p. 155-181.

Herak, M. and Stringfield, V. T. (Editors), 1972. "Karst", Elsevier Publishing Company, Amsterdam. 551 pp.

Hermanrud, C., 1993. Basin modelling techniques - an overview: p. 1-34 in "Basin Modeling: Advances and Applications" (Edited by A. G. Dore et al), Norwegian Petroleum Society Special Publication 3, Elsevier, Amsterdam.

Hermanrud, C. and Shen, P. Y., 1989. Virgin rock temperatures from well logs - accuracy analysis for some advanced inversion models: Marine and Petroleum Geology, v. 6, p. 360-363.

Hermanrud, C., Eggen, S., Jacobsen, T., Carlsen, E. M., and Pallesen, S., 1990. On the accuracy of modelling hydrocarbon generation and migration: The Egersund Basin oil find, Norway: Organic Geochemistry, v. 16, p. 389-399.

Hermansen, D., 1993. Optimization of temperature history - aspects of vitrinite reflectance and sterane isomerization: p. 119-126 in "Basin Modeling: Advances and Applications" (Edited by A. G. Dore et al), Norwegian Petroleum Society Special Publication 3, Elsevier, Amsterdam.

Heroux, Y., Chagnon, A., and Bertrand, R., 1979. Compilation and correlation of major thermal maturation indicators: AAPG Bulletin, v. 63, p. 2128-2144.

Herron, S. L., 1991. In situ evaluation of potential source rocks by wireline logs: p. 127-134 in "Source and Migration Processes and Evaluation Techniques" (Edited by R. K. Merrill), AAPG Treatise of Petroleum Geology/Handbook of Petroleum Geology.

Hess, J., Bender, M. L., and Schilling, J. G., 1986. Evolution of the ratio of strontium-87 to strontium-86 in seawater from Cretceous to present: Science, v. 231, p. 979-984.

Hill, K. C. and Gleadow, A.J.W., 1989. Uplift and thermal history of the Papuan fold belt, Papua New Guinea: apatite fission track analysis: Australain Journal of Earth Sciences, v. 36, p. 515-539.

Hillis, R. R., 1995. Quantification of Tertiary exhumation in the United Kingdom southern North Sea using sonic velocity data: AAPG Bulletin, v. 79, p. 130-152.

Hills, J. M., 1984. Sedimentation, tectonism, and hydrocarbon generation in Delaware Basin, West Texas and southeastern New Mexico: AAPG Bulletin, v. 68, p. 250-267.

Hitchon, B., 1984. Geothermal gradients, hydrodynamics and hydrocarbon occurrences, Alberta, Canada: AAPG Bulletin, v. 68, p. 713-743.

Ho, T. Y., Jensen, R. P., Sahai, S. K., Leadholm, R. H., and Senneseth, O., 1996. Comparative studies of pre- and post-drilling modelled thermal conductivity and maturity data with post-drilling results: Implications to basin modelling and hydrocarbon exploration: Geological Society of London, Special Publication, in press.

Hoefs, J., 1986. Isotope geothermometers: p. 45-52 in "Paleogeothermics" (Edited by G. Buntebarth and L. Stegena), Springer-Verlag, Berlin.

Hoffman, J. and Hower, J., 1979. Clay mineral assemblages as low grade metamorphic geothermometers: applications to the thrust faulted disturbed belt of Montana, U.S.A.: p. 55-79 in "Aspects of Diagenesis" (Edited by P. A. Scholle and P. R. Schluger), Society of Economic Paleontologists and Mineralogists Special Publication No. 26.

Hoffman, J. J., Hower, J., and Aronson, J. L., 1976. Radiometric dating of time of thrusting in the disturbed belt of Montana: Geology, v. 4, p. 16-20.

Hofmann, P., Leythaeuser, D., and Carpentier, B., 1993. Paleoclimatic controlled accumulation of organic matter in Oligocene evaporite sediments of the Mulhouse basin: Organic Geochemistry, v. 20, p. 125-138.

Holder, G. D., Malone, R. D., and Lawson, W. F., 1987. Effects of gas composition and geothermal properties on the thickness and depth of natural-gas-hydrate zones: Journal of Petroleum Technology, v. 39, no. 9, p. 1147-1152.

Hong, Z.-H., Li, H.-X., Rullkotter, J., and Mackenzie, A. S., 1986. Geochemical application of sterane and triterpane biological marker compounds in the Linyi Basin: Organic Geochemistry, v. 10, p. 433-439.

Hood, A., Gutjahr, C. C. M., and Heacock, R. L., 1975. Organic metamorphism and the generation of petroleum : AAPG Bulletin, v. 59, p. 986-996.

Horai, K., 1971. Thermal conductivity of rock-forming minerals: Journal of Geophysical Research, v. 76, p. 1278-1308.

Horibe, Y. and Oba, T., 1972. Temperature scales of aragonite-water and calcite-water systems (in Japanese): Fossils, v. 23/24, p. 69-79.

Horner, D. R., 1951. Pressure build-up in wells: Proceedings of Third World Petroleum Congress, v. 2, p. 503-521.

Horsfield, B. and McClimans, R. L., 1984. Geothermometry and geochemistry of aqueous and oil-bearing fluids inclusions from Fateh Field, Dubai: p. 733-740 in "Advances in Organic Geochemistry 1983" (Edited by P. A. Schenck, J. W. deLeeuw, and G. S. M. Lijmbach), Pergamon Press, New York.

Horsfield, B., 1984. Pyrolysis studies and petroleum exploration: p. 247-298 in "Advances in Petroleum Geochemistry, v. 1" (Edited by J. Brooks and D. Welte), Academic Press, London.

Horsfield, B., Duppenbecker, S J., Schenk, H. J., and Schaefer, R. G., 1993. Kerogen typing concepts designed for the quantitative geochemical evaluation of petroleum potential: p. 243-249 in "Basin Modeling: Advances and Applications" (Edited by A. G. Dore et al), Norwegian Petroleum Society Special Publication 3, Elsevier, Amsterdam.

Horstman, E. L., 1984. Evidence for post-Permian epeirogenic uplift in the Canning Basin from vitrinite reflectance data: p. 401-409 in "The Canning Basin, W. A." (Edited by P. G. Purcell), Proceedings of the Canning Basin Symposium, Perth, 1984.

Horvath, F., Dovenyi, P., and Laczo, I., 1986. Geothermal effect of magmatism and its contribution to the maturation of organic matter in sedimentary rocks: p. 173-183 in "Paleogeothermics" (Edited by G. Buntebarth and L. Stegena), Springer-Verlag, Berlin.

Horvath, F., Szalay, A., Dovenyi, P., and Rumpler, J., 1986. Structural and thermal evolution of the Pannonian Basin: an overview: p. 339-358 in "Thermal Modeling in Sedimentary Basins" (Edited by J. Burrus), Editions Technip, Paris.

Houseknecht, D. W. and Matthews, S. M. 1985. Thermal maturity of Carboniferous strata, Ouachita Mountains: AAPG Bulletin, v. 69, p. 335-345.

Houseknecht, D. W., 1987. Assessing the relative importance of compaction and cementation to reduction of porosity in sandstones: AAPG Bulletin, v. 71, p. 633-642.

Houzay, J. P., Ferret, J. C., Harang, C., LeTendre, L., and Schein, F., 1993. Identification and evaluation of source rocks on wireline logs: p. 344-354 in "Applied Petroleum Geochemistry" (Edited by M. L. Bordenave), Editions Technip, Paris.

Huc, A. Y. and Sinninghe Damste, J. S. (Editors), 1993. Origin and significance of the organic matter in evaporites: Application to the Mulhouse Basin: Organic Geochemistry, Volume 20.

Hughes, M. and Davison, I., 1993. Geometry and growth kinematics of salt pillows in the southern North Sea: Tectonophysics, v. 228, p. 239-254.

Hulen, J. B., Goff, F., Ross, J. R., Bortz, L. C., and Robert, S., 1993. Geology and geothermal origin of Grant Canyon and Bacon Flat oil fields, Railroad Valley, Nevada: AAPG Bulletin, v. 78, p. 596-623.

Hunt, J. M., 1995. "Petroleum Geochemistry and Geology" (Second Edition), W. H. Freeman, San Francisco, 743 pp.

Hunt, J. M., Lewan, M. D., and Hennet, R. J-C., 1991. Modeling oil generation with time-temperature index graphs based on the Arrhenius equation: AAPG Bulletin, v. 75, p. 795-807.

Hwang, R.J., Ahmed, A.S., and Moldowan, J.M., 1994. Oil composition variation and reservoir continuity: Unity Field, Sudan: Organic Geochemistry, v. 21, p. 171-188.

Ibach, L. E. J., 1982. Relationship between sedimentation rate and total organic carbon content in ancient marine sediments: AAPG Bulletin, v. 66, p. 170-188.

Iijima, A., 1988. Diagenetic transformations of minerals as exemplified by zeolites and silica minerals - a Japanese view: Chapter 3, p. 147-211 in "Diagenesis II" (Edited by G. V. Chilingarian and K. H. Wolf), Elsevier, Amsterdam.

Ilchik, R. B., Brimhall, G. H., and Schull, W. H., 1996. Hydrothermal maturation of indigenous organic matter at the Alligator Ridge fold deposits, Nevada: Economic Geology, v. 81, p. 113-130.

Iliffe, J. E., Lerche, I., and Cao, S., 1991. Basin analysis predictions of known hydrocarbon occurrences: the North Sea Viking Graben as a test case: Earth-Science Reviews, v. 30, p. 51-80.

Illiffe, J. E., Lerche, I., and DeBuyl, M., 1991. Basin analysis and hydrocarbon generation of the South Mozambique graben using extensional models of heat flow: Marine and Petroleum Geology, v. 8, p. 151-162.

Ingle, J. C., Jr., 1980. Cenozoic paleobathymetry and depositional history of selected sequences within the southern California continental borderland: Cushman Foundation Special Publication, v. 19, p. 163-195.

Issler, D. R., 1984. Calculation of organic maturation levels for offshore eastern Canada - implications for general application of Lopatin's method: Canadian Journal of Earth Sciences, v. 21, p. 477-488.

Issler, D. R. and Beaumont, C., 1987. Thermal and subsidence history of the Labrador and West Greenland continental margin: p. 45-69 in "Sedimentary Basins and Basin-Forming Mechanisms" (Edited by C. Beaumont and A. J. Tankard), Canadian Society of Petroleum Geologists, Memoir 12.

Issler, D. R. and Snowdon, L. R., 1990. Hydrocarbon generation kinetics and thermal modeling, Beaufort-Mackenzie basin: Bulletin of Canadian Petroleum Geology, v. 38, p. 1-16.

Jabour, H. and Nakayama, K., 1988. Basin modelling of Tadla Basin, Morocco, for hydrocarbon potential: AAPG Bulletin, v. 72, p. 1059-1073.

Jackson, K. S., McKirdy, D. M., and Deckelman, J. A., 1984. Hydrocarbon generation in the Amadeus Basin, Central Australia: Australian Petroleum Exploration Association (APEA) Journal, v. 24, p. 42-65.

Jaeger, J. C., 1964. Thermal effects of intrusions: Reviews of Geophysics, v. 2, p. 443-466.

Jaeger, T. S., 1968. Cooling and solidification of igneous rocks: p. 503-536 in "Basalts" (Edited by H. H. Hess), Wiley, New York.

James, N. P. and Choquette, P. W. (Editors), 1988. "Paleokarst", Springer-Verlag, New York, 416 pp.

Jansa, L. F. and Urrea, V.H.N., 1990. Geology and diagenetic history of overpressured sandstone reservoirs, Venture gas field, offshore Nova Scotia, Canada: AAPG Bulletin, v. 74, p. 1640-1658.

Jarvie, D. M., 1991. Factors affecting Rock-Eval derived kinetic parameters: Chemical Geology, v. 93, p. 79-99.

Jarvis, G. T. and McKenzie, D. P., 1980. Sedimentary basin formation with finite extension rates: Earth and Planetary Science Letters, v. 48, p. 42-58.

Jenkins, D. G., 1967. Recent distribution, origin and coiling ratio changes in *Globorotalia pachyderma* (Ehrenberg): Micropaleontology, v. 13(2), p. 195-203.

Jenkins, D. G., 1968. Planktonic foraminifera as indicators of New Zealand Tertiary paleotemperatures: Tauatara, v. 16(1), p. 32-37.

Jennings, S. and Thompson, G. R., 1986. Diagenesis of Plio-Pleistocene sediments of the Colorado River delta, southern California: Journal of Sedimentary Petrology, v. 56, p. 89-98.

Jensen, P. K., 1983. Calculations on the thermal conditions around a salt diapir: Geophysical Prospecting, v. 31, p. 481-489.

Jensen, P.K., Holm, L., and Thomsen, E., 1985. Modelling burial history, temperature and maturation: p. 145-152 in "Petroleum Geochemistry in Exploration of the Norwegian Shelf" (Edited by B. M. Thomas et al), Graham and Trotman, for the Norweigian Petroleum Society, London.

Jensenius, J., 1987. High-temperature diagenesis in shallow chalk reservoir, Skjold oil field, Danish North Sea: Evidence from fluid inclusions and oxygen isotopes: AAPG Bulletin, v. 71, p. 1378-1386.

Johansen, H., Fjeldskaar, W., and Mykkeltveit, J., 1996. The influence of glaciation on the basin temperature regime: Global and Planetary Change, v. 12, p. 437-448.

Johnson, V. G., Graham, D. L., and Reidel, S. P., 1993. Methane in Columbia River Basalt aquifers: Isotopic and geohydrologic evidence for a deep coal-bed gas source in the Columbia Basin, Washington: AAPG Bulletin, v. 77, p. 1192-1207.

Johnsson, M. J., 1986. Distribution of maximum burial temperatures across northern Appalachian basin and implications for Carboniferous sedimentation patterns: Geology, v. 14, p. 384-387.

Johnsson, M.J., Howell, D. G., and Bird, K. J., 1993. Thermal maturity patterns in Alaska: Implications for tectonics evolution and hydrocarbon potential: AAPG Bulletin, v.77, p. 1874-1903.

Jones, F. W. and Majorowicz, J. A., 1987. Some aspects of the thermal regime and hydrodynamics of the western Canadian sedimentary basin: p. 79-85 in "Fluid Flow in Sedimentary Basins and Aquifers" (Edited by J. C. Goff and B. P. J. Williams), Geological Society Special Publication No. 34.

Jones, F. W., Majorowicz, J. A., Linville, A., and Osadetz, K. G., 1986. The relationship of hydrocarbon occurrences to geothermal gradients and time-temperature indices in Mesozoic formations of southern Alberta: Bulletin of Canadian Petroleum Geology, v. 34, p. 226-239.

Jones, P. H., 1975. Geothermal and hydrocarbon regimes, northern Gulf of Mexico Basin: presented at the First Geopressured Geothermal Energy Conference, May 1975.

Jones, R. and Edison, T., 1978. Microscopic observations of kerogen related to geochemical parameters with emphasis on thermal maturation: p. 1-12 in "Symposium in Geochemistry, Low Temperature Metamorphism of Kerogen and Clay Minerals," Society of Economic Paleontologists and Mineralogists (Pacific Section).

Jones, R. W., 1985. Hydrocarbon habitat in main producing areas, Saudi Arabia: Reply: AAPG Bulletin, v. 69, p. 2031-2033.

Jowett, E. C., Cathles, L. M., III, and Davis, B. W., 1993. Predicting depths of gypsum dehydration in evaporitic sedimentary basins: AAPG Bulletin, v. 77, p. 402-413.

Juntgen, H. and Klein, J., 1975. Origin of natural gas from coaly sediments: Erdol und Kohle, Erdgas-Petrochemie, v. 17, p. 180-186.

Kablanow, R. I., II and Surdam, R. C., 1983. Diagnesis and hydrocarbon generation in the Monterey formation, Huasna Basin, California: p 53-68 in "Petroleum Generation and Occurrence in the Miocene Monterey Formation, California" (Edited by C. M. Isaacs and R. E. Garrison), Society of Economic Paleontologists and Mineralogists (Pacific Section).

Kalkreuth, W. and McMechan, M., 1988. Burial history and thermal maturity, Rocky Mountain Front Ranges, Foothills, and Foreland, East-Central British Columbia and adjacent Alberta, Canada: AAPG Bulletin, v. 72, p. 1395-1410.

Kamp, P. J. J. and Green, P. F., 1990. Thermal and tectonic history of selected Taranaki basin (New Zealand) wells assessed by apatite fission track analysis: AAPG Bulletin, v. 74, p. 1401-1419.

Kantsler, A. J., Prudence, T. J. C., Cook, A.C., and Zwigulis, M., 1984. Hydrocarbon habitat of the Cooper/Eromanga Basins, Australia: AAPG Memoir 35, p. 373-390.

Karweil, J., 1956. Die metamorphose der Kohlen vom Standpunkt der physikalischen Chemie: Z. Deutsch. Geol. Gesell., v. 107, p. 132-139.

Kastner, M., Keene, J. B., and Gieskes, J. M., 1977. Diagenesis of siliceous oozes - I. Chemical control on the rate of opal-A to opal-CT transformation - an experimental study: Geochimica et Cosmochimica Acta, v. 41, p. 1041-1059.

Katz, B. J., 1995. A survey of rift basin source rocks: p. 213-242 in "Hydrocarbon Habitat in Rift Basins" (Edited by J. J. Lambiase), Geological Society (London) Special Publication No. 80.

Katz, B. J. and Pratt, L. M., 1993. "Source Rocks in a Sequence Stratigraphic Framework", AAPG Studies in Geology No. 37. 246 pp.

Katz, B. J., Liro, L. M., Lacey, J. E., White, J. W., and Waples, D. W., 1982. Time and temperature in petroleum formation; application of Lopatin's method to petroleum exploration discussion and reply: AAPG Bulletin, v. 66, p. 1150-1152.

Katz, B. J., Pheifer, R. N., and Schunk, D. J., 1988. Interpretation of discontinuous vitrinite reflectance profiles: AAPG Bulletin, v. 72, p. 926-931.

Katz, B. J., Schunk, D. J., and Pheifer, R. N., 1990. Interpretation of discontinuous vitrinite reflectance profiles: Reply: AAPG Bulletin, v. 74, p. 94.

Keen, C. E. and Lewis, T., 1982. Measured radiogenic heat production in sediments from continental margin of eastern North America: Implications for petroleum generation: AAPG Bulletin, v. 66, p. 1402-1407.

Keen, C. E., 1983. Salt diapirs and thermal maturity: Scotian Basin: Bulletin of Canadian Petroleum Geology, v. 31, p. 101-108.

Keen, C. R., 1978. Thermal history and subsidence of rifted continental margins – evidence from wells on the Nova Scotian and Labrador shelves: Canadian Journal of Earth Sciences, v. 16, p. 505-522.

Kehle, R. O., 1971. Geothermal survey of North America, 1971 Annual Progress Report, Research Committee AAPG (unpublished), 31 pp.

Keller, M. A. and Isaacs, C. M., 1985. An evaluation of temperature scales for silica diagenesis in diatomaceous sequences including a new approach based on the Miocene Monterey Formation, California: Geo-Marine Letters, v. 5, p. 31-35.

Kelley, S., Duncan, I., and Blackwell, D., 1983. Use of fission-track annealing systematics in constraining the thermal evolution of sedimentary basins (Abstr): AAPG Bulletin, v. 67, p. 494.

Kennett, J. P. and Barker, P. F., 1990. Latest Cretaceous to Cenozoic climate and oceanographic development in the Weddell sea, Antarctica: An ocean-drilling perspective: Proceedings of the Ocean Drilling Program, Scientific Results, 113, College Station, Texas.

Kennicutt, M. C., II, Barker, C., Brooks, J. M., DeFeitas, D. A., and Zhu, G. H., 1987. Selected organic matter source indicators in the Orinoco, Nile and Changjiang deltas: Organic Geochemistry, v. 11, p. 41-51.

Kerans, C., 1988. Karst-controlled reservoir heterogeneity in Ellenburger Group carbonates of West Texas: AAPG Bulletin, v. 72, p. 1160-1183.

Kettel, D., 1983. The East Groningen Massif - detection of an intrusive body by means of coalification: Geologie en Mijnbouw, v. 62, p. 203-210.

Kharaka, Y. K. and Mariner, R. H., 1989. Chemical geothermometers and their application to formation waters from sedimentary basins: p. 99-117 in "Thermal History of Sedimentary Basins" (Edited by N. D. Naeser and T. H. McCulloh), Springer-Verlag, New York.

Kharaka, Y. K., Callender, E., and Carothers, W. W., 1977. Geochemistry of geopressured geothermal waters from the Texas Gulf Coast: Proceedings Third Geopressure-Geothermal Energy Conference, University of Southwestern Louisiana, vol. 1, p. G1121-G1165.

Khorasani, G. K., 1989. Factors controlling source rock potential of the Mesozoic coal-bearing strata from offshore central Norway: Application to petroleum exploration: Bulletin of Canadian Petroleum Geology, v. 37, p. 417-427.

Killingley, J. S., 1983. Effects of diagenetic recrystallization of $^{18}O/^{16}O$ values of deep-sea sediments: Nature, v. 301, p. 594-597.

Killops, S. D., Woolhouse, A. D., Weston, R. J., and Cook, R. A., 1994. A geochemical appraisal of oil generation in the Taranaki Basin, New Zealand: AAPG Bulletin, v. 78, p. 1560-1585.

King, J. D. and Claypool, G. E., 1983. Biological marker compounds and implications for generation and migration of petroleum in rocks of the Point Conception Deep-Stratigraphic Test Well OCS-CAL 78-164 No. 1, Offshore California: p. 191-200 in "Petroleum Generation and Occurrence in the Miocene Monterey Formation, California" (Edited by C. M. Isaacs and R. E. Garrison), Society of Economic Paleontologists and Mineralogists (Pacific Section).

King, J. D., Yang, J., and Pu, F., 1994. Thermal history of the periphery of the Junggar Basin, northwest China: Organic Geochemistry, v. 21, p. 393-405.

Kingston, D. R., 1996. Postmortem should be *de rigueur* : AAPG Explorer, May 1996, p. 68-69.

Kirby, G. A., Smith, K., Smith, N.J.P., and Swallow, P. W., 1987. Oil and gas generation in eastern England: p. 171-180 in "Petroleum Geology of North West Europe" (Edited by J. Brooks and K. Glennie), Graham and Trotman, London.

Kisch, H. J., 1987. Correlation between indicators of very low-grade metamorphism: p. 227-246 in "Low Temperature Metamorphism" (Edited by M. Frey), Chapman and Hall, New York.

Kleinspehn, K. L. and Paola, C. (Editors), 1988. "New Perspectives in Basin Analysis", Springer-Verlag, New York. 453 pp.

Klemme, H. D. and Ulmishek, G. F., 1991. Effective petroleum source rocks of the World: Stratigraphic distribution and controlling depositional factors: AAPG Bulletin, v. 75, p. 1809-1851.

Klemme, H. D., 1975. Geothermal gradients, heat flow, and hydrocarbon recovery: p. 251-304 in "Petroleum and Global Tectonics" (Edited by A. G. Fischer and S. Judson), Princeton University Press.

Klemme, H. D., 1980. Types of petroliferous basins: p. 101-115 in "Petroleum Geology in China" (Edited by J. F. Mason), PennWell Publishing, Tulsa.

Klomp, U. C. and Wright, P. A., 1990. A new method for the measurement of kinetic parameters of hydrocarbon generation from source rocks: Organic Geochemistry, v. 16, p. 49-60.

Kontrovitz, M. and Myers, J. H., 1988. Ostracode eyes as paleoenvironmental indicators: Physical limits of vision of some podocopids: Geology, v. 16, p. 293-295.

Kruge, M. A., 1983. Diagenesis of Miocene biogenic sediments in Lost Hills Oil Field, San Joaquin Basin, California: p. 39-51 in "Petroleum Generation and Occurrence in the Miocene Monterey Formation, California" (Edited by C. M. Isaacs and R. E. Garrison), Society of Economic Paleontologists and Mienralogists (Pacific Section).

Kuo, L. C. and Michael, G. E., 1994. A multicomponent oil-cracking kinetics model for modelling preservation and composition of reservoired oils: Organic Geochemistry, v.21, p. 911-925.

Kuo, L. C., 1994. Lower Cretaceous lacustrine source rocks in northern Gabon: effect of organic facies and thermal maturity on crude oil quality: Organic Geochemistry, v. 22, p. 257-273.

Kupecz, J. E. and Land, L. S., 1991. Late-stage dolomitzation of the Lower Ordovician Ellenburger Group, West Texas: Journal of Sedimentary Petrology, v. 61, p. 551-574.

Kvenvolden, K. A., 1980. Molecular architecture: p. 178-185 in "Geochemistry of Organic Molecules," Benchmark Papers in Geology, v. 52, Dowden, Hutchinson and Ross, Inc.

Kvenvolden, K. A., 1993. A primer on gas hydrates: p. 279-291 in "The Future of Energy Gases" (Edited by D. G. Howell), U.S. Geological Survey Professional Paper 1570.

Kvenvolden, K. A., 1995. A review of the geochemistry of methane in natural gas hydrate: Organic Geochemistry, v. 23, p. 997-1008.

Kvenvolden, K. A., Ginsburg, G. D., and Soloviev, V.A., 1993. Worldwide distribution of subaquatic gas hydrates; Geo-Marine Letters, v. 13, p. 32-40.

Laczo, I. and Jambor, A., 1988. Secondary heating of vitrinite: Some geological implications: AAPG Memoir 45, p. 311-317.

Lafargue, E. and Barker, C., 1988. Effect of water washing on crude oil compositions: AAPG Bulletin, v. 72, p. 263-276.

Lafargue, E., Espitalie, J., Broks, T. M., and Nyland, B., 1994. Experimental simulation of primary migration: Organic Geochemistry, v. 22, p. 575-586.

Lakshmanan, C. C., Bennett, M. L., and White, N., 1991. Implications of multiplicity in kinetic parameters to petroleum exploration: distributed activation energy models: Energy and Fuels, v. 5, p. 110-117.

Lam, H. L. and Jones, F. W., 1984. Geothermal gradients of Alberta in western Canada: Geothermics, v. 13, p. 181-192.

Lambiase, J. J., 1990. A model for tectonic control of lacustrine stratigraphic sequence in continental rift basins: AAPG Memoir 50, p. 265-176.

Lambiase, J. J. (Editor), 1995. "Hydrocarbon Habitat in Rift Basins", Geological Society (London) Special Publication No. 80.

Land, L. S. and MacPherson, G. L., 1992. Geothermometry from brine analyses: Lessons learned from the Gulf Coast, USA: Applied Geochemistry, v. 7, p. 333-340.

Land, L. S., 1985. The origin of massive dolomite: Journal of Geological Education, v. 33, p. 112-125.

Lane, F. H. and Jackson, K. S., 1980. Controls on occurrence of oil and gas in the Beaufort-Mackenzie basin: Canadian Society of Petroleum Geologists, Memoir No. 6, p. 489-507.

LaPlante, R. E., 1974. Hydrocarbon generation in Gulf Coast Tertiary sediments: AAPG Bulletin, v. 58, p. 1281-1289.

Larrieu, T.L., 1995. Basin analysis with a spreadsheet: Journal of Geological Education, v. 43, p. 107.

Larson, K. W., Waples, D. W., Fu, H., and Kodama, K., 1993. Predicting tectonic fractures and fluid flow through fractures in basin modelling: p. 373-383 in "Basin Modeling: Advances and Applications" (Edited by A. G. Dore et al), Norwegian Petroleum Society Special Publication 3, Elsevier, Amsterdam.

Larter, S. R., 1984. Application of analytical pyrolysis techniques to kerogen characterization and fossil fuel exploration exploitation: p. 212-275 in "Analytical Pyrolysis Techniques and Explorations" (Edited by K. J. Vorhees), Butterworth, London.

Larter, S. R., 1988. Some pragmatic perspectives in source rock geochemistry: Marine and Petroleum Geology, v. 5, p. 194-204.

Larter, S., 1989. Chemical models of vitrinite reflectance evolution: Geologische Rundshau, v. 78, p. 349-359.

Larue, D. K., Schoonmaker, J., Schneider, R., Torrini, R., and Lucas-Clark, J., 1985. Barbados: maturation, source rock potential and burial history within a Cenozoic accretionary complex: Marine and Petroleum Geology, v. 2, p. 96-110

Laschet, C. A., 1984. On the origin of cherts: Erlangen, v. 10, p. 257-290.

Law, B. E. and Dickinson, W. W., 1985. Conceptual model for origin of abnormally pressured gas accumulations in low-permeability reservoirs: AAPG Bulletin, v. 69, p. 1295-1304.

Law, B. E., 1992. Thermal maturity patterns of Cretaceous and Tertiary rocks, San Juan basin, Colorado and New Mexico: Geological Society of America Bulletin, v. 104, p. 192-207.

Law, B. E., Lickus, M. R., and Pawlewicz, M. J., 1986. Fluid migration pathways - evidence from thermal maturity mapping in southwestern Wyoming: United States Geological Survey Circular 974.

Law, B. E., Nuccio, V. F., and Barker, C. E., 1989. Kinky vitrinite reflectance well profiles: evidence of paleopore pressure in low-permeability, gas-bearing sequences in Rocky Mountain foreland basins: AAPG Bulletin, v. 73, p. 999-1010. (see also Discussion, AAPG Bulletin, v. 74, p. 946).

Lawrence, D. T., Doyle, M., and Aigner, T., 1990. Stratigraphic simulation of sedimentary basins: concepts and calibration: AAPG Bulletin, v. 74, p. 273-295.

Leadholm, R. H., Ho, T.T.Y., and Sahai, S.K., 1985. Heat flow, geothermal gradients and maturation modelling on the Norwegian continental shelf using computer methods: p. 131-143 in "Petroleum Geochemistry in Exploration of the Norwegian Continental Shelf (Edited by B.M. Thomas et al), Graham and Trotman, London.

Leblanc, Y., Lam, H. L., Pascoe, L. J., and Jones, F. W., 1982. A comparison of two methods of estimating static formation temperature from well logs: Geophysical Prospecting, v. 30, p. 348-357.

Leblanc, Y., Pascoe, L. J., and Jones, F. W., 1981. The temperature stablization of a borehole: Geophysics, v. 46, p. 1301-1303.

Leder, F. and Park, W. C., 1986. Porosity reduction in sandstones by quartz overgrowth: AAPG Bulletin, v. 70, p. 1713-1728.

Lee, H. K., and Uyeda, S., 1965. Review of heat flow data, Terrestrial Heat Flow, Geophysics Monograph, v. 8, p. 87-190.

Lee, M. K. and Bethke, C. M., 1994. Groundwater flow, late cementation, and petroleum accumulation in the Permian Lyons Standstone: AAPG Bulletin, v. 78, p. 217-237.

Lee, M., Aronson, J. L., and Savin, S. M., 1985. K/Ar dating of time of gas emplacement in Rotliegendes sandstone, Netherlands: AAPG Bulletin, v. 69, p. 1381-1385.

Lee, M., Aronson, J. L., and Savin, S. M., 1989. Timing and conditions of Permian Rotliegende sandstone diagenesis, southern North Sea: K/Ar and oxygen isotope data: AAPG Bulletin, v. 73, p. 195-215.

Legall, F. D., Barnes, C. R., and MacQueen, R. W., 1981. Thermal maturation, burial history and hotspot development, paleozoic strata of southern Ontario, Quebec, from conodont and acritarch colour alteration studies: Bulletin of Canadian Petroleum Geology, v. 29, p. 492-539.

Lehner, F. K., Marsal, D., Hermans, L., and vanKuyk, A., 1987. A model of secondary hydrocarbon migration as a buoyancy-driven separate phase flow: p. 457-471 in "Migration of Hydrocarbons in Sedimentary Basins" (Edited by B. Doligez), Editions Technip, Paris.

Leischner, K., Welte, D. H., and Littke, R., 1993. Fluid inclusions and organic maturity parameters as calibration tools in basin modelling: p. 161-172 in "Basin Modeling: Advances and Applications" (Edited by A. G. Dore et al), Norwegian Petroleum Society Special Publication 3, Elsevier, Amsterdam.

Leith, M. J., and Rowsell, D. M., 1979. Burial history and temperature-depth conditions for hydrocarbon generation and migration on the Agulhas Bank, South Africa: Geological Society South Africa Special Publication No. 6, p. 205-217.

Lerche, I., 1990. "Basin Analysis. Quantitative Methods; Volume 1," Academic Press, San Diego. 562 pp.

Lerche, I., 1990. "Basin Analysis. Quantitative Methods; Volume 2," Academic Press, San Diego. 570 pp.
Lerche, I., Yarkzab, R. F., and Kendall, C G. St.C., 1984. Determination of paleoheat flux from vitrinite reflectance data: AAPG Bulletin v. 68, p. 1704-1717. (see Discussion AAPG Bulletin, v. 69, p. 2148-2149).
Lerche, I. and O'Brien, J. J., 1987 (Editors). "Dynamical Geology of Salt and Related Structures," Academic Press, Orlando. 832 pp.
Lerche, I. and Petersen, K., 1995. "Salt and Sediment Dynamics", CRC Press, Boca Raton. 320 pp.
Lerche, I. and Torudbakken, B. O., 1996. The Nordkapp basin, Norway: Development of salt and sediment interplays for hydrocarbon exploration: Proceedings of the 1996 Offshore Technology Conference, Houston, Texas, v. 1, p. 547-560.
Levorsen, A. I., 1967. "Geology of Petroleum" (Second Edition), W. H. Freeman, San Francisco. 724 pp.
Lewan, M. D., 1988. Petrographic study of primary petroleum migration in the Woodford Shale and related rock units: p. 113-130 in "Migration of Hydrocarbons in Sedimentary Basins" (Edited by B. Doligez), Editions Technip, Paris.
Lewan, M. D., 1993. Laboratory simulation of petroleum formation: hydrous pyrolysis: p.419-442 in "Organic Geochemistry" (Edited by M. H. Engel and S. A. Macko), Plenum Publishing, New York.
Lewan, M. D., Winter, J. C., and McDonald, J. H., 1979. Generation of oil-like pyrolyzates from organic-shales: Science, v. 203, p. 897-899.
Lewis, S. E. and Hower, J. C., 1990. Implications of thermal events on thrust emplacement sequence in the Appalachian fold and thrust belt: some new vitrinite reflectance data: Journal of Geology, v. 98, p. 927-942.
Leythaeuser, D., 1973. Effects of weathering on organic matter in shales: Geochimica et Cosmochimica Acta, v. 37, p. 113-120.
Leythaeuser, D., Radke, M., and Schaefer, R. G., 1984. Efficiency of petroleum expulsion from shale source rocks: Nature, v. 311, p. 745-748.
Leythaeuser, D., Schaefer, R. G., and Radke, M., 1988. Geochemical effects of primary migration of petroleum in Kimmeridge source rocks from Brae field area, North Sea, I. Gross composition of C_{15+} soluble organic matter and molecular composition of C_{15+}-saturated hydrocarbons: Geochimica et Cosmocimica Acta, v, 52, p. 701-713.
Leythaeuser, D., Schaefer, R. G., and Yukler A., 1982. Role of diffusion in primary migration of hydrocarbons: AAPG Bulletin, v. 66, p. 408-429.
Leythaeuser, D., Schaefer, R. G., and Yukler, A., 1980. Diffusion of light hydrocarbons through near surface rocks: Nature, v. 284. p. 522-525.
Lickus, M. R., Pawlewicz, M. J., Law, B. E., and Dickinson, W. W., 1984. Thermal maturity map, northern Green River Basin, Wyoming: U.S. Geological Survey Open-File Rept No. 84-753.
Liewig, N., Clauer, N., and Sommer, F., 1987. Rb-Sr and K-Ar dating of clay diagenesis in Jurassic sandstone oil reservoir, North Sea: AAPG Bulletin, v. 71, p. 1467-1474.

Link, C. M. and Bustin, R. M., 1989. Organic maturation and thermal history of Phanerozoic strata in northern Yukon and northwestern District of MacKenzie: Bulletin of Canadian Petroleum Geology, v. 37, p. 266-292.

Liu, G. and Roaldset, E., 1994. A new decompaction model and its application to the northern North Sea: First Break, v. 12, p. 81-89.

Liu, J., Lerche, I., Bissada, K. K., and Lacey, J., 1992. An end-member sensitivity analysis of vitrinite reflectance distributions for the COST G-2 well in the Georges Bank Basin: Organic Geochemistry, v. 18, p. 385-396.

Lo, H. B. and Cardott, B. J., 1995. Detection of natural weathering of Upper McAlester coal and Woodford Shale, Oklahoma, U.S.A.: Organic Geochemistry, v. 22, p. 73-83.

Lomando, A. J., 1992. The influence of solid reservoir bitumen on reservoir quality: AAPG Bulletin, v. 76, p. 1127-1152.

Lonsdale, P., 1978. Ecuadorian subduction system: AAPG Bulletin, v. 62, p. 2454-2477.

Lopatin, N. V., 1971. Temperature and geologic time as a factor in coalification (in Russian): Izveotiya Akademii. Nauk SSSR, Seriya Geolicheskya, v. 3, p. 95-106.

Lovering, T. S. and Goode, H. D., 1963. Measuring geothermal gradients in drill holes less than 60 feet deep, East Tintic District, Utah: U.S. Geological Survey Bulletin 1172. 48 pp.

Ludvigsen, A., Gram, K., Palm, E., and Bjorlykke, K., 1993. Effects of convection-currents on heat transfer in sedimentary basins: p. 353-359 in "Basin Modeling: Advances and Applications" (Edited by A. G. Dore et al), Norwegian Petroleum Society Special Publication 3, Elsevier, Amsterdam.

Luo, M., Baker, M. R., and LeMone, D. V., 1994. Distribution and generation of the overpressure system, eastern Delaware basin, western Texas and southern New Mexico: AAPG Bulletin, v. 78, p. 1386-1405.

Luo, X., Brigaud, F., and Vasseur, G., 1993. Compaction coefficients of argillaceous sediments: their implications, significance and determination: p. 321-332 in "Basin Modeling: Advances and Applications" (Edited by A. G. Dore et al), Norwegian Petroleum Society Special Publication 3, Elsevier, Amsterdam.

Luo, X. and Vasseur, G., 1996. Geopressuring mechanism of organic matter cracking: numerical modeling: AAPG Bulletin, v. 80, p. 856-874.

Macchi, L., Curtis, C. D., Levison, A., Woodward, K., and Hughes, C. R., 1990. Chemistry, morphology, and distribution of illites from Morecambe gas fields, Irish Sea, offshore United Kingdom: AAPG Bulletin, v. 74, p. 296-308.

Macintyre, I. G., Burke, R. B., and Stuckenrath, R., 1977. Thickest recorded Holocene reef section, Isla Perez core hole, Alacran Reef, Mexico: Geology, v. 5, p. 749-754.

Mack, G. H. and James, W. C., 1994. Paleoclimate and the global distribution of paleosols: Journal of Geology, v. 102, p. 360-366.

Mack, G. H., Giordano, T. H., Cole, D. R., James, W.C., and Salyards, S. L., 1994. Stable oxygen and carbon isotopes of pedogenic carbonate as indicators of Plio-Pleistocene paleoclimate in the southern Rio Grande rift, south-central New Mexico: American Journal of Science, v. 294, p. 621-640.

Mackenzie, A. S. and Maxwell, J. R., 1981. Assessment of thermal maturation in sediments by molecular measurements: p. 239-2532 in "Organic Maturation Studies and Fossil Fuel Exploration" (Edited by J. Brooks), Academic Press, New York.

Mackenzie, A. S. and McKenzie, D., 1983. Isomerization and aromatization of hydrocarbons in sedimentary basins formed by extension: Geological Magazine, v. 120, p. 417-470.

Mackenzie, A. S., 1984a. Applications of biological markers in petroleum geochemistry: p. 115-214 in "Advances in Petroleum Geochemistry, Vol. 1" (Edited by J. Brooks and D. Welte), Academic Press, London.

Mackenzie, A. S., 1984b. Estimation of kinetics of geochemical reactions with geophysical models and applications: p. 135-145 in "Thermal Phenomena in Sedimentary Basins" (Edited by B. Durand), Editions Technip, Paris.

Mackenzie, A. S., and Quigley, T. M., 1988. Principles of geochemical prospect appraisal: AAPG Bulletin v. 72, p. 399-415.

Mackenzie, A. S., Beaumont, C., and McKenzie, D. P., 1984. Estimation of the kinetics of geochemical reactions with geophysical models of sedimentary basins and applications: Organic Geochemistry, v. 6, p. 875-884.

Mackenzie, A. S., Beaumont, C., Boutilier, R., and Rullkotter, J., 1985. The aromatization and isomerization of hydrocarbons and the thermal and subsidence history of the Nova Scotia margin: Philosophical Transactions of the Royal Society London, v. A315, p. 203-232.

Mackenzie, A. S., Leythaeuser, D., Altebaumer, F. J., Disko, U., and Rullkotter, J., 1988. Molecular measurements of maturity for Lias δ shales in N. W. Germany: Geochimica et Cosmochimica Acta, v.52, p. 1145-1154.

Mackenzie, A. S., Price, I., Leythaeuser, D., Muller, P., Radke, M., and Schaefer, R. G., 1987. The expulsion of petroleum from Kimmeridge clay source-rocks in the area of the Brae Oilfield, UK continental shelf: p. 865-877 in "Petroleum Geology of North West Europe" (Edited by J. Brooks and K. Glennie), Graham and Trotman, London.

MacLeod, M. K., 1982. Gas hydrates in ocean bottom sediments: AAPG Bulletin, v. 66, p. 2649-2662.

MacMillan, L., 1980. Oil and gas of Colorado: A conceptual view: p.191-197 Rocky Mountain Association of Geologists, Denver, Colorado.

Magara, K., 1976. Thickness of removed sedimentary rocks, paleopore pressure, and paleotemperature, southwestern part of Western Canadian Basin: AAPG Bulletin, v. 60, p. 554-565.

Magara, K., 1977. Petroleum migration and accumulation: p.83-126 in "Developments in Petroleum Geology, v. 1" (Edited by G. D. Hobson), Applied Science Publishers, London.

Magoon, L. B. and Dow, W. G., 1994. "Petroleum Systems - From Source to Trap," AAPG Memoir 60. 654 pp.

Majorowicz, J. A. and Jessop, A. M., 1981a. Present heat flow and a preliminary paleogeothermal history of the central Prairies Basin, Canada: Geothermics, v. 10, p. 81-93.

Majorowicz, J. A. and Jessop, A. M., 1981b. Regional heat flow patterns in the western Canadian sedimentary basin: Tectonophysics, v. 74, p. 209-238.

Majorowicz, J. A., Jones, F. W., and Jessop, A. M., 1986. Geothermics of the Williston Basin in Canada in relation to hydrodynamics and hydrocarbon occurrences: Geophysics, v. 51, p. 767-779.

Majorowicz, J. A., Jones, F. W., Ertman, M. E., Osadetz, K. G., and Stasiuk, L. D., 1990. Relationships between thermal maturation gradients, geothermal gradients and estimates of the thickness of the eroded foreland section, southern Alberta Plains, Canada: Marine and Petroleum Geology, v. 7, p. 138-152.

Majorowicz, J. A., Jones, F. W., Lam, H. L., and Jessop, A. M., 1985. Terrestrial heat flow and geothermal gradients in relation to hydrodynamics in the Alberta Basin, Canada: Journal of Geodynamics, v. 4, p. 265-283.

Makagon, Y. F., 1981. "Hydrates of Natural Gas," PennWell Publishing Co., Tulsa. 237 pp.

Mani, K. S., Hussain, A., Basu, B., and Arunachalam, A., 1988. Maturation studies of source sediments based on burial history analysis in Upper Assam Shelf: p. 241-255 in "Petroleum Geochemistry and Exploration in the Afro-Asian Region" (Edited by R. K. Kumar, P. D. Wivedi, V. Banerjie, and V. Gupta), A. Balkema, Rotterdam.

Mann, D. M. and Mackenzie, A. S., 1990. Prediction of pore fluid pressures in sedimentary basins: Marine and Petroleum Geology, v. 7, p. 55-65.

Mann, U. and Muller, P. J., 1988. Source rock evaluation by well log analysis (Lower Toarcian, Hils Syncline): Advances in Organic Geochemistry 1987, v. 13, p. 109-119.

Mann, U., Leythaeuser, D., and Muller, P. J., 1986. Relation between source rock properties and wireline log parameters. An example from Lower Jurassic Posidonia shale, NW German: Advances in Organic Geochemistry 1985, v. 10, p. 1105-1112.

Maragna, B., Zaro, G., and Pessina, P., 1985. Block 33/6 geochemical evaluation: p. 197-204 in "Petroleum Geochemistry in Exploration of the Norwegian Continental Shelf" (Edited by B.M. Thomas et al), Graham and Trotman, London.

March, J., 1992. "Advanced Organic Chemistry, Reactions, Mechanisms and Structure" (Fourth Edition), Wiley-Interscience. 1495 pp.

Marchand, A. and Conard, J., 1980. Electron paramagnetic resonance in kerogen studies: p. 243-270 in "Kerogen" (Edited by B. Durand), Editions Technips, Paris.

Marie, J. P. P., 1975. Rotliegendes stratigraphy and diagenesis: p. 205-210 in "Petroleum and the Continental Shelf of North West Europe" (Edited by A. Woodland), Institute of Petroleum, London.

Marquez, X. M. and Mountjoy, E. W., 1996. Microfractures due to overpressures caused by thermal cracking in well-sealed Upper Devonian reservoirs, Deep Alberta basin: AAPG Bulletin, v. 80 , p. 570-588.

Martin, K. R., Baker, J. C., Hamilton, P. J., and Thrasher, G. P., 1994. Diagenesis and reservoir quality of Paleocene sandstones in the Kupe South Field, Taranaki basin, New Zealand: AAPG Bulletin, v. 78, p. 624-643.

Marzi, R. and Rullkotter, J., 1992. Qualitative and quantitative evolution and kinetics of biological marker transformations – Laboratory experiments and application to the Michigan Basin: p. 18-41 in "Biological Markers in Sediments and Petroleum" (Edited by J. M. Moldowan, P. Albrecht, and R. P. Philp), Prentice Hall, Englewood Cliffs, New Jersey.

Marzi, R., Rullkotter, J., and Perriman, W. S., 1990. Application of the change of sterane isomer ratios to the reconstruction of geothermal histories: Implications of the results of hydrous pyrolysis experiments: Organic Geochemistry, v. 16, p. 91-102.

Marzi, R. W. and Waples, D.W., 1996. The relationship between vitrinite reflectance and transformation ratio in basin modeling (Abstract): AAPG Annual Meeting, Abstracts with Program, p. A92.

Mattavelli, L. and Novelli, L., 1990. Geochemistry and habitat of the oils in Italy: AAPG Bulletin, v. 74, p. 1623-1639.

Mattavellie, L., Ricchiuto, T., Grignani, D., and Schoell, M., 1983. Geochemistry and habitat of natural gases in Po Basin, Northern Italy: AAPG Bulletin, v. 67, p. 2239-2254.

Mattick, R. E., Rumpler, J., Uifalusy, A., Szanyi, B., and Nagy, I., 1994. Sequence stratrigraphy of the Bekes Basin: p. 39-65 in "Basin Analysis in Petroleum Exploration. A Case Study from the Bekes Basin, Hungary" (Edited by P. G. Teleki, R. E. Mattick, and J. Kokai), Kluwer Academic Publishers, Dordrecht.

Maubeuge, F. and Lerche, I., 1993. A north Indonesian basin: geo, thermal and hydrocarbon histories: Marine and Petroleum Geology, v. 10, p. 231-245.

Maubeuge, F. and Lerche, I., 1994. Geopressure evolution and hydrocarbon generation in a north Indonesian basin: two-dimensional quantitative modelling: Marine and Petroleum Geology, v. 11, p. 104-116.

McCaffrey, M. A., Dahl, J. E., Sundararaman, P., Moldowan, J. M., and Schoell, M., 1994. Source rock quality determination from oil biomarkers II - A case study using Tertiary-reservoired Beaufort Sea oils: AAPG Bulletin, v. 78, p. 1527-1540.

McCulloch, A. A., 1993. Apatite fission track results from Ireland and the Porcupine basin and their significance for the evolution of the North Atlantic: Marine and Petroleum Geology, v. 10, p. 572-590.

McCulloh, T. H., Cashman, S. M., and Stewart, R. J., 1978. Diagenetic baselines for interpretive reconstructions of maximum burial depths and paleotemperatures in clastic sedimentary rocks: p. 18-46 in "Symposium in Geochemistry: Low Temperature Metamorphism of Kerogen and Clay Minerals" (Edited by D. F. Oltz). Society of Economic Paleontologists and Mineralogists, Pacific Section.

McDougall, I., and Harrison, T. M., 1988. "Geochronology and Thermochronology by the $^{40}Ar/^{39}Ar$ Method," Oxford University Press, New York, 212 pp.

McGee, H. W., Meyer, H. J., and Pringle, T. R., 1989. Shallow geothermal anomalies overlying deeper oil and gas deposits in Rocky Mountain region: AAPG Bulletin, v. 73, p. 576-697.

McKenna, T. E., Sharp, J. M., Jr., and Lynch, F. L., 1996. Thermal conductivity of Wilcox and Frio sandstones in south Texas (Gulf of Mexico basin): AAPG Bulletin, v. 80, p. 1203-1215.

McKenzie, D. P., 1981. The variation of temperature with time and hydrocarbon maturation in sedimentary basins formed by extension: Earth and Planetary Science Letters, v. 55, p. 87-98.

McKenzie, D., 1978. Some remarks on the development of sedimentary basins: Earth and Planetary Science Letters, v. 40, p. 25-32.

McLimans, R. K., 1987. The application of fluid inclusions to migration of oil and diagenesis in petroleum reservoirs: Applied Geochemistry, v. 2, p. 585-603.

McLimans, R. K. and Videtich, P. E., 1989. Diagenesis and burial history of Great Oolite Limestone, southern England: AAPG Bulletin, v. 73, p 1195-1205.

Meissner, F. F., 1978. Petroleum geology of the Bakken Formation Williston basin, North Dakota and Montana: p. 207-227 in "The Economic Geology of the Williston Basin" 1978 Williston Basin Symposium. Montana Geological Society.

Mello, M. R. and Maxwell, J. R., 1990. Organic geochemical and biological marker characterization of source rocks and oils derived from lacustrine environments in the Brazilian continental margin: AAPG Memoir 50, p. 77-95.

Mello, U. T. and Karner, G. D., 1996. Development of sediment overpressure and its effect on thermal maturation: Application to the Gulf of Mexico basin: AAPG Bulletin, v. 80, p. 1367-1396.

Mello, U. T., Karner, G. D., and Anderson, R. N., 1995. Role of salt in restricting the maturation of subsalt source rocks: Marine and Petroleum Geology, v. 12, p. 697-716.

Meshri, I. and Ortoleva, P. J. (Editors), 1991. "Prediction of reservoir quality through chemical modeling", AAPG Memoir No. 49.

Meyer, B. L. and Nederlof, M. H., 1984. Identification of source rocks on wireline logs by density/resistivity and sonic transit time/resistivity cross plots: AAPG Bulletin, v. 68, p. 121-129.

Meyer, H. J. and McGee, H. W., 1985. Oil and gas fields accompanied by geothermal anomalies in Rocky Mountain region: AAPG Bulletin, v. 69, p. 933-945.

Miall, A. D., 1984. "Principles of Sedimentary Basin Analysis", Springer-Verlag, New York. 490 pp.

Michelsen, J. K. and Khorasani, G. K., 1995. The kinetics of thermal degradation of individual oil-generating macerals: calibration with microscopical fluorescence spectrometry and bulk flow pyrolysis: Organic Geochemistry, v. 22, p. 179-189.

Middleton, M. F. and Falvey, D. A., 1983. Maturation modelling in Otway Basin, Australia: AAPG Bulletin, v. 67, p. 271-279.

Middleton, M. F. and Schmidt, P. W., 1982. Paleothermometry of the Sydney Basin: Journal of Geophysical Research, v. 87, p. 5351-5359.

Middleton, M. F., 1979. A model for bottom-hole temperature stabilization: Geophysics, v. 44, p. 1458-1462.

Middleton, M. F., 1982a. Bottom-hole temperature stabilization with continued circulation of drilling mud: Geophysics, v. 47, p. 1716-1723.

Middleton, M. F., 1982b. Tectonic history from vitrinite reflectance: Geophysical Journal of the Royal Astronomical Society, v. 68, p. 121-132.

Middleton, M. F., 1982c. The subsidence and thermal history of the Bass Basin, southeastern Australia: Tectonophysics, v. 87, p. 383-397.

Middleton, M. F., 1993. A transient method of measuring the thermal properties of rocks: Geophysics, v. 58, p. 357-365.

Miles, J. A., 1989. "Illustrated Glossary of Petroleum Geochemistry", Clarendon Press, Oxford. 137 pp.

Miles, J. A., 1990. Secondary migration routes in the Brent sandstones of the Viking Graben and East Shetland Basin: Evidence from oil residues and subsurface pressure data: AAPG Bulletin, v. 74, p. 1718-1735.

Miller, D. S. and Duddy, I. R., 1989. Early Cretaceous uplift and erosion of the North Apalachian Basin, New York, based on apatite fission track analysis: Earth and Planetary Science Letters, v. 78, p. 255-270.

Miller, K. G., Fairbanks, R. G., and Mountain, G. S., 1987. Tertiary oxygen isotope synthesis, sea level history, and continental margin erosion: Paleoceanography, v. 2, p. 1-19.

Mitsui, K. and Taguchi, K., 1977. Silica mineral diagenesis in Neogene Tertiary shales in the Tempoku district, Hokkaido, Japan: Journal of Sedimentary Petrology, v. 47, p. 158-167.

Mohan, M., 1985. Geohistory analysis of Bombay High region: Marine and Petroleum Geology, v. 2, p. 350-360.

Moldowan, J. M., Albrecht, P., and Philp, R. P., (Editors) 1992. "Biological Markers in Sediments and Petroleum", Prentice Hall, Englewood Cliffs, New Jersey. 411pp.

Moldowan, J. M., Dahl, J., Fago, F. J., Huizinga, B. T., Hickey, L. J., Peakman, T.M., and Taylor, D.W., 1994 The molecular fossil record of oleanane and its relation to angiosperms: Science, v. 265, p. 768-771.

Momper, J. A., 1978. Oil migration limitations suggested by geological and geochemical considerations: p.B1-B60 in "Physical and Chemical Controls on Petroleum Migration", AAPG Continuing Education Course Note Series, No. 8.

Momper, J. A., and Williams, J. A., 1984. Geochemical exploration in the Powder River basin: AAPG Memoir 35, p. 181-191.

Monnier, F., 1982. Thermal diagenesis in the Swiss Molasse Basin: implications for oil generation: Canadian Journal of Earth Sciences, v. 19, p. 328-342.

Montanez, I. P., 1994. Late diagenetic dolomitization of Lower Ordovician Upper Knox carbonates: A record of the hydrodynamic evolution of the southern Appalachian basin: AAPG Bulletin, v. 78, p. 1210-1239.

Moore, M. E., Gleadow, A. J. W., and Lovering, J. F., 1986. Thermal evolution of rifted continental margins: new evidence from fission tracks in basement apatites from southeastern Australia: Earth and Planetary Science Letters, v. 78, p. 255-270.

Morin, R. and Silva, A. J., 1984. The effects of high pressure and high temperature on physical properties of ocean sediments: Journal of Geophysical Research, v. 89, p. 5611-526.

Morrow, D. W. and Issler, D. R., 1993. Calculation of vitrinite reflectance from thermal histories: A comparison of some methods: AAPG Bulletin, v. 77, p. 610-624.

Morton, J. P., 1985. Rb-Sr dating of diagenesis and source rock age of clays in Upper Devonian black shales of Texas: Geological Society of America Bulletin, v. 96, p. 1043-1049.

Moshier, S. O. and Waples, D. W., 1985. Quantitative evaluation of Lower Cretaceous Mannville Group as source rock for Alberta's oil sands: AAPG Bulletin, v. 69, p. 161-172.

Mossop, G. D., 1972. Origin of the peripheral rim, Redwater Reef, Alberta: Bulletin of Canadian Petroleum Geology, v. 20, p. 238-280.

Mudford, B. S. and Best, M. E., 1989. Venture gas field, offshore Nova Scotia: Case study of overpressuring in region of low sedimentation rate: AAPG Bulletin v. 73, p. 1383-1396.

Mudford, B. S., 1990. A one-dimensional, two phase model of overpressure generation in the Venture gas field, offshore Nova Scotia: Bulletin of Canadian Petroleum Geology, v. 38, p. 246-258.

Mullan, H. S., 1984. Deep gas potential of Aquitaine basin, France: AAPG Bulletin, v. 68, p. 1857-1869.

Mullis, J., 1987. Fluid inclusion studies during very low-grade metamorphism: p. 162-199 in "Low Temperature Metamorphism" (Edited by M. Frey), Blackie, Glasgow, U.K.

Mussman, W. J., Montanez, I. P., and Read, J. F., 1988. Ordovician Knox paleokarst unconformity, Appalachians: p. 211-228 in "Paleokarst" (Edited by N. P. James and P. W. Choquette), Springer-Verlag, New York.

Nadeau, P. H. and Reynolds, R. C., Jr., 1981. Burial and contact metamorphism in the Mancos Shale: Clays and Clay Minerals, v. 29, p. 249-259.

Naeser, C. W., 1979. Thermal history of sedimentary basins: fission-track dating of subsurface rocks: p. 109-112 in "Aspects of Diagenesis: (Edited by P. A. Scholle and P. R. Schluger), Society of Economic Paleontologists and Mineralogists Special Publication No. 26.

Naeser, N. D., 1986. Neogene thermal history of the northern Green River Basin, Wyoming - evidence from fission-track dating: p.65-72 in Society of Economic Paleontologists and Mineralogists Special Publication No. 38.

Naeser, N. D., 1993. Apatite fission-track analysis in sedimentary basins - a critical appraisal: p. 147-160 in "Basin Modeling: Advances and Applications" (Edited by A. G. Dore et al), Norwegian Petroleum Society Special Publication 3, Elsevier, Amsterdam.

Naeser, N. D. and McCulloh, T. H. (Editors), 1989. "Thermal History of Sedimentary Basins. Methods and Case Histories", Springer-Verlag, New York. 500 pp.

Naeser, N. D., Naeser, C. W., and McCulloh, T. H., 1990. Thermal history of rocks in southern San Joaquin Valley, California: evidence from fission-track analysis: AAPG Bulletin, v. 74, p. 13-29.

Nairn, A. E. M., Lerche,I., and Iliffe, J. E., 1991. Geology, basin analysis, and hydrocarbon potential of Mozambique and the Mozambique Channel: Earth-Science Reviews, v. 30, p. 81-124.

Nakayama, K. and Cao, S., 1993. An inverse method for determining kinetic parameters for kerogen from routine pyrolysis data and application to the Peace River arch area, Alberta, Canada: p. 251-263 in "Basin Modeling: Advances and Applications" (Edited by A. G. Dore et al), Norwegian Petroleum Society Special Publication 3, Elsevier, Amsterdam.

Nakayama, K. and Van Siclen, D. C., 1981. Simulation model for petroleum exploration: AAPG Bulletin, v. 65, p. 1230-1255.

Nakayama, K., 1987. Hydrocarbon-explusion model and its application to Niigata area, Japan: AAPG Bulletin, v. 71, p. 810-821.

Narr, W. and Currie, J. B., 1982. Origin of fracture porosity - example from Altamont Field, Utah: AAPG Bulletin, v. 66, p. 1231-1247.

Narr, W. M. and Burruss, R. C., 1984. Origin of reservoir fractures in Little Knife Field, North Dakota: AAPG Bulletin, v. 68, p. 1087-1100.

Nelson, R. A., Patton, T. L., and Morley, C.K., 1992. Rift-segment interaction and its relation to hydrocarbon exploration in continental rift systems: AAPG Bulletin, v. 76, p. 1153-1169.

Newell, K. D., Burruss, R. C., and Palacas, J. G., 1993. Thermal maturation and organic richness of potential petroleum source rocks in Proterozoic Rice Formation, North American Mid-continent Rift System, northeastern Kansas: AAPG Bulletin, v. 77, p. 1922-1941.

Nielsen, S. B. and Barth, T., 1991. An application of least-squares inverse analysis in kinetic interpretations of hydrous pyrolysis experiments: Mathematical Geology, v. 23, p. 565-582.

Nielsen, S. B. and Barth, T., 1991. Vitrinite reflectance: Comments on "A chemical kinetic model of vitrinite maturation and reflectance" by Alan K. Burnham and Jerry J. Sweeney: Geochimica et Cosmochimica Acta, v. 55, p. 639-641.

Nielsen, S. B. and Dahl, B., 1991. Confidence limits on kinetic models of primary cracking and implications for the modelling of hydrocarbon generation: Marine and Petroleum Geology, v. 8, p. 483-492.

Nielsen, S. B., 1993. Fast Monte Carlo simulation of hydrocarbon generation: p. 265-276 in "Basin Modeling: Advances and Applications" (Edited by A. G. Dore et al), Norwegian Petroleum Society Special Publication 3, Elsevier, Amsterdam.

Nixon, J. F., 1986. Thermal simulation of subsea saline permafrost: Canadian Journal of Earth Sciences, v. 23, p. 2039-2046.

North, F. K., 1985. "Petroleum Geology," Allen and Unwin, Boston. 607 pp.

Novelli, L., Welte, D. H., Mattavelli, L., Yalcin, M.N., Cinelli, D., and Schmitt, K. J., 1987. Hydrocarbon generation in Southern Sicily - A three dimensional computer aided basin modeling study: p. 153-164 in "Advances in Organic Geochemistry 1987" (Edited by L. Mattavelli and L. Novelli). Pergamon Press, Oxford.

Nowland, G.S. and Barnes, C.R., 1987. Thermal maturation of Paleozoic strata in eastern Canada from conodont colour alteration index (CAI) data with implications for burial history, tectonic evolution, hotspot tracks and mineral and hydrocarbon exploration: Canadian Geological Survey Bulletin, No. 367. 51pp.

Nuccio, V. F. and Barker, C. E. (Editors), 1990. "Applications of thermal maturity studies to energy exploration," Rocky Mountain Section, The Society of Economic Paleontologists and Mineralogists (Society of Sedimentary Geologists). 175 pp.

Nuccio, V. F. and Johnson, R. C., 1984a. Retardation of vitrinite reflectance in Green River oil shales, Piceance Creek Basin, Northwestern Colorado (Abstr.): AAPG Bulletin, v. 68, p. 513.

Nuccio, V. F. and Johnson, R. C., 1984b. Thermal maturation and burial history of the Upper Cretaceous Mesaverde Group, including the multiwell experiment (MWX), Piceance Creek Basin, Colorado: U.S. Geological Survey Open-File Report No. 84-757.

Nunn, J. A. and Deming, D., 1991. Thermal constraints on basin-scale flow systems: Geophysical Research Letters, v. 18, p. 967-970.

Nunn, J. A. and Sassen, R., 1986. The framework of hydrocarbon generation and migration, Gulf of Mexico continental slope: Gulf Coast Association of Geological Societies Transactions, v. 36, p. 257-262.

Nunn, J. A., Sleep, N. H., and Moore, W. E., 1984. Thermal subsidence and generation of hydrocarbons in Michigan Basin: AAPG Bulletin, v. 68, p. 296-315.

Nurkowski, J. R., 1984. Coal quality, coal rank variation and its relation to reconstructed overburden, Upper Cretaceous and Tertiary Plains coals, Alberta, Canada: AAPG Bulletin, v. 68, p. 285-295.

Nwachukwu, J. I. and Barker, C., 1985. Variations in kerogen densities of sediments from the Orinoco Delta, Venezuela: Chemical Geology, v. 51, p. 193-198.

Nwachukwu, S. O., 1976. Approximate geothermal gradients in Niger delta sedimentary basin: AAPG Bulletin, v. 60, p. 1073-1077.

Nyland, B., 1993. Application of basin modelling on the Norwegian Shelf (extended abstract): p. 415-418 in "Basin Modeling: Advances and Applications" (Edited by A. G. Dore et al), Norwegian Petroleum Society Special Publication 3, Elsevier, Amsterdam.

O'Brien, J. J. and Lerche, I., 1984. Influence of salt domes on paleotemperature distributions: Geophysics, v. 49, p. 2032-2043.

O'Brien, J. J. and Lerche, I., 1987. Heat flow and thermal maturation near salt diapirs: p. 711-750 in "Dynamical Geology of Salt and Related Structures" (Edited by I. Lerche and J. J. O'Brien), Academic Press.

O'Hara, K., Hower, J. C., and Rimmer, S. M., 1990. Constraints on the emplacement and uplift history of the Pine Mountain thrust sheet, Eastern Kentucky: evidence from coal rank trends: Journal of Geology, v. 98, p. 43-51.

O'Neil, J. R., Clayton, R. N., and Mayeda, T. K., 1969. Oxygen isotope fractionation in divalent metal carbonates: Journal of Chemical Physics, v. 51, p. 5547-5558.

O'Sullivan, P. B., Green, P. R., Bergman, S. C., Decker, J., Duddy, I. R., Gleadow, A. J. W., and Turner, D. L., 1993. Multiple phases of Tertiary uplift and erosion in the Arctic National Wildlife Refuge, Alaska, revealed by apatite fission track analysis: AAPG Bulletin, v. 77, p. 359-385.

Ogura, N., 1983. Estimation of thermal maturation of sedimentary rocks in the Niigata Basin, Northeast Japan: Japanese Association of Petroleum Technology, v. 48, p. 341-354.

Okui, A. and Waples, D. W., 1993. Relative permeabilities and hydrocarbon expulsions from source rocks: p. 293-301 in "Basin Modeling: Advances and Applications" (Edited by A. G. Dore et al), Norwegian Petroleum Society Special Publication 3, Elsevier, Amsterdam.

Oliver, J., 1986. Fluids expelled tectonically from orogenic belts: Geology, v. 14, p. 99-102.

Omar, G. I., Lutz, T. M., and Giegengack, R., 1994. Apatite fission-track evidence for Laramide and post-Laramide uplift and anomalous thermal regime at the Beartooth overthrust, Montana-Wyoming: Geological Society of America Bulletin, v 106, p. 74-85.

Omar, G. I., Streckler, M. S., Buck, W. R., and Kohn, B. P., 1989. Fission-track analysis of basement apatites at the western margin of the Gulf of Suez rift, Egypt: evidence for synchroneity of uplift and subsidence: Earth and Planetary Science Letters, v. 94. p. 316-328.

Oxburgh, E. H. and Andrews-Speed, C. P., 1981. Temperature, thermal gradients and heat flow in the southwestern North Sea: p. 141-151 in "Petroleum Geology of the Continental Shelf of Northwestern Europe" (Edited by L. V. Illing and G. D. Hobson), Heyden Press, London.

Oxburgh, E. R. and Turcotte, D. L., 1974. Thermal gradients and regional metamorphism in overthrust terrains with special reference to the eastern Alps: Schweizerische Mineralogische und Petrogphische Mitteilungen, v. 54, p. 641-662.

Ozkaya, I., 1991. Computer simulation of primary oil migration in Kuwait: Journal of Petroleum Geology, v. 14, p. 37-48.

Pagel, M. and Poty, B., 1984. The evolution of composition temperature and pressure of sedimentary fluids over time: a fluid inclusion reconstruction: p. 71-88 in "Thermal Phenomena in Sedimentary Basins" (Edited by B. Durand), Editions Technip, Paris.

Pagel, M., 1975. Determination des conditions physico-chimiques de la silicification diagenetique des gres Athabasca (Canada) au moyen des inclusions fluides: C. R. Acad. Science Paris, v. 280 D, p. 2301-2304.

Pagel, M., Walgenwitz, F., and Dubessy, J., 1986. Fluid inclusions in oil and gas-bearing sedimentary formations: p. 565-583 in "Thermal Modeling in Sedimentary Basins" (Edited by J. Burrus), Editions Technip, Paris.

Palmer, S. E., 1993. Effect of biodegradation and water washing on crude oil composition: p. 511-533 in "Organic Geochemistry" (Edited by M.H. Engel and S.A. Macko), Plenum Press, New York.

Pande, D. K., Misra, K. N., and Sharma, S., 1984. Source bed history of Bombay Offshore Region: Proceedings of the 27th International Geological Congress, v. 13, p. 211-244.

Pantano, J. and Lerche, I., 1989. Stratigraphic age determination from inversion of vitrinite reflectance data and sterane isomerization data: correlation at its best and worst: p. 33-43 in "Correlation in Hydrocarbon Exploration" (Edited by J. D. Collinson), Graham and Trotman, London.

Parasnic, D. S., 1971. Temperature extrapolation to infinite time in geothermal measurements: Geophysical Prospecting, v. 19, p. 612-614.

Parnell, J. and Carey, P.F., 1995. Emplacement of bitumen (asphaltite) veins in the Neuquen basin, Argentina: AAPG Bulletin, v. 79, p. 1798-1816.

Parrish, R. and Roddick, J. C., 1985. "Geochronology and Isotope Geology for the Geologist and Explorationist." Geological Association Canada, Cordilleran Section, Short Course 4.

Pasley, M. A., Riley, G. W., and Nummedal, D., 1993. Sequence stratigraphic significance of organic matter variations: Example from the Upper Cretaceous Mancos shale of the San Juan Basin, New Mexico: AAPG Studies in Geology, No. 37, p. 221-241.

Passey, Q. R., Creaney, S., Kulla, J. B., Moretti, F. J., and Stroud, J. D., 1990. A practical model for organic richness from porosity and resistivity logs: AAPG Bulletin, v. 74, p. 1777-1794.

Pearson, M. J. and Small, J. S., 1988. Illite-smectite diagenesis and paleotemperatures in northern North Sea Quaternary to Mesozoic shale sequences: Clay Minerals, v. 23, p. 109-132.

Pelet, R. and Dembicki, H., 1994. Comment and reply on "The effects of the mineral matrix on the determination of kinetic parameters using modified Rock-Eval pyrolysis": Organic Geochemistry, v. 21, p. 979-986.

Penn, I. E., Chadwick, P. A., Holloway, S., Roberts, G., Pharaoh, T. C., Allsop, J. M., Hulbert, A. G., and Burns, I.M., 1987. Principle features of the hydrocarbon prospectivity of the Wessex-Channel Basin, UK: p. 109-118 in "Petroleum Geology of North West Europe" (Edited by J. Brooks and K. Glennie), Graham and Trotman, London.
Pepper, A. S. and Corvi, P. J., 1995. Simple kinetic models of petroleum formation. Part I: oil and gas generation from kerogen: Marine and Petroleum Geology, v.12, p. 291-319.
Perrier, R. and Quiblier, J., 1974. Thickness changes in sedimentary layers during compaction history; methods for quantitative evaluation: AAPG Bulletin, v. 58, p. 507-520.
Perrodon, A., 1983. "Dynamics of Oil and Gas Accumulations," Bulletin des Centres de Recherches Exploration-Production Elf-Aquitaine, Memoir 5, Pau. 368 pp.
Perry, E. A., Jr., 1974. Diagenesis and the K-Ar dating of shales and clay minerals: Geological Society of America Bulletin, v. 85, p. 827-830.
Person, M. and Garven, G., 1992. Hydrologic constraints on petroleum generation within continental rift basins: Theory and application to the Rhine graben: AAPG Bulletin, v. 76, p. 468-488.
Person, M., Toupin, D., and Eadington, P., 1995. One-dimensional models of groundwater flow, sediment thermal history and petroleum generation within continental rift basins: Basin Research, v. 7, p. 81-96.
Peters, K. E. and Moldowan, J. M., 1993. "The Biomarker Guide. Interpreting Molecular Fossils in Petroleum and Ancient Sediments," Prentice Hall. 352 pp.
Peters, K. E., 1986. Guidelines for evaluating petroleum source rock using programmed pyrolysis: AAPG Bulletin, v. 70, p. 318-329.
Petersen, K. and Lerche, I., 1995. Quantification of thermal anomalies in sediments around salt structures: Geothermics, v. 24, p.253-268.
Pettijohn, F. J., 1957. "Sedimentary Rocks" (Second Edition), Harper, New York. 718 pp.
Philippi, G. T., 1965. On the depth, time and mechanism of petroleum generation: Geochimica et Cosmochimica Acta, v. 22, p. 1021-1040.
Philp, R. P. and Lewis, C. A., 1987. Organic geochemistry of biomarkers: Annual Reviews Earth Planetary Science, v. 15, p. 363-395.
Philp, R. P., 1985. "Fossil Fuel Biomarkers: Applications and Spectra," Methods in Geochemistry and Geophysics, No. 23, Elsevier, Amsterdam. 294 pp.
Pigott, J. D., 1985. Assessing source maturity in frontier basins: Importance of time, temperature, and tectonics: AAPG Bulletin, v. 69, p. 1269-1274.
Pilaar, W. F. H. and Wakefield, L. L., 1984. Hydrocarbon generation in the Taranaki Basin, New Zealand: AAPG Memoir No. 35, p. 405-423.
Pirrie, D. and Marshall, J. D., 1990. High-paleolatitude Late Cretaceous paleotemperatures: New data from James Ross Island, Antarctica: Geology, v. 18, p. 31-34.

Pisciotto, K. A., 1981. Diagenetic trends in the siliceous facies of the Monterey Shale in the Santa Maria region, California: Sedimentology, v. 28, p. 547-571.

Pitman, J. K., Franczyk, K. J. and Anders, D. E., 1987. Marine and non-marine gas-bearing rocks in Upper Cretaceous Blackhawk and Neslen Formations, Eastern Uinta Basin, Utah. Sedimentology, diagenesis, and source rock potential: AAPG Bulletin, v. 71, p. 76-94.

Pitt, G. J., 1961. The kinetics of the evolution of volatile products from coal: 4th International Conference on Coal Science, Le Touquet, May 30-June 2, 1961.

Pitt, G. M., 1986. Geothermal gradients, geothermal histories and the timing of thermal maturation in the Eromanga-Cooper Bains: p. 323-351 in "Contributions to the Geology and Hydrocarbon Potential of the Eromanga Basin" (Edited by D. L. Gravestock, P. S. Moore, and G. M. Pitt), Geological Society of Australia, Special Publication No. 12.

Pollack, H.N and Cercone, K. R., 1994. Anomalous thermal maturities caused by carbonaceous sediments: Basin Research, v. 6, p. 47-51.

Pollastro, R. M. and Barker, C. E., 1986. Application of clay-mineral, vitrinite reflectance, and fluid inclusion studies to the thermal and burial history of the Pinedale Anticline, Green River basin, Wyoming: p.73-83 in Society of Economic Paleontologists and Mineralogists Special Publication No. 38.

Pollastro, R. M. and Schmoker, J., 1989. Relationship of clay-minerals diagenesis to temperature, age, and hydrocarbon generation - an example from the Anadarko basin, Oklahoma: p. 257-261 in "Anadarko Basin Symposium, 1988" (Edited by K. S. Johnson), Oklahoma Geological Survey Circular 90.

Pollastro, R. M. and Scholle, P. A., 1986. Exploration and development of hydrocarbons from low-permeability chalks - an example from the Upper Cretaceous Niobrara Formation, Rocky Mountain region: p.129-142 in AAPG Studies in Geology, No. 24.

Pollastro, R. M., 1990. The illite/smectite geothermometer - concepts, methodology, and application to basin history and hydrocarbon generation: p. 1-18 in "Applications of Thermal Maturity Studies to Energy Exploration" (Edited by V. F. Nuccio and C. E. Barker), RMS-Society of Economic Paleontologists and Mineralogists.

Potter, R. W., 1977. Pressure correction for fluid inclusion temperature based on the volumetric properties of the system NaCl-H_2O: U.S. Geological Survey Journal of Research, p. 603-607.

Poulet, M. and Roucache, J., 1969. Geochemical study of the North Sahara reservoirs of Algeria (in French): Review of the French Petroleum Institute, May, p. 615-644.

Prezbindowski, D. R. and Tapp, J. B., 1991. Dynamics of fluid inclusion alteration in sedimentary rocks: a review and discussion: Organic Geochemistry, v. 17, p. 131-142.

Prezbindowski, D. R., and Larese, R. E., 1987. Experimental stretching of fluid inclusions in calcite - Implications for diagentic studies: Geology, v. 15, p. 333-336.

Price, K. L., Huntoon, J. E., and McDowell, S. D., 1996. Thermal history of the 1.1-Ga Nonesuch Formation, North American mid-continent rift, White Pine, Michigan: AAPG Bulletin, v. 80, p. 1-15.

Price, L. C. and Barker, C. E., 1985. Suppression of vitrinite reflectance in amorphous rich kerogen - a major unrecognized problem: Journal of Petroleum Geology, v. 8, p. 59-84.

Price, L. C., 1983. Geologic time as a parameter in organic metamorphism and vitrinite reflectance as an absolute paleogeothermometer: Journal of Petroleum Geology, v. 6, p. 5-38.

Price, L. C., Daws, T., and Pawlewicz, M., 1986. Organic metamorphism in the Lower Mississippian-Upper Devonian Bakken shales. Part 1: Rock-Eval pyrolysis and vitrinite reflectance: Journal of Petroleum Geology, v. 9, p. 125-162.

Puckett, T. M., 1991. Absolute paleobathymetry of Upper Cretaceous chalks based on ostracodes - Evidence from the Demopolis Chalk (Campanian and Maastrichtian) of the northern Gulf Coastal Plain: Geology, v. 19, p. 449-452.

Pusey, W. C., 1973. How to evaluate potential oil and gas source rocks: World Oil, April, p. 71-74.

Pytte, A. M. and Reynolds, R. C., 1989. The thermal transformation of smectite to illite: p. 133-140 in "Thermal History of Sedimentary Basins" (Edited by N. D. Naeser and T. H. McCulloh), Springer-Verlag, New York.

Qing, H. and Mountjoy, E. W., 1994. Formation of coarsely crystalline, hydrothermal dolomite reservoirs in the Presqu'ile barrier, western Canada sedimentary basin: AAPG Bulletin, v. 78, p. 55-77.

Qing, H. and Mountjoy, E., 1992. Large-scale fluid flow in the Middle Devonian Presqu'ile barrier, western Canada sedimentary basin: Geology, v. 20, p. 903-906.

Quigley, T. M. and Mackenzie, A. S., 1988. The temperatures of oil and gas formation in the subsurface: Nature, v. 333, p. 549-552.

Quigley, T. M., Mackenzie, A. S., and Gray, J. R., 1987. Kinetic theory of petroleum generation: p. 649-665 in "Migration of Hydrocarbons in Sedimentary Basins," (Edited by B. Doligez), Editions Technip, Paris.

Radke, M., Welte, D. H., and Willsch, H., 1984. Geochemical study on a well in the Western Canada basin: relation of the aromatic distribution pattern to maturity of organic matter: Geochimica et Cosmochimica Acta, v. 46, p. 1-10.

Rahman, M., Majorowicz, J. A., Bosse, V., and Jones, F. W., 1985. Some implications of organic maturation modelling in south-central Alberta: Bulletin of Canadian Petroleum Geology, v. 33, p. 396-409.

Ramseyer, K. and Boles, J. R., 1986. Mixed-layer illite/smectite minerals in Tertiary sandstones and shales, San Joaquin basin, California: Clays and Clay Minerals, v. 34, p. 115-124.

Rashid, M. A. and McAlary, J. D., 1977. Early maturation of organic matter and genesis of hydrocarbons as a result of heat from a shallow piercement salt dome: Journal of Geochemical Exploration, v. 8, p. 549-569.

Rashid, M. A., 1978. The influence of a salt dome on the diagenesis of organic matter in the Jeanne d'Arc Subbasin of the northeast Grand Banks of Newfoundland: Organic Geochemistry, v. 1, p. 67-77.

Reiter, M. and Clarkson, G., 1983. Relationships between heat flow, paleotemperatures, coalification and petroleum maturation in the San Juan basin, northwestern New Mexico and southwest Colorado: Geothermics, v. 12, p. 323-339.

Reminton, C. J. and Pranyotao, U., 1985. A hydrocarbon generation analysis in Northwest Java Basin using Lopatin's method: Proceedings of the Annual Convention - Indonesian Petroleum Association, (IDN) 14, v. 1, p. 121-141.

Ren, Z., Zhao, Z., Zhang, J., and Yu, Z., 1994. Research on paleotemperature in the Ordos basin: Acta Sedimentologica Sinica, v. 12, p. 56-65.

Reynolds, L. A. and Thunell, R. C., 1986. Seasonal production and morphologic variation of *Neogloboquadrina pachyderma* (Ehrenberg) in the northeast Pacific: Micropaleontology, v. 32(1), p. 1-18.

Rhea, L., Person, M., deMarsily, G., Ledoux, E., and Galli, A., 1994. Geostatistical models of secondary oil migration within heterogeneous carrier beds: a theoretical example: AAPG Bulletin, v. 78, p. 1679-1691.

Rice, D. D. and Claypool, G. E., 1981. Generation, accumulation, and resource potential of biogenic gas: AAPG Bulletin, v. 65, p. 5-25.

Rice, D. D., 1983. Relation of natural gas composition to thermal maturity and source rock type in San Juan basin, northwestern New Mexico and southwestern Colorado: AAPG Bulletin, v. 67, p. 1199-1218.

Rice, D. D., 1993. Biogenic gas: controls, habitats, and resource potential: p. 583-606 in "The Future of Energy Gases" (Edited by D. G. Howell), U.S. Geological Survey Professional Paper 1570.

Rieke, H. H., III and Chilingarian, C. V., 1974. "Compaction of Argillaceous Sediments," Elsevier, Amsterdam. 474 pp.

Risley, D. E., Martin, G. C., Lynch, M. B., Flett, T. L., Larson, J. A., and Horowitz, W. L., 1992. Geologic report for the Gulf of Alaska planning area: Minerals Management Service, OCS Report MMS 92-0065.

Ritter, U., 1984. The influence of time and temperature on vitrinite reflectance: Organic Geochemistry, v. 6, p. 473-480.

Ritter, U., 1988. Modelling of hydrocarbon generation patterns in the Egersund Sub-Basin, North Sea: Organic Geochemistry, v. 13, p. 165-174.

Ritter, U., Aareskjold, K., and Schou, L., 1993. Distributed activation energy models of isomerisation reactions from hydrous pyrolysis: Organic Geochemistry, v. 20, p. 511-520.

Ritter, U., Leith, T. L., Griffiths, C. M., and Schou, L., 1987. Hydrocarbon generation and thermal evolution in parts of the Egersund Basin, Northern North Sea: p. 75-85 in "Sedimentary Basins and Basin-Forming Mechanisms" (Edited by C. Beaumont and A. J. Tankard), Canadian Society of Petroleum Geologists, Memoir 12.

Ritter, U., Myhr, M. B., Aareskjold, K., Ganz, H.H., Mostafa, A., and Schou, L., 1993. Distributed activated energy models from hydrous pyrolysis: towards a second-generation tool of thermal history analysis: p. 201-208 in "Basin Modeling: Advances and Applications" (Edited by A. G. Dore et al), Norwegian Petroleum Society Special Publication 3, Elsevier, Amsterdam.

Ritter, U., Myhr, M.B., Aareskjold, K., and Schou, L., 1993. Validation and first application of palaeotemperature models of isomerization and NMR aromatisation parameters: p. 185-199 in "Basin Modeling: Advances and Applications" (Edited by A. G. Dore et al), Norwegian Petroleum Society Special Publication 3, Elsevier, Amsterdam.

Robert, P., 1980. The optical evolution of kerogen and geothermal histories applied to oil and gas exploration: Chapter 12 in "Kerogen" (Edited by B. Durand), Editions Technip, Paris.

Robert, P., 1988. "Organic Metamorphism and Geothermal History," Dordrecht, Netherlands, Reidel. 311 pp.

Roberts, S. V. and Burbank, D. W., 1993. Uplift and thermal history of the Teton Range (northwestern Wyoming) defined by apatite fission-track dating: Earth and Planetary Science Letters, v. 118, p. 295-309.

Roberts, W. H., 1981. Some uses of temperature data in petroleum exploration: p. 8-49 in "Symposium II. Unconventional Methods in Exploration of Petroleum and Natural Gas" (Edited by B. M. Gittlieb), Southern Methodist University Press, Dallas.

Robinson, A. G., Coleman, M. L., and Gluyas, J. G., 1993. The age of illite cement growth, Village Fields area, southern North Sea: Evidence from K-Ar ages and $^{18}O/^{16}O$ ratios: AAPG Bulletin, v. 77, p. 68-80.

Roden, M. K., 1990. Apatite fission-track thermochronology of the southwestern Appalachian basin: Maryland, West Virginia, and Virginia: Journal of Geology, v. 99, p. 41-53.

Roedder, E., 1979. Fluid inclusion evidence on the environments of sedimentary diagenesis, a review: p.89-107 in "Aspects of Diagenesis" (Edited by P. A. Scholle and P. R. Schluger), Society of Economic Paleontologists and Mineralogists Special Publication No. 26.

Roedder, E., 1984. "Fluid Inclusions." Reviews in Mineralogy, Vol. 12, Mineralogical Society of America, Washington, D.C. 644 pp.

Rogers, M. A., McAlary, J. D., and Bailey, N. J. L., 1974. Significance of reservoir bitumens to thermal maturation studies, Western Canadian basin: AAPG Bulletin, v. 58, p. 1806-1824.

Roksandic, M. M., 1986. Dynamic interpretation of organic-matter maturation and evolution of oil-generative window: Discussion: AAPG Bulletin, v. 70, p. 1008-1010.

Ronov, A. B., 1958. Organic carbon in sedimentary rocks (in relation to the presence of petroleum): Geochemistry (Transl. from Russian), v. 5, p. 510-536.

Rose, P. R., 1986. Hydrocarbon resources of the Palo Duro Basin, Texas Panhandle: Office of Nuclear Waste Isolation, Report No. BMI/ONWI-590. 100 pp.

Ross, L. M. and Ames, R. L., 1988. Stratification of oils in Columbus Basin off Trinidad: Oil and Gas Journal, v. 86:39, p. 72-76.

Roufosse, M. C., 1987. The formation and evolution of sedimentary basins in the western Barents Sea: p. 1149-1161 in "Petroleum Geology of North West Europe" (Edited by J. Brooks and K. Glennie), Graham and Trotman, London.

Royden, L. and Keen, C. E., 1980. Rifting process and thermal evolution of the continental margin of eastern Canada determined from subsidence curves: Earth and Planetary Science Letters, v. 51, p. 343-361.

Royden, L. H., 1988. Late Cenozoic tectonics of the Pannonian basin system: AAPG Memoir 45, p. 27-48.

Royden, L., Sclater, J. G., and von Herzen, R. P., 1980. Continental margin subsidence and heat flow: important parameters in formation of petroleum hydrocarbons: AAPG Bulletin, v. 64, p. 173-187.

Rudkiewicz, J. L. and Behar, F., 1994. Influence of kerogen type and TOC content on multiphase primary migration: Organic Geochemistry, v. 21, p. 121-133.

Rullkotter, J. and Marzi, R., 1989. New aspects of the application of sterane isomerization and steroid aromatization to petroleum exploration and the reconstruction of geothermal histories of sedimentary basins: American Chemical Society Preprint, Division of Petroleum Chemistry, p. 126-131.

Rullkotter, J., Marzi, R., and Meyers, P.A., 1992. Biological markers in Paleozoic sedimentary rocks and crude oils from the Michigan basin: Reassessment of sources and thermal history of organic matter: p. 324-335 in "Early Organic Evolution: Implications for Mineral and Energy Resources" (Edited by M. Schidlowski et al), Springer-Verlag, Berlin.

Rullkotter, J., Meyers, P. A., Schaefer, R. G., and Dunham, K. W., 1986. Oil generation in the Michigan Basin: A biological marker and carbon isotope approach: p. 359-375 in "Advances in Organic Geochemistry 1985" (Edited by D. Leythaeuser and J. Rullkotter). Pergamon Press.

Rundberg, Y. and Smalley, P. C., 1989. High-resolution dating of Cenozic sediments from northern North Sea using $^{87}Sr/^{86}Sr$ stratigraphy: AAPG Bulletin, v. 73, p. 298-308.

Russell, N. J. and Baillie, P. W., 1989. Vitrinite paleothermometry of offshore exploration wells, Tasmania: Australian Petroleum Exploration Association (APEA) Journal., v. 29, p. 130-156.

Russell, N. J. and Bone, Y., 1989. Palaeogeothermometry of the Cooper and Eromanga Basins, South Australia: p. 559-581 in "The Cooper and Eromanga Basins, Australia" (Edited by B. J. O'Neil), Proceedings Petroleum Exploration Society of Australia, Society of Petroleum Engineers, Australian Society of Exploration Geophysics (S. A. Branch), Adelaide.

Rybach, L. and Cermak, V., 1982. Radioactive heat generation in rocks: p. 353-371 in "Physical Properties of Rocks, Volume 5" (Edited by G. Angenheister), Springer-Verlag, New York.

Ryer, T. A. and Langer, A. W., 1980. Thickness change involved in the peat-to-coal transformation for a bituminous coal of Cretaceous age in central Utah: Journal of Sedimentary Petrology, v. 50, p. 987-992.

Sachsenhofer, R. F., 1994. Petroleum generation and migration in the Styrian Basin (Pannonian Basin system, Austria): an integrated geochemical and numerical modelling study: Marine and Petroleum Geology, v. 11, p. 684-701.

Sadler, P. M., 1981. Sediment accumulation rates and the completeness of stratigraphic sections: Journal of Geology, v. 89, p. 569-584.

Sahay, B. and Fertl, W. H., 1989. "Origin and Evaluation of Formation Pressures", Kluwer Academic Publishers, Dordrecht. 292 pp.

Saigal, G. C., Bjorlykke, K., and Larter, S., 1992. The effects of oil emplacement on diagenetic processes - Examples from the Fulmar Reservoir sandstones, central North Sea: AAPG Bulletin, v. 76, p. 1034-1033.

Sajgo, C. and Lefler, J., 1986. A reaction kinetic approach to the temperature-time history of sedimentary basins: p. 119-151 in "Paleogeothermics" (Edited by G. Buntebarth and L. Stegena), Springer-Verlag, Berlin.

Sarg, J. F., 1988. Carbonate sequence stratigraphy: p. 155-181 in "Sea-Level Changes- An Integrated Approach", Society of Economic Paleontologists and Mineralogists Special Publication No. 42.

Sarwar, G. and Friedman, G.M., 1994. Late Paleozoic sediment cover on the Adirondacks, New York: evidence from fluid inclusions and clay diagenesis: Northeastern Geology, v. 16, p. 18-36.

Sass, J. H., Blackwell, D. D., Chapman, D. S., Costain, J. K., Decker, E. R., Lawver, L. A., and Swanberg, C. A., 1989. Heat flow from the crust of the United States: p. 503-548 in "Physical Properties of Rocks and Minerals Volume II-2" (Edited by Y. S. Touloukian, W. R. Judd, and R. F. Roy). Hemisphere Publishing, New York.

Sassen, R. and Moore, C.H., 1988. The framework of hydrocarbon generation and destruction in the eastern Smackover trend: AAPG Bulletin, v. 72, p. 649-663.

Savin, S. M., 1977. The history of the earth's surface temperature during the past 100 million years: Annual Reviews of Earth and Planetary Sciences 1977, v. 5, p. 319-355.

Savin, S. M., Abel, L., Barrera, E., Hodell, D., Kennett, J. P., Murphy, M., Keller, G., Killingley J., and Vincent, E., 1985. The evolution of Miocene surface and near surface marine temperatures: Oxygen isotopic evidence: Geological Society of America Memoir 163, p. 49-82.

Sawyer, D. S., Toksoz, M. N., Sclater, J. G., and Swift, B. A., 1982. Thermal evolution of the Baltimore Canyon Trough and Georges Bank basin: AAPG Memoir No. 34, p. 743-762.

Schaefer, R. G., Schenk, H. J., Hardelauf, H., and Harms, R., 1990. Determination of gross kinetic parameters for petroleum formation from Jurassic source rocks of different maturity levels by means of laboratory experiments: Organic Geochemistry, v. 16, p. 115-120.

Schanmugam, G. and Higgins, J. B., 1988. Porosity enhancement from chert dissolution beneath Neocomian unconformity: Ivishak Formation, North Slope, Alaska: AAPG Bulletin, v. 72, p. 523-535.

Schlager, W., 1981. The paradox of drowned reefs and carbonate platforms: Geological Society of America Bulletin, Part I, v. 92, p. 197-211.

Schlanger, S. O. and Combs, J., 1975. Hydrocarbon potential of marginal basins bounded by an island arc: Geology, July 1975, p. 397-400.

Schmoker, J. W. and Gautier, D. L., 1988. Sandstone porosity as a function of thermal maturity: Geology, v. 16, p. 1007-1010.

Schmoker, J. W. and Halley, R. B., 1982. Carbonate porosity versus depth: a predictable relation for South Florida: AAPG Bulletin, v. 66, p. 2561-2570.

Schmoker, J. W. and Hester, T. C., 1983. Organic carbon in Bakken Formation, United States portion of Williston basin: AAPG Bulletin, v. 67, p. 2165-2174.

Schmoker, J. W. and Hester, T. C., 1989. Oil generation inferred from formation resistivity - Bakken Formation, Williston basin, North Dakota: Transactions of the Thirtieth Society of Petroleum Well Log Analysts Annual Logging Symposium, paper H.

Schmoker, J. W., 1981. Determination of organic matter content of Appalachian Devonian shales from gamma-ray logs: AAPG Bulletin, v. 65, p. 1285-1298.

Schmoker, J. W., 1984. Empirical relation between carbonate porosity and thermal maturity: an approach to regional porosity prediction: AAPG Bulletin, v. 68, p. 1697-1703.

Schmoker, J. W., 1986. Oil generation in the Anadarko basin, Oklahoma and Texas: Modeling using Lopatin's Method: Oklahoma Geological Survey, Special Publication 86-3. 40 pp.

Schoell, M., 1980. The hydrogen and carbon isotopic composition of methane from natural gases of various origins: Geochimica et Cosmochimica Acta, v. 44, p. 649-661.

Scholle, P. A., 1977. Chalk diagenesis and its relationship to oil exploration: oil from chalks a modern miracle: AAPG Bulletin, v. 61, p. 982-1009.

Scholle, P. A., Stemmerik, L., and Ulmer, D. S., 1991. Diagenetic history and hydrocarbon potential of Upper Permian carbonate buildups, Wegener Halvo Area, Jameson Land Basin, East Greenland: AAPG Bulletin, v. 75, p. 701-725.

Schowalter, T., 1979. Mechanics of secondary hydrocarbon migration, and entrapment: AAPG Bulletin, v. 63, p. 723-760.

Schroeder, F. W. and Sylta, O., 1993. Modelling the hydrocarbon system of the North Viking Graben: a case study: p. 469-484 in "Basin Modeling: Advances and Applications" (Edited by A. G. Dore et al), Norwegian Petroleum Society Special Publication 3, Elsevier, Amsterdam.

Schull, T. T., 1988. Rift basins of interior Sudan: Petroleum exploration and discovery: AAPG Bulletin, v. 72, p. 1128-1142.

Schwab, F. L., 1976. Modern and ancient sedimentary basins, comparative accumulation rates: Geology, v. 4, p. 723-727.

Schwarz, E. J. and Buchan, K. L., 1985. Post-Aphebian uplift deduced from remanent magnetization, Yellowknife area of Slave Province: Canadian Journal of Earth Science, v. 22, p. 1793-1802.

Sclater, J. G. and Christie, P. A. F., 1980. Continental stretching: an explanation of the post mid-Cretaceous subsidence of the Central North Sea Basin: Journal of Geophysical Research, v. 85, p. 3711-3739.

Scott, A. C. and Fleet, A. J. (Editors), 1994. "Coal and Coal Bearing Strata as Oil-Prove Source Rocks?", The Geological Society (London), Special Publication 77. 213pp.

Seifert, W. K. and Moldowan, J. M., 1978. Applications of steranes, terpanes and monoaromatics to the maturation, migration and source of crude oils: Geochimica et Cosmochimica Acta, v. 42, p. 77-95.

Selig, F. and Wallick, G. C., 1966. Temperature distribution in salt domes and surrounding sediments: Geophysics, v. 31, p. 346-361.

Seni, S. J. and Jackson, M. P. A., 1983. Evolution of salt structures, East Texas diapir province, Part 2: Patterns and rates of halokinesis: AAPG Bulletin, v. 67, p. 1245-1274.

Sestini, G., 1989. Nile Delta: a review of depositional environments and geologic history: p. 99-127 in "Deltas: Sites and Traps for Fossil Fuels" (Edited by M. K.G. Whateley and K. T. Pickering), Geological Society (London) Special Publicaiton No. 41.

Shackleton, N. J., 1974. Attainment of isotopic equilibrium between ocean water and the benthonic foraminifera genus *Uvigerina:* Isotopic changes in the ocean during the last glacial: Colloq. Int. C.N.R.S., v. 219, p. 203-210.

Shepherd, T. J., Rankin, A. H., and Alderton, D. H.M., 1985. "A Practical Guide to Fluid Inclusion Studies", Blackie, London. 239 pp.

Shipley, T. H., Houston, M. H., Buffler, R. T. , Shaub, F. J., McMillen, K. J., Ladd, J. W., and Worzel, J. L., 1979. Seismic evidence for widespread possible gas hydrate horizons on continental slopes and rises: AAPG Bulletin, v. 63, p. 2204-2213.

Siever, R. and Hager, J. L., 1981. Paleogeography, tectonics and thermal history of some Atlantic margin sediments: p. 95-117 in "Geology of North Atlantic Borderlands" (Edited by J. W. Kerr and A. J. Fergguson), Canadian Society of Petroleum Geologists, Memoir 7.

Siever, R., 1983. Burial history and diagenetic reaction kinetics: AAPG Bulletin, v. 67, p. 684-691.

Skjervoy, A. and Sylta, O., 1993. Modelling of expulsion and secondary migration along the south western margin of the Horda Platform: p. 499-537 in "Basin Modeling: Advances and Applications" (Edited by A. G. Dore et al), Norwegian Petroleum Society Special Publication 3, Elsevier, Amsterdam.

Sloan, E. D., Jr., 1990. "Clathrate Hydrates of Natural Gases", Marcel Dekker, New York. 640 pp.

Sloan, E. D., Jr., Happel, J., and Hnatow, M. (Editors), 1994. Natural gas hydrates: Annals of the New York Academy of Sciences, v. 715. 580 pp.

Smith, L. and Chapman, D. S., 1983. On the thermal effects of groundwater flow: 1. Regional scale systems: Journal of Geophysical Research, v. 88, p. 593-608.

Smosna, R., 1989. Compaction law for Cretaceous sandstones of Alaska's North Slope: Journal of Sedimentary Petrology, v. 59, p. 572-584.

Snowdon, L. R., 1979. Errors in extrapolation of experimental kinetic parameters to organic geochemical systems: AAPG Bulletin, v. 63, p. 1128-1134.

Sofer, Z., 1984. Stable carbon isotope compositions of crude oils: application to source depositional environments and petroleum alterations: AAPG Bulletin, v. 68, p. 31-49.

Somerton, W. H., 1992. "Thermal Properties and Temperature-Related Behaviour of Rock/Fluid Systems", Developments in Petroleum Science, 37. Elsevier, Amsterdam. 257 pp.

Sorensen, S., 1985. Basin analysis and maturation modelling, onshore Denmark: a case study from the Danish First Round: p. 153-160 in "Petroleum Geochemistry in Exploration of the Norwegian Shelf" (Edited by B. M. Thomas et al), Graham and Trotman, for the Norwegian Petroleum Society, London.

Speece, M. A., Bowen, T. D., Folcik, J. L., and Pollack, H. N., 1985. Analysis of temperatures in sedimentary basins: the Michigan Basin: Geophysics, v. 50, p. 1318-1334.

Spencer, C. W., 1987. Hydrocarbon generation as a mechanism for over-pressuring in Rocky Mountain region: AAPG Bulletin, v. 71, p. 368-388.

Spencer, C. W., 1994. Abnormal formation pressures caused by hydrocarbon generation – examples from the Rocky Mountain region: p. 343-375 in “Studies in Abnormal Pressures”, Elsevier Developments in Petroleum Science No. 38.

Springer, J., 1993. Decompaction and backstripping with regard to erosion, salt movement, and interlayered bedding: Computers and Geosciences, v. 19, p. 1115-1125.

Stam, B., Gradstein, F. M., Lloyd, P., and Gillis, D., 1987. Algorithms for porosity and subsidence history: Computers and Geosciences, v. 13, p. 317-349.

Stegena, L., Horvath, F., Sclater, J. G., and Royden, L., 1981. Determination of paleotemperature by vitrinite reflectance data: Earth Evolut. Science, v. 1, p. 292-300.

Stein, R., 1990. Organic carbon content/sedimentation rate relationship and its paleoenvironmental significance for marine sediments: Geo-Marine Letters, v. 10, p. 37-44.

Stein, R., 1991. "Accumulation of organic carbon in marine sediments", Lecture Notes in Earth Sciences, Volume 34, Springer-Verlag, Berlin. 217pp.

Stockmal, G. S. and Beaumont, C., 1987. Geodynamic models of convergent margin tectonics: the southern Canadian cordillera and the Swiss Alps: p. 393-411 in "Sedimentary Basins and Basin-forming Mechanisms" (Edited by C. Beaumont and A. J. Tankard), Canadian Society of Petroleum Geologists, Memoir 12.

Stoll, R. D. and Bryan, G. M., 1979. Physical properties of sediments containing gas hydrates: Journal of Geophysical Research, v. 84, p. 1629-1634.

Storzer, D. and Selo, M., 1984. Toward a new tool in hydrocarbon resource evaluation: the potential of the apatite fission track chronothermometer: p. 89-110 in "Thermal Phenomena in Sedimentary Basins" (Edited by B. Durand), Editions Technip, Paris.

Strachan, M. G., Alexander, R., van Bronswijk, W., and Kagi, R. I., 1989. Source and heating rate effects upon maturity parameters based on ratios of 24-ethylcholestane diastereomers: Journal of Geochemical Exploration, v. 31, p. 285-294.

Strauch, F., 1968. Determination of Cenozoic sea-temperatures using *Hiatella artica* (Linne): Palaeogeography, Palaeoclimatology, Palaeoecology, v. 5, p. 213-233.

Streckler, M. S. and Watts, A. B., 1978. Subsidence of the Atlantic-type continental margin off New York: Earth and Planetary Science Letters, v. 41, p. 1-13.

Summa, L. L., Pottorf, R. J., Schwarzer, T.F., and Harrison, W. J., 1993. Palaeohydrology of the Gulf of Mexico Basin: development of compaction overpressure and timing of hydrocarbon migration relative to cementation: p. 641-656 in "Basin Modeling: Advances and Applications" (Edited by A. G. Dore et al), Norwegian Petroleum Society Special Publication 3, Elsevier, Amsterdam.

Summer, N. S. and Verosub, K. L., 1989. A low-temperature hydrothermal maturation mechanism for sedimentary basins associated with volcanic rocks: p. 129-136 in "Origin and Evolution of Sedimentary Basins and Their Energy and Mineral Resources" (Edited by R. A. Price), Geophysical Monograph 48, IUGC Volume 3 American Geophysical Union and the International Union of Geodesy and Geophysics.

Surdam, R. C., Crossey, L. J., Hagen, E. S., and Heasler, H. P., 1989. Organic-inorganic interactions and sandstone diagenesis: AAPG Bulletin, v. 73, p. 1-23.

Sute, M., Schlosser, P., Clark, J. F., and Broecker, W. S., 1992. Paleotemperatures in the southwestern United States derived from noble gases in ground water: Science, v. 256, p. 1000-1003.

Suzuki, N., 1984. Estimation of maximum temperature of mudstone by two kinetic parameters: epimerization of sterane and hopane: Geochimica et Cosmochimica Acta, v. 48, p. 2273-2282.

Suzuki, N., 1990. Application of sterane epimerization to evaluation of Yoshii gas and condensate reservoir, Niigata basin, Japan: AAPG Bulletin, v. 74, p. 1571-1589.

Suzuki, N., Matsubayashi, H., and Waples, D. W., 1993. A simpler kinetic model of vitrinite reflectance: AAPG Bulletin, v 77, p. 1502-1508.

Swanberg, C. A. and Morgan, P., 1979. The linear relation between heat flow based on silica content of groundwater and regional heat flow: a new heat flow map of the United States: Pure and Applied Geophysics v. 117, p. 227-241.

Swanberg, C. A. and Morgan, P., 1980. The silica heat flow interpretation technique: assumption and applications: Journal of Geophysical Research, v. 85, p. 7206-7214.

Sweeney, J. J. and Burnham, A. K., 1990. Evaluation of a simple model of vitrinite reflectance based on chemical kinetics: AAPG Bulletin, v. 74, p. 1559-1570. (see also Errata, AAPG Bulletin, v. 75, p. 848.)

Sweeney, J. J., 1990. BASINMAT. Fortran program calculates oil and gas generation using a distribution of discrete activation energies: Geobyte, April, p. 37-43.

Sweeney, J. J., Braun, R. L., Burnham, A. K., Talukdar, S., and Vallejos, C., 1995. Chemical kinetic model of hydrocarbon generation, expulsion, and destruction applied to the Maracaibo basin, Venezuela: AAPG Bulletin, v. 79, p. 1515-1532.

Sweeney, J. J., Burnham, A. K., and Braun, R. L. 1987. A model of hydrocarbon generation from Type I kerogen: application to the Uinta Basin, Utah: AAPG Bulletin, v. 71, p. 967-985.

Sweeney, J. J., Burnham, A. K., and Braun, R. L., 1986. A model of hydrocarbon maturation in the Uinta Basin, Utah, U.S.A.: p. 547-561 in "Thermal Modeling in Sedimentary Basins" (Edited by J. Burrus), Editions Technip, Paris.

Sweeney, J., Talukdar, S., Burnham, A., and Vallejos, C., 1990. Pyrolysis kinetics applied to prediction of oil generation in the Maracaibo Basin, Venezuela: Organic Geochemistry, v. 16, p. 189-196.

Sweeney, J., Braun, R. L., Burnham, A. K., Talukdar, S., and Vallejos, C., 1995. Chemical and kinetic model of hydrocarbon generation, expulsion, and destruction applied to the Maracaibo basin, Venezuela: AAPG Bulletin, v. 79, p. 1515-1532.

Swift, B. A., Sawyer, D. S., Grow, J. A., and Klitgard, K. D., 1987. Subsidence, crustal structure, and thermal evolution of Georges Bank basin: AAPG Bulletin, v. 71, p. 702-718.

Swift, J. H. and Williams, J. A., 1980. Petroleum source rocks, Grand Banks area: Canadian Society Petroleum Geologist Memoir 6, p. 567-587.

Sylta, O., 1987. SEMI – a program for the modelling of buoyancy-driven, secondary migration of oil and gas by means of a ray-tracing technique: Continental Shelf Petroleum Technology Research Institute (IKU) Rep 25.1403.00//01/87. 51pp.

Sylta, O., 1993. New techniques and their applications in the analysis of secondary migration: p. 385-398 in "Basin Modeling: Advances and Applications" (Edited by A. G. Dore et al), Norwegian Petroleum Society Special Publication 3, Elsevier, Amsterdam.

Tainter, P. A., 1984. Stratigraphic and paleostructural controls on hydrocarbon migration in Cretaceous D and J sandstones of the Denver-Basin: p. 339-354 in "Hydrocarbon Source Rocks of the Greater Rocky Mountain Region" (Edited by J. Woodward, F. F. Meissner, and J. L. Clayton), Rocky Mountain Association of Geologists, Denver.

Talukdar, S., Gallango, O., Vallejos, C., and Ruggiero, A., 1987. Observations on the primary migration of oil in the La Luna source rocks of the Maracaibo Basin, Venezuela: p. 59-77 in "Migration of Hydrocarbons in Sedimentary Basins" (Edited by B. Doligez), Editions Technip, Paris.

Tang, J. and Lerche, I., 1993. Analysis of the Beaufort-Mackenzie Basin, Canada: burial, thermal and hydrocarbon histories: p. 623-640 in "Basin Modeling: Advances and Applications" (Edited by A. G. Dore et al), Norwegian Petroleum Society Special Publication 3, Elsevier, Amsterdam.

Teichmuller, R. and Teichmuller, M., 1986. Relations between coalification and palaeogeothermics in Variscan and Alpidic foredeeps of western Europe: p. 53-78 in "Paleogeothermics" (Edited by G. Buntebarth and L. Stegena), Springer-Verlag, Berlin.

Theis, N.J., Nielsen, H. M., Sales, T. K., and Gail, G. J., 1993. Impact of data integration on basin modelling in the Barents Sea: p. 433-444 in "Basin Modeling: Advances and Applications" (Edited by A. G. Dore et al), Norwegian Petroleum Society Special Publication 3, Elsevier, Amsterdam.

Thomas, B. H., Moller-Pedersen, P., Whitaker, M. F., and Shaw, N. D., 1985. Organic facies and hydrocarbon distributions in the Norwegian North Sea: p. 3-26 in "Petroleum Geochemistry in Exploration of the Norwegian Shelf" (Edited by B. M. Thomas, et al), Graham and Trotman, London.

Thomas, B. M., 1984, Hydrocarbons, source rocks, and maturity trends in the northern Perth Basin, Australia: AAPG Memoir No. 35, p. 391-403.

Thomas, J. and Thomas, N. J., 1988. Use of TTI to evaluate hydrocarbon prospects in West Bengal Basin: p. 275-282 in "Petroleum Geochemistry and Exploration in the Afro-Asian Region" (Edited by R. K. Kumar, P. D. Wivedi, V. Banerjie, and V. Gupta), A. Balkema, Rotterdam.

Thomas, M. M. and Clouse, J. A., 1995. Scaled physical model of secondary oil migration: AAPG Bulletin, v. 79, p. 19-29.

Thorne, J. A. and Watts, A. B., 1989. Quantitative analysis of North Sea subsidence: AAPG Bulletin, v. 73, p. 88-116.

Throndsen, R., Andresen, B., and Unander, A., 1993. Comparison of different models for vitrinite reflectance evolution using laboratory calibration and modelling of well data: p. 127-133 in "Basin Modeling: Advances and Applications" (Edited by A. G. Dore et al), Norwegian Petroleum Society Special Publication 3, Elsevier, Amsterdam.

Tilley, B. J., Nesbitt, B. E., and Longstaffe, F. J., 1989. Thermal history of Alberta Deep Basin: Comparative study of fluid inclusion and vitrinite reflectance data: AAPG Bulletin, v. 73, p. 1206-1222.

Tillman, J. E. and Barnes, H. L., 1983. Deciphering fracturing and fluid migration histories in northern Appalachian basin: AAPG Bulletin, v. 67, p. 692-705.

Tillman, J. E., 1983. Exploration for reservoirs with fracture-enhanced permeability: Oil and Gas Journal, Feb. 21st, p. 165-180.

Timko, D. J. and Fertl, W. H., 1972. How down-hole temperature pressure affect drilling: World Oil, v. 175, p. 73.

Tissot, B. and Espitalie, J., 1975. L'evolution thermique de la matiere organique des sediments: Applications d'une simulation mathematique: Review Institute French Petroleum, v. 30, p. 743-777.

Tissot, B. P. and Welte, D. H., 1984. "Petroleum Formation and Occurrence" (Second Edition), Springer-Verlag, New York. 699 pp.

Tissot, B. P., 1969. Premieres donnees sur les mecanismes et la cinetique de la formation du petrole dans les sediments. Simulation d'un schema reactionnel sur ordinateur: Review French Petroleum Institute, v. 24, p. 470-501.

Tissot, B. P., 1987. The role of geochemistry in exploration risk evaluation and decision making: Proceedings Twelfth World Petroleum Congress, v. 2, p. 99-112.

Tissot, B. P., Pelet, R., and Ungerer, P., 1987. Thermal history of sedimentary basins, maturation indices, and kinetics of oil and gas generation: AAPG Bulletin, v. 71, p. 1445-1466.

Tissot, B., and Pelet, R., 1971. Nouvelle donnees sur les mechanismes de genese et de migration du petroleum simulation mathematique et application a la prospection: Proceedings of the 8th World Petroleum Congress, v. 2, p. 35-46, Applied Science Publishers, London.

Tissot, B., Califet-Debyser, Y., Deroo, G., and Oudin, J. L., 1971. Origin and evolution of hydrocarbons in early Toarcian shales, Paris Basin, France: AAPG Bulletin, v. 55, p. 2177-2193.

Tissot, B., Durand, B., Espitalie, J., and Combaz, A., 1974. Influence of nature and diagenesis of organic matter in formation of petroleum: AAPG Bulletin, v. 58, p. 499-506.

Tocco, R., Alberdi, M., Ruggiero, A., and Jordan, N., 1994. Organic geochemistry of the Carapita Formation and terrestrial crude oils in the Maturine subbasin, Eastern Venezuelan Basin: Organic Geochemistry, v. 21, p. 1107-1119.

Touret, J. L. R., 1994. Fluid inclusions in sedimentary and diagenetic environments: p. 251-269 in "Fluid Inclusions in Minerals" Methods and Applications" (Edited by B. DeVito and M. L. Frezzotti), Short Course Notes of the Working Group "Inclusions in Minerals", International Mineralogical Association.

Trindade, L.A.F., Brassell, S.C., and Neto, E.V.S., 1992. Petroleum migration and mixing in the Potiguar basin, Brazil: AAPG Bulletin, v. 76, p. 1903-1924.

Turcotte, D. L. and Ahern, J. L., 1977. On the thermal and subsidence history of sedimentary basins: Journal of Geophysical Research, v. 82, p. 3762-3766.

Turcotte, D. L. and Angevine, C. L., 1982. Thermal mechanisms of basin formation: Philosophical Transactions Royal Society London, v. A305, p. 283-294.

Turcotte, D. L. and McAdoo, D. C., 1979. Thermal subsidence and petroleum generation in the southwestern block of the Los Angeles Basin, California: Journal of Geophysical Research, v. 84, p. 3460-3464.

Turcotte, D. L., 1980. On the thermal evolution of the earth: Earth and Planetary Letters, v. 48, p. 53-58.

Underwood, M. B., Fulton, D. A., and McDonald, K. W., 1988. Thrust control on thermal maturity of the frontal Ouachita Mountains, central Arkansas, U.S.A.: Journal of Petroleum Geology, v. 11, p. 325-340.

Ungerer, P., 1990. State of the art of research in the kinetic modeling of oil formation and expulsion: Organic Geochemistry, v. 16, p. 1-25.

Ungerer, P., 1993. Modelling of petroleum generation and explusion - an update to recent reviews: p. 219-232 in "Basin Modeling: Advances and Applications" (Edited by A. G. Dore et al), Norwegian Petroleum Society Special Publication 3, Elsevier, Amsterdam.

Ungerer, P. and R. Pelet, 1987. Extrapolation of oil and gas formation kinetics from laboratory experiments to sedimentary basins: Nature, v. 327, p. 52-54.

Ungerer, P., Behar, F., and Discamps, D., 1983. Tentative calculation of overall volume of organic matter: implications for primary migration: p. 129-135 in "Advances in Organic Geochemistry 1981" (Edited by M. Bjoroy et al), Wiley, Chichester.

Ungerer, P., Bessis, F., Chenet, P. Y., Durand, B., Nogaret, E., Chiarelli, A., Oudin, J. L., and Perrin, J. F., 1984. Geological and geochemical models in oil exploration; principles and practical examples: AAPG Memoir 35, p. 53-77.

Ungerer, P., Burrus, J., Doligez, B., Chenet, P. Y., and Bessis, F., 1990. Basin evaluation by integrated two-dimensional modeling of heat transfer, fluid flow, hydrocarbon generation, and migration: AAPG Bulletin, v. 74, p. 309-335.

Ungerer, P., Chenet, P. Y., Moretti, I., Chiarelli, A., and Oudin, J. L., 1986. Modelling oil formation and migration in the southern part of the Suez rift, Egypt: Organic Geochemistry, v. 10, p. 247-260.

Ungerer, P., Chiarelli, A., and Oudin, J. L., 1985. Modelling of petroleum genesis and migration with a bidimensional computer model in the Frigg sector, Viking Graben: p. 121-130 in "Petroleum Geochemistry in Exploration of the Norwegian Continental Shelf" (Edited by B. M. Thomas, et al), Graham and Trotman, London.

Unnerve, P., Espitalie, J., Behar, F., and Egging, S., 1988. Modelisation mathematique des interactions entre craquage thermique et migration lors de la formation du petrole et du gaz: Compte Rendu a l'Academie des Sciences, serie II, p. 927-934.

Vacquier, V., 1984. Oil fields - a source of heat flow data: Tectonophysics, v. 103, p. 81-98.

Vacquier, V., Mathieu, Y., Legendre, E., and Blondin, E., 1988. Experiment on estimating thermal conductivity of sedimentary rocks from oil well logging: AAPG Bulletin, v. 72, p. 758-765.

Valentine, P. C., 1976. Zoogeography of Holocene Ostracoda off western North America and paleoclimatic implications: U. S. Geological Survey Professional Paper No. 916. 47 pp.

van Aarssen, B.G.K., Cox, H. C., Hoogendoorn, P., and deLeeuw, J. W., 1990. A cadinene biopolymer present in fossil and extant dammar resins as a source for cadinenes and bicadinanes in crude oils from south east Asia. Geochimica et Cosmochimica Acta, v. 54, p. 3021-3031.

van Engelen, G.P., Kaul, C.L., Vos, B., and Aranha, H.P., 1981. Study of flow improvers for transportation of Bombay High crude oil through submarine pipelines: Journal of Petroleum Technology, December, p. 2539-2544.

Van Gijel, P., 1985. Organic indicators of thermal maturity, a review (Abstr.): AAPG Bulletin, v. 69, p. 314.

van Hinte, J. E., 1978. Geohistory analysis-applications of micropaleontology in exploration geology: AAPG Bulletin, v. 62, p. 201-222.

van Wijhe, D. H., 1987. The structural evolution of the Broad Fourteens Basin: p. 315-323 in "Petroleum Geology of North West Europe" (Edited by J. Brooks and K. W. Glennie), Graham and Trotman, London.

Vandenbrouke, M., 1980. Structure of kerogens as seen by investigations on soluble extracts: p. 415-443 in “Kerogen” (Edited by B. Durand), Editions Technip, Paris.

Vandenbrouke, M., 1993. Migration of hydrocarbons: p. 124-148 in "Applied Petroleum Geochemistry" (Edited by M. L. Bordenave), Editions Technip, Paris.

Van Kooten, G. K., 1988. Structure and hydrocarbon potential beneath the Iron Springs laccolith, southwestern Utah: Geological Society of America Bulletin, v. 100, p. 1533-1540.

Vasseur, G. and Velde, B., 1993. A kinetic interpretation of the smectite-to-illite transformation: p. 173-184 in "Basin Modeling: Advances and Applications" (Edited by A. G. Dore et al), Norwegian Petroleum Society Special Publication 3, Elsevier, Amsterdam.

Veto, I., 1980. An examination of the timing of catagenesis of organic matter using three published methods: p. 163-169 in "Advances in Organic Geochemistry, 1979" (Edited by A. G. Douglas and J. R. Maxwell), Pergamon Press, Oxford, England.

Videtich, P. E., McClimans, R. K., Watson, H. K. S., and Nagy, R. M., 1988. Depositional diagenetic, thermal, and maturation histories of Cretaceous Mishrif Formation, Fateh Field, Dubai: AAPG Bulletin, v. 72, p. 1143-1159.

Vik, E. and Hermanrud, C., 1993. Transient thermal effects of rapid subsidence in the Haltenbanken area: p. 107-117 in "Basin Modeling: Advances and Applications" (Edited by A. G. Dore et al), Norwegian Petroleum Society Special Publication 3, Elsevier, Amsterdam.

Vincelette, R. R., and Chittum, W. E., 1981. Exploration for oil accumulations in Entrada Sandstone, San Juan Basin, New Mexico: AAPG Bulletin, v. 65, p. 2546-1570.

Vincent, P. W., Montimore, I. R., and McKirdy, D. M., 1985. Hydrocarbon generation, migration, and entrapment in the Jackson-Naccowlah Area, ATP 259P, southwestern Queensland: Australian Petroleum Exploration Association (APEA) Journal, v. 25, p. 62-84.

Visser, W., 1982. Maximum diagenetic temperature in a petroleum source-rock from Venezuela by fluid inclusion geothermometry: Chemical Geology, v. 37, p. 95-101.

Vizgirda, J., O'Brien, J. J., and Lerche, I., 1985. Thermal anomalies on the flanks of a salt dome: Geothermics, v. 14, p. 553-565.

Vlierboom, F. W., 1991. Glossary of terms applicable to petroleum geochemistry: p. 159-185 in "Source and Migration Processes and Evaluation Techniques" (Edited by R. K. Merrill), AAPG Treatise of Petroleum Geology/Handbook of Petroleum Geology.

Volkman, J. K., Alexander, R., Kagi, R. I., and Woodhouse, G. W., 1983. Demethylated hopanes in crude oils and their applications in petroleum geochemistry: Geochimica et Cosmochimica Acta, v. 47, p. 785-794.

Vugrinovich, R., 1989. Subsurface temperatures and surface heat flow in the Michigan basin and their relationships to regional subsurface fluid movement: Marine and Petroleum Geology, v. 6, p. 60-70.

Wagner, W. and van den Haute, P., 1992. "Fission Track Dating", Kluwer Academic Publishers, Dordrechet, Netherlands.

Wahab, A. and Martono, D., 1985. Application of oil geochemistry for hydrocarbon exploration in Northwest Java: Proceedings of the Annual Convention - Indonesian Petroleum Association, (IDN) 14; v. 2, p. 657-682a.

Walgenwitz, F., Pagel, M., Meyer, A., Maluski, H., and Monie, P., 1990. Thermochronological approach to reservoir diagenesis in the offshore Angola basin: a fluid inclusion, ^{40}Ar-^{39}Ar and K-Ar investigation: AAPG Bulletin, v. 74, p. 547-563.

Walker, C. T., Francis, R. D., Dennis, J. G., and Lumsden, W. W., 1992. Cenozoic attenuation detachment faulting: A possible control on oil and gas accumulation in east-central Nevada: AAPG Bulletin, v. 76, p. 165-1686.

Wallem, D. B., Surdam, R. C., and Steidtmann, J. R., 1981. Source rock analysis of the Lower Cretaceous Bear River Formation, western Wyoming overthrust belt: p. 41-64 in "Energy Resources of Wyoming" (Edited by S. G. Reid and D. D. Miller), 32nd Annual Field Conference Guidebook, Wyoming Geological Association.

Walters, C. C. and Kotra, R. K., 1990. Thermal maturity of Jurassic shales from the Newark Basin, U.S.A.: influence of hydrothermal fluids and implications to basin modeling: Applied Geochemistry, v. 5, p. 211-225.

Wang, S.-C. and Fu, S.-T., 1986. Geomathematical research on the thermal evolution of Paleozoic coal-series of Shaanganning basin: Experimental Petroleum Geology, v. 8, p. 50-57.

Wang, X., Lerche, I., and Walters, C., 1989. The effect of igneous intrusive bodies on sedimentary thermal maturity: Organic Geochemistry, 4. 14, p. 571-584.

Wanli, Y., Yongkang, L., Ruiqi, G., 1985. Formation and evolution of nonmarine petroleum in Songliao basin, China: AAPG Bulletin, v. 69, p. 1112-1122.

Waples, D. W. and Machihara, T., 1991. "Biomarkers for geologists - a practical guide to the application of steranes and triterpanes in petroleum exploration": AAPG Methods in Exploration No. 9. 90 pp.

Waples, D. W., 1980. Time and temperature in petroleum formation: application of Lopatin's method to petroleum exploration: AAPG Bulletin, v. 64, p. 916-926.

Waples, D. W., 1984. Thermal models for oil generation: p. 7-67 in "Advances in Petroleum Geochemistry, Vol. 1" (Edited by J. Brooks and D. Welte), Academic Press, London.

Waples, D. W., 1985. "Geochemistry in Petroleum Exploration", International Human Resources Development Corporation, Boston. 232 pp.

Waples, D. W., Kamata, H., and Suizu, M., 1992. The art of maturity modeling. Part 1: Finding a satisfactory geologic model: AAPG Bulletin, v. 76, p. 31-46.

Waples, D. W., Suizu, M., and Kamata, H., 1992. The art of maturity modeling. Part 2: Alternative models and sensitivity analysis: AAPG Bulletin, v. 76, p. 47-66.

Warner, M. A. and Royse, F., 1987. Thrust faulting and hydrocarbon generation: discussion: AAPG Bulletin, v. 71, p. 882-889.

Warner, M. A., 1982. Source and time of generation of hydrocarbons in the Fossil basin, western Wyoming thrust belt: p. 805-815 in "Geologic Studies of the Cordilleran Thrust Belt, Vol II" (Edited by R. B. Powers), Rocky Mountain Association of Geologists, Denver.

Wavrek, D. A., 1992. Role of sulfur in altering maturity-dependent biomarker transformations - a quantitative approach: unpublished Ph.D. dissertation, The University of Tulsa, 851 pp.

Weaver, C. E., 1979. Geothermal alteration of clay minerals and shales: Office of Nuclear Waste Isolation Technical Report 21. 176 pp.

Weaver, C. E., 1989. "Clays, Muds, and Shales". Elsevier, Amsterdam. 819 pp.

Weaver, C. E., Eslinger, E. V., and Yeh, H. W., 1984. Oxygen isotopes: p. 141-152 in "Shale-Slate Metamorphism in Southern Appalachians" (Edited by C. E. Weaver and Associates), Developments in Petrology 10, Elsevier, Amsterdam.

Weber, K. J. and Daukoru, E. M., 1975. Petroleum geology of the Niger Delta: Proceedings of the 9th World Petroleum Congress, v. 2, p. 209-222.

Weber, K. J., 1987. Hydrocarbon distribution patterns in Nigerian growth fault structures controlled by structural style and stratigraphy: Journal of Petroleum Science and Engineering, v. 1, p. 91-104.

Webster, R. L., 1984. Petroleum source rocks and stratigraphy of the Bakken formation in North Dakota: p. 57-81 in "Hydrocarbon Source Rocks of the Greater Rocky Mountain Region" (Edited by J. Woodward, F. F. Meissner, and J. L. Clayton), Rocky Mountain Association of Geologists, Denver.

Weir, L. A. and Furlong, K. P., 1987. Thermal regimes of small basins: effects of intrabasinal conductive and advective heat transport: p. 351-362 in "Sedimentary Basins and Basin-forming Mechanisms" (Edited by C. Beaumont and A. J. Tankarad), Canadian Society of Petroleum Geologists, Memoir 12.

Welte, D. H. and Yukler, M. A., 1981. Petroleum origin and accumulation in basin evolution - a quantitative model: AAPG Bulletin, v. 65, p. 1387-1396.

Welte, D. H., and Yalcin, M. N., 1987. Basin modeling - A new methodology in petroleum geology: p. 141-151 in "Advances in Organic Geochemistry 1987" (Edited by L. Mattavelli and L. Novelli), Pergamon Press, Oxford.

Welte, D. H., Schaefer, R. G., Stoessinger, W., and Radke, M., 1984. Gas generation and migration in the Deep Basin of western Canada: AAPG Memoir 38, p. 35-47.

Welte, D. H., Yukler, M. A., Radke, M., Leythaeuser, D., Mann, U., and Ritter, U., 1983. Organic geochemistry and basin modelling - important tools in petroleum exploration: p. 237-252 in "Petroleum Geochemistry and Exploration in Europe" (Edited by J. Brooks), Geological Society (London) Special Publication No. 12.

Welte, D.H., Schaefer, R. G., and Yalcin, M.N., 1988. Gas generation from source rocks: aspects of a quantitative treatment: Chemical Geology, v. 71, p. 105-116.

Werner, D., 1986. Paleotemperatures in the Central Alps - an attempt at interpretation: p. 185-193 in "Paleogeothermics" (Edited by G. Buntebarth and L. Stegena), Springer-Verlag, Berlin.

Westcott, W. A. and Hood, W. C., 1994. Hydrocarbon generation and migration routes in the East Texas basin: AAPG Bulletin, v. 78, p. 287-307.

Whateley, M.K.G. and Pickering, K. T., 1989. "Deltas: Sites and Traps for Fossil Fuels", Geological Society (U.K.) Special Publication No. 41, Blackwell Scientific Publication, Oxford. 360 pp.

Wilcoxon, B. R., Ferrell, R. E., Sassen, R., and Wade, W. J., 1990. Illite polytype distribution as an inorganic indicator of thermal maturity in the Smackover Formation of the Manila Embayment, southwest Alabama: Organic Geochemistry, v. 15, p. 1-8.

Wilgus, C. K., Hastings, B. S., Kendall, C. G. St. C., Posamentier, H. W., Ross, C. A., and Van Wagoner, J. C. (Editors), 1988. "Sea-level changes: an integrated approach", Special Publication No. 42, Society of Economic Paleontologists and Mineralogists, Tulsa, Oklahoma.

Wilkinson, J. J., 1994. A new system for high-magnification thermometric studies of fluid inclusions in diagentic minerals: Journal of Sedimentary Research, Section A, v. 64A, p. 701-703.

Willett, S. D. and Chapman, D. S., 1987. Analysis of temperatures and thermal processes in the Uinta Basin: p. 447-461 in "Sedimentary Basins and Basin-Forming Mechanisms" (Edited by C. Beaumont and A. J. Tankard), Canadian Society of Petroleum Geologists, Memoir 12.

Williams, H. H. and Eubank, R. T., 1995. Hydrocarbon habitat in the rift graben of the Central Sumatra Basin, Indonesia: p. 331-371 in "Hydrocarbon Habitat in Rift Basins" (Edited by J. J. Lambiase), Geological Society (London) Special Publication No. 80.

Williams, J. A., 1974. Application of oil-correlation and source rock data to exploration in the Williston Basin: AAPG Bulletin, v. 58, p. 1243-1252.

Williams, P. F. V., 1986. Petroleum geochemistry of the Kimmeridge Clay of onshore southern and eastern England: Marine and Petroleum Geology, v. 3, p. 258-281.

Williamson, M. A., 1992. The subsidence, compaction, thermal and maturation history of the Egret Member source rock, Jeanne D'Arc Basin, offshore Newfoundland: Bulletin of Canadian Petroleum Geology, v. 40, p. 136-150.

Williamson, M. A. and Smyth, C., 1992. Timing of gas and overpressure generation in the Sable Basin offshore Nova Scotia: implications for gas migration dynamics: Bulletin of Canadian Petroleum Geology, v. 40, p. 151-169.

Williamson, M., Coflin, K., King, S., Desroches, K., Moir, P., McAlpine, D., and Wray, A., 1994. A hydrocarbon charge model of the Hibernia Drainage Area, Jeanne d'Arc Basin, Offshore, Newfoundland: Energy Exploration and Exploitation, v. 12, p.295-323.

Williamson, P. E., Pigram, C. J., Colwell, J. C., Scherl, A. S., Lockwood, K. L., and Branson, J. C., 1986. New deeper exploration frontiers in the Bass basin, Australia: Oil and Gas Journal, Sept. 8th, p. 121-126.

Williamson, P. E., Swift, M. G., O'Brien, G. W., and Falvey, D. A., 1990. Two-stage Early Cretaceous rifting of the Otway Basin margin of southeastern Australia: Implications for rifting of the Australian southern margin: Geology, v. 18, p. 75-78.

Wilson, J. C. and McBride, E. F., 1988. Compaction and porosity evaluation of Pliocene sandstones, Ventura Basin, California: AAPG Bulletin, v. 72, p. 664-681.

Wilson, J. L., 1975. "Carbonate Facies in Geologic History", Springer-Verlag, New York. 471pp.

Wilson, N. P. and Luheshi, M. N., 1987. Thermal aspects of the East Midlands aquifer system: p. 157-169 in "Fluid Flow in Sedimentary Basins and Aquifers" (Edited by J. C. Goff and B. P. J. Williams), Geological Society Special Publication No. 34.

Wood, D. A., 1988. Relationships between thermal maturity indices calculated using Arrhenius equation and Lopatin method: implications for petroleum exploration: AAPG Bulletin, v. 72, p. 115-134.

Wood, D. A., 1990. Thermal maturation modeling using spreadsheets: Geobyte, February, p. 56-61.

Wood, J. R. and Hewett, T. A., 1982. Fluid convection and mass transfer in porous sandstones - a theoretical model: Geochimica Cosmochimica Acta, v. 46, p. 1707-1713.

Wright, N. J. R., 1980. Time, temperature and organic maturation - the evolution of rank within a sedimentary pile: Journal of Petroleum Geology, v. 2, p. 411-425.

Wright, V. P. and Smart, P. L., 1994. Paleokarst (dissolution diagenesis): its occurrence and hydrocarbon exploration significance: p. 477-517 in "Diagenesis, IV", Elsevier, New York.

Wygrala, B. P., Yalcin, M. N., and Dohmen, L., 1990. Thermal histories and overthrusting - Application of numerical simulation techniques: Organic Geochemistry, v. 16, p. 267-285.

Xiao, H-B. and Suppe, J., 1989. Role of compaction in listric shape of growth normal faults: AAPG Bulletin, v. 73, p. 777-786.

Xue, L. and Galloway, W. E., 1993. Genetic sequence stratigraphic framework, depositional style, and hydrocarbon occurrence of the Upper Cretaceous QYN Formations in the Songliao lacustrine basin, northern China: AAPG Bulletin, v. 77, p. 1792-1808.

Yalcin, M. N. and Welte, D. H., 1988. The thermal evolution of sedimentary basins and significance for hydrocarbon generation: Bulletin of Turkish Petroleum Geology, v. 1, p. 12-26.

Yalcin, M. N., Welte, D. H., Misra, K. N., Mandal, S. K., Balan, K.C., Mehrotra, K. L., Lohar, B. L.,Kuma, S. P., and Misra, G.S., 1988. Three-dimensional computer aided basin modeling of Cambay Basin, India. A case history of hydrocarbon generation: p. 417-450 in "Petroleum Geochemistry and Exploration in the Afro-Asian Region" (Edited by R.K. Kumar, P. Dwivedi, V. Banerjie, and V. Gupta), A. A. Balkema, Rotterdam.

Yamano, M., Uyeda, S., Aoki, Y., and Shipley, T. H., 1982. Estimates of heat flow derived from gas hydrates: Geology, v. 10, p. 339-343.

Yang, C. and Hesse, R., 1993. Diagnesis and anchimetamorphism in an overthrust belt, external domain of the Taconian Orogen, southern Canadian Appalachians - II. Paleogeothermal gradients derived from maturation of different types of organic matter: Organic Geochemistry, v. 20, p. 381-403.

Yonkee, W. A., Parry, W. T., Bruhn, R. L., and Cashman, P. H., 1989. Thermal models of thrust faulting: constraints from fluid-inclusions observations, Willard thrust sheet, Idaho-Utah-Wyoming thrust belt: Geological Society of America Bulletin, v. 101, p. 304-313.

Yu, Z., 1992. Quantitative basin modeling: An application in the Arctic National Wildlife Refuge of Alaska, and salt influences on thermal anomalies and fracturing patterns of sediments in Gulf of Mexico: Ph.D. Dissertation, University of South Carolina. 175pp.

Yukler, M. A. and Kokesh, F., 1984. A review of models used in petroleum resource estimation and organic geochemistry: p. 69-113 in "Advances in Petroleum Geochemistry, Vol. 1" (Edited by J. Brooks and D. Welte), Academic Press, London.

Yukler, M. A. and Welte, D. H., 1980. A three-dimensional deteriministic dynamic model to determine geologic history and hydrocarbon generation and accumulation in a sedimentary basin: p. 267-285 in Fossils Fuels, 26th International Geological Congress, Technips, Paris.

Zhao, K. and Lerche, I., 1993. Thermal maturation and burial history of the Multi well Experiment Site, Piceance Basin, Colorado: application of thermal indicator tomography: p. 135-145 in "Basin Modeling: Advances and Applications" (Edited by A. G. Dore et al), Norwegian Petroleum Society Special Publication 3, Elsevier, Amsterdam.

Zhao, Z.-Y., 1990. History of structural evolution, mechanism of formation, and the possible petroliferous area of the Ordos basin: p. 93-102 in "The Formation and Evolution of the Sedimentary Basins and their Hydrocarbon Occurrence in the North China Craton" (Edited by Z.-Y. Zhao et al), Northwest University Press, Xi'an.

Zhao, M.-W., Behr, H.-J., Ahrendt, H., Wemmer, K., Ren, Z.-L., and Zhao, Z.-Y., 1996. Thermal and tectonic history of the Ordos basin, China: Evidence from apatite fission track analysis, vitrinite reflectance, and K-Ar dating: AAPG Bulletin, v. 80, p. 1110-1134.

Zhou, S., 1989. Subsidence history of the Eromanga Basin, Australia: p. 329-335 in "The Cooper and Eromanga Basins Australia" (Edited by B. J. O'Neil), Proceedings of Petroleum Exploration Society of Australia, Society of Petroleum Engineers, Australian Society of Exploration Geophysicsts, Adelaide.

Ziegler, D. L. and Spotts, J. H., 1978. Reservoir and source-bed history of Great Valley, California: AAPG Bulletin, v. 62, p. 813-826.

Ziegler, K., Sellwood, B. W., and Fallick, A. E., 1994. Radiogenic and stable isotope evidence for age and origin of authigenic illites in the Rotliegend, southern North Sea: Clay Minerals, v. 29, p. 555-565.

Zielinski, G. W. and Bruchhausen, P. M., 1983. Shallow temperatures and thermal regime in the hydrocarbon province of Tierra Del Fuego: AAPG Bulletin, v. 67, p. 166-177.

Zielinski, G. W., Drahovzal, J. A., Decoursey, G. M., and Ruperto, J. M., 1985. Hydrothermics in the Wyoming Overthrust belt: AAPG Bulletin, v. 69, p. 699-709.

Zimmerman, R. W., 1989. Thermal conductivity of fluid-saturated rocks: Journal of Petroleum Science and Engineering, v. 3, p. 219-227.

SUBJECT INDEX

DEVELOPMENTS IN PETROLEUM SCIENCE

Volumes 1-5, 7, 10, 11, 13-17A, 18A-B, 21, 24-26 are out of print

6. D.W. PEACEMAN – Fundamentals of Numerical Reservoir Simulation
8. L.P. DAKE – Fundamentals of Reservoir Engineering
9. K. MAGARA – Compaction and Fluid Migration
12. T.D. VAN GOLF-RACHT – Fundamentals of Fractured Reservoir Engineering
17B. E.C. DONALDSON, G.V. CHILINGARIAN and T.F. YEN (Editors) – Enhanced Oil Recovery, II. Processes and operations
19A. G.V. CHILINGARIAN, J.O. ROBERTSON Jr. and S. KUMAR – Surface Operations in Petroleum Production, I
19B. G.V. CHILINGARIAN, J.O. ROBERTSON Jr. and S. KUMAR – Surface Operations in Petroleum Production, II
20. A.J. DIKKERS – Geology in Petroleum Production
22. E.C. DONALDSON, G.V. CHILINGARIAN and T.F. YEN – Microbial Enhanced Oil Recovery
23. J. HAGOORT – Fundamentals of Gas Reservoir Engineering
27. G. DA PRAT – Well Test Analysis for Naturally Fractured Reservoirs
28. E.B. NELSON (Editor) – Well Cementing
29. R.W. ZIMMERMAN – Compressibility of Sandstones
30. G.V. CHILINGARIAN, S.J. MAZZULLO and H.H. RIEKE – Carbonate Reservoir Characterization: A Geologic-Engineering Analysis, Part I
31. E.C. DONALDSON (Editor) – Microbial Enhancement of Oil Recovery – Recent Advances
32. E. BOBOK – Fluid Mechanics for Petroleum Engineers
33. E. FJÆR, R.M. HOLT, P. HORSRUD, A.M. RAAEN and R. RISNES – Petroleum Related Rock Mechanics
34. M.J. ECONOMIDES – A Practical Companion to Reservoir Stimulation
35. J.M. VERWEIJ – Hydrocarbon Migration Systems Analysis
36. L. DAKE – The Practice of Reservoir Engineering
37. W.H. SOMERTON – Thermal Properties and Temperature related Behavior of Rock/fluid Systems
38. W.H. FERTL, R.E. CHAPMAN and R.F. HOTZ (Editors) – Studies in Abnormal Pressures
39. E. PREMUZIC and A. WOODHEAD (Editors) – Microbial Enhancement of Oil Recovery – Recent Advances – Proceedings of the 1992 International Conference on Microbial Enhanced Oil Recovery
40A. T. F. YEN and G.V. CHILINGARIAN (Editors) – Asphaltenes and Asphalts, 1
41. E.C. DONALDSON, G. CHILINGARIAN and T.F. YEN (Editors) – Subsidence due to Fluid Withdrawal
42. S.S. RAHMAN and G.V. CHILINGARIAN – Casing Design – Theory and Practice
43. B. ZEMEL – Tracers in the Oil Field
44. G.V. CHILINGARIAN, S.J. MAZULLO and H.H. RIEKE – Carbonate Reservoir Charaterization: A Geologic – Engineering Analysis, Part II
45. C. BARKER – Thermal Modeling of Petroleum Generation: Theory and Applications

ACKNOWLEDGMENTS

This book, like so many others, started out as a set of lecture notes and has benefited from the input of all the students who had early versions of the course inflicted on them. The course was taught to university graduate students, and also internationally to a wide variety of petroleum industry geologists. Their questions, comments, and insights have helped mold this text. I especially appreciate the time spent by Ted Beaumont, Tom Ho, Mike Horn, Indu Meshri, and Radomir Petrovich in reviewing early drafts of the manuscript. The final version is much improved by their contributions – but I take responsibility for any remaining errors!

I sincerely appreciate permission from the following organizations to use Figures and Tables for which they hold the copyright: *Academic Press*: Figures 1.40, 1.41, 4.19, 4.20, 7.13, 7.15, and 10.49, and Table 1.4; *American Association of Petroleum Geologists (AAPG)*: Figures 1.9, 1.14, 1.15, 1.20, 1.26, 1.32, 1.34, 1.42, 1.43, 1.45, 1.49, 2.12, 2.13, 3.6, 3.8, 3.9, 3.10, 3.12, 3.17, 3.19, 3.20, 3.21, 3.35, 4.2, 4.5, 4.10, 4.13, 4.14, 4.15, 4.18, 5.5, 5.9, 5.10, 5.11, 5.20, 5.21, 5.23, 5.24, 5.31, 5.32, 5.33, 6.2, 6.4, 6.6, 7.2, 7.4, 7.5, 7.8, 7.9, 7.12, 7.17, 7.19, 8.2, 8.3, 8.4, 8.5, 8.11, 8.12, 8.13, 9.1, 9.2, 9.4, 9.5, 9.6, 9.8, 9.10, 10.1, 10.5, 10.6, 10.8, 10.9, 10.10, 10.11, 10.12, 10.13, 10.15, 10.16, 10.17, 10.31, 10.32, 10.33, 10.34, 10.35, 10.38, 10.39, 10.40, 10.41, 10.50, 10.51, and 10.52, and Tables 1.5, 2.2, 8.2, 10.3, 10.4, 10.6, and 10.8; *American Chemical Society*: Figure 2.10; *American Geophysical Union*: Figures 10.37 and 10.44; *American Journal of Science*: Figure 10.36; *Australian Petroleum Exploration Association*: Figures 5.19 and 8.7; *A. A. Balkema Publishers*: Figures 10.21, 10.22, and 10.23; *Blackwell Scientific Publications*: Figure 4.1; *Canadian Society of Petroleum Geologists*: Figures 10.24, 10.25, 10.26, and 10.27; *Cushman Foundation*: Figure 3.30; *Editions Technip*: Figures 1.48, 7.11, and 9.9; *Elf Aquitaine*: Figure 6.1; *Erdol und Kohle*: Figure 2.11; *Geochimica et Cosmochimica Acta*: Figure 1.18; *Geological Society of America*: Figures 3.36, 5.12, 5.13, 5.16, 10.36, and Appendix III; *Geological Society (London)*: Figures 5.35, 8.1, 8.8, and 10.42; *Geological Survey of Canada*: Figures 1.30 and 3.39; *Graham and Trotman*: Figures 1.28, 1.35, 3.15, and 10.14; *Institute Francais du Petrole*: Figure 1.44; *Journal of Geochemical Exploration*: Figures 1.17, 3.16, and 5.14; *The*

Log Analyst: Figure 9.3; *Offshore Technology Conference*: Figure 10.30; *Oklahoma Geological Survey*: Figures 3.3, 4.3, 5.34, and 7.16; *Platte River Associates*: Table 4.1; *Prentice Hall*: Figure 10.7; *Rocky Mountain Association of Geologists*: Figures 3.31, 5.4, 5.15, 6.5, 6.8, 10.3, and 10.4; *SEPM (Society for Sedimentary Geology)*: Figures 1.31, 5.25, 5.26, and 5.27; *Society of Exploration Geophysicists*: Figure 10.48; *Springer-Verlag*: Figures 1.19, 2.9, 4.9, 5.6, 5.18, 5.22, 5.28, 5.29, and 7.1; *Tulsa Geological Society*: Figure 10.29; *John Wiley*: Figure 3.34.

ERRATUM

The caption to figure 10.52 (page 415) is a repeat of the caption from figure 10.51.
The correct text is as follows:

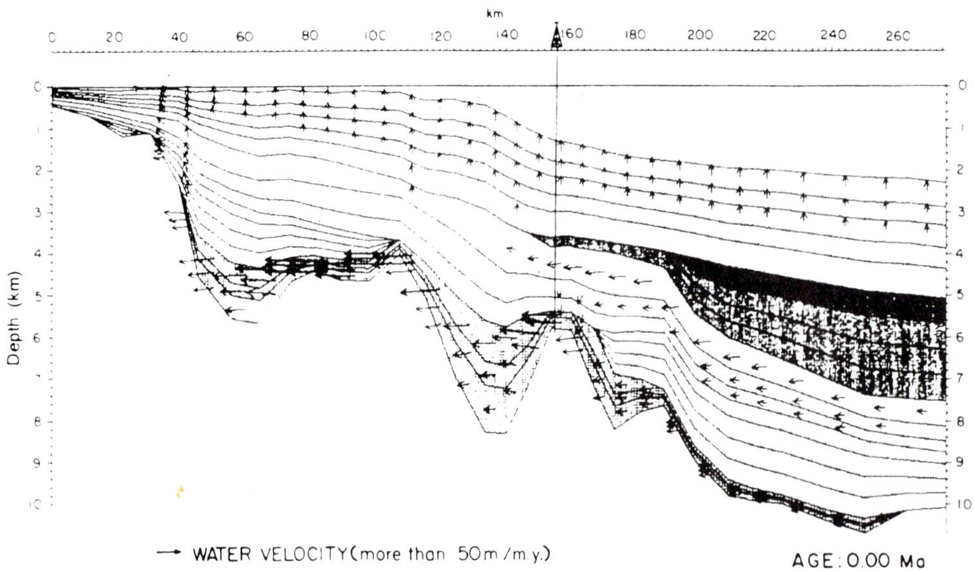

Figure 10.52. Predicted compaction-driven water flow at present. Flow is vertical in the upper kilometers, where most of the porosity is lost, but is horizontal below the impermeable evaporitic sequence and in the more sandy synrift sequence. The largest arrows correspond to 3000 m/m.y. (From Burrus and Audebert, 1990).